Surface-Wave Devices
for Signal Processing

STUDIES IN ELECTRICAL AND
ELECTRONIC ENGINEERING 19

Surface-Wave Devices for Signal Processing

DAVID P. MORGAN

Plessey Research (Caswell) Ltd., Allen Clark Research Centre, Towcester, U.K.

ELSEVIER
Amsterdam – Oxford – New York – Tokyo   1991

ELSEVIER SCIENCE PUBLISHERS B.V.
Sara Burgerhartstraat 25
P.O. Box 211, 1000 AE Amsterdam, The Netherlands

Distributors for the United States and Canada:

ELSEVIER SCIENCE PUBLISHING COMPANY, INC.
655, Avenue of the Americas
New York, NY 10010, U.S.A.

First Edition 1985
Paperback Edition 1991

Library of Congress Cataloging in Publication Data

Morgan, David P.
    Surface-wave devices for signal processing.

    (Studies in electrical and electronic engineering;
vol. 19)
    Bibliography: p.
    Includes index.
    1. Acoustic surface wave devices. 2. Signal process-
ing. I. Title. II. Series: Studies in electrical and
electronic engineering; 19.
TK5981.M63   1985   621.38'043   85-10-330
ISBN 0-444-42511-X (U.S.)

ISBN 0-444-42511-X (Vol. 19, hard cover)
ISBN 0-444-88845-4 (Vol. 19, soft cover)

© ELSEVIER SCIENCE PUBLISHERS B.V., 1991

All rights reserved. No part of this publication may be reproduced, stored in a retrieval system, or transmitted, in any form or by any means, electronic, mechanical, photocopying, recording or otherwise, without the prior written permission of the publisher, Elsevier Science Publishers B.V., P.O. Box 521, 1000 AN Amsterdam, The Netherlands.

Special regulations for readers in the U.S.A. - This publication has been registered with the Copyright Clearance Center Inc. (CCC), 27 Congress Street, Salem, MA 01970, U.S.A. Information can be obtained from the CCC about conditions under which photocopies of parts of this publication may be made in the U.S.A. All other copyright questions, including photocopying outside of the U.S.A., should be referred to the publisher.

Printed in The Netherlands

STUDIES IN ELECTRICAL AND ELECTRONIC ENGINEERING

Vol. 1 Solar Energy Conversion: The Solar Cell (Neville)
Vol. 3 Integrated Functional Blocks (Novák)
Vol. 4 Operational Amplifiers (Dostál)
Vol. 5 High-Frequency Application of Semiconductor Devices (F. Kovács)
Vol. 6 Electromagnetic Compatibility in Radio Engineering (Rotkiewicz)
Vol. 7 Design of High-Performance Negative-Feedback Amplifiers (Nordholt)
Vol. 8 Discrete Fourier Transformation and its Applications to Power Spectra Estimation (Geçkinli and Yavuz)
Vol. 9 Transient Phenomena in Electrical Machines (P. K. Kovács)
Vol. 10 Theory of Static Converter Systems: Mathematical Analysis and Interpretation. Part A: Steady-State Processes (Slonim)
Vol. 11 Power Sources for Electric Vehicles (edited by McNicol and Rand)
Vol. 12 Classical Electrodynamics (Ingarden and Jamiotkowski)
Vol. 13 Reliability of Analogue Electronic Systems (Klaassen)
Vol. 14 Electrets (Hilczer and Małecki)
Vol. 15 Graph Theory: Application to the Calculation of Electrical Networks (Vágó)
Vol. 16 Eddy Currents in Linear Conducting Media (Tegopoulos and Kriezis)
Vol. 17 Electrical Measurements in Engineering (Boros)
Vol. 18 Active RC Filters (Herpy and Berka)
Vol. 19 Surface-Wave Devices for Signal Processing (Morgan)
Vol. 20 Micromachining and Micropackaging of Transducers (edited by Fung, Cheung, Ko and Fleming)
Vol. 21 Nonlinear and Environmental Electromagnetics (edited by Kikuchi)
Vol. 22 Microwave Measurements of Complex Permittivity by Free Space Methods and their Applications (Musil and Žáček)
Vol. 24 Piezoelectric Resonators and their Applications (Zelenka)
Vol. 25 Large Power Transformers (Karsai, Kerényi and Kiss)
Vol. 26 Power Supplies (Ferenczi)
Part A: Linear Power Supplies, DC-DC Converters
Part B: Switched-mode Power Supplies
Vol. 27 Proceedings of the Eighth Colloquium on Microwave Communication, Budapest, Hungary, August 25-29, 1986 (edited by Berceli)
Vol. 28 U.R.S.I. International Symposium on Electromagnetic Theory, Budapest, Hungary, August 25-29, 1986 (edited by Berceli)
Vol. 29 Nonlinear Active Microwave Circuits (Berceli)
Vol. 30 Power System Stability (Rácz and Bókay)
Vol. 31 Analysis and Synthesis of Translinear Integrated Circuits (Seevinck)
Vol. 32 Microwave Measurements by Comparison Methods (Kneppo)
Vol. 33 Nodal Analysis of Electrical Networks (Fodor)
Vol. 34 Noise and Vibration of Electrical Machines (edited by Timár)
Vol. 35 Adaptive Arrays (Nicolau and Zaharia)
Vol. 36 Transient Stability Analysis of Synchronous Motors (Ĉemus and Hamata)
Vol. 37 Electric Drive Systems Dynamics (Szklarski, Jaracz and Horodecki)
Vol. 38 Digital Microwave Transmission (Frigyes, Szabó and Ványai)
Vol. 39 Magnetic Heads for Digital Recording (Ciureanu and Gavrilă)

Contents

1. **Introductory survey** — 1

2. **Acoustic waves in elastic solids** — 15
   2.1 Elasticity in anisotropic materials — 15
      2.1.1 Non-piezoelectric materials — 15
      2.1.2 Piezoelectric materials — 17
   2.2 Waves in isotropic materials — 19
      2.2.1 Plane waves — 19
      2.2.2 Rayleigh waves in a half-space — 21
      2.2.3 Shear-horizontal waves in a half-space — 25
      2.2.4 Waves in a layered half-space — 25
      2.2.5 Waves in a parallel-sided plate — 28
   2.3 Waves in anisotropic materials — 29
      2.3.1 Plane waves in an infinite medium — 30
      2.3.2 Theory for a piezoelectric half-space — 30
      2.3.3 Surface-wave solutions — 32
      2.3.4 Other solutions — 34
      2.3.5 Materials for devices — 36

3. **Electrical excitation at a plane surface** — 39
   3.1 Non-piezoelectric half-space — 39
   3.2 Piezoelectric half-space — 42
   3.3 Some properties of the effective permittivity — 45
   3.4 Green's function — 50
   3.5 Other applications of the effective permittivity — 52

4. **Analysis of interdigital transducers** — 57
   4.1 Delta-function model — 57
   4.2 Discussion of second-order effects and methods of analysis — 64
   4.3 Analysis for a general array of electrodes — 67
      4.3.1 The quasi-static approximation — 67

	4.3.2 Electrostatic equations and charge superposition	70
	4.3.3 Current entering one electrode	72
	4.3.4 Evaluation of the acoustic potential	74
4.4	Quasi-static analysis of transducers	75
	4.4.1 Launching transducer	76
	4.4.2 Transducer admittance	77
	4.4.3 Receiving transducer	78
	4.4.4 Scattering coefficients	80
4.5	Transducers with regular electrodes	83
	4.5.1 Electrostatic charge density and element factor	83
	4.5.2 End effects	88
	4.5.3 Transducer response in terms of gap elements	88
4.6	Admittance of uniform transducers	89
	4.6.1 Acoustic conductance and susceptance	89
	4.6.2 Capacitance	93
	4.6.3 Comparative performance	94
4.7	Two-transducer devices	95
	4.7.1 Devices using unapodised transducers	96
	4.7.2 Device using one apodised transducer	97
	4.7.3 Apodised transducer with regular electrodes	99
4.8	Device resonse allowing for terminating circuits	100
	4.8.1 Main response	102
	4.8.2 Multiple-transit responses	103

5. The multi-strip coupler and its applications — 107

5.1	Analysis for an infinite array of electrodes	109
5.2	Basic coupler behaviour	113
5.3	Interdigital devices using couplers	118
5.4	Unidirectional transducer	120
5.5	Other applications of 3-dB couplers	123
5.6	Bandpass filtering using multi-strip couplers	124
5.7	Beam compression	126

6. Propagation effects and materials — 129

6.1	Surface-wave probing	129
6.2	Diffraction and beam steering	132
	6.2.1 Formulation using angular spectrum of plane waves	133
	6.2.2 Beam steering in the near field	134
	6.2.3 Minimal-diffraction orientations	136
	6.2.4 Diffracted field in the parabolic approximation: scaling	137
	6.2.5 Two-transducer devices	140
6.3	Propagation loss and non-linear effects	144
6.4	Temperature effects and velocity errors	147
6.5	Materials for surface-wave devices	151

CONTENTS

7. **Delay lines and multi-phase transducers** — 157
   - 7.1 Delay lines — 157
     - 7.1.1 Bandwidth and conversion loss of uniform transducers — 158
     - 7.1.2 Parasitic components — 165
     - 7.1.3 Triple-transit signal — 168
     - 7.1.4 Delay line types and performance — 171
   - 7.2 Multi-phase unidirectional transducers — 173
     - 7.2.1 Transducer types and performance — 173
     - 7.2.2 Analysis of multi-phase transducers — 176

8. **Bandpass filters** — 183
   - 8.1 Apodised transducer as a transversal filter — 184
     - 8.1.1 Transversal filter analogy — 185
     - 8.1.2 Sampling and surface-wave transducers — 187
     - 8.1.3 Examples of particular cases — 189
   - 8.2 Design of apodised transducers — 192
     - 8.2.1 Use of window functions — 193
     - 8.2.2 Optimised design methods — 197
     - 8.2.3 Minimum-phase filters — 199
   - 8.3 Thinning and withdrawal weighting — 200
   - 8.4 Filter design and performance — 203
     - 8.4.1 Basic types of bandpass filter — 203
     - 8.4.2 Circuit effect — 204
     - 8.4.3 Second-order effects and design — 205
     - 8.4.4 Performance — 206
     - 8.4.5 Other types of bandpass filter — 208
   - 8.5 Filter banks — 209

9. **Chirp filters and their applications** — 213
   - 9.1 Principles of pulse compression radar — 214
   - 9.2 Waveform characteristics and design — 219
     - 9.2.1 Stationary-phase approximation — 221
     - 9.2.2 Linear chirp waveforms — 222
     - 9.2.3 Weighting of linear-chirp filters — 224
     - 9.2.4 Weighting using non-linear chirps — 229
   - 9.3 Chirp transducers — 231
     - 9.3.1 Transducer analysis — 233
     - 9.3.2 Admittance in the stationary phase approximation — 236
     - 9.3.3 Transducer design — 239
   - 9.4 Design and performance of interdigital devices — 241
   - 9.5 Second-order effects in chirp filters — 247
     - 9.5.1 Effect of phase errors on compressor output waveform — 247
     - 9.5.2 Velocity errors and temperature effects — 250
     - 9.5.3 Doppler shifts — 252
     - 9.5.4 Other second-order effects in interdigital devices — 255

CONTENTS

9.6 Reflective array compressors	256
9.6.1 Basic principles	256
9.6.2 Analysis and performance	259
9.6.3 Other types of RAC	267
9.7 Spectrum analysis and other types of signal processing	268
9.7.1 Compressive receiver principles	268
9.7.2 Analysis of compressive receivers and Fourier transform systems	270
9.7.3 Experimental compressive receivers and Fourier transform systems	275
9.7.4 Other types of signal processing	277

10. Devices for spread-spectrum communications — 281

10.1 Principles of spread-spectrum systems	282
10.2 Linear devices	285
10.2.1 Matched filters for PSK waveforms	286
10.2.2 Output waveform and effect of phase errors	288
10.2.3 Devices for MSK waveforms	290
10.2.4 Frequency hopping	293
10.3 Acoustic convolvers	294
10.3.1 Principles of non-linear convolvers	295
10.3.2 Performance of basic convolvers	300
10.3.3 Waveguide convolvers	305
10.3.4 Convolver fidelity and frequency response	309
10.4 Other non-linear devices	312
10.4.1 Semiconductor convolvers	312
10.4.2 Storage convolvers	314
10.4.3 Correlation of long waveforms	316
10.5 Oscillators	318
10.5.1 Delay-line oscillator	318
10.5.2 Resonators	319
10.5.3 Oscillator performance	321

Appendix A. Fourier transforms and linear filters — 325

A.1 Fourier transforms	325
A.2 Linear filters	330
A.3 Matched filtering	332
A.4 Non-uniform sampling	334
A.5 Some properties of bandpass waveforms	337

Appendix B. Reciprocity — 343

B.1 General relation for a mechanically free surface	343
B.2 Reciprocity for two-terminal transducers	344
B.3 Symmetry of the Green's function	347
B.4 Reciprocity for surface excitation of a half-space	348

B.5	Reciprocity for surface wave transducers	348
B.6	Surface wave generation	351

Appendix C. Elemental charge density for regular electrodes — 355
    C.1 Some properties of Legendre functions — 355
    C.2 Elemental charge density — 358
    C.3 Net charges on electrodes — 360

Appendix D. Floquet analysis for an infinite array of regular electrodes — 363
    D.1 General solution for low frequencies — 363
    D.2 Propagation outside the stop band — 367
    D.3 Stop bands — 370
    D.4 Solution at higher frequencies — 372

Appendix E. Electrode interactions in transducers — 375
    E.1 Reflective array model — 376
    E.2 Electrode reflection coefficient due to electrical loading — 381
    E.3 Electrical loading in transducers — 386
    E.4 Mechanical loading — 390

Appendix F. Bulk waves — 393
    F.1 Bulk wave generation by interdigital transducers — 394
    F.2 Mode conversion in arrays — 397
    F.3 Bulk wave devices — 399

References — 403

Index — 423

Preface to the Second Edition (Student Edition)

Since this book was first published in 1985 the topic of Surface Acoustic Wave Devices has continued to develop, confirming its niche as the major technology for passive filters in the 50 to 2000 MHz region. In addition to widespread development of commercial products, continuing research and development efforts have yielded further advances. For example, a variety of new techniques for low-loss filtering make front-end filtering feasible, thus opening up the possibility of substantial new markets. Stabilities obtainable from SAW oscillators have improved markedly, and fabrication techniques can now give 0.5 micron linewidths in production, increasing the SAW frequencies obtainable.

While maintaining its original function as a reference work, it is hoped that this soft-covered second Edition will also be useful for students on university courses and others new to the field. I am delighted that, with some support from colleagues, the publishers have decided to issue this edition. The subject continues to fascinate its practitioners, with its seemingly endless variety of devices, the wide range of technical topics, and the challenge of meeting exacting practical requirements – more exacting as time goes on. I hope that new readers will find the book both illuminating and enjoyable.

Some minor mathematical errors have been corrected in the new edition.

DAVID MORGAN

Preface

Devices using acoustic waves have been employed in electronic systems for many years, notable examples being the quartz crystal oscillator and the acoustic delay line, both of which use acoustic wave propagation in the bulk of a material. In contrast, the use of *surface* acoustic waves, in which the wave motion is bound to a plane surface of a solid, has developed quite recently, though the existence of the wave itself was established by Lord Rayleigh in the 19th Century. The use of surface waves introduces several attractions, notably a considerable degree of versatility due to the accessibility of the wave in two dimensions, and the prior existence of a variety of suitable fabrication techniques. These attractions were first recognised in the 1960's, and since then there have been substantial developments in understanding the wave behaviour and a wide variety of electronic devices has emerged. Today, surface-wave devices are used in many practical systems, particularly in communications, radar and broadcasting.

In this book I have chosen to concentrate on the devices most commonly found in electronic systems, and the principles underlying them. Most of these devices perform signal processing operations – for example, a bandpass filter is used to select some required frequency band, while chirp filters and PSK filters perform correlation of complex waveforms. To appreciate the function of the devices some knowledge of signal processing is necessary, and this is included in the appropriate parts of the book.

Chapter 1 gives a descriptive survey, intended to introduce the subject to those unfamiliar with it, and Chapters 2 to 5 give the theoretical background needed to appreciate the operation of the devices considered later. The devices here use surface waves in piezoelectric materials, which enable the waves to be generated or detected by means of metal electrodes on the surface. Chapter 2 considers basic properties of acoustic waves and emphasises surface waves in piezoelectric materials, though some other relevant cases are also included. Chapter 3 covers electrical excitation of surface waves, introducing the effective permittivity and the Green's function, and these concepts are applied to the analysis of interdigital transducers in Chapter 4 and to multi-strip couplers in Chapter 5. Interdigital transducers are used in all of the devices considered in this book, and in many devices the response is determined mainly by the transducer behaviour, which is therefore treated in detail. The analysis for transducers and multi-strip couplers makes use of the quasi-state approximation, described in Section 4.3. This simplifies the results considerably since it neglects electrode interactions, which are not very significant in most practical devices; however, interaction effects are considered in Appendices D and E. Generation of bulk waves in surface-wave devices is another complication ignored in Chapters 4 and 5, but this is considered in Appendix F.

Chapter 6 describes several surface-wave propagation effects, particularly diffraction, and gives a comparative assessment of materials commonly used for practical devices. The remaining chapters are mainly concerned with the design and performance of devices. Chapter 7 describes delay lines, including some practical aspects of transducer performance, while bandpass filters are covered in Chapter 8. Chapter 9 describes chirp filters, commonly used in pulse-compression radar systems, including interdigital devices and reflective array compressors. This chapter also includes the characteristics and design of chirp waveforms. Finally, Chapter 10 is mainly concerned with devices for spread-spectrum communication systems, including the PSK filter and the non-linear convolver which are used to correlate phase-shift-keyed waveforms. The surface-wave oscillator and resonator are also considered briefly.

It should be noted that the coverage here is quite selective. The literature includes substantial material on topics hardly mentioned, for example interaction with light and with semiconductors, and the behaviour of surface-wave waveguides. There is also a very considerable variety of devices in addition to those mentioned above, and these are omitted apart from some brief comments.

The book is intended to appeal mainly to engineers developing surface-wave devices and to those developing systems using the devices, though it should also be helpful in connection with university course or research work. The reader will not need a prior knowledge of acoustics, as the concepts required are included in Chapter 2. However, an undergraduate-level knowledge of network analysis, and of some basic concepts of crystallography, are assumed. The extensive use of Fourier transforms arises quite naturally, since the time- and frequency-domain representations of a device response both correspond to common laboratory measurements, and both domains occur in device specifications. In addition, Fourier transforms are used in the analysis of transducers and other structures. The relationships needed are summarised in Appendix A, which also gives some basic relationships for analysis of linear filters. However, the reader unfamiliar with these topics will probably find that further reading, from the references quoted in Appendix A for example, will be helpful.

The material in this book arises from experience in several laboratories, and has benefited substantially from cooperation and discussion with many colleagues. It is a pleasure to acknowledge the past involvement with colleagues in University College London, the Central Research Laboratories of the Nippon Electric Company Ltd. (Kawasaki), and the University of Edinburgh. I am especially indebted to the surface-wave group in Plessey Research (Caswell) Ltd. Many of the ideas in the book arose from the work of this group, and most of the experimental results shown refer to devices developed by the group. In particular, I wish to mention R. W. Allen, R. Almar, R. Arnold, R. E. Chapman, R. K. Chapman, J. M. Deacon, R. M. Gibbs (who fabricated most of the devices), W. Gibson, J. Heighway, J. A. Jenkins, P. M. Jordan, B. Lewis, J. G. Metcalfe (who also computed some theoretical figures for the book), R. F. Milsom, J. J. Purcell, D. Selriah and D. H. Warne. Many helpful discussions were contributed by E. G. S. Paige and M. F. Lewis. During the writing of the book the initial drafts were reviewed by E. G. S. Paige, B. Lewis and, in part, by J. J. Purcell, and I am greatly indebted to these gentlemen for their thorough and painstaking efforts; their comments have been of immense value throughout the writing process. Much of the work reported was supported by the Procurement Executive of the U.K. Ministry of Defence, sponsored by DCVD, and this applies in particular to the development of reflective array compressors and convolvers.

Thanks are also due to the management of Plessey Research (Caswell) Ltd. for some of the time involved and for funding the typesetting, and in particular to the Director, Dr. J. C. Bass, for his encouragement and for kindly contributing the Foreword. I am also grateful to the staff of The Alden Press (London and Northampton) Ltd. for their fine work in typesetting the script and preparing the figures.

DAVID MORGAN
Caswell, 1985

Foreword

It is a source of amazement that, given a mask with which to perform the photolithography, all you need is a single crystal with a thin metal film deposited on its surface in order to make a surface acoustic wave device with outstanding performance. It can function as a band-pass filter, passing most of the signal in the pass band but providing 60 dB or more rejection out-of-band. It can be made into a matched filter capable of extracting a signal for noise when the strength of the signal is four orders of magnitude below the prevailing noise. How is it that such a seemingly simple device can achieve so much? An important part of the answer lies in the fact that it is dependent for its performance on *intrinsic* properties of the single crystal-surface acoustic wave velocity and piezo-electric constant. But a vital ingredient is the design of the mask for here is where the subtlety and sophistication enters. It is the design of the mask which makes the difference between a SAW device having either excellent or mediocre performance; the design of the lithography mask can be equated to the design of the device.

A major strength of Dr. Morgan's book is the logical and coherent development of those topics which underpin or are relevant to the design of SAW devices. This book provides a clear and careful treatment of topics ranging from the basic theory of bulk and surface acoustic waves, the electrical excitation and detection of surface waves, material properties, through to the design, performance and application of a range of key SAW devices. Though originally planned as a reference book for the SAW device design engineer, the development of the material is well suited to undergraduate and MSc courses. Students will appreciate not only the clear exposition of the central subject matter but also the linkage with such topics as the application of Fourier Transform techniques and signal processing as applied to radar, and to video and audio telecommunication systems.

Dr. Morgan's book, making its first appearance exactly one century after Rayleigh's classic paper recording his discovery of elastic surface waves, is now five years old. It says much for the original choice and treatment of the subject that virtually all is as relevant today as the day it was written.

E.G.S. PAIGE
November 1990

Chapter 1

Introductory Survey

This book is concerned with a variety of surface acoustic wave devices and their applications in electronic systems. In subsequent chapters a number of theoretical topics are developed and are then, starting at Chapter 7, applied to the analysis of practical devices. However, in view of the breadth of the subject it is helpful to first survey the entire field briefly, thus clarifying the objectives of the later analysis. The survey, given in this chapter, also serves to introduce some of the terminology. Acoustic waves are described briefly, followed by a discussion of some bulk acoustic wave devices used in electronics. Some principles used in surface acoustic wave devices are then given, followed by an account of the devices most commonly found in electronic systems. Finally, the fabrication of the devices, and their applications, are discussed. The coverage given is necessarily very selective.

(a) *Acoustic Waves in Solids.* In a solid, an acoustic wave is a form of disturbance involving deformations of the material. Deformation occurs when the motions of individual atoms are such that the distances between them change, and this is accompanied by internal restoring forces which tend to return the material to its equilibrium state. If the deformation is time-variant, the motion of each atom is determined by these restoring forces and by inertial effects, and this can give rise to propagating wave motion with each atom oscillating about its equilibrium position. In most materials the restoring forces are proportional to the amount of deformation, provided the latter is small, and this can be assumed for most practical purposes. The material is then described as "elastic", and the waves are often called "elastic waves", though the term "acoustic waves" is used here. In an ideal elastic material, acoustic waves can propagate with no attenuation.

The simplest types of wave are the plane waves that can propagate in an infinite homogeneous medium. The deformation is harmonic in space and time, and all the atoms on a particular plane, normal to the propagation direction, have the same motion. There are two types of plane waves: longitudinal waves, in which the atoms vibrate in the propagation direction, and shear waves, in which the atoms vibrate in the plane normal to the propagation direction. These are directly analogous to the longitudinal and transverse waves that can propagate on an elastic string. The waves are non-dispersive at the frequencies of interest here, with velocities usually between 1000 and 10,000 m/s.

If the propagation medium is bounded, the boundary conditions can substantially alter the character of the waves. The case of primary interest here is the *surface acoustic wave*, whose existence was first shown by Lord Rayleigh. This type of wave can exist in a homogeneous material with a plane surface. It is guided along the surface, with its amplitude decaying exponentially with depth. The wave is strongly confined, with typically 90% of the energy propagating within one wavelength of the surface. It is non-dispersive, with a velocity of typically 3000 m/s. A bounded medium also supports many other types of acoustic waves, and the boundary conditions can substantially affect the nature of the waves. For example, in a plate with two plane parallel boundaries, a series of dispersive modes with different velocities can propagate. On the other hand, a medium with dimensions much larger than the wavelength can support waves with characteristics similar to those of waves in an infinite medium. The term *bulk waves* is often used to describe waves which are not bound to a surface.

Acoustic waves have a practical significance in many different contexts. A particular example is seismology. The motion of earthquakes involves both bulk and surface acoustic waves, and surface waves often contribute a major part of the motion because they are guided along the surface, spreading in two dimensions rather than three. The substantial seismological literature, extending back into the 19th century, established many of the important properties of acoustic waves, and has had an impact on many later developments. There are also many industrial uses of acoustic waves, in particular nondestructive testing, in which invisible defects such as cracks are detected without damaging the material.

(b) Bulk Wave Devices. Electronic applications are of prime concern here. Bulk waves have been used in several ways, taking advantage of two particular features. Firstly, acoustic velocities are very much less than electromagnetic velocities. Secondly, the attenuation can be low, though this depends on the choice of propagation medium, particularly at high frequencies. A device taking advantage of these features is the bulk-wave delay line [1, 2], which can take the form illustrated in Figure 1.1(a). This device consists of a solid propagation medium with a transducer at each end. The transducer at one end generates an acoustic wave when an oscillatory voltage is applied to it, and the transducer at the other end generates a voltage in response to the incident wave. The output voltage waveform is thus a delayed replica of the input waveform, with the delay determined by the acoustic path length and velocity. The low velocity enables large delays to be obtained compactly, typically a few microseconds for each cm of the propagation path.

For frequencies below about 50 MHz the propagation medium is usually fused quartz or glass, and the transducers are usually parallel-sided plates of a ceramic material. Sometimes the device is made more compact by using plane facets on the propagation medium to reflect the waves, thus folding the propagation path. In this

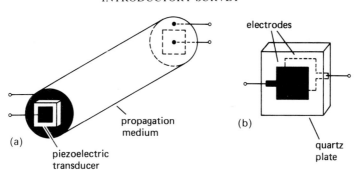

FIGURE 1.1. Devices using bulk acoustic waves. (a) delay line, (b) resonator.

way, delays up to 1 ms are achievable. Applications include radar systems requiring delay lines for moving target indication, and television receivers which require a delay corresponding to one line scan, about 60 μs. The acoustic wave can also be used to diffract a beam of light in a manner similar to a diffraction grating, and this phenomenon may be used to detect the acoustic wave and to measure its frequency.

At higher frequencies a crystalline propagation medium is used in order to obtain acceptably low attenuation. For example, at 1 GHz sapphire (Al_2O_3) gives about 0.3 dB attenuation per microsecond of delay, at room temperature. The transducers are usually plates of zinc oxide or lithium niobate. Such devices can operate at frequencies up to about 5 GHz, with delays up to about 10 μs.

The transducers in these devices make use of the piezoelectric effect [3]. This phenomenon is a property of many materials, and couples acoustic deformations of the material to electric fields. A transducer consists of a parallel-sided plate of piezoelectric material, firmly bonded on to the propagation medium. An oscillatory voltage is applied to the plate by means of electrodes (shown solid in Figure 1.1(a)), causing the plate to vibrate and thus generate acoustic waves. The piezoelectric effect is used in a wide variety of acoustic devices.

Another common acoustic device is the crystal resonator [4] shown in Figure 1.1(b). This consists of a parallel-sided plate of crystalline quartz with electrodes on both sides. If the major dimensions are much larger than the thickness, the plate resonates at a frequency such that its thickness equals half the acoustic wavelength, and at harmonics of this frequency. Quartz is piezoelectric, so the acoustic resonances can be excited electrically. In the familiar crystal-controlled oscillator the resonator is incorporated in an electrical oscillator circuit to control its frequency. The quartz resonator gives a very high Q-factor, up to 10^6, and excellent temperature stability. It is very widely used in electronic systems, particularly for communications. Frequencies up to about 50 MHz are obtainable, the limitation being that higher frequencies require thinner, more fragile, crystals.

The crystal resonator is also used in bandpass filters, designed to pass signals in some specified band of frequencies and reject signals at other frequencies. These may take the form of ladder circuits, incorporating a number of resonators coupled electrically. Alternatively, acoustic coupling may be obtained by fabricating the electrodes for several resonators on a single plate of quartz. With an appropriate spacing between the electrodes, a controlled amount of coupling is obtained by means of the evanescent acoustic fields which surround each resonator. These devices are known as "monolithic crystal filters". They are practicable up to about 50 MHz, giving bandwidths of typically a few kHz [5], and are used in telecommunications systems.

(c) Surface Wave Technology. In this book we are concerned with applications of surface acoustic waves in electronics. The use of surface waves in electronic devices was first considered in the early 1960's, and since then there has been a substantial growth of research into methods of generating and manipulating the waves, and in developing practical devices for use in a wide range of electronic applications. Some general literature on the subject is listed in Refs [6–22]. These include books, special issues of journals, and conference proceedings. A considerable amount of literature appears in the Proceedings of the annual IEEE Ultrasonics Symposium, and papers from the 1970–1977 Proceedings are published in a collected edition [14].

As for bulk waves, surface waves are attractive for electronics applications since they offer low velocity non-dispersive propagation, with low attenuation up to microwave frequencies. There is however a significant additional advantage since the propagation path, at the surface of the material, is accessible. This implies, at least in principle, a considerable degree of versatility. Because two dimensions are available rather than one, there is much more scope to exploit methods of generating and detecting the waves, or of modifying them as they propagate, and considerable structural complexity is feasible. A similar argument applies in the field of semiconductor devices, where the use of planar technology for integrated circuits has led to a remarkable growth in sophistication and complexity. In fact, the technology of integrated circuits has had a very direct bearing on the development of surface-wave devices, because of the range of fabrication techniques that it has made available. Established techniques of particular relevance include the deposition of thin films of various materials, etching of the propagation medium itself, and lithography for defining complex geometries with high precision. These techniques enable structures of considerable complexity to be made quite conveniently; moreover, in many cases they are also economically effective and suitable for large-scale production.

In the past twenty years a wide variety of techniques has been developed for use in surface-wave devices. Methods have been developed for electrically generating and detecting the waves (that is, for transduction), for reflecting, guiding, focussing and amplifying the waves, and for introducing controlled dispersion. These methods employ a variety of physical principles. As in bulk wave devices an important factor is the use of piezoelectric materials, though for surface waves the usage is somewhat different in that the propagation medium itself is piezoelectric. Some of the uses of piezoelectricity in surface-wave devices are illustrated in Figure 1.2. For a piezoelectric material, a propagating surface wave is accompanied by an electric field localised at

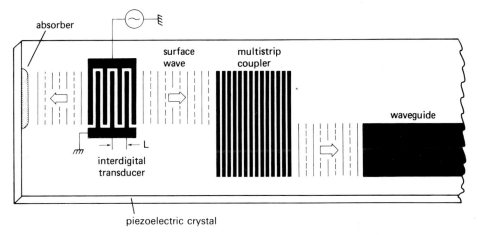

FIGURE 1.2. Metal film components using the piezoelectric effect.

the surface, and this enables the wave to be generated by applying a voltage to an array of metal electrodes on the surface. The electrode array is known as an *interdigital transducer*, and will be considered later in more detail. The transducer can also be used to detect surface waves, producing an electrical output waveform, and is used in all the devices under consideration here. In another application of piezoelectricity, a set of metal strips in the path of a surface wave can be used to generate a secondary surface wave, which may be displaced laterally with respect to the input wave (Figure 1.2), or may propagate in a different direction. This principle is used in the *multi-strip coupler*, which has a variety of forms with many different applications. An array of metal strips may also be used to reflect surface waves, and with two such arrays a resonant cavity can be formed. Another consequence of piezoelectricity is that a metal strip on the surface may be used as a *waveguide* for surface waves, enabling a narrow beam to propagate long distances without diffraction spreading. However, this method of controlling diffraction is necessary only for beam widths less than about five wavelengths, and in most practical cases larger widths are used.

In all the above examples the structure is simply a piezoelectric medium with a metal film on the surface, etched to give an appropriate geometry. Owing to the simplicity of the structure, and the availability of convenient fabrication methods, nearly all surface-wave devices use piezoelectric materials. Crystalline materials are usually chosen in order to obtain low attenuation of the waves, and the commonest choices are quartz and lithium niobate.

In addition to the direct use of piezoelectricity, there are several other principles that can be employed. Some devices make use of grooves etched in the surface of the substrate, in order to reflect surface waves, or to guide them. Dielectric films can be deposited, and can be used to introduce dispersion or to guide the wave. It is also possible to deposit a piezoelectric film, such as zinc oxide (ZnO), and then deposit metal electrodes on top. This enables an interdigital transducer to be fabricated on a non-piezoelectric substrate. Some devices make use of non-linear effects associated with the propagating surface wave. The non-linearity is weak, but in some materials, notably lithium niobate, is strong enough for useful interactions to be obtained. The prime example is the surface-wave convolver in which two surface waves are mixed, giving an output at the sum frequency; this device is used to correlate coded waveforms.

Interaction of surface waves with light has received much attention. The waves scatter light in a manner similar to a diffraction grating. This effect can be used to measure the distribution of surface waves over the surface, a procedure known as probing. Since the light is diffracted through an angle dependent on the frequency of the surface wave, the frequency can be measured. This principle may be used for electronic frequency measurement, in a device known as a Bragg cell; the electrical signal, whose frequency is required, is converted into a surface wave by an interdigital transducer.

The techniques described above have been used in a wide variety of surface-wave devices with applications in many electronic systems, notably in radar, communications and broadcasting. In many cases the function of a device is that of linear *signal processing*, that is, an electrical input waveform is applied to the device, which then produces an electrical output waveform linearly related to the input in a prescribed manner. In the terminology of systems analysis, such a device is called a linear filter. Examples are delay lines, bandpass filters and filters for correlating complex waveforms. To appreciate the operation of these devices, we first need to consider the interdigital transducer in more detail.

(d) Transducers and Delay Lines. A key feature of all the devices for electronics applications is the interdigital transducer for generating and receiving surface waves, illustrated in Figure 1.2. This transducer was first used for surface wave excitation by White and Voltmer [23] in 1965, though it is also referred to in earlier patents [24]. There are in fact many other types of transducer for surface waves [9], but these will not be considered here; most of them are not compatible with planar technology, and are not used in devices for electronics applications.

The interdigital transducer generates surface waves by exploiting the piezoelectric effect. As shown in Figure 1.2, the transducer has a set of identical electrodes connected alternately to two metal bus-bars. When an oscillatory voltage is applied, the transducer generates an electric field which is spatially periodic, with its period, L, equal to the spacing of the electrodes connected to one of the bus-bars. Owing to the piezoelectric effect, a corresponding pattern of mechanical displacements is also produced. Efficient coupling to surface waves occurs if the transducer period L is equal to or close to the surface-wave wavelength, and this requires an appropriate frequency for the applied voltage. Typically, the transducer will be designed for operation at, say, 100 MHz, where the wavelength is about 32 μm. The width of each electrode, equal to one quarter of the wavelength, is then about 8 μm. Owing to the symmetry, the transducer generates surface waves equally in two opposite directions,

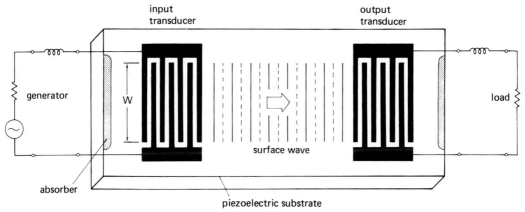

FIGURE 1.3. Interdigital delay line.

so that it is *bidirectional*. Usually, the waves in one direction are not required, and are eliminated by an absorber comprising a lossy material applied to the surface.

The simplest type of surface-wave device is a delay line employing two such transducers, one to generate the waves and one to receive them, as shown in Figure 1.3. The propagation medium, often called the *substrate*, is a piezoelectric crystal typically 1 mm thick. An electrical signal applied to the input transducer is converted to a corresponding surface wave, which causes a voltage to appear on the output transducer after a delay determined by the transducer separation and the surface wave velocity. Provided the input signal is confined to a frequency band in which the transducers are effective, there is little distortion because the wave is non-dispersive. Typical delays are from 1 to 50 μs.

Practical transducers can be quite efficient, converting most of the available electrical power into surface wave power. However, half of the power is radiated in an unwanted direction, giving a loss of 3 dB, and in a delay line with two transducers this factor contributes 6 dB to the total insertion loss. Losses due to other causes can be small. The surface wave propagates with little attenuation, and diffraction spreading can be minimised by using a sufficiently wide aperture (W in Figure 1.3), so that the output transducer is in the near field of the input transducer. For low loss one or more lumped components are usually added to match the transducer electrically to the source or load. The transducer impedance is largely capacitive, and often it is sufficient to tune it using a series inductor, as in Figure 1.3. The aperture W influences the transducer impedance and the diffraction spreading, but can often be chosen such that minimal diffraction spreading and an impedance convenient for matching are both obtained. Typical apertures are 20 to 100 wavelengths, or a few mm, and are convenient for fabrication.

With appropriate design, practical delay lines can give insertion losses of 10 dB or less. However, the devices are usually designed to give larger losses in order to reduce reflections. It is a consequence of the bidirectional nature of the transducer that, when it is well matched to an electrical source or load, it reflects incident surface waves quite strongly. This gives rise to an unwanted additional output signal known as the *triple-transit* signal, due to surface waves traversing the device three times. The triple-transit signal is often suppressed by deliberately avoiding a good electrical match to the source and load, and in consequence the insertion loss usually exceeds 15 dB. However, some more complex types of transducer are unidirectional, generating surface waves in only one direction, and these enable low losses to be obtained while still suppressing the unwanted reflections.

(e) Main Surface-Wave Devices. So far, we have considered only the simplest form of interdigital transducer. The transducer design can however be modified in a variety of ways, enabling the device to process an applied electrical signal in a prescribed manner, for example, to reject unwanted frequency components. Signal processing is one of the commonest uses of surface wave devices, and the versatility of the interdigital transducer is a crucial factor in this context. The two commonest modifications are to vary the electrode lengths and to vary the pitch. Transducers which do not use these modifications, such as the transducers in Figure 1.3, are described as *uniform*.

The technique of varying the electrode lengths is known as *apodisation* and is illustrated in Figure 1.4, which shows a device with one apodised and one uniform transducer. For convenience, it is assumed here that the uniform transducer is much shorter than the apodised transducer; for this case, the response of the device as a whole is essentially determined by the apodised transducer. The effect of apodisation can be appreciated by supposing that a short electrical pulse is applied to the uniform transducer. A short packet of surface-wave energy is produced, and this travels along the surface of the substrate, scanning the apodised transducer. At any instant, the output voltage produced by the apodised transducer depends on the amount by which its electrodes overlap at the location of the scanning surface wave packet. Thus the

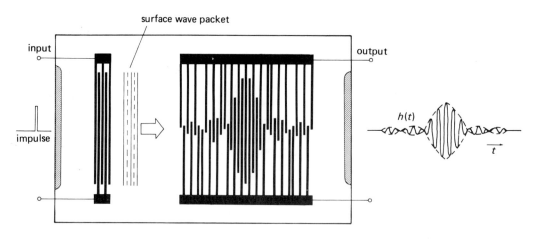

FIGURE 1.4. Bandpass filter using apodised transducer.

output waveform has an amplitude, as a function of time, directly related to the overlap of the electrodes as a function of position, with the time-scale related to the position-scale by the surface wave velocity. The output amplitude is approximately proportional to the electrode overlaps.

The output waveform produced in response to a short pulse is known as the *impulse response*, designated $h(t)$. The Fourier transform of $h(t)$ is the frequency response of the device, giving the output amplitude and phase when a c.w. signal is applied to the device, as functions of frequency. The frequency response can therefore be calculated quite straightforwardly from the electrode overlaps. More significantly, this procedure can be reversed in order to design a transducer to give some specified frequency response: Fourier transformation gives the required impulse response $h(t)$, and the amplitude of this function then gives the required apodisation for the electrodes. This demonstrates a very high degree of versatility, since the method may be used for *any* specified frequency response, provided it is consistent with the limitations of the technology.

This principle is commonly used in surface-wave *bandpass filters*, where the usual requirement is that the device should pass c.w. signals with frequencies within a specified band and reject signals with other frequencies. For this case the required impulse response has an amplitude of the form $(\sin \alpha t)/(\alpha t)$, with the constant α determined by the bandwidth, and the geometry is typified by Figure 1.4. In practice there are many complications affecting the performance and different approaches to the design may be adopted, depending on the requirements.

The direct relationship between the transducer geometry and the impulse response also applies to transducers whose electrode pitch varies. Figure 1.5 shows a device using a transducer of this type and a short uniform transducer. When the uniform transducer is impulsed a short surface wave packet is produced, and this scans along the output transducer. At any one time, the frequency of the output voltage depends on the pitch of the electrodes at the location of the scanning pulse, so that a frequency-swept output pulse is produced. If the periodicities at the two ends of the transducer are L_1 and L_2, the frequency sweeps from $f_1 = v/L_1$ to $f_2 = v/L_2$, where v is the surface wave velocity. The duration of the output pulse, T, corresponds to the length of the transducer, vT. A frequency-swept pulse of this type is often called a *chirp* pulse, and the surface-wave device is called a *chirp filter*. It is also described as a dispersive delay line, because the group delay of the device varies with frequency; the delays at frequencies f_1 and f_2 differ by an amount T.

The main application of the chirp filter occurs in radar systems, where the device is used to perform *pulse compression*. This process is illustrated in the lower part of Figure 1.5. A chirp waveform is applied to the device, the waveform being similar to the device impulse response but reversed in time, so that the frequency starts at f_2 and ends at f_1. The short uniform transducer generates a surface wave pulse with a corresponding form, and this propagates along the surface. At a particular instant, the peaks and troughs of the surface wave pulse match the electrode positions of the long transducer, and the output voltage peaks, producing a narrow pulse. The width of the output pulse is approximately $1/B$, where $B = f_1 - f_2$ is the bandwidth. The ratio of the input and output pulse widths is called the compression ratio, and is equal

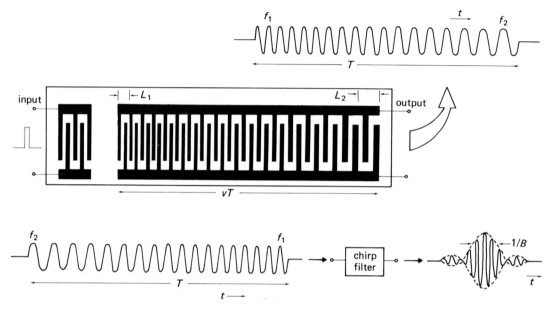

FIGURE 1.5. Upper: interdigital chirp filter and its impulse response. Lower: pulse compression.

to the time-bandwidth product of the device, TB. Typical time-bandwidth products are in the range 50 to 500.

The chirp filter is one example of a device designed to have its impulse response corresponding to the time-reverse of a specified waveform. Such a device is called a *matched filter*, and the process of compressing the waveform in the filter is also called "correlation". The filter responds most strongly to the waveform that it is matched to, discriminating against other waveforms and, in particular, against noise. This feature can be used to improve the sensitivity of a radar system. In a pulse-compression radar, the transmitter emits a frequency-swept chirp pulse. The echo received from a target has the same form, and is compressed by a chirp filter in the receiver. As in all electronic systems, noise is also present, but the filter discriminates against it. Thus a weak echo, initially obscured by noise, is processed to produce a pulse exceeding the noise level, and can therefore be detected. This principle is frequently used in present-day radar systems.

An alternative type of chirp filter is the *Reflective Array Compressor*, or RAC, shown in Figure 1.6. This device has two arrays of inclined shallow grooves, with graded periodicity, arranged to reflect surface waves through 90°. The surface waves are generated by a uniform interdigital transducer at one end of the device, and are then reflected twice by the grooves so that they reach the interdigital output transducer, located at the same end as the input transducer. Because the grooves are shallow, with a depth typically 1% of the wavelength, the reflection coefficient of any one groove is small. However, at any one frequency the reflected waves from many grooves add coherently, producing a much larger reflected wave amplitude. This

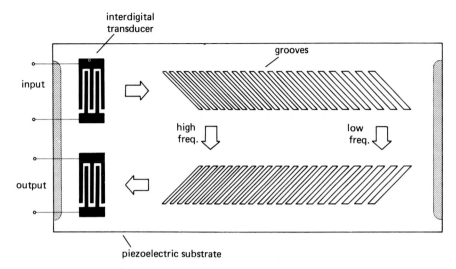

FIGURE 1.6. Reflective Array Compressor (RAC).

requires the pitch of the grooves to correspond with the surface-wave wavelength. Because the pitch varies along the length of the device, different frequencies are reflected at different locations, giving different path lengths. The delay therefore depends on frequency, and the device is a type of dispersive delay line. The RAC can be used for pulse compression of chirp waveforms, and enables very large compression ratios, up to 10,000, to be obtained. In contrast, the interdigital chirp filter described above is, for technical reasons, limited to compression ratios below 1,000.

Matched filtering is also applicable to other types of waveform, notably to phase-shift-keyed, or PSK, waveforms. For such waveforms, an appropriate matched filter is the surface-wave *PSK filter*, which is essentially a form of tapped delay line using interdigital transducers. This device is applicable to spread-spectrum communication systems, and to some radar systems.

Surface wave techniques can also be used to produce several types of stable *oscillator*, generally using quartz as the propagation medium because of its good temperature stability. One method is to use an interdigital delay line with an amplifier connected between the output and the input, forming a loop. The amplifier small-signal gain exceeds the loss of the delay line, so that the loop oscillates at a frequency related to the surface wave velocity. Alternatively a surface-wave resonator may be used. The resonator is basically two reflectors forming a surface wave cavity, the reflectors being periodic arrays of either metal strips or grooves. The resonator can give very high Q-factors, up to 10,000, giving good stability. The delay-line oscillator gives lower Q-factors, but has the advantage that the frequency can be made adjustable by incorporating a phase shifter in the loop external to the delay line. These techniques give highly stable c.w. sources with frequencies up to about 2 GHz, in contrast to the parallel plate resonator using bulk waves, which is limited to about 50 MHz.

(f) Fabrication. The manufacturing methods used for surface-wave devices have a strong bearing on device performance. In particular, the minimum line width obtainable determines the maximum frequency of operation, and for long delays the ability to process long substrates is required. A brief outline of the method is given here, and the reader is referred elsewhere [25] for further details.

Most devices consist essentially of metal patterns, such as transducers, deposited on crystalline piezoelectric substrates. These devices are made by photolithography, using a procedure exemplified by Figure 1.7. The substrate is carefully polished and cleaned to give a flat smooth surface of optical quality, free of extraneous particles or grease. A metal film is then deposited, usually by vacuum evaporation (Figure 1.7(a)). The film is usually of aluminium and is typically 0.1 to 0.3 μm thick. A thin underlay of chromium is often used to improve the adhesion. The sample is then coated with photo-resist, a solution of a photo-sensitive polymer, and is spun at high speed so that the resist becomes a thin uniform layer. Subsequent baking solidifies the resist, which is then exposed to ultraviolet light through a mask, as in Figure 1.7(b). The mask has opaque areas, usually of photographic emulsion or chromium film, corresponding to the areas to be metallised on the final device. The exposed areas of resist undergo a chemical change, and can then be removed by a developing solution, as in Figure 1.7(c). This exposes areas of metal which are removed chemically, and finally the remaining resist is dissolved away, leaving a metal pattern on the substrate corresponding to the pattern on the mask, as in Figure 1.7(d). With this process, line widths down to 0.5μm can be replicated with care, giving interdigital transducers operating at about 1.5 GHz. The most critical stage is the optical exposure, where the mask must be in contact with the sample, and a commercial mask aligner is normally used here. Owing to anisotropy, the angular alignment of the mask relative to the substrate is important, and an accuracy of 1° or better is often necessary. There are many variants to this basic process, and for very fine lines there are techniques using X-rays or electron beams instead of optical exposure, though these are not generally in commercial usage.

For the smaller devices, up to a few cm long, the process is very similar to standard techniques used in the semiconductor industry and is well suited to large-scale

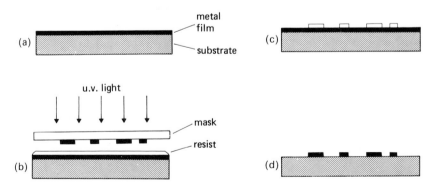

FIGURE 1.7. Fabrication using optical lithography.

production. The crystal substrate is often in the form of a disc of diameter 7.5 cm, with a straight edge cut as a reference for orientation. The required pattern is repeated many times on the mask so that many devices are fabricated simultaneously, and the disc is subsequently sawn to separate the individual devices. For larger devices, which are usually chirp filters or delay lines, the process is more specialised. Crystal substrates are available commercially with lengths up to about 25 cm, and present-day pattern generation machines can produce masks of this length with sub-micron accuracy. This enables interdigital devices to have delays up to about 50 μs. The devices are usually made individually, using specially adapted equipment for resist coating and mask alignment.

Devices using grooves, such as RAC's, require an additional fabrication stage. The grooves are usually cut by exposing the substrate to an ion beam, using a metal masking pattern on the surface to define the geometry.

The final stage of fabrication is packaging. Production devices are hermetically sealed in an inert atmosphere or vacuum, to exclude moisture (which may cause erosion of the metal) and surface contaminants (which cause attenuation of the surface waves). Standard packages, such as the circular TO-series or the dual-in-line types, are often used for the smaller devices, but for longer devices custom-designed packages are necessary. The substrates are mounted using adhesives, often of the elastomer type to minimise mechanical stress, since stress can affect the performance somewhat. Electrical connections are made by means of thin wires, typically of gold and with 25 μm diameter, attached to the substrate metallisation by thermo-compression or ultrasonic bonding. In some cases temperature regulation is necessary, so the package is mounted within a thermally insulating enclosure.

(g) Performance and Applications. The applicability of surface-wave devices to practical electronic systems is basically determined by the centre frequencies, bandwidths and delays obtainable, together with the performance and cost effectiveness. The upper limit for the centre frequency is determined by fabrication techniques, and for current photolithography is about 1.5 GHz. At low frequencies surface-wave devices become more bulky and expensive and other technologies become more suitable, for example bulk acoustic wave resonators for bandpass filters or digital techniques for signal processing. Consequently, surface-wave devices are not normally used below a few MHz. In most applications, this frequency range implies that the devices are used in the I.F. section of the system, and this has the important consequence that low insertion loss is not generally a priority. For example, bandpass filters have typically 15 to 30 dB loss, and this is acceptable for many applications. Bandwidths generally range from a minimum of about 100 kHz to a maximum of about 50% of the centre frequency. The delay obtainable ranges from about 0.1 to 50 μs, with the upper limit determined by the lengths of available crystal substrates. For the RAC, in which the surface-wave traverses the length of the substrate twice, the maximum delay becomes 100 μs.

These figures are well suited to a wide range of system requirements, particularly in radar and communication systems. In terms of quantity, the most widely used surface-wave device is the bandpass filter used in the I.F. section of colour television receivers. Bandpass filters are also used extensively in radar and communication systems, and in television broadcasting equipment. Chirp filters (interdigital and RAC) are widely used in radar systems. They can also serve a number of other functions, notably in the compressive receiver, a sub-system for frequency measurement with applications in electronic surveillance. Some radar systems make use of delay lines or PSK filters. In spread-spectrum communication systems PSK filters are used as matched filters, and surface-wave non-linear convolvers are also beginning to impact this area. It is emphasised that this description is selective, and more details are given elsewhere [19–22]. For example, Williamson [21] lists 45 separate types of surface-wave device, and gives an extensive list of systems using them. Some of the applications are discussed further in later chapters of this book.

The success of surface-wave devices in meeting these requirements can be ascribed to a number of factors. Foremost among these is the substantial degree of versatility due to the accessibility of the propagation path, a feature well illustrated by the variety of types of interdigital transducer. Accurate reproducibility is another key feature. The device characteristics are mainly determined by the geometry of the pattern on the mask, and by the acoustic properties of the substrate. Modern pattern generation machines produce masks of very high accuracy, and good reproducibility of the substrate properties is obtained by the use of single-crystal materials. This is particularly true for quartz substrates, which can also give excellent temperature stability. For this material the surface wave velocity, for example, can be controlled to within 50 parts per million or better. In addition, design techniques are sophisticated enough to exploit this accuracy, so that the design procedure does not limit the device accuracy obtainable. The accuracy is an important factor in the success of matched filters and oscillators. Other contributing factors are the simplicity of the fabrication procedure, using equipment already developed for semiconductor fabrication, and the low surface wave velocity, which enables long delays to be obtained in relatively short devices.

Chapter 2

Acoustic Waves in Elastic Solids

Many different types of acoustic wave can propagate in solid materials. Here we are particularly concerned with surface waves, though several other kinds of acoustic wave are also relevant to surface wave devices. This chapter gives a brief account of the analysis and properties of the waves. Propagation in piezoelectric materials is emphasised, because such materials are used in most surface-wave devices.

2.1. ELASTICITY IN ANISOTROPIC MATERIALS

We first describe the elastic behaviour of anisotropic materials, summarising the development given in more detail elsewhere [26–31]. It is convenient to consider the non-piezoelectric case first, and then consider piezoelectric materials later.

2.1.1. Non-piezoelectric materials

Elasticity is concerned with the internal forces within a solid and the related displacement of the solid from its equilibrium, or force-free, configuration. It is assumed here that the solid is homogeneous. The forces will be expressed in terms of the *stress*, T, while the displacements are expressed in terms of the *strain*, S.

We first consider the strain. Suppose that, in the equilibrium state, a particle in the material is located at the point $\mathbf{x} = (x_1, x_2, x_3)$. When the material is not in its equilibrium state, this particle is displaced by an amount $\mathbf{u} = (u_1, u_2, u_3)$, where the components u_1, u_2 and u_3 are in general functions of the coordinates x_1, x_2, x_3. Thus a particle with equilibrium position $\mathbf{x}$ has been displaced to a new position $\mathbf{x} + \mathbf{u}$. For the present, the displacement $\mathbf{u}$ is taken to be independent of time, t. Now clearly there will be no internal forces if $\mathbf{u}$ is independent of $\mathbf{x}$, since this simply denotes a displacement of the material as a whole. There will also be no forces if the material is rotated. To avoid these cases the strain at each point is defined by

$$S_{ij}(x_1, x_2, x_3) = \frac{1}{2}\left(\frac{\partial u_i}{\partial x_j} + \frac{\partial u_j}{\partial x_i}\right), \quad i, j = 1, 2, 3. \tag{2.1}$$

With this definition, any displacements or rotations of the material as a whole cause no strain, and the strain is related to the internal forces. The strain is a second-rank tensor and is clearly symmetrical:

$$S_{ij} = S_{ji} \tag{2.2}$$

so that only six of the nine components are independent.

The internal forces are described by a stress tensor T_{ij}. To define this, consider the plane $x_1 = x'_1$ within the material, where x'_1 is a constant. If the material is strained, the material on one side of the plane exerts a force on the material on the other side. The force may be in any direction, and may vary with the coordinates (x_2, x_3) in the plane. The stress is defined such that the force per unit area has an x_i-component equal to $T_{i1}(x'_1, x_2, x_3)$, with $i = 1, 2, 3$. This is the force exerted on the material at $x_1 < x'_1$. The force exerted on the material at $x_1 > x'_1$ is the negative of this. The definition applies for any value of x'_1, so we can write the stress as $T_{i1}(x_1, x_2, x_3)$. Similarly, we may consider forces on planes perpendicular to the x_2 and x_3 axes, defining the stresses in the same way, to arrive at the second-rank stress tensor $T_{ij}(x_1, x_2, x_3)$. Although we have only considered planes perpendicular to the coordinate axes, it can be shown that the forces acting on any plane can be deduced from this tensor. It can also be shown that the stress tensor is symmetric, that is,

$$T_{ij} = T_{ji}. \tag{2.3}$$

In most materials the stresses can be taken to be proportional to the strains, provided the strains are sufficiently small. If this is true, the material is said to be *elastic*. The relationship is a generalisation of Hooke's law, which states that stress is proportional to strain for the one-dimensional case. Unless stated otherwise, it will be assumed throughout that the material is elastic, and hence each component of the stress is given by a linear combination of the strain components. The coefficients required are given by the *stiffness tensor*, c_{ijkl}, defined such that

$$T_{ij} = \sum_k \sum_l c_{ijkl} S_{kl}, \quad i, j, k, l = 1, 2, 3. \tag{2.4}$$

The stiffness is a fourth rank tensor, with 81 elements. However many of these elements are related. The symmetry of S_{ij} and T_{ij}, equations (2.2) and (2.3), implies that the stiffness is unaltered if i and j are interchanged, or if k and l are interchanged, that is,

$$c_{jikl} = c_{ijkl}, \tag{2.5}$$

$$c_{ijlk} = c_{ijkl}, \tag{2.6}$$

Thus only 36 of the 81 elements are independent. It can also be shown, from thermodynamic considerations, that the second pair of indices can be interchanged with the first pair:

$$c_{klij} = c_{ijkl}. \tag{2.7}$$

This reduces the number of independent elements to 21. These elements are of course physical properties of the material under consideration, so that the number of independent components may well be reduced further by the symmetry of the material. For example, a crystalline material with cubic symmetry has only three independent elements. It should be noted that the coordinate axes x_1, x_2, x_3 will not in general be parallel to the axes of the crystal lattice.

Equation of motion. If the stress and strain are functions of time as well as position, the motion is subject to Newton's laws in addition to the above equations, and these constraints can be combined in the form of an equation of motion. Consider an elementary cube within the material, centred at $\mathbf{x}' = (x_1', x_2', x_3')$. The edges are parallel to the x_1, x_2, and x_3 axes, and each edge is of length δ. The material surrounding the cube exerts forces on all six faces. For the faces at $x_1 = x_1' \pm \delta/2$, the components of force in the x_i direction are $\pm \delta^2 T_{i1}(x_1' \pm \delta/2, x_2', x_3')$. The forces on the faces normal to x_2 and x_3 are obtained in the same way, and we add these to obtain the total force on the cube. Noting that δ is small, the total force has an x_i component

$$\delta^3 \left[\sum_j \frac{\partial T_{ij}}{\partial x_j} \right]_{\mathbf{x}'}.$$

This must be equal to the acceleration $\partial^2 u_i(\mathbf{x}')/\partial t^2$, multiplied by the mass $\varrho \delta^3$, where ϱ is the density. This is valid for all points $\mathbf{x}'$, and hence

$$\varrho \frac{\partial^2 u_i}{\partial t^2} = \sum_j \frac{\partial T_{ij}}{\partial x_j}, \quad i, j = 1, 2, 3, \tag{2.8}$$

which is the equation of motion.

2.1.2. Piezoelectric Materials

Piezoelectricity is the phenomenon which, in many materials, couples elastic stresses and strains to electric fields and displacements. It occurs only in anisotropic materials whose internal structure lacks a centre of symmetry. It occurs in many crystal classes but is often weak, thus having little effect on the elastic behaviour. However, here we are concerned with devices that make crucial use of piezoelectricity, so it is necessary to take account of the effect in the analysis. Only insulating materials will be considered here.

In a homogeneous piezoelectric insulator, the stress components T_{ij} at each point are dependent on the electric field $\mathbf{E}$ (or, equivalently, the electric displacement $\mathbf{D}$) in addition to the strain components S_{ij}. Assuming all these quantities are small enough we can take the relationship to be linear, so that T_{ij} is given by the linear relation

$$T_{ij} = \sum_k \sum_l c^E_{ijkl} S_{kl} - \sum_k e_{kij} E_k. \tag{2.9}$$

Here, the superscript on c^E_{ijkl} identifies this as the stiffness tensor for constant electric field; that is, if $\mathbf{E}$ is held constant this tensor relates changes of T_{ij} to changes of S_{kl}. Similarly, the electric displacement $\mathbf{D}$ is usually determined by the field $\mathbf{E}$ and the permittivity tensor ε_{ij}, but in a piezoelectric material it is also related to the strain:

$$D_i = \sum_j \varepsilon^S_{ij} E_j + \sum_j \sum_k e_{ijk} S_{jk}, \tag{2.10}$$

where ε^S_{ij} is the permittivity tensor for constant strain. The forms of these equations are justified by thermodynamic arguments which are not considered here. The tensor

e_{ijk}, relating elastic to electric fields, is called the piezoelectric tensor. From equation (2.9) and the symmetry of T_{ij}, this tensor has the symmetry

$$e_{ijk} = e_{ikj}. \tag{2.11}$$

It is equally valid to relate **D** to the stress instead of the strain, and this can be done by eliminating S_{ij} from equations (2.9) and (2.10). The result is expressed in the form

$$D_i = \sum_j \varepsilon_{ij}^T E_j + \sum_j \sum_k d_{ijk} T_{jk}, \tag{2.12}$$

where the new tensors ε_{ij}^T and d_{ijk} are related in a rather complicated manner to the tensors in equations (2.9) and (2.10). The tensor ε_{ij}^T is the permittivity tensor for constant stress. We can also eliminate **E** between equations (2.9) and (2.10) to obtain an equation giving T_{ij} in terms of S_{kl} and **D**; the coefficients of S_{kl} then give a stiffness tensor for constant electric displacement.

The mechanical equation of motion, equation (2.8), is valid for a piezoelectric material. It is convenient to express this in terms of the displacements u_i and the electric potential Φ. Since elastic disturbances travel much more slowly than electromagnetic ones the electric field can be taken to be quasi-static, that is, it is given by the gradient of the potential, so that

$$E_i = -\partial \Phi / \partial x_i. \tag{2.13}$$

Using this relation in equation (2.9), and equation (2.1) for the strain, the equation of motion becomes

$$\varrho \frac{\partial^2 u_i}{\partial t^2} = \sum_j \sum_k \left[e_{kij} \frac{\partial^2 \Phi}{\partial x_j \partial x_k} + \sum_l c_{ijkl}^E \frac{\partial^2 u_k}{\partial x_j \partial x_l} \right]. \tag{2.14a}$$

In addition there are no free charges, since the material is assumed to be an insulator. Hence div **D** $= 0$, and using equation (2.10) this gives

$$\sum_i \sum_j \left[\varepsilon_{ij}^S \frac{\partial^2 \Phi}{\partial x_i \partial x_j} - \sum_k e_{ijk} \frac{\partial^2 u_j}{\partial x_i \partial x_k} \right] = 0. \tag{2.14b}$$

Equations (2.14) give four equations relating the four quantities u_i and Φ, and hence the motion is determined if appropriate boundary conditions are specified.

When specifying the stiffness and piezoelectric tensors for a particular material, it is usual to adopt a special notation known as the matrix notation. This is convenient because it reduces the number of elements to be specified. The stiffness tensor c_{ijkl}^E has only 36 independent components because of its symmetry, equations (2.5) and (2.6), and is expressed in terms of a stiffness matrix c_{mn}^E. This is defined by

$$c_{mn}^E = c_{ijkl}^E, \quad m, n = 1, 2, \ldots, 6. \tag{2.15}$$

where m is related to i and j by

$$m = i \qquad \text{for } i = j$$
$$m = 9 - i - j \quad \text{for } i \neq j, \qquad i, j = 1, 2, 3.$$

A similar definition relates n to k and l. A simplified piezoelectric matrix is also used, defined by

$$e_{km} = e_{kij}, \quad k = 1, 2, 3, \quad m = 1, 2, \ldots, 6 \tag{2.16}$$

with m related to i and j as above.

2.2. WAVES IN ISOTROPIC MATERIALS

In this book we are concerned mainly with wave motion in anisotropic materials. The complexity of the equations of elasticity, described in the previous section, is such that the properties of the waves can usually be found only by numerical techniques. In contrast, solutions for isotropic materials are much easier to obtain, and since they have many features in common with the solutions for anisotropic materials it is helpful to consider the isotropic case first [30–39]. Numerical examples are given here for fused quartz which has acoustic properties somewhat similar to crystalline quartz, used in many surface wave devices.

In an isotropic material the stiffness tensor c_{ijkl} has only two independent components. From the symmetry it can be shown [30] that the stiffness can be written in the form

$$c_{ijkl} = \lambda \delta_{ij} \delta_{kl} + \mu(\delta_{ik}\delta_{jl} + \delta_{il}\delta_{jk}), \tag{2.17}$$

where $\delta_{ij} = 1$ for $i = j$ and $\delta_{ij} = 0$ for $i \neq j$. The constants λ and μ are known as Lamé constants and in practice are always positive; μ is also called the rigidity. Substituting into equation (2.4), the stress can be written in the form

$$T_{ij} = \lambda \delta_{ij} \Delta + 2\mu S_{ij} \tag{2.18}$$

where

$$\Delta = \sum_i S_{ii} = \sum_i \frac{\partial u_i}{\partial x_i}. \tag{2.19}$$

The equation of motion, equation (2.8), becomes, on substituting equation (2.18),

$$\varrho \frac{\partial^2 u_j}{\partial t^2} = (\lambda + \mu) \frac{\partial \Delta}{\partial x_j} + \mu \nabla^2 u_j \tag{2.20}$$

where

$$\nabla^2 = \sum_i \frac{\partial^2}{\partial x_i^2}.$$

2.2.1. Plane Waves

We first consider an infinite medium supporting plane waves, with frequency ω, in which the displacement **u** takes the form

$$\mathbf{u} = \mathbf{u}_0 \exp[j(\omega t - \mathbf{k} \cdot \mathbf{x})], \tag{2.21}$$

where $\mathbf{u}_0$ is a constant vector, independent of $\mathbf{x}$ and t. The actual displacement is the real part of equation (2.21), but the complex form can be used throughout the analysis because the equations are linear. The wave vector is $\mathbf{k} = (k_1, k_2, k_3)$, which gives the direction of propagation. The wavefronts are solutions of $\mathbf{k} \cdot \mathbf{x} =$ constant, and are perpendicular to $\mathbf{k}$. The phase velocity of the wave is $V = \omega/|\mathbf{k}|$. With this form for $\mathbf{u}$, we have $\partial u/\partial x_j = -jk_j u$, and on substituting into equation (2.20) we obtain

$$\omega^2 \rho u_j = (\lambda + \mu)(\mathbf{k} \cdot \mathbf{u})k_j + \mu |\mathbf{k}|^2 u_j, \quad j = 1, 2, 3$$

where

$$|\mathbf{k}|^2 = k_1^2 + k_2^2 + k_3^2.$$

Substituting for u_j using equation (2.21), and writing the result in vector form, gives

$$\omega^2 \rho \mathbf{u}_0 = (\lambda + \mu)(\mathbf{k} \cdot \mathbf{u}_0)\mathbf{k} + \mu |\mathbf{k}|^2 \mathbf{u}_0. \tag{2.22}$$

Here there are two terms parallel to $\mathbf{u}_0$ and one term parallel to $\mathbf{k}$, with the latter including the scalar product $\mathbf{k} \cdot \mathbf{u}_0$. There are therefore two cases to consider. Firstly, if $\mathbf{u}_0$ is perpendicular to $\mathbf{k}$ the scalar product $\mathbf{k} \cdot \mathbf{u}_0$ is zero, and the remaining terms in the equation are parallel. Secondly, if $\mathbf{u}_0$ is not perpendicular to $\mathbf{k}$ the product $\mathbf{k} \cdot \mathbf{u}_0$ is non-zero, so that for non-trivial solutions we must have $\mathbf{u}_0$ parallel to $\mathbf{k}$. These two cases give shear wave solutions and longitudinal wave solutions, respectively.

Taking $\mathbf{u}_0$ to be perpendicular to $\mathbf{k}$ gives *shear*, or transverse, waves. For these the wave vector is denoted by $\mathbf{k}_t$, and equation (2.22) gives

$$|\mathbf{k}_t|^2 = \omega^2 \rho/\mu.$$

The phase velocity for shear waves is denoted by V_t, equal to $\omega/|\mathbf{k}_t|$, so that

$$V_t = \sqrt{\mu/\rho}, \tag{2.23}$$

taking V_t to be positive. Since this is independent of the frequency ω, the wave is non-dispersive. The displacement $\mathbf{u}_0$ can have any direction in the plane of the wavefront, perpendicular to $\mathbf{k}_t$.

For *longitudinal* waves we consider solutions of equation (2.22) in which $\mathbf{u}_0$ is parallel, or anti-parallel, to $\mathbf{k}$. Thus $\mathbf{k}$ is given by

$$\mathbf{k} = \pm \mathbf{u}_0 \frac{|\mathbf{k}|}{|\mathbf{u}_0|}. \tag{2.24}$$

With this relation we find

$$(\mathbf{k} \cdot \mathbf{u}_0)\mathbf{k} = \mathbf{u}_0 |\mathbf{k}|^2$$

irrespective of the sign in equation (2.24). In this case the wave vector is denoted $\mathbf{k}_l$ and substitution into equation (2.22) gives

$$|\mathbf{k}_l|^2 = \omega^2 \rho/(\lambda + 2\mu).$$

The velocity for this case is denoted by V_l and is thus given by

$$V_l = \sqrt{\frac{\lambda + 2\mu}{\rho}} \tag{2.25}$$

and hence the wave is non-dispersive. Since λ and μ are always positive, the velocity of longitudinal waves is always greater than the velocity of shear waves.

The velocities are typically in the region of 3000 m/s for shear waves, and 6000 m/s for longitudinal waves. For example, in fused quartz [44], $V_t = 4100$ m/s and $V_l = 6050$ m/s. The displacements involved are usually very small. In fused quartz, a shear wave with a power density of 1 mW/mm^2 and a frequency of 10 MHz has maximum displacements of about 0.3 nm, and a similar figure applies for longitudinal waves. This displacement is some six orders of magnitude smaller than the wavelengths.

We now consider several configurations involving plane boundaries. The solutions are obtained by summing plane wave solutions of the above types. We consider solutions in which the displacements are proportional to $\exp(-j\beta x_1)$, with the x_1 axis parallel to the boundaries. In each case this gives one or more characteristic solutions, and these are often called modes. It should however be noted that these modes are not the only solutions that can exist.

In most cases we are concerned with waves propagating on a half-space of some material, which may have a layer of another material on the surface. The solutions of most interest are described as *surface acoustic waves* (SAW) or, more simply, as surface waves. These are solutions for which the displacements in the half-space decay rapidly in the direction normal to the surface, and the energy of the wave is transported parallel to the surface. The Rayleigh wave to be considered next is one type of surface acoustic wave.

2.2.2. Rayleigh Waves in a Half-Space

Consider an isotropic medium with infinite extent in the x_1 and x_2 directions but with a boundary at $x_3 = 0$, so that the medium occupies the space $x_3 < 0$. The space $x_3 > 0$ is a vacuum. The surface wave solution for this case is named after Lord Rayleigh, who first discovered it [37]. We assume propagation in the x_1 direction, so that the wavefronts are parallel to x_2, as depicted in Figure 2.1. The (x_1, x_3) plane, which contains the surface normal and the propagation direction, is known as the *sagittal plane*. The solution must satisfy the equation of motion,

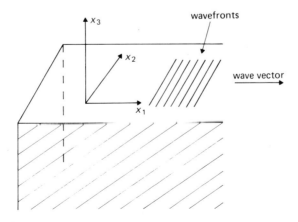

FIGURE 2.1. Axes for surface wave analysis.

equation (2.20), and the boundary conditions that there should be no forces on the free surface at $x_3 = 0$.

The solution is found by adding two components corresponding to plane shear and longitudinal waves, and these components are described as *partial* waves. The wave vectors, $\mathbf{k}_t$ and $\mathbf{k}_l$, have magnitudes given by

$$|\mathbf{k}_t|^2 = \omega^2/V_t^2; \qquad |\mathbf{k}_l|^2 = \omega^2/V_l^2.$$

Now since the Rayleigh wave has no variation in the x_2 direction these vectors must have no x_2-components, and are therefore in the sagittal plane. In the x_1-direction the displacements are assumed to vary as $\exp(-j\beta x_1)$, where β is the wavenumber of the Rayleigh wave. The x_1-components of $\mathbf{k}_t$ and $\mathbf{k}_l$ must therefore be equal to β. The x_3-components are denoted respectively by T and L, and we thus have

$$T^2 = \omega^2/V_t^2 - \beta^2, \qquad (2.26)$$

$$L^2 = \omega^2/V_l^2 - \beta^2. \qquad (2.27)$$

The displacement of the longitudinal wave, $\mathbf{u}_l$, must be parallel to the wave vector $\mathbf{k}_l = (\beta, 0, L)$. Thus, ommitting a factor $\exp(j\omega t)$, we can write

$$\mathbf{u}_l = A(1, 0, L/\beta) \exp[-j(\beta x_1 + L x_3)], \qquad (2.28)$$

where A is a constant. The displacement of the shear wave, $\mathbf{u}_t$, is perpendicular to the wave vector $\mathbf{k}_t = (\beta, 0, T)$. This does not determine the direction of $\mathbf{u}_t$, but we assume for the present that $\mathbf{u}_t$ is the sagittal plane (as is $\mathbf{u}_l$). Thus $\mathbf{u}_t$ is given by

$$\mathbf{u}_t = B(1, 0, -\beta/T) \exp[-j(\beta x_1 + T x_3)], \qquad (2.29)$$

where B is a constant. The total displacement $\mathbf{u}$ is the sum

$$\mathbf{u} = \mathbf{u}_t + \mathbf{u}_l. \qquad (2.30)$$

Now, in the x_3-direction, the shear wave displacement $\mathbf{u}_t$ varies as $\exp(-jTx_3)$. For surface wave solution the displacement must decay for negative x_3, and hence the value of T must be positive imaginary. Thus β must be large enough to make the right hand side of equation (2.26) negative, and since the Rayleigh wave velocity V_R will be given by $V_R = \omega/\beta$ we must have

$$V_R < V_t. \qquad (2.31)$$

Similarly, L must be positive imaginary and this implies $V_R < V_l$, but this is already implied by equation (2.31) because V_l is greater than V_t. With T and L imaginary, the partial wave solutions, $\mathbf{u}_l$ of equation (2.28) and $\mathbf{u}_t$ of equation (2.29), are no longer plane waves; however, they are still valid as solutions for an infinite medium, satisfying the equation of motion, equation (2.20). If the total displacement $\mathbf{u}$ of equation (2.30) is to be a valid solution for the half-space, it must also satisfy the boundary conditions that $T_{13} = T_{23} = T_{33} = 0$ at the surface $x_3 = 0$, where the stresses T_{i3} are given by equation (2.18). This gives two linear homogeneous equations

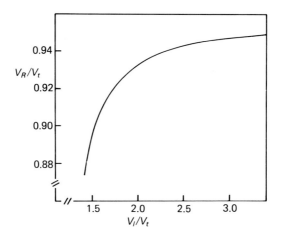

FIGURE 2.2. Normalised Rayleigh wave velocity for isotropic materials.

relating the constants A and B. For non-trivial solutions the determinant of the coefficients must be zero, and this gives

$$(T^2 - \beta^2)^2 + 4\beta^2 LT = 0. \qquad (2.32)$$

Using equations (2.26) and (2.27) for T and L, and defining the phase velocity $V = \omega/\beta$, this gives

$$[2 - V^2/V_t^2]^2 = 4[1 - V^2/V_l^2]^{1/2}[1 - V^2/V_t^2]^{1/2}. \qquad (2.33)$$

For Rayleigh waves we require a solution giving V^2 a real positive value, less than V_t^2. The equation has only one such solution, which is denoted V_R^2, and we take V_R to be positive. The ratio V_R/V_t is determined by the ratio V_l/V_t of the plane wave velocities and is shown in Figure 2.2, which thus gives the Rayleigh velocity for any isotropic material. V_R is usually quite close to V_t, and is independent of frequency.

The above analysis also gives the displacements. Omitting a factor $\exp[j(\omega t - \beta x_1)]$ and an arbitrary multiplier, the displacements are

$$u_1 = \gamma \exp(a\beta x_3) - \exp(b\beta x_3),$$
$$u_3 = j[\gamma a \exp(a\beta x_3) - b^{-1} \exp(b\beta x_3)], \qquad (2.34)$$

where a, b and γ are real positive quantities given by $a = -jL/\beta$, $b = -jT/\beta$ and $\gamma = (2 - V_R^2/V_t^2)/(2ab)$, and $\beta = \omega/V_R$. These displacements are shown in Figure 2.3, as functions of the depth normalised to the Rayleigh wavelength $\lambda_R = 2\pi V_R/\omega$. Since u_3 is in phase quadrature with u_1, the motion of each particle is an ellipse. Because of the change of sign of u_1 at a depth of about 0.2 wavelengths, the ellipse is described in different directions above and below this point; at the surface the motion is retrograde, while lower down it is prograde. The distortion of the material at one instant is shown in Figure 2.4, with the displacements exaggerated. The dots in this figure represent the equilibrium positions of particles within the

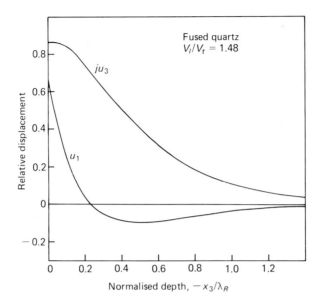

FIGURE 2.3. Rayleigh wave displacements for isotropic material.

material, while the lines show the displacements when a Rayleigh wave is present. Note that there is little motion at depths greater than one wavelength.

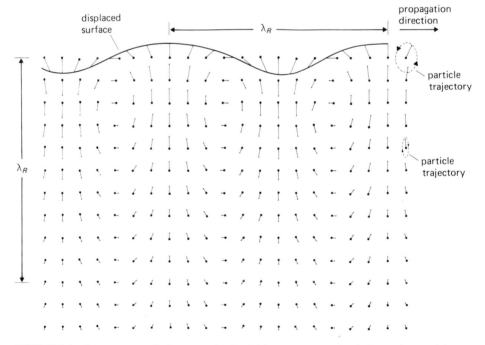

FIGURE 2.4. Instantaneous displacements for Rayleigh wave propagation in isotropic material.

2.2.3. Shear-horizontal Waves in a Half-space

It was assumed above that for surface wave solutions the displacements would be confined to the sagittal plane (x_1, x_3). We now consider whether a valid solution can be obtained with a perpendicular component, in the x_2-direction. As before, the wave vector of any partial wave must be confined to the sagittal plane, and hence a displacement normal to this plane can only be produced by shear waves. The wave vector must therefore be $\mathbf{k}_t = (\beta, 0, T)$, as before, and the displacement must be

$$\mathbf{u} = A(0, 1, 0) \exp[-j(\beta x_1 + T x_3)],$$

where A is a constant. For this wave the stresses T_{13} and T_{33} are zero, while $T_{23} = -j\mu T \cdot u_2$. The boundary conditions, $T_{i3} = 0$ at $x_3 = 0$, cannot therefore be satisfied if T is finite, and hence there is no solution representing a wave bound to the surface. However, if $T = 0$ the stress components T_{i3} are zero everywhere, and this satisfies the boundary conditions. This solution is simply a plane shear wave propagating parallel to the surface, with its amplitude independent of x_3 within the material. It is called a shear-horizontal, or SH, wave, since its displacements are parallel to the surface. The phase velocity is equal to V_t.

2.2.4. Waves in a Layered Half-space

Now consider a half space of material with a layer of another material, of thickness d, on top, as shown in Figure 2.5. Structures of this type are common in surface wave devices, where the layer may for example be a metal film. Usually, one is concerned with minimising the perturbing effect of the film, for example to minimise the dispersion. In seismology the structure is of considerable interest because the layering of rocks influences the propagation of surface waves, and consequently the solutions have been studied in detail [32, 38–44].

We first consider waves with their displacements confined to the sagittal plane. If the layer thickness is small, the solution will be similar to the Rayleigh wave for a half-space, described in Section 2.2.2 above, so the solutions here are described as

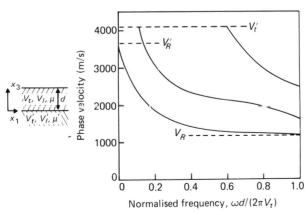

FIGURE 2.5. Velocities for layered Rayleigh waves, gold layer on fused quartz substrate. After Farnell and Adler [44], with permission.

"layered Rayleigh waves". The solutions may be found by summing partial waves, as before, though the calculation is more complex. The layer material is taken to have plane wave velocities V_t and V_l. Assuming the displacements are proportional to $\exp(-j\beta x_1)$, a partial shear wave in the layer gives displacements with the form of equation (2.29), with T given by equation (2.26). However, there are two solutions for T. To account for this, we define T as the positive solution of equation (2.26), so that the partial shear waves have wavenumbers $\mathbf{k}_t = (\beta, 0, \pm T)$. The displacements of these two waves then give

$$\mathbf{u}_t = A(1, 0, -\beta/T) \exp(-jTx_3) + B(1, 0, \beta/T) \exp(jTx_3), \quad (2.35)$$

where A and B are constants, and the variations with x_1 and t have been omitted. Similarly, there are two longitudinal partial waves in the layer, with displacements

$$\mathbf{u}_l = C(1, 0, L/\beta) \exp(-jLx_3) + D(1, 0, -L/\beta) \exp(jLx_3), \quad (2.36)$$

where L is the positive solution of equation (2.27), and C and D are constants. The total displacement in the layer is $\mathbf{u} = \mathbf{u}_t + \mathbf{u}_l$.

We define V_t' and V_l' as the velocities of plane shear and longitudinal waves in the half-space material. The displacement $\mathbf{u}'$ in the half-space is obtained as before; comparing with equations (2.28) and (2.29) we can write

$$\mathbf{u}' = E(1, 0, -\beta/T') \exp(-jT'x_3) + F(1, 0, L'/\beta) \exp(-jL'x_3), \quad (2.37)$$

where E and F are constants. T' and L' are the x_3-components of the wavenumbers, given by equations (2.26) and (2.27) but with V_t' and V_l' replacing V_t and V_l. For a surface wave solution, with displacements decaying in the half-space, T' and L' must be positive imaginary, and hence the phase velocity must be less than V_t'.

The boundary conditions require that the stresses T_{i3} should be zero on the free upper surface, while on the lower surface the stresses T_{i3} and the displacements should be continuous. This gives six homogeneous equations relating the constants A, $B, \ldots, F$, and the determinant of coefficients is set to zero to give the dispersion relation. This in turn gives the allowed values for β, and hence the velocities.

Figure 2.5 shows the calculated phase velocities, for a gold layer on a fused quartz substrate, after Farnell and Adler [44]. The horizontal axis here gives normalised frequency, but may also be read as the thickness of the layer divided by the wavelength of plane shear waves in the layer material. The result is typical of cases in which the layer material has acoustic velocities much less than those of the half-space material. The fundamental mode, that is, the solution with lowest velocity, is of primary importance. At low frequencies the layer thickness is much less than the wavelength, so the velocity approaches V_R', the Rayleigh velocity for the half-space material. At high frequencies the structure can support a Rayleigh wave with its energy concentrated near the upper surface, so the velocity approaches the Rayleigh velocity for the layer material, denoted V_R. The velocity thus varies from V_R' to V_R. Clearly, if we wish to minimise the dispersion, the materials should be chosen such that V_R is not substantially less than V_R'. For this reason the metal film used for the electrodes in surface wave devices is usually aluminium, which has a Rayleigh velocity similar to those of the common substrate materials quartz and lithium niobate. In addition to the fundamental, there is a series of higher modes, which are named after Sezawa [40].

If the layer material has acoustic velocities greater than those of the substrate, the velocity of the fundamental is V_R' at zero frequency, rising to a value V_t', at which point there is a cut-off. There is therefore little dispersion in this case.

In addition to layered Rayleigh waves, the layered system can also support surface waves with the displacements normal to the sagittal plane. These are known as Love waves [43]. In this case the partial waves are shear waves, with wave vectors $\mathbf{k}_t = (\beta, 0, \pm T)$ in the layer and $\mathbf{k}_t' = (\beta, 0, T')$ in the substrate. Thus the displacements in the layer can be written

$$\mathbf{u}_t = A(0, 1, 0) \exp(-jTx_3) + B(0, 1, 0) \exp(jTx_3) \tag{2.38}$$

and the displacement in the substrate is

$$\mathbf{u}_t' = C(0, 1, 0) \exp(-jT'x_3),$$

where A, B and C are constants. The boundary conditions are the same as for the Rayleigh wave case, and applying these gives the dispersion relation

$$\tan(Td) = j\mu'T'/(\mu T), \tag{2.39}$$

where μ and μ' are respectively the rigidities of the layer material and the half-space material. Solving for β gives in general a number of modes, and the velocities for gold on fused quartz are shown in Figure 2.6. Solutions are obtainable only for $V_t < V_t'$, and the solutions must have velocities less than V_t', so that T' is imaginary and the displacement decays in the half-space. At zero frequency the Love wave solution becomes identical to the SH plane wave solution for a half-space. Thus Love waves can be regarded as modified forms of the SH plane wave, where the presence of a layer with low acoustic velocity converts the plane wave into a surface wave and causes dispersion.

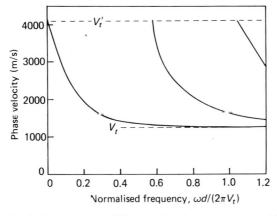

FIGURE 2.6. Velocities for Love waves, gold layer on fused quartz substrate. After Farnell and Adler [44], with permission.

2.2.5. Waves in a Parallel-sided Plate

We now consider a plate of material with boundaries at $x_3 = \pm d/2$ and with infinite extent in the x_1 and x_2 directions. As for the layered substrate there are two types of solution [32, 34]. For displacements confined to the sagittal plane (x_1, x_3) the solutions are called Lamb waves, while there are also SH-wave solutions with displacements perpendicular to the sagittal plane. Both types have some relevance to surface wave devices.

The parallel plate is rather similar to the layered half-space considered above, with the half-space omitted. Thus, for Lamb waves, the partial waves have the same form as those for layered Rayleigh waves, given by equations (2.35) and (2.36). The allowed values for β, and the relative values of the constants A, B, C and D, are obtained by applying the boundary conditions $T_{i3} = 0$ at the two surfaces, $x_3 = \pm d/2$. This gives two families of dispersive solutions, known as symmetric and antisymmetric modes. The dispersion relation is

$$\left[\frac{\tan(Ld/2)}{\tan(Td/2)}\right]^{\pm 1} = -\frac{(T^2 - \beta^2)^2}{4LT\beta^2}, \tag{2.40}$$

taking the upper sign for symmetric modes and the lower sign for antisymmetric modes

At high frequencies, Td and Ld are large, and if we also assume T and L to be imaginary the left side of equation (2.40) approaches unity. Comparison with equation (2.32) shows that the velocity approaches the Rayleigh velocity, V_R. Thus the two Lamb modes, one symmetric and one antisymmetric, each become equivalent to a Rayleigh wave on the upper surface plus a Rayleigh wave on the lower surface. This gives some insight into the behaviour of a wave generated on one surface of a parallel-sided plate. The wave is equivalent to the sum of the two Lamb waves, with equal amplitudes and phases; owing to the symmetry the Lamb wave displacements are additive near the upper surface but cancel near the lower surface, so that if d is large enough the wave is essentially a Rayleigh wave on the upper surface. However, for finite d the Lamb waves have different wave numbers, β_s and β_a, say, so that the disturbance at the upper surface has the form $[\exp(-j\beta_a x_1) + \exp(-j\beta_s x_1)]$, with an amplitude proportional to $\cos[(\beta_a - \beta_s)x_1/2]$. Thus, after travelling a distance

$$x_c = \frac{\pi}{|\beta_a - \beta_s|}$$

the amplitude at the top surface is zero. At this point, the two Lamb waves are in anti-phase, so that they reinforce at the lower surface; in effect, the Rayleigh wave has been transferred from one surface to the other. The distance x_c is therefore called the coupling distance. At a distance $2x_c$, the amplitude on the upper surface is again maximised, so that the Rayleigh wave has been transferred back again.

The coupling distance may be evaluated by solving equation (2.40) for β_a and β_s, giving the result shown in Figure 2.7, where x_c is normalised to the Rayleigh wavelength. The coupling length is many thousand wavelengths, even when the plate

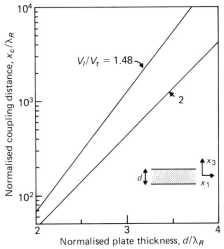

FIGURE 2.7. Coupling distance for transfer of Rayleigh waves between the two sides of a parallel-sided plate.

is only a few wavelengths thick. In surface-wave devices the wave is generated on a rectangular substrate, and it can be concluded that a substrate thickness of a few wavelengths should be sufficient to prevent the rear surface from having any significant effect. This conclusion remains valid if the substrate is mounted on a carrier using an adhesive (as is usually the case).

Setting $\beta = 0$ in equation (2.40) gives the resonant frequencies of the plate. These are the frequencies at which $2d$ is a multiple of the shear wavelength $2\pi V_t/\omega$, or of the longitudinal wavelength $2\pi V_l/\omega$. Resonances of this type are used in the parallel-plate crystal resonator. They are sometimes observed in surface wave devices, though usually most of them are damped out because of the mounting of the substrate.

The parallel-sided plate also supports shear horizontal (SH) modes, with the displacements in the x_2-direction. These solutions can be obtained by assuming partial waves of the form given by equation (2.38). This gives a series of modes, most of which are dispersive. However, at low frequencies there is only one solution, a non-dispersive mode with velocity V_t; this is simply a plane shear wave propagating in the x_1 direction. The same solution was found in Section 2.2.3 for propagation in a half-space, and it was shown that this wave gives no stresses on planes perpendicular to x_3.

The non-dispersive SH mode of the parallel-sided plate is used in a type of dispersive delay line known as the IMCON, in which the wave is reflected by arrays of grooves. Dispersive modes (Lamb waves and SH modes) have also been used in dispersive delay lines.

2.3. WAVES IN ANISOTROPIC MATERIALS

This section is concerned with acoustic waves in piezoelectric materials, which must of course be anisotropic. Because of the complexity of the equations for this case, the solutions can usually be found only by using numerical techniques. The account here is therefore mainly descriptive.

2.3.1. Plane Waves in an Infinite Medium

For a piezoelectric material, the equation of motion takes the form of equations (2.14), in terms of the displacements **u** and the potential Φ. We consider plane wave solutions with frequency ω and wave vector **k**, with **u** and Φ having the forms

$$\mathbf{u} = \mathbf{u}_0 \exp[j(\omega t - \mathbf{k} \cdot \mathbf{x})],$$

$$\Phi = \Phi_0 \exp[j(\omega t - \mathbf{k} \cdot \mathbf{x})],$$

where $\mathbf{u}_0$ and Φ_0 are constants, independent of **x** and t, and **k** is real. For isotropic materials, Section 2.2.1, there are two solutions, the shear wave and the longitudinal wave, with $\mathbf{u}_0$ respectively perpendicular to and parallel to **k**.

To find the solutions for anisotropic materials, the above functions **u** and Φ are substituted into equations (2.14), giving four equations in the four variables $\mathbf{u}_0$, Φ_0. Setting the determinant of coefficients to zero then gives four solutions. One of these solutions is essentially electrostatic in nature — it corresponds to the electrostatic solution for a non-piezoelectric material, for which $e_{ijk} = 0$ so that equation (2.14b) reduces to Laplace's equation. This solution is of little interest here. The other three solutions are non-dispersive acoustic waves. Usually, one solution has the displacement $\mathbf{u}_0$ almost parallel to **k**, and is called the "quasi-longitudinal", or simply longitudinal, wave. The other two solutions (with different velocities) usually have $\mathbf{u}_0$ almost perpendicular to **k**, are are called "quasi-shear", or shear, waves. For particular propagation directions the longitudinal wave has $\mathbf{u}_0$ parallel to **k**, and is then called a "pure longitudinal" wave. Similarly, shear waves may have $\mathbf{u}_0$ perpendicular to **k**, and are then called "pure shear" waves. Owing to anisotropy, each of the three waves has a phase velocity (equal to $\omega/|\mathbf{k}|$) dependent on the propagation direction. In addition, all three solutions may have associated electric potentials, though for particular propagation directions the potential may disappear.

2.3.2. Theory for a Piezoelectric Half-space

In Section 2.2.2 we saw that the Rayleigh wave solution for an isotropic half-space can be obtained by adding two partial waves, corresponding to plane shear and longitudinal waves, with the x_1-components of their wave vectors equal. For anisotropic materials the method [30, 45–50] is essentially the same, though a numerical procedure must be adopted to obtain the solutions.

Care is needed in specifying the orientation of the material. For most crystalline materials the internal structure is referenced to an orthogonal set of axes denoted by upper-case symbols X, Y, Z, with directions defined in relation to the crystal lattice [52]. The surface orientation and the wave propagation direction must be defined in relation to these axes. The convention usually adopted is to define the surface normal x_3, followed by the propagation direction x_1. For example "Y, Z lithium niobate" indicates that x_3 is parallel to the crystal Y-axis, and x_1 is parallel to the crystal Z-axis.

The orientation of x_3 is also referred to as the cut, so that for Y, Z lithium niobate the crystal is Y-cut. The material tensors, the stiffness, permittivity and piezoelectric tensor, are specified in relation to the internal axes X, Y and Z, so for the analysis they must be rotated into the frame defined by x_1, x_2, x_3. For cubic crystals, the orientation is usually defined directly in relation to the lattice by using Miller indices.

For a piezoelectric material it is necessary to use an electrical boundary condition at the surface, in addition to the stress-free condition which applies for isotropic materials. Two cases are usually considered. In the first case the space above the surface is a vacuum and conductors are excluded, so that there are no free charges. This is known as the *free-surface* case. In general there will be a potential in the vacuum above the surface. In the second case the surface is assumed to be covered with a thin metal layer with infinite conductivity, which shorts out the horizontal component of **E** at the surface but does not affect the mechanical boundary conditions. This is called the *metallised* case. These two cases generally give different velocities. The velocity difference is a measure of the coupling between the wave and electrical perturbations at the surface, and will be seen to be of crucial importance to the performance of surface wave transducers.

For the free-surface case the potential in the vacuum satisfies Laplace's equation $\nabla^2 \Phi = 0$. If the wavenumber of the surface wave is β, the potential Φ in the vacuum can be written

$$\Phi = f(x_3) \exp[j(\omega t - \beta x_1)].$$

Using Laplace's equation shows that the function $f(x_3)$ has the form $\exp(\pm \beta x_3)$, and since Φ must vanish at $x_3 = +\infty$ the potential is given for $x_3 \geq 0$ by

$$\Phi = \Phi_0 \exp(-|\beta|x_3) \exp[j(\omega t - \beta x_1)], \tag{2.41}$$

where Φ_0 is a constant. Since there are no free charges D_3 must be continuous, so that in both the piezoelectric and the vacuum we have

$$D_3 = \varepsilon_0 |\beta| \Phi, \quad \text{at } x_3 = 0. \tag{2.42}$$

For the metallised case the potential at the surface is zero:

$$\Phi = 0, \quad \text{at } x_3 = 0. \tag{2.43}$$

In addition, for either case there are no forces on the surface, so

$$T_{13} = T_{23} = T_{33} = 0, \quad \text{at } x_3 = 0. \tag{2.44}$$

To find the surface wave solutions, we first consider partial waves in which the displacements and potential, denoted by $\mathbf{u}'$ and Φ', take the form

$$\mathbf{u}' = \mathbf{u}_0' \exp(j\gamma x_3) \exp[j(\omega t - \beta x_1)],$$

$$\Phi' = \Phi_0' \exp(j\gamma x_3) \exp[j(\omega t - \beta x_1)], \tag{2.45}$$

where β is the wave number of the surface wave, assumed to be real. These expressions are to satisfy the equations of motion, equations (2.14), for an infinite material. As in the isotropic case, if β is fixed there are a number of specific solutions for γ, the

x_3-component of the wave vector. We assume a particular real value of β and substitute equations (2.45) into equations (2.14). These can then be solved numerically, giving eight solutions for γ, and for each solution the relative values of $\mathbf{u}'_0$ and Φ'_0 are obtained. The values of γ are generally complex, and we can only allow values whose imaginary parts are negative, so that $\mathbf{u}'$ and Φ' vanish at $x_3 = -\infty$. There are four such values of γ in general, and these are denoted $\gamma_1, \ldots, \gamma_4$. The partial waves are therefore

$$\mathbf{u}'_m = \mathbf{u}'_{0m} \exp(j\gamma_m x_3) \exp[j(\omega t - \beta x_1)],$$
$$\Phi'_m = \Phi'_{0m} \exp(j\gamma_m x_3) \exp[j(\omega t - \beta x_1)], \quad m = 1, 2, 3, 4, \quad (2.46)$$

where $\mathbf{u}'_{0m}$ and Φ'_{0m} are the displacement and potential corresponding to γ_m.

In the half-space it is assumed that the solution is a linear sum of these partial waves, so that

$$\mathbf{u} = \sum_{m=1}^{4} A_m \mathbf{u}'_m,$$
$$\Phi = \sum_{m=1}^{4} A_m \Phi'_m. \quad (2.47)$$

The coefficients A_m are to be such that the solution satisfies the boundary conditions, given by equations (2.44) and either equation (2.42) (for the free-surface case) or equation (2.43) (for the metallised case). These conditions give a determinant which must be zero for a valid solution, and when zero gives the relative values of the constants A_m. However, the determinant will only be zero if the correct value for β has been chosen. Thus, to find the solution the entire procedure is iterated using different values for β until the boundary condition determinant vanishes. The velocity of the wave is then ω/β, and the displacements and potential are given by equation (2.47).

2.3.3. Surface-wave Solutions

The solutions obtained by the above procedure are of course significantly affected by the anisotropy of the material and by its orientation. The determination of surface wave characteristics in general is a considerable task because of the variety of crystal symmetries, and because for any one symmetry the orientation depends on three angular variables. However, a very extensive range of cases has been studied, as shown for example by Refs. [45–51]. It appears that one or more surface wave solutions can always be found, whatever the symmetry and orientation. In general the solution may involve all three components of the displacement, so that the motion is not confined to the sagittal plane. However, since there is no variation in the x_2-direction the electric field $\mathbf{E}$, given by the negative gradient of Φ, is always in the sagittal plane. For a metallised surface the parallel component, E_1, is always zero at the surface, though the normal components, E_3, is not necessarily zero. In some orientations the solution has no associated electric field, and can thus be described as non-piezoelectric. In this case the solution is not affected by metallisation of the surface.

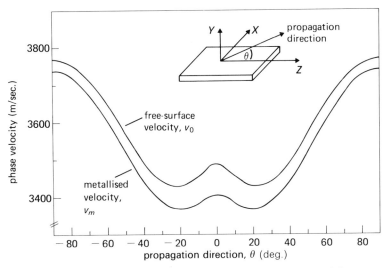

FIGURE 2.8. Rayleigh-wave velocities for Y-cut lithium niobate.

In general, the velocity of a surface wave must be less than the velocities of plane waves propagating in the x_1 direction in an infinite material. The reason for this is that, as for isotropic materials, the wave vectors of the partial waves must not have real x_3-components. Of the three plane waves (longitudinal, fast shear, slow shear) the slow shear wave has the lowest velocity, so the surface wave velocity must be less than this. In practice, the surface wave velocity is usually quite close to the slow shear velocity.

The solution encountered most frequently has its displacement **u** directed parallel, or almost parallel, to the sagittal plane, and has an associated electric field. This solution is known as a *piezoelectric Rayleigh wave*. It is similar to the Rayleigh wave for an isotropic material, with its behaviour modified somewhat by anisotropy and piezoelectricity. The penetration depth is typically about one wavelength. This type of solution is found in, for example, Y-cut lithium niobate. For this material the velocities are shown in Figure 2.8, as functions of the propagation direction. The free-surface velocity is denoted v_0 and the metallised velocity is denoted v_m. Note that v_m is less than v_0, which is always the case for a piezoelectrically coupled wave. For Y-cut lithium niobate the marked difference between the two velocities shows that the piezoelectric coupling is strong for this case. The coupling is strongest for propagation in the Z direction, and this orientation, described as Y, Z, is often used for surface-wave devices. The displacements and potential are shown in Figure 2.9, where the scales are appropriate for a power density of 1 mW/mm and a frequency of 100 MHz.

For some particular orientations the piezoelectric Rayleigh wave can have its displacement confined to the sagittal plane. This occurs if the sagittal plane is a plane of mirror symmetry for the crystal. The wave is then called a *pure* mode, and the propagation direction is called a pure mode direction. For a given surface orientation, the wave velocity is symmetrical with respect to a pure mode direction. For Y-cut

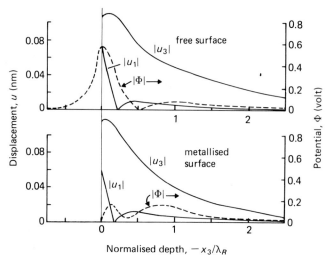

FIGURE 2.9. Displacements and potential for surface waves on Y, Z lithium niobate. After Farnell [49], with permission.

lithium niobate the X and Z directions are pure mode directions, and the symmetry can be seen in Figure 2.8. Some authors use the term "pure mode directions" for directions about which the velocity is symmetrical, and for these the displacements are not necessarily confined to the sagittal plane.

A quite different solution, the *Bleustein–Gulyaev wave* [53–56], occurs if the sagittal plane is normal to a two-fold axis of the crystal, or an axis of higher even order. This wave has its displacement normal to the sagittal plane, and has an associated electric field. It is closely related to the plane SH wave that can propagate in an isotropic half-space; in effect, piezoelectricity has caused the plane wave to become bound to the surface. However the wave is not very strongly bound, even in a strongly piezoelectric material; for example, in cadmium sulphide the penetration depth is typically 4 wavelengths for a metallised surface, changing to 44 wavelengths for a free surface. The velocity is very close to the velocity of slow shear waves. For the same orientation there is also a separate non-piezoelectric Rayleigh wave solution, with its displacements in the sagittal plane. For a given surface orientation the Bleustein–Gulyaev solution is found over a range of propagation directions, though not usually for all directions in the surface [56].

The Rayleigh wave and Bleustein–Gulyaev wave solutions are strongly related to the Rayleigh and SH wave solutions for an isotropic half-space, and these are in turn related to the solutions for a layered half-space. These relationships are summarised in Table 2.1, which also includes other solutions described below.

2.3.4. Other Solutions

In addition to the above surface wave solutions, the boundary conditions for a piezoelectric half-space can sometimes be satisfied by a plane shear wave propagating parallel to the surface, as in the isotropic case.

ACOUSTIC WAVES IN ELASTIC SOLIDS

TABLE 2.1

Isotropic half-space		Anisotropic half-space
Layered	Non-layered	
Rayleigh waves (dispersive)	Rayleigh wave	Rayleigh wave Pseudo-surface wave Leaky surface wave
Love waves (dispersive)	SH plane wave	Bleustein–Gulyaev wave Plane wave

There can also be in some cases yet another type of surface wave solution, known as a *pseudo-surface wave*. Rather surprisingly, the pseudo-surface wave has a phase velocity higher than that of the slow shear plane wave, though it is less than that of the fast shear wave. Despite this, the pseudo-surface wave is a true surface wave, since its displacements decay exponentially with depth. This behaviour is possible because the displacement of the pseudo-surface wave is perpendicular to the displacement of the slow shear wave, so that the surface wave does not have a partial wave component corresponding to the slow shear wave. An example is the Y–Z plane of quartz [48, 57] which gives the velocities shown in Figure 2.10. For all propagation directions in this plane there is a Rayleigh wave solution, but in addition a pseudo-surface wave exists for propagation at an angle 153° away from the Z-axis. The Rayleigh wave and pseudo-surface wave are both piezoelectric. The plane shear wave velocities for an unbounded medium are also shown.

For propagation in a slightly different direction it is found that the boundary conditions cannot be satisfied without including a partial wave corresponding to the slow shear wave. This implies that a surface wave solution cannot exist, since the slow

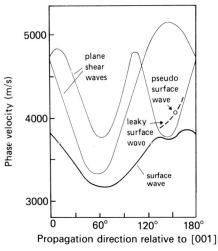

FIGURE 2.10. Velocities for surface waves and leaky waves on the YZ plane of quartz. After Farnell [48], with permission.

shear wave component will carry energy away from the surface. However, if the slow shear component only contributes a small fraction of the total displacement, the energy lost in this way will be small. In this case there is a solution with characteristics very similar to a true surface wave, except that it has a small amount of attenuation. This is known as a *leaky surface wave*. The solution may be found by the procedure outlined above (Section 2.3.2) for surface waves, allowing the wave number β to become complex. For propagation on the Y–Z plane of quartz, the velocity of the leaky surface wave is shown in Figure 2.10. For a range of propagation directions covering about 30°, the attenuation of the wave is less than 0.006 dB per wavelength. Experimentally, an attenuation as low as this is difficult to detect, and the wave appears to behave as a true surface wave.

Pseudo-surface waves and leaky waves are known to exist in many materials, including lithium niobate [59, 60], bismuth germanium oxide [60], and many non-piezoelectric anisotropic materials. In most cases the attenuation of the leaky wave is substantially larger than that of the quartz case described above.

2.3.5. Materials for Devices

For a surface-wave device the choice of an appropriate material is a vital part of the design procedure. Nearly always, the material and orientation are chosen such that only one surface wave mode can be excited. Bleustein–Gulyaev waves are usually excluded because their large penetration depths would require thicker substrates in order to avoid coupling to the rear surface.

It is usual to choose a material and orientation such that the one surface wave mode is a piezoelectric Rayleigh wave. Strong piezoelectric coupling, as shown by a relatively large difference between the free-surface and metallised velocities, is sometimes, but not always, desirable. A pure mode direction is usually chosen as the axis of the device. There are also many other factors such as temperature effects, diffraction and attenuation, and these will be considered later, in Chapter 6. The commonest materials used in devices are quartz and lithium niobate.

Perturbations Due to Thin Films. The effect of a layer on the surface of a half-space was considered for isotropic materials in Section 2.2.4, where it was shown that the layer causes Rayleigh waves to become dispersive, and a series of modes can exist. For anisotropic materials the behaviour is much more complex, though there are strong parallels with the isotropic case.

The case of most interest concerns a conducting metal layer on a piezoelectric substrate, since a metal layer is used for transducers and other structures. We consider a half-space with a uniform layer, as in Figure 2.5. The layer material can be taken to be isotropic, and it is assumed that the wave is of the piezoelectric Rayleigh type. A very thin layer has the effect of reducing the wave velocity because of the change of electrical boundary conditions, as discussed above. This effect is known as *electrical loading*. If the layer thickness is finite its elastic properties become relevant, causing an additional dispersive change of velocity, and this is known as *mass loading*. At high frequencies a series of higher modes can sometimes exist, as in the isotropic case (Figure 2.5), but these are not usually relevant in surface-wave devices.

In practical cases the velocity changes are small; electrical loading changes the velocity by at most a few percent, while mass loading must be controlled to avoid undue dispersion. For such cases a perturbation theory has been developed to enable the velocity change to be evaluated approximately, using a calculation much simpler than the exact analysis. An important result of this theory is a relation giving the velocity change due to electrical loading. If v_0 is the free-surface Rayleigh wave velocity and v_m the metallised velocity, the relation is [32, 49, 61]

$$\frac{v_m - v_0}{v_0} \approx - (\varepsilon_0 + \varepsilon_p^T)|\Phi(0)|^2 \omega/(4P_s), \qquad (2.48)$$

where

$$\varepsilon_p^T = [\varepsilon_{33}^T \varepsilon_{11}^T - (\varepsilon_{13}^T)^2]^{1/2}.$$

The permittivity components ε_{ij}^T are measured at constant stress, as in equation (2.12). Here $\Phi(0)$ is the potential at the surface $x_3 = 0$ for the free-surface case, and P_s is the power carried by this wave per unit length in the x_2 direction. The perturbation theory can also be applied to many other problems, including mass loading due to an isotropic film [32, 49].

Some special properties of the layered system are sometimes exploited in devices. For example, surface waves can be generated on a non-piezoelectric substrate by first depositing a piezoelectric film and then fabricating an interdigital transducer in the usual way. An example is a zinc oxide layer on silicon. Thin films are also used to guide surface waves, or to deliberately introduce dispersion. For these reasons, waves in layered anisotropic media have been studied quite extensively [44, 60].

Chapter 3

Electrical Excitation at a Plane Surface

For a piezoelectric material, the surface waves described in Chapter 2 can be generated and detected electrically by means of metal electrodes on the surface. This principle is used in interdigital transducers and multistrip couplers. A basic concept used in the analysis of these components is the effective permittivity, which gives a description of the electrical behaviour of the surface taking account of the acoustic behaviour of the material. A Green's function derived from the effective permittivity is also used. This chapter gives the basic theory. Sections 3.1 to 3.4 describe the effective permittivity and the Green's function for excitation of a half-space, including some approximations which will be used for analysis of transducers and multistrip couplers in the following chapters. In Section 3.5 some other applications of the basic concepts are described briefly.

It is assumed throughout this chapter that the electrodes do not cause any mechanical perturbations.

3.1. NON-PIEZOELECTRIC HALF-SPACE

We first consider electrical excitation at the surface of a half-space, assuming the material to be non-piezoelectric. This will be referred to as the electrostatic case. In the cases of practical interest the material is always piezoelectric, of course. However, the electrostatic solution is of considerable importance because it can be used as a first-order approximation in the piezoelectric case, and it will be used extensively in later chapters. Some previous analysis is given by Ingebrigtsen [62] and by Hartmann and Secrest [63].

Coordinate axes are defined as in Figure 2.1, so that a homogeneous anisotropic dielectric occupies the half-space $x_3 < 0$, with a vacuum in the region $x_3 > 0$. The potential is assumed to be invariant in the x_2-direction. We shall be concerned with cases where electrodes are present at the surface, so that there will in general be free charges present in the plane $x_3 = 0$. However, the electrodes are not considered explicitly at this stage; they will be allowed for later by applying appropriate boundary conditions at the surface. Since the potential does not vary in the x_2-direction, the edges of the electrodes must be parallel to the x_2-axis.

It is assumed that there are no free charges except at the plane $x_3 = 0$, so that the potential $\Phi(x_1, x_3)$ must satisfy Laplace's equation in the vacuum and in the dielectric. In the dielectric we have div $\mathbf{D} = 0$, and $\mathbf{E} = -\,\mathrm{grad}\,\Phi$, assuming the electric field to be quasi-static. Using $\partial/\partial x_2 = 0$ and $\mathbf{D} = \varepsilon \cdot \mathbf{E}$, we find

$$\varepsilon_{11}\frac{\partial^2 \Phi}{\partial x_1^2} + 2\varepsilon_{13}\frac{\partial^2 \Phi}{\partial x_1 \partial x_3} + \varepsilon_{33}\frac{\partial^2 \Phi}{\partial x_3^2} = 0 \qquad (3.1)$$

for $x_3 < 0$. In the vacuum, $x_3 > 0$, the same relation applies, with ε_{11} and ε_{33} replaced by ε_0 and ε_{13} set to zero, so that

$$\frac{\partial^2 \Phi}{\partial x_1^2} + \frac{\partial^2 \Phi}{\partial x_3^2} = 0 \qquad (3.2)$$

for $x_3 > 0$.

We consider first a harmonic solution in which the potential varies as $\exp(j\beta x_1)$. The potential is required to vanish at $x_3 = \pm \infty$ and must also be continuous at $x_3 = 0$. Taking β to be real, a harmonic solution that satisfies these requirements and also satisfies equations (3.1) and (3.2) is given by

$$\tilde{\Phi}(x_1, x_3) = \exp[j\beta x_1 - |\beta|x_3], \quad \text{for } x_3 > 0, \qquad (3.3\mathrm{a})$$

$$\tilde{\Phi}(x_1, x_3) = \exp[j\beta(x_1 - x_3 \cdot \varepsilon_{13}/\varepsilon_{33}) + |\beta|x_3 \cdot \varepsilon_p/\varepsilon_{33}], \quad \text{for } x_3 < 0 \qquad (3.3\mathrm{b})$$

where

$$\varepsilon_p = (\varepsilon_{11}\varepsilon_{33} - \varepsilon_{13}^2)^{1/2} \qquad (3.4)$$

and the tilde is used to indicate the harmonic solution. With these expressions for the potential, the component of the displacement normal to the surface is, in the vacuum,

$$\tilde{D}_3(x_1, x_3) = \varepsilon_0 |\beta| \cdot \tilde{\Phi}(x_1, x_3), \quad \text{for } x_3 > 0 \qquad (3.5\mathrm{a})$$

and, in the dielectric,

$$\tilde{D}_3(x_1, x_3) = -\varepsilon_p |\beta| \cdot \tilde{\Phi}(x_1, x_3), \quad \text{for } x_3 < 0. \qquad (3.5\mathrm{b})$$

The displacements can also be written in terms of the electric field component parallel to the surface:

$$\tilde{D}_3(x_1, x_3) = j\varepsilon_0 \,\mathrm{sgn}(\beta) \cdot \tilde{E}_1(x_1, x_3), \quad \text{for } x_3 > 0, \qquad (3.6\mathrm{a})$$

$$\tilde{D}_3(x_1, x_3) = -j\varepsilon_p \,\mathrm{sgn}(\beta) \cdot \tilde{E}_1(x_1, x_3), \quad \text{for } x_3 < 0, \qquad (3.6\mathrm{b})$$

where $\mathrm{sgn}(\beta) = +1$ for $\beta > 0$ and $\mathrm{sgn}(\beta) = -1$ for $\beta < 0$.

Any discontinuity in $\tilde{D}_3$ at the surface $x_3 = 0$ implies, by Gauss's law, that there must be free charges there. Define $\tilde{D}_3(+)$ as the value of $\tilde{D}_3$ in the vacuum at $x_3 = 0$, and $\tilde{D}_3(-)$ as the value of $\tilde{D}_3$ in the dielectric at $x_3 = 0$. The density of free charges, denoted $\tilde{\sigma}(x_1)$, is then given by

$$\tilde{\sigma}(x_1) = \tilde{D}_3(+) - \tilde{D}_3(-),$$

where $\tilde{\sigma}(x_1)$ includes charges on both sides of the electrodes. Using equations (3.5),

$$\tilde{\sigma}(x_1) = (\varepsilon_0 + \varepsilon_p)|\beta|\tilde{\Phi}(x_1, 0). \tag{3.7}$$

The quantity $(\varepsilon_0 + \varepsilon_p)$ is an effective permittivity which characterises the interface between the vacuum and the dielectric. It gives the discontinuity in $\tilde{D}_3$ in terms of the parallel field $\tilde{E}_1$ at the surface:

$$\tilde{D}_3(+) - \tilde{D}_3(-) = j \operatorname{sgn}(\beta)(\varepsilon_0 + \varepsilon_p)\tilde{E}_1(x_1, 0). \tag{3.8}$$

We now consider a more general solution. It is convenient to write this in terms of the potential at the surface which, for the harmonic solution, is defined as $\tilde{\phi}(x_1) \equiv \tilde{\Phi}(x_1, 0)$. Thus equation (3.7) becomes

$$\tilde{\sigma}(x_1) = (\varepsilon_0 + \varepsilon_p)|\beta|\tilde{\phi}(x_1), \tag{3.9}$$

where $\tilde{\sigma}(x_1)$ and $\tilde{\phi}(x_1)$ are both proportional to $\exp(j\beta x)$. The general solution, with charge density $\sigma(x_1)$ and surface potential $\phi(x_1)$, is obtained by Fourier synthesis. We have

$$\sigma(x_1) = \frac{1}{2\pi}\int_{-\infty}^{\infty} \bar{\sigma}(\beta) \exp(j\beta x_1) \, d\beta, \tag{3.10}$$

and

$$\phi(x_1) = \frac{1}{2\pi}\int_{-\infty}^{\infty} \bar{\phi}(\beta) \exp(j\beta x_1) \, d\beta, \tag{3.11}$$

where $\bar{\sigma}(\beta)$ and $\bar{\phi}(\beta)$ are respectively the Fourier transforms of $\sigma(x_1)$ and $\phi(x_1)$. The general solution is an infinite sum of harmonic solutions with different values of β. For each β, equation (3.9) applies, and we thus have

$$\bar{\sigma}(\beta) = (\varepsilon_0 + \varepsilon_p)|\beta|\bar{\phi}(\beta). \tag{3.12}$$

Thus the charge density $\sigma(x_1)$ may be obtained if any surface potential $\phi(x_1)$ is specified.

It should be noted that $\bar{\phi}(\beta)$ must be zero at $\beta = 0$, because if this were not so the potential would be finite at $x_3 = \pm \infty$, as can be seen by considering equations (3.3) for $\beta = 0$. It follows from equation (3.12) that $\bar{\sigma}(\beta)$ is also zero at $\beta = 0$, and from the definition of the Fourier transform this implies that

$$\int_{-\infty}^{\infty} \sigma(x_1) \, dx_1 = 0, \tag{3.13}$$

that is, the sum total of all the charges at the surface $x_3 = 0$ is equal to zero.

Corresponding relationships in the spatial domain can be obtained using Fourier analysis. The tangential electric field $E_1(x_1)$ at the surface is the negative differential of $\phi(x_1)$, and hence its Fourier transform is $\bar{E}_1(\beta) = -j\beta\bar{\phi}(\beta)$. Equation (3.12) thus gives

$$(\varepsilon_0 + \varepsilon_p)\bar{E}_1(\beta) = -j \operatorname{sgn}(\beta)\bar{\sigma}(\beta)$$

This is transformed to the x_1-domain by using the convolution theorem, equation (A.19), noting that the inverse transform of $\operatorname{sgn}(\beta)$ is $j/(\pi x)$ as shown by equation

(A.35). This gives

$$(\varepsilon_0 + \varepsilon_p)E_1(x_1) = \sigma(x_1) * \frac{1}{\pi x_1}. \qquad (3.14)$$

The surface potential $\phi(x_1)$ can be obtained by integrating $E_1(x_1)$, and with the aid of equation (3.13) this gives

$$(\varepsilon_0 + \varepsilon_p)\phi(x_1) = -\sigma(x_1) * \ln|x_1|/\pi. \qquad (3.15)$$

To obtain solutions for $\phi(x_1)$ and $\sigma(x_1)$, it is also necessary to use the boundary conditions that $\phi(x_1)$ must be constant on any electrode and $\sigma(x_1)$ must be zero at all unmetallised locations. Methods for obtaining solutions will be considered later, in Chapter 4.

3.2. PIEZOELECTRIC HALF-SPACE

In Section 3.1 it was found that the potential and charge density at the surface of a non-piezoelectric half-space are related by an effective permittivity $(\varepsilon_0 + \varepsilon_p)$. Here we consider the effective permittivity for a piezoelectric half-space. The method was first given by Ingebrigtsen [62] and developed later by Greebe et al. [64] and by Milsom et al. [65, 66]. The theory given here follows most closely that of Milsom. Other approaches, which will not be considered here, are the perturbation theory [67, 68] and normal mode theory [69, 70]; these give results which are essentially the same as the results of the effective permittivity approach.

It is assumed that the potential and the acoustic displacements are proportional to $\exp(j\omega t)$, with the frequency ω positive. As in Section 3.1 we consider initially a harmonic solution with variables proportional to $\exp(j\beta x_1)$, with β real, and generalise later using Fourier synthesis. The procedure is similar to that used for calculating surface wave velocities, Section 2.3.2, but here the electric boundary condition at the surface is not specified. This enables a solution to be obtained for any value of β. Another difference is that positive values of β refer to wave motion propagating in the $-x_1$ direction instead of the $+x_1$ direction. This is done for convenience when using Fourier synthesis.

As in Section 2.3.2, we consider partial waves in which the displacements $\mathbf{u}'$ and potential Φ' have the form

$$\mathbf{u}' = \mathbf{u}'_0 \exp(j\gamma x_3) \exp[j(\omega t + \beta x_1)],$$
$$\Phi' = \Phi'_0 \exp(j\gamma x_3) \exp[j(\omega t + \beta x_1)], \qquad (3.16)$$

where $\mathbf{u}'_0$ and Φ'_0 are constants and γ is the x_3-component of the wave vector, which by definition has no x_2-component. These expressions are required to satisfy the equations of motion, equations (2.14), for an infinite medium, with the material tensors rotated into the frame of the axes x_1, x_2, x_3. Substitution into equations (2.14) gives four linear homogeneous equations in the four variables $\mathbf{u}'_0$, Φ'_0, and for non-trivial solutions the determinant of coefficients is set to zero. The determinant is an eight-order polynomial in γ and thus gives eight roots, and for each root the equations also give the relative values of $\mathbf{u}'_0$ and Φ'_0. Four of the roots are unacceptable however, because they do not correspond to excitation at the surface. Care is needed in choosing acceptable roots, because the solution will not in general be a surface wave solution. Complex or imaginary values of γ are acceptable if the imaginary part is negative, so that $\mathbf{u}'$ and Φ' decay away from the surface. Real values of γ give plane wave solutions, and are acceptable only if they carry energy away from the surface. Usually, this requires γ to have its sign opposite to that of β. This is not always the case, however, because for an anisotropic material the power flow direction is not in general collinear with the wave vector, an effect known as beam steering. The power flow direction may be found by examining the variation of phase velocity with propagation direction, as will be shown later for surface waves in Chapter 6.

The four acceptable partial wave solutions are written

$$\mathbf{u}'_m = \mathbf{u}'_{0m} \exp(j\gamma_m x_3) \exp[j(\omega t + \beta x_1)],$$

$$\Phi'_m = \Phi'_{0m} \exp(j\gamma_m x_3) \exp[j(\omega t + \beta x_1)], \quad m = 1, 2, 3, 4. \quad (3.17)$$

The total solution in the half-space has displacements $\tilde{\mathbf{u}}$ and $\tilde{\Phi}$, where the tilde indicates that the solution is harmonic, with variables proportional to $\exp(j\beta x_1)$. The total solution is taken to be a linear combination of the partial waves, so that

$$\tilde{\mathbf{u}} = \sum_{m=1}^{4} A_m \mathbf{u}'_m,$$

$$\tilde{\Phi} = \sum_{m=1}^{4} A_m \Phi'_m. \quad (3.18)$$

The relative values of the constants A_m are determined by the boundary condition that there must be no force on the free surface $x_3 = 0$, that is,

$$T_{13} = T_{23} = T_{33} = 0, \quad \text{at } x_3 = 0, \quad (3.19)$$

with the stresses given by equation (2.9). The electrical boundary conditions are not specified here. We thus have three equations relating the four constants A_m, and hence the relative values of these constants can be found. The relative values of the displacements $\tilde{\mathbf{u}}$ and potential $\tilde{\Phi}$ for the harmonic solution can then be obtained from equation (3.18), giving a solution for any value of β.

For problems concerning electrical excitation at the surface, the variables of interest, constrained by boundary conditions, are the potential and the normal component of the electric displacement $\tilde{D}$. The electric displacement can be calculated from the potential and the acoustic displacements by using equation (2.10). At the surface, the normal component $\tilde{D}_3$ in the piezoelectric is denoted $\tilde{D}_3(-)$. The potential at the surface is denoted $\tilde{\phi}(x_1)$, so that $\tilde{\phi}(x_1) = \tilde{\Phi}(x_1, 0)$. The ratio $\tilde{D}_3(-)/\tilde{\phi}(x_1)$ is determined by the solution described above, and will in general be a function of β.

In the vacuum, $x_3 > 0$, the potential $\tilde{\Phi}(x_1, x_3)$ must satisfy Laplace's equation $\nabla^2 \tilde{\Phi} = 0$. Since $\tilde{\Phi}$ is proportional to $\exp(j\beta x_1)$ and it must vanish at $x_3 = \infty$, the x_3 dependence is $\exp(-|\beta|x_3)$, and so

$$\tilde{\Phi}(x_1, x_3) = \tilde{\phi}(x_1) \exp(-|\beta|x_3), \quad (3.20)$$

for $x_3 > 0$. At the surface $x_3 = 0$, the normal displacement in the vacuum is denoted $\tilde{D}_3(+)$, and this is given by

$$\tilde{D}_3(+) = \varepsilon_0 |\beta| \tilde{\phi}(x_1). \quad (3.21)$$

The surface potential $\tilde{\phi}(x_1)$ must of course be the same on both sides of the boundary. However the normal component of displacement can be different. The discontinuity is related to the potential by the *effective permittivity* $\varepsilon_s(\beta)$, defined by

$$\varepsilon_s(\beta) = \frac{\tilde{D}_3(+) - \tilde{D}_3(-)}{|\beta|\tilde{\phi}(x_1)}. \tag{3.22}$$

Here the x_1-dependence cancels on the right side, so that $\varepsilon_s(\beta)$ is not dependent on x_1. Thus the effective permittivity gives the electrical behaviour of the interface between the vacuum and the piezoelectric half-space.

If $\tilde{D}_3(+)$ and $\tilde{D}_3(-)$ differ, there must be free charges present at the surface, implying the presence of electrodes. $\tilde{D}_3(+)$ will then be equal to the charge density on the vacuum side of the electrodes, and $\tilde{D}_3(-)$ is equal to the negative of the charge density on the piezoelectric side. Thus, if the total charge density at x_1, including both sides, is denoted $\tilde{\sigma}(x_1)$, we have

$$\varepsilon_s(\beta) = \frac{\tilde{\sigma}(x_1)}{|\beta|\tilde{\phi}(x_1)}, \tag{3.23}$$

where $\tilde{\sigma}(x_1)$ and $\tilde{\phi}(x_1)$ are both proportional to $\exp[j(\omega t + \beta x_1)]$. If $\tilde{D}_3(+) = \tilde{D}_3(-)$ there may be electrodes present with equal and opposite charges on the two sides, giving $\tilde{\sigma}(x_1) = 0$. Alternatively there may be no electrodes and hence no free charges. Some authors exclude the charges on the vacuum side when defining $\varepsilon_s(\beta)$, that is, they omit the term $\tilde{D}_3(+)$ in equation (3.22). This reduces the value of $\varepsilon_s(\beta)$ by an amount ε_0, as can be seen from equation (3.21).

In the above equations the potential $\tilde{\phi}(x_1)$ and charge density $\tilde{\sigma}(x_1)$ are proportional to $\exp(j\omega t)$, and the frequency ω was taken to be constant throughout. If ω is changed, the value of $\varepsilon_s(\beta)$ changes, so $\varepsilon_s(\beta)$ is a function of ω as well as β. However, since $\varepsilon_s(\beta)$ is essentially the ratio of $\tilde{D}_3$ to $\tilde{E}_1$, as shown by equation (3.22), it can be seen that it remains unchanged if ω and β are changed in proportion. Thus $\varepsilon_s(\beta)$ is a function of the normalised variable β/ω. This has dimensions the same as the reciprocal of velocity, and is often termed the "slowness". In this chapter the analysis applies for constant frequency, and for brevity the effective permittivity is written as $\varepsilon_s(\beta)$, without showing the frequency dependence explicity.

A more general solution, with surface potential $\phi(x_1)$ and charge density $\sigma(x_1)$, is readily obtained by Fourier synthesis. The method is the same as in the electrostatic case of Section 3.1, so the result follows directly. Thus, comparing with equation (3.12), the general solution obtained from equation (3.23) gives

$$\varepsilon_s(\beta) = \frac{\bar{\sigma}(\beta)}{|\beta|\bar{\phi}(\beta)}, \tag{3.24}$$

where $\bar{\sigma}(\beta)$ and $\bar{\phi}(\beta)$ are respectively the Fourier transforms of $\sigma(x_1)$ and $\phi(x_1)$. Thus, given some general potential function $\phi(x_1)$, the corresponding charge density may be obtained by transforming to obtain $\bar{\phi}(\beta)$, using $\varepsilon_s(\beta)$ to obtain $\bar{\sigma}(\beta)$, and then transforming back to the x_1-domain.

For the general solution the potential $\phi(x_1)$ and charge density $\sigma(x_1)$ are proportional to exp ($j\omega t$), with the frequency ω regarded as a constant in the above equations. In solving a particular problem it is usually found that the potential and charge density are functions of frequency, so their transforms $\bar{\phi}(\beta)$ and $\bar{\sigma}(\beta)$ will also be functions of frequency. In the Fourier transform, the frequency ω is held constant during the integration. The relationship given by the effective permittivity, equation (3.24), applies for all values of ω.

3.3. SOME PROPERTIES OF THE EFFECTIVE PERMITTIVITY

The effective permittivity described above is a powerful tool for solving problems concerning a one-dimensional set of electrodes on the surface of a piezoelectric half-space. For variables proportional to exp ($j\omega t$), the surface potential $\phi(x_1)$ and charge density $\sigma(x_1)$ are related by the effective permittivity, and the solution is then determined if appropriate boundary conditions are applied. Usually, $\phi(x_1)$ is specified at the electrode locations, while $\sigma(x_1)$ must be zero on all unmetallised regions. Acoustic wave excitation is allowed for implicitly by the definition of the effective permittivity. This includes all forms of acoustic wave that can be excited; thus, in addition to the usual excitation of piezoelectric Rayleigh waves, the effective permittivity will when appropriate include the effects of Bleustein–Gulyaev waves, pseudo-surface waves and bulk waves. In fact, many of the properties of these waves in the material under consideration may be deduced by examining the effective permittivity. However, it should be noted that the permittivity does not show the effect of any acoustic waves which are not piezoelectrically coupled at the surface. Such waves, which may occur in a piezoelectric material, cannot of course be excited by electrodes on the surface; nevertheless, they may be present in a practical device owing to mode conversion at a discontinuity, for example an edge of the substrate.

An important limitation of the method follows from the assumption, used in the derivation, that there are no mechanical forces on the surface. This implies that any electrodes on the surface must be sufficiently thin that they can be assumed to cause negligible mechanical perturbations, that is, mechanical loading is neglected. In practice, this is usually a good approximation.

Generally, the effective permittivity is a complicated function of β, and must be found numerically using the method described in Section 3.2 above. There are however a number of important properties which are readily deduced. Firstly, the function is symmetrical, so that $\varepsilon_s(-\beta) = \varepsilon_s(\beta)$. This follows from the general reciprocity relation, as shown in Appendix B, Section B.4. Secondly, if $\varepsilon_s(\beta)$ is complex this indicates that energy is being radiated away from the surface into the bulk of the material, in the form of acoustic waves. This can be seen from the definition involving the harmonic solution, equation (3.23). The quantity $\tilde{\sigma}(x_1)$ is the charge density on some set of electrodes with edges parallel to the x_2-axis. Since the charge density is time-variant there must in general be currents entering these electrodes from outside. For unit length in the x_2-direction, the current in an interval dx_1 is $J(x_1)\,dx_1$, where $J(x_1) = j\omega\tilde{\sigma}(x_1)$ is the current density. Now, if there is no power transferred into the system from outside, the current density $J(x_1)$ and the potential $\tilde{\phi}(x_1)$ must be in phase

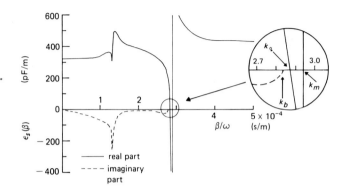

FIGURE 3.1. Effective permittivity for Y, Z lithium niobate. After Milsom et al. [65], copyright © 1977 IEEE.

quadrature, and hence, from equation (3.23), $\varepsilon_s(\beta)$ must be real. In the steady-state condition, a net transfer of power can only occur if bulk acoustic waves are radiated away from the surface, so a complex value of $\varepsilon_s(\beta)$ indicates that bulk wave radiation is occurring. It is usually found that $\varepsilon_s(\beta)$ is complex for some values of β and real for other values.

The form of $\varepsilon_s(\beta)$ depends markedly on the type of acoustic wave involved. Usually we shall be concerned with excitation of piezoelectric Rayleigh waves. However, bulk wave excitation has been investigated quite extensively because it occurs in most surface-wave devices to some extent, and in fact some devices use bulk waves as the main form of acoustic propagation. For excitation of Bleustein–Gulyaev waves the effective permittivity can be expressed as an analytic formula [64, 71].

In the description to follow it is assumed that the surface wave involved is a piezoelectric Rayleigh wave, as is usually the case in practice. An example is Y, Z lithium niobate, and the effective permittivity is shown for this material in Figure 3.1. The permittivity has a zero and a pole at two values of β close together. From Equation (3.23), the zero corresponds to a surface wave solution for a free surface, since the charge density is zero. The wavenumber here is denoted k_0, which is taken to be positive, so that the zeros of $\varepsilon_s(\beta)$ occur at $\beta = \pm k_0$. The pole of $\varepsilon_s(\beta)$ indicates a surface-wave solution for a metallised surface, since it gives a finite charge density and zero potential. In this case the wavenumber is $k_m > 0$, so that the poles occur at $\beta = \pm k_m$. The surface wave velocities for these two cases are v_0 and v_m, so that $k_0 = \omega/v_0$ and $k_m = \omega/v_m$. At a lower value of β, denoted by k_b in Figure 3.1, the permittivity becomes complex, and it remains complex for all smaller values of β. In this region, bulk wave excitation is occurring. For most practical purposes it is not necessary to consider bulk wave excitation in any detail.

An important parameter is the differential of the permittivity at the free-surface wavenumber k_0. This quantity is directly related to the amplitude of the surface waves generated by a transducer. The relationship is derived in Appendix B, Section B.6. The same quantity also relates the power flow of a surface wave to its associated surface potential, as shown below.

Surface-wave Power Flow.
The power flow is derived by a method similar to that of Ingebrigtsen [62]. We consider the harmonic solution for the half space, with the surface potential $\tilde{\phi}(x_1)$ and charge density $\tilde{\sigma}(x_1)$ both proportional to $\exp(j\beta x_1)$. If β is assumed to have a small imaginary part the amplitude of the wave rises or falls exponentially with x_1, and the surface wave power density may be obtained by considering the electrical power applied to the system from outside.

We consider the power transferred in a width W in the x_2-direction. If $J(x_1)$ is the current density at the surface, the power flowing into the system, in a small interval Δx_1, is given by

$$\Delta P(x_1) = \tfrac{1}{2} W \operatorname{Re}[\tilde{\phi}^*(x_1) J(x_1)] \Delta x_1, \qquad (3.25)$$

where the asterisk indicates a complex conjugate and the current density $J(x_1) = j\omega \tilde{\sigma}(x_1)$. The potential and charge density are written, omitting a term $\exp(j\omega t)$, as

$$\tilde{\phi}(x_1) = \tilde{\phi}_0 \exp(j\beta x_1),$$

$$\tilde{\sigma}(x_1) = \tilde{\sigma}_0 \exp(j\beta x_1),$$

where $\tilde{\phi}_0$ and $\tilde{\sigma}_0$ are constants. Setting $\beta = \beta_r + j\beta_i$, this gives

$$\Delta P(x_1) = -\tfrac{1}{2} \omega W \exp(-2\beta_i x_1) \operatorname{Im}[\tilde{\phi}_0^* \tilde{\sigma}_0] \Delta x_1. \qquad (3.26)$$

Now, for real β we have $\tilde{\sigma}_0 = f(\beta) \tilde{\phi}_0$ with, from equation (3.23),

$$f(\beta) = |\beta| \varepsilon_s(\beta). \qquad (3.27)$$

It is assumed that $f(\beta)$ can be continued analytically for complex values of β near the real axis, that is, for small β_i. Equation (3.26) thus becomes

$$\Delta P(x_1) = -\tfrac{1}{2} \omega W |\tilde{\phi}_0|^2 \exp(-2\beta_i x_1) \operatorname{Im}[f(\beta)] \Delta x_1. \qquad (3.28)$$

Using the Cauchy–Riemann equations and noting that $f(\beta)$ is real for $\beta_i = 0$, we have for small β_i

$$\operatorname{Im}[f(\beta)] = \beta_i \left[\frac{\partial f(\beta)}{\partial \beta}\right]_{\beta_i=0}. \qquad (3.29)$$

Now, since the potential $\tilde{\phi}(x_1)$ is proportional to $\tilde{\phi}_0 \exp(-\beta_i x_1)$ the power of the wave, $P_s(x_1)$, must have the form

$$P_s(x_1) = C |\tilde{\phi}_0|^2 W \exp(-2\beta_i x_1), \qquad (3.30)$$

where C is a constant. The change of $P_s(x_1)$ over a distance Δx_1 is equated with $\Delta P(x_1)$ of equation (3.28), giving

$$C = -\tfrac{1}{4} \omega \left[\frac{\partial f(\beta)}{\partial \beta}\right]_{\beta_i=0}, \qquad (3.31)$$

where equation (3.29) has also been used.

In the limit $\beta_i \to 0$ the power of the wave, P_s, is independent of x_1. We consider a wave propagating on a free surface, so that $\beta = k_0$. For this case, using equation

(3.27) and noting that $\varepsilon_s(k_0) = 0$, we have

$$P_s = -\tfrac{1}{4}\omega W |\tilde{\phi}_0|^2 k_0 \left[\frac{d\varepsilon_s(\beta)}{d\beta}\right]_{k_0}. \qquad (3.32)$$

It is convenient to define a real positive constant Γ_s by

$$\frac{1}{\Gamma_s} = -k_0 \left[\frac{d\varepsilon_s(\beta)}{d\beta}\right]_{k_0}. \qquad (3.33)$$

This is independent of the frequency ω; it depends only on the properties of the material and the orientation. The power flow of the surface wave is thus

$$P_s = \tfrac{1}{4}\omega W |\tilde{\phi}_0|^2/\Gamma_s. \qquad (3.34)$$

Milsom et al. [65] give an alternative derivation.

Ingebrigtsen's Approximation. A convenient approximate form for $\varepsilon_s(\beta)$, suitable when the main acoustic wave present is a piezoelectric Rayleigh wave, has been given by Ingebrigtsen [62, 72]. Since $\varepsilon_s(\beta)$ has a zero at $\beta = k_0$, it must be proportional to $(\beta - k_0)$ when β is close to k_0, by Taylor's theorem. Similarly, $1/\varepsilon_s(\beta)$ is zero at $\beta = k_m$, the wavenumber for the metallised case, so $\varepsilon_s(\beta)$ is proportional to $(\beta - k_m)^{-1}$ for β close to k_m. Thus $\varepsilon_s(\beta)$ must be proportional to $(\beta - k_0)/(\beta - k_m)$. This is modified a little to make it an even function of β, giving the approximate formula

$$\varepsilon_s(\beta) \approx A \frac{\beta^2 - k_0^2}{\beta^2 - k_m^2}, \qquad (3.35)$$

where A is a constant. Differentiating this equation with respect to β, the constant Γ_s, defined by equation (3.33), is given by

$$\Gamma_s = \frac{v_0^2/v_m^2 - 1}{2A} \approx \frac{1}{A}\frac{v_0 - v_m}{v_0}, \qquad (3.36)$$

where the approximate form is obtained by using $v_0 \approx v_m$.

The value of the constant A cannot be given unambiguously, because equation (3.35) is only an approximate form for the permittivity. However, for a non-piezoelectric material the permittivity is equal to $(\varepsilon_0 + \varepsilon_p)$, as shown by equation (3.7). In this case $k_m = k_0$ and hence equation (3.35) gives $A = \varepsilon_0 + \varepsilon_p$, with ε_p defined by

$$\varepsilon_p = (\varepsilon_{11}\varepsilon_{33} - \varepsilon_{13}^2)^{1/2}. \qquad (3.37)$$

For a piezoelectric material, an appropriate value for A may be obtained by comparing the power flow formula, equation (3.34), with the corresponding result

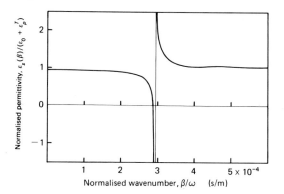

FIGURE 3.2. Effective permittivity for Y, Z lithium niobate, using Ingebrigtsen's approximation.

obtained by perturbation theory, given in Chapter 2, equation (2.48). Using equation (3.36) for Γ_s, this gives

$$A \approx \varepsilon_0 + \varepsilon_p^T,$$

where ε_p^T is defined as in equation (3.37), with the permittivity elements ε_{ij} measured at constant stress. This value for A is adequate for most practical purposes, and has the merit that it is easily evaluated for many materials, using readily available data. A further justification is given in Section 3.5 below. Thus, Ingebrigtsen's approximation, equation (3.35), becomes

$$\varepsilon_s(\beta) \approx (\varepsilon_0 + \varepsilon_p^T) \frac{\beta^2 - k_0^2}{\beta^2 - k_m^2} \qquad (3.38)$$

and the constant Γ_s is

$$\Gamma_s \approx \frac{1}{\varepsilon_0 + \varepsilon_p^T} \frac{v_0 - v_m}{v_0}. \qquad (3.39)$$

A relation almost identical to equation (3.38) can be obtained by using normal mode theory [70].

The permittivity is plotted on Figure 3.2, using equation (3.38) with data appropriate for lithium niobate (v_0 = 3488 m/s, v_m = 3404 m/s). The exact solution for this case is shown in Figure 3.1.

The approximate form of the permittivity is convenient for analysis of practical devices, because it depends on only three variables, and numerical data for these are readily available. The main limitation is that bulk wave excitation is excluded, since the function is real for all values of β. It should be noted that the approximation is only valid for piezoelectric Rayleigh waves; for example, it cannot be used for leaky waves or Bleustein–Gulyaev waves.

3.4. GREEN'S FUNCTION

The effective permittivity $\varepsilon_s(\beta)$ relates the charge density to the surface potential in the β-domain. However, here we are concerned with problems in which the boundary conditions are expressed in the x_1-domain. The relationship in this domain can be expressed by using a Green's function [65, 66].

From equation (3.24) the surface potential and charge density, in the β-domain, are related by

$$\bar{\phi}(\beta) = \bar{G}(\beta, \omega)\bar{\sigma}(\beta), \qquad (3.40)$$

where

$$\bar{G}(\beta, \omega) = [|\beta|\varepsilon_s(\beta)]^{-1}. \qquad (3.41)$$

In the x_1-domain the surface potential $\phi(x_1)$ and the charge density $\sigma(x_1)$ are the inverse transforms of $\bar{\phi}(\beta)$ and $\bar{\sigma}(\beta)$. Generally, these functions also depend on the frequency ω, but here ω is taken to be a constant. Now, by the convolution theorem of Fourier analysis [Appendix A, equation (A.19)] we may transform equation (3.40) from the β-domain to the x_1-domain by transforming the two functions on the right side individually and the convolving. Thus, if $G(x_1, \omega)$ is the inverse transform of $\bar{G}(\beta, \omega)$, we have

$$\phi(x_1) = G(x_1, \omega) * \sigma(x_1)$$
$$= \int_{-\infty}^{\infty} G(x_1 - x_1', \omega)\sigma(x_1')\,dx_1', \qquad (3.42)$$

where the asterisk indicates convolution. The function $G(x_1, \omega)$ is the *Green's function*. It can be interpreted as the surface potential produced by a line charge at $x_1 = 0$, as can be seen by putting $\sigma(x_1) = \delta(x_1)$ in equation (3.42). Transforming equation (3.41) we have

$$G(x_1, \omega) = \frac{1}{2\pi}\int_{-\infty}^{\infty} \frac{\exp(j\beta x_1)}{|\beta|\varepsilon_s(\beta)}\,d\beta. \qquad (3.43)$$

Using reciprocity, it is shown in Appendix B that this is an even function of x_1, that is $G(-x_1, \omega) = G(x_1, \omega)$.

Green's Function for Piezoelectric Rayleigh Waves. We consider the form of the Green's function for the particular case of a substrate supporting propagation of piezoelectric Rayleigh waves, as in for example Y, Z lithium niobate. In addition to surface waves, excitation of bulk waves can occur, as shown by the fact that $\varepsilon_s(\beta)$ is complex for some values of β (Figure 3.1). Milsom et al. [65] have shown that the Green's function $G(x_1, \omega)$ may be regarded as a sum of three terms, giving contributions due to surface wave excitation, electrostatic effects, and bulk wave excitation.

The surface wave term is deduced by considering the generation of surface waves by a transducer, consisting of a set of electrodes occupying a finite region of x_1, with voltages applied to them. The charge density on the electrodes is $\sigma(x_1)$, with Fourier transform $\bar{\sigma}(\beta)$. The potential associated with these waves is derived in Appendix B, Section B.6, and is given by

$$\phi(x_1) = j\Gamma_s \bar{\sigma}(\mp k_0, \omega) \exp(\mp jk_0 x_1), \qquad (3.44)$$

where the upper signs refer to waves radiated in the $+x_1$ direction, and the lower signs to waves radiated in the $-x_1$ direction. This potential is considered to arise from a surface-wave component $G_s(x_1, \omega)$ of the Green's function. By equation (3.42), this is the surface-wave potential obtained when $\sigma(x_1) = \delta(x_1)$, that is, for $\bar{\sigma}(\beta) = 1$. Thus, from equation (3.44)

$$G_s(x_1, \omega) = j\Gamma_s \exp(-jk_0|x_1|). \qquad (3.45)$$

The surface-wave component of $G(x_1, \omega)$ arises because $\varepsilon_s(\beta)$ is zero at $\beta = k_0$, so that the integrand of equation (3.43) has a pole at this point. There is also a pole at $\beta = 0$. If the variation of $\varepsilon_s(\beta)$ is ignored in this region, then from equation (3.40), $\bar{\phi}(\beta)$ is proportional to $\bar{\sigma}(\beta)/|\beta|$. A relation of this form applies for the electrostatic case, as shown by equation (3.12) of Section 3.1. In the x_1-domain this contribution to the potential is given by an electrostatic Green's function $G_e(x_1)$, and by comparing with equation (3.15) this can be written

$$G_e(x_1) = -\frac{\ln|x_1|}{\pi(\varepsilon_0 + \varepsilon_p^T)}. \qquad (3.46)$$

The use of the constant $(\varepsilon_0 + \varepsilon_p^T)$ will be justified later in this section, though equation (3.15) shows that it is correct if the material is not piezoelectric. It should be noted that $G_e(x_1)$ is independent of the frequency ω.

In addition to surface wave effects and electrostatic effects, the total Green's function $G(x_1, \omega)$ must also include the effects of bulk waves. It is assumed that these can be accounted for by adding a further term $G_b(x_1, \omega)$, so that the total Green's function of equation (3.43) is given by

$$G(x_1, \omega) = G_e(x_1) + G_s(x_1, \omega) + G_b(x_1, \omega), \qquad (3.47)$$

with $G_e(x_1)$ and $G_s(x_1, \omega)$ given by equations (3.46) and (3.45) respectively. The bulk wave term $G_b(x_1, \omega)$ is more difficult to obtain analytically, though it can be obtained numerically from the effective permittivity [65]. For most purposes the details of this function will not be required.

Approximate Form for the Green's Function.

An approximate Green's function, convenient for solving many practical problems, is obtained by omitting the bulk wave contribution from the exact expression of equation (3.47). Thus, using equations (3.46) and (3.45) for the electrostatic and surface wave terms, the approximate Green's function is

$$G(x_1, \omega) \approx G_e(x_1) + G_s(x_1, \omega)$$
$$= -\frac{\ln|x_1|}{\pi(\varepsilon_0 + \varepsilon_p^T)} + j\Gamma_s \exp(-jk_0|x_1|). \qquad (3.48)$$

To confirm the validity of this, we deduce the corresponding effective permittivity and compare it with Ingebrigtsen's approximation. From equation (3.41), the permittivity is the reciprocal of $|\beta| \cdot \bar{G}(\beta, \omega)$, where $\bar{G}(\beta, \omega)$ is the transform of the right side of equation (3.48). The transform of the exponential term is given by equation (A.41), and the transform on $\ln |x_1|$ can be taken as $-\pi/|\beta|$, which follows by comparing equation (3.15) with equation (3.12). We thus find

$$\bar{G}(\beta, \omega) \approx [(\varepsilon_0 + \varepsilon_p^T)|\beta|]^{-1} + j\Gamma_s[\pi\delta(\beta - k_0) + \pi\delta(\beta + k_0)$$
$$+ 2jk_0/(\beta^2 - k_0^2)]. \tag{3.49}$$

This function is infinite at $\beta = \pm k_0$, so that $\varepsilon_s(\beta)$ is zero at these points, as required. It is also necessary that $\bar{G}(\beta, \omega)$ should be zero at $\beta = \pm k_m$, so that $\varepsilon_s(\beta)$ is infinite at these points. This condition requires an appropriate value for the constant Γ_s. Setting $\bar{G}(\pm k_m, \omega) = 0$ in equation (3.49) gives

$$\Gamma_s = \frac{k_m^2 - k_0^2}{2(\varepsilon_0 + \varepsilon_p^T) k_0 k_m} \approx \frac{1}{\varepsilon_0 + \varepsilon_p^T} \frac{v_0 - v_m}{v_0}, \tag{3.50}$$

in good agreement with the expression deduced previously, equation (3.39).

In equation (3.49), the term $1/(\beta^2 - k_0^2)$ makes $\bar{G}(\beta, \omega)$ infinite at $\beta = \pm k_0$ and hence the reciprocal of $\bar{G}(\beta, \omega)$ is zero at these points. It follows that the reciprocal of $\bar{G}(\beta, \omega)$ is not affected by the presence of the delta-functions, which can therefore be omitted when calculating $\varepsilon_s(\beta)$. Using equation (3.50) for Γ_s, the effectivity permittivity is

$$\varepsilon_s(\beta) = \frac{1}{|\beta|\bar{G}(\beta, \omega)} = (\varepsilon_0 + \varepsilon_p^T) \frac{\beta^2 - k_0^2}{(|\beta| - k_m)(|\beta| + k_0^2/k_m)}. \tag{3.51}$$

This is almost identical with Ingebrigtsen's approximation, equation (3.38); it becomes exactly the same if the term k_0^2/k_m is replaced by k_m, which is quantitatively similar. The agreement confirms the validity of the approximate Green's function, equation (3.48), and also justifies the use of the constant $(\varepsilon_0 + \varepsilon_p^T)$ in the electrostatic Green's function, equation (3.46).

3.5. OTHER APPLICATIONS OF THE EFFECTIVE PERMITTIVITY

The above development shows how the effective permittivity $\varepsilon_s(\beta)$ may be used to relate the surface charge density and potential for a piezoelectric half-space, with the region above the piezoelectric assumed to be a vacuum. In Chapter 4 this method will be used for the analysis of surface-wave transducers on a half-space. However, the concept of the effective permittivity may be generalised to analyse a number of other problems, and here we digress briefly to discuss these.

We first consider the coupling between a piezoelectric half-space and a plane at a height h above the surface [62, 67, 68, 70]. As shown in Figure 3.3(a), the surface of

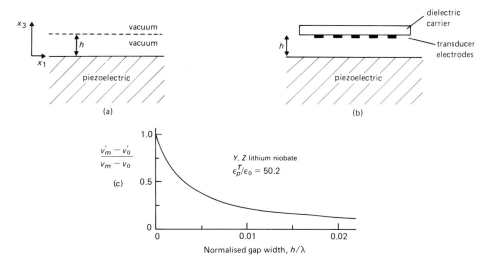

FIGURE 3.3. Piezoelectric coupling across a gap.

the piezoelectric is at $x_3 = 0$, and free charges are allowed to exist only at the plane $x_3 = h$. Considering a harmonic solution, with the relevant quantities proportional to $\exp(j\beta x_1)$, we can define an effective permittivity $\varepsilon'_s(\beta)$ for the plane $x_3 = h$:

$$\varepsilon'_s(\beta) = \frac{\tilde{\sigma}}{|\beta|\tilde{\Phi}(h)}, \qquad (3.52)$$

where $\tilde{\Phi}(h)$ and $\tilde{\sigma}$ are the potential and charge density at $x_3 = h$. This function may be expressed in terms of the surface permittivity $\varepsilon_s(\beta)$, defined in equation (3.23) above. At the plane $x_3 = 0$, the value of $\tilde{D}_3$ in the piezoelectric is, from equations (3.21) and (3.22),

$$\tilde{D}_3(-) = \tilde{\phi}|\beta|[\varepsilon_0 - \varepsilon_s(\beta)], \qquad (3.53)$$

where $\tilde{\phi}$ is the potential at $x_3 = 0$. This relation is valid irrespective of the electrical conditions above the surface $x_3 = 0$. In the vacuum region $0 < x_3 < h$ the potential $\tilde{\Phi}$ has terms proportional to $\exp(\pm \beta x_3)$, from Laplace's equation, and in the region $x_3 > h$ the potential is proportional to $\exp(-|\beta|x_3)$. The potential must be continuous everywhere, and $\tilde{D}_3$ must be continuous at $x_3 = 0$. The discontinuity of $\tilde{D}_3$ at $x_3 = h$ gives the charge density, $\tilde{\sigma}$. Using these relations, the effective permittivity at $x_3 = h$ is found to be

$$\varepsilon'_s(\beta) = \frac{\varepsilon_s(\beta)[1 + \tanh(|\beta|h)]}{1 - [1 - \varepsilon_s(\beta)/\varepsilon_0]\tanh(|\beta|h)}. \qquad (3.54)$$

This function may be used to analyse excitation by a transducer located at the plane $x_3 = h$, as in Figure 3.3(b). In practice the transducer would be supported by a dielectric, but this does not appreciably alter the effective permittivity. As for excitation at the surface, the permittivity has a zero at $\beta = k'_0$, say, and a pole at $\beta = k'_m$, say, with corresponding surface wave velocities $v'_0 = \omega/k'_0$ and $v'_m = \omega/k'_m$. The difference between these velocities is a measure of the coupling strength. Equation (3.54) shows that $\varepsilon'_s(\beta)$ is zero when $\varepsilon_s(\beta)$ is zero, so that $v'_0 = v_0$. The metallised-surface velocity v'_m depends on h, approaching v'_0 for large h. Assuming that the surface wave is a piezoelectric Rayleigh wave, Ingebrigtsen's approximation, equation (3.38), may be used for $\varepsilon_s(\beta)$ in equation (3.54) to evaluate v'_m. Noting that the four velocities v'_m, v'_0, v_m and v_0 are numerically similar, this gives

$$\frac{v'_m - v'_0}{v_m - v_0} \approx \left[1 - \frac{1 + \varepsilon_p^T/\varepsilon_0}{1 - \coth(k_0 h)}\right]^{-1}. \tag{3.55}$$

Here the left side is a measure of the coupling strength for the plane at $x_3 = h$, dividied by the coupling strength obtained at the surface $x_3 = 0$. The function is shown in Figure 3.3(c), assuming a Y, Z lithium niobate half-space, with the height h normalised to the wavelength $\lambda = 2\pi/k_0$. The coupling strength decreases very rapidly with h. It can be concluded that, to be practically effective, a transducer held above the surface would need to be very close, with a gap width much less than the wavelength. In practice, such a small gap is usually awkward to obtain, and so transducers are usually deposited directly on the substrate. However, coupling across a gap has the advantage that the transducer is movable, and this has been exploited as a rapid method of assessing the surface-wave properties of materials [73].

The quantitative results obtained from equation (3.55) have been found to give excellent agreement with accurate calculations, obtained without the use of approximations [67, 70]. This gives further confirmation of the validity of Ingebrigtsen's approximation, equation (3.38), and in particular confirms the use of the coefficient $(\varepsilon_0 + \varepsilon_p^T)$.

Another type of problem which can be analysed using the effective permittivity is that of coupling to a semiconductor above the piezoelectric surface [64, 70, 71, 74]. Usually the semiconductor is considered to be separated from the surface by a small gap, as in Figure 3.4(a). As before, the ratio of $\tilde{D}_3/\tilde{\Phi}$ at the piezoelectric surface is determined by the effective permittivity, while the same ratio at the semiconductor surface is determined by the semiconductor equations. Taken together, these relations give the wavenumber β, which in this case is complex. As in the case of a transducer held above the surface, the gap has to be very small if significant interactions are to be obtained. Coupling to a semiconductor has been exploited in several ways, which are reviewed by Kino [75]. Amplification of the surface wave can be obtained by applying a drift field in the x_1-direction, so that the carrier drift velocity in the semiconductor exceeds the surface wave velocity. Typically, a gain of 50 dB/cm can be obtained, though a large drift field, typically a few kV/cm, is required. Coupling to a semiconductor is also used in a type of convolver, where a non-linear effect in the semiconductor is used to mix two surface-wave signals.

The effective permittivity concept can also be applied to the analysis of transducers using piezoelectric layers. As illustrated in Figure 3.4(b), a piezoelectric film is

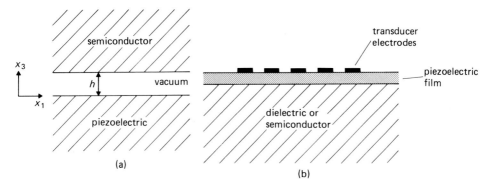

FIGURE 3.4. Other excitation problems.

deposited on a non-piezoelectric substrate, and this enables an interdigitial transducer to be used to generate surface waves. The transducer may be on top of the film, as in Figure 3.4(b), or at the interface between the film and the substrate. Experimentally, the commonest material for the film is zinc oxide and, with a glass substrate, this is used in some bandpass filters for television receivers [76]. Alternatively, a zinc oxide film can be used to generate surface waves on a silicon substrate, enabling a surface wave device to be integrated with circuitry on the same substrate [77].

For the layered system, an effective permittivity can be defined in the usual way, relating the charge density to the potential at the plane of the transducer. As before, two key parameters are the surface wave velocities obtained when there is no metallisation at the transducer plane, and when a uniform conducting sheet is present at this plane. The velocity difference gives a measure of the coupling strength. For simplicity, Ingebrigtsen's approximation [equation (3.38)] can be used for the permittivity, with the wavenumbers k_0 and k_m obtained from the two surface wave velocities. Much of the transducer analysis of Chapter 4 can then be used. However, there is the complication that for the layered system the surface wave is dispersive, as already noted in Section 2.3.6, so that k_0 and k_m are not proportional to ω. An analysis based on normal mode theory is given by Kino and Wagers [78].

Chapter 4

Analysis of Interdigital Transducers

As noted in Chapter 1, interdigital transducers are used in all practical surface-wave devices, and in many devices the performance of the transducers is the main factor determining the device performance. A detailed understanding of the behaviour of transducers is therefore crucial for both analysis and design of the devices. This chapter describes methods for transducer analysis, and for analysis of devices comprising two transducers. Most of the analysis is based on the Green's function derived in Chapter 3, using an approximation called the "quasi-static approximation" for simplicity.

Section 4.1 describes the delta-function model, which gives a straightforward approximate analysis. Despite being over-simplified, this model is very convenient to use, and illustrates some important features directly. It is particularly useful for transducer design, as will be considered in later chapters. Section 4.2 discusses some important second-order effects that are neglected in the delta-function model, and also gives a comparative discussion of several other methods for transducer analysis.

The quasi-static approximation is introduced in Section 4.3, which derives results applicable to both transducers and multi-strip couplers, and the main properties of transducers are then derived in Section 4.4. The remaining sections are concerned with some applications of this basic analysis. In Sections 4.5 and 4.6 some particular restrictions are imposed on the transducer geometry; these restrictions enable some useful results, applicable to most interdigital devices, to be derived. Finally, Sections 4.7 and 4.8 are concerned with analysis of devices comprising two transducers.

4.1. DELTA-FUNCTION MODEL

This model was introduced by Tancrell and Holland [79, 80], and a closely related approach called the "impulse model" was described later by Hartmann *et al.* [81].

(a) *Launching Transducer.* We first consider generation of surface waves by a uniform transducer, as shown in Figure 4.1(a). The transducer electrodes are all identical, and are connected alternately to the two bus-bars. When a voltage is applied

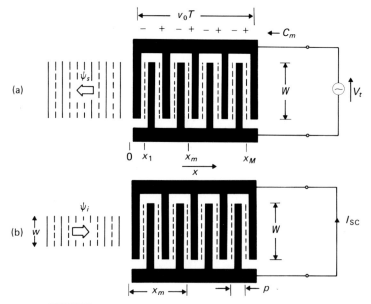

FIGURE 4.1. Uniform transducer: launching and reception.

an electric field is set up in each inter-electrode gap, and surface waves are generated. It is assumed that each gap can be regarded as an independent source, launching surface waves with amplitudes proportional to the voltage difference between the adjacent electrodes. We also assume that an adequate approximation is obtained by taking each source to be strongly localised, so that the sources can be represented by the broken lines in Figure 4.1(a), midway between the electrodes. We consider waves generated to the left, though the transducer will also generate waves to the right.

The wave amplitude is denoted by $\psi_s(x)$, and this will be proportional to the potential $\phi_s(x)$ associated with the wave. A precise physical definition for $\psi_s(x)$ is not needed in this model, though its relation to the surface wave power flow is given later. For the source located at x_m, the wave generated to the left, at frequency ω, is written as

$$\psi_s(x) = EV_t C_m \exp[jk_0(x - x_m)], \qquad (4.1)$$

where a factor $\exp(j\omega t)$ is implicit and the real part is to be taken. Here the factor $C_m = \pm 1$ accounts for the alternating polarities of the fields in the gaps, as shown in Figure 4.1(a), and E is a constant. The wavenumber k_0 is taken to be ω/v_0, where v_0 is the free-surface velocity. Equation (4.1) thus assumes that each source generates waves which propagate out of the transducer, unaffected by the electrodes that they pass under. In fact, the electrodes reflect the waves to some extent, and they also perturb the surface-wave velocity. It is assumed that these effects can be neglected, as is often found to be the case, particularly when the piezoelectric coupling of the material is weak.

The material can be taken to be linear, so that the total amplitude of the wave generated by the transducer is the sum of the contributions due to individual sources.

Thus the total wave amplitude is

$$\psi_s(x) = V_t E A(\omega) \exp(jk_0 x), \qquad (4.2)$$

where

$$A(\omega) = \sum_{m=1}^{M} C_m \exp(-jk_0 x_m) \qquad (4.3)$$

and M is the number of sources. The product $EA(\omega)$ is essentially the *frequency response* of the transducer, giving the surface wave amplitude as a function of frequency. The function $A(\omega)$, determined by the relative positions and the polarities of the elements, is the *array factor*, and E is the *element factor*. A detailed analysis, given in later sections, shows that E varies a little with frequency. However, $A(\omega)$ is found to vary much more rapidly than E and, in the delta-function model, a reasonable approximation is obtained by taking E to be constant. Thus the frequency response is essentially given by the array factor alone, and since this is readily found from the transducer geometry the model is convenient to use. The element factor is evaluated in Section 4.5.3.

For the uniform transducer considered here, the summation in equation (4.3) can be written in a more convenient form. Since C_m alternates between $+1$ and -1 we may write $C_m = -C_1 \exp(jm\pi)$. Also, taking the spacing of the sources to be p, their locations may be written $x_m = mp$. With these substitutions, equation (4.3) is readily summed as a geometric progression. Since the phase is of little interest here, we consider the magnitude of the array factor, which is found to be

$$|A(\omega)| = \left|\frac{\sin M\theta/2}{\sin \theta/2}\right|, \qquad (4.4)$$

where $\theta = k_0 p - \pi$. We also define ω_c as the frequency at which the transducer periodicity, $2p$, equals the surface-wave wavelength, so that $\omega_c = \pi v_0/p$. We then have

$$\theta = \pi(\omega - \omega_c)/\omega_c. \qquad (4.5)$$

The magnitude of the array factor is shown, as a function of ω, in Figure 4.2. The first main peak occurs at frequency ω_c, and the response in this region is called the fundamental response. Harmonic responses of equal magnitude occur at odd multiples of ω_c.

The fundamental response is of most interest. If ω is near ω_c, θ is small and the magnitude of the array factor is, approximately,

$$|A(\omega)| \approx M\left|\frac{\sin M\theta/2}{M\theta/2}\right|. \qquad (4.6)$$

For this function, the zeros nearest to ω_c are at $\omega = \omega_c \pm 2\omega_c/M$. Roughly speaking, the transducer responds effectively over a band of frequencies bounded by the points $\omega = \omega_c \pm \omega_c/M$, where the amplitude is about 4 dB below its maximum value. Thus the bandwidth can be taken to be $\Delta\omega = 2\omega_c/M$. Now, the length of the transducer

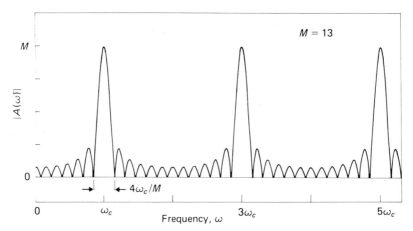

FIGURE 4.2. Magnitude of array factor for a uniform transducer.

is approximately Mp, which corresponds to a surface-wave propagation time $T = Mp/v_0$. Thus the bandwidth can be written as

$$\Delta\omega = 2\pi/T. \qquad (4.7)$$

Thus the bandwidth, in Hz, is simply the reciprocal of the transducer length, measured in time units.

It was stated earlier that the element factor E in equation (4.2) can be taken to be independent of frequency. This is valid for the fundamental response, where ω is close to ω_c, and for a harmonic response, where ω is close to an odd multiple of ω_c. However, the element factor has a strong effect on the relative levels of the harmonics. In particular, it is shown in a later section that the third harmonic, centred at $\omega = 3\omega_c$, is absent altogether if the electrode widths are equal to $p/2$.

(b) Reception. We now consider the same transducer receiving a surface-wave beam, as shown on Figure 4.1(b). The transducer is taken to be shorted, and generates a current I_{sc}. The incident beam has a width, w, less than or equal to the transducer aperture, W. The amplitude of the incident beam is written as

$$\psi_i(x) = \psi_{i0} \exp(-jk_0 x), \qquad (4.8)$$

where ψ_{i0} is a constant. As before, it is assumed that the surface wave is not perturbed by the presence of the electrodes, so that equation (4.8) gives the wave amplitude at all x.

For generation of surface waves, the transducer was regarded as an array of localised sources, shown by the broken lines in Figure 4.1(a). For a receiving transducer, each of these sources becomes a receiving element, making a contribution to the current I_{sc}. Clearly, the element at x_m contributes a current proportional to $C_m \psi_i(x_m)$. The total current is simply the sum of these contributions, with a proportionality constant Ew which is justified below. Thus,

ANALYSIS OF INTERDIGITAL TRANSDUCERS

$$I_{sc} = \psi_{i0} w E \sum_{m=1}^{M} C_m \exp(-jk_0 x_m), \quad \text{for } w \leqslant W. \tag{4.9}$$

Comparing with equation (4.3), we have $I_{sc} = \psi_{i0} w E A(\omega)$. Thus the frequency response, $EA(\omega)$, applies to the reception process as well as to the launching process, as expected by reciprocity.

In equation (4.9), I_{sc} is taken to be proportional to w, and this is justified by the following argument. Physically, the current arises because the surface wave induces charges on the electrodes. The charge density is proportional to the wave amplitude ψ_{i0}, and is uniform over the width of the beam. However, the current flowing out of any one electrode is given by the total charge on it, and must therefore be proportional to the beamwidth w. Hence the transducer current I_{sc} is also proportional to w.

The use of the coefficient E in equation (4.9) is justified by the observation that the magnitudes of the surface wave amplitudes, ψ_s and ψ_i, have not so far been specified. In equation (4.9) E is proportional to $1/\psi_{i0}$, while in equation (4.2) for the launching process E is proportional to the magnitude of $\psi_s(x)$. Thus, if a suitable definition is used for ψ_s and ψ_i, the same element factor, E, may be used in equations (4.2) and (4.9). This is assumed to be the case, and consequently the magnitudes of ψ_s and ψ_i are determined. It is found that, for the launching transducer, Figure 4.1(a), the power of the surface wave beam radiated to the left is given by

$$P_s = \tfrac{1}{4} W |\psi_s|^2, \tag{4.10}$$

where W is the beam width. A similar relation applies for the incident beam in Figure 4.1(b). This relation can be proved by considering the power conversion efficiency of the transducer, for launching and reception of surface waves, making use of equations (4.2) and (4.9). From the analysis in later sections of this chapter, it also follows that $\psi_s(x)$ is proportional to the potential $\phi_s(x)$ associated with the propagating wave. Comparing equation (4.10) with the power flow given by equation (3.34) shows that the magnitude of $\psi_s(x)$ is given by

$$|\psi_s(x)| = |\phi_s(x)|(\omega/\Gamma_s)^{1/2}, \tag{4.11}$$

where the constant Γ_s, given by equation (3.39), is a measure of the piezoelectric coupling of the substrate material.

(c) Apodised Transducers. To appreciate the behaviour of an apodised transducer, it is necessary to consider the response of a surface-wave device having two transducers. Here, the device of Figure 4.3 is considered, where transducer A is apodised and transducer B is unapodised. To preserve the symmetry, the source locations in the two transducers are referred to different axes, x for transducer A and x' for transducer B; in each transducer, the axis is directed away from the other transducer. A voltage V_t, with frequency ω, is applied to transducer A. Transducer B is shorted, and produces a current I_{sc}.

In transducer A, the sources are taken to exist only where electrodes of differing polarity overlap. The length of source m will be denoted a_m, and it is assumed that the

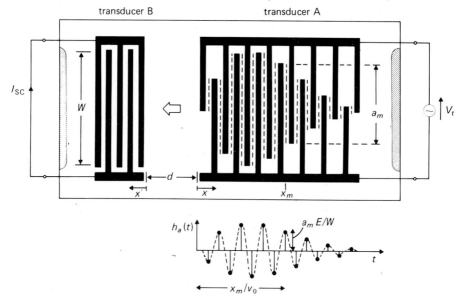

FIGURE 4.3. Apodised transducer and its impulse response.

longest source has a length W, equal to the aperture of transducer B. Consider first the surface wave beam generated by source m of transducer A. Assuming diffraction to be negligible, this is given by equation (4.1), so that $\psi_s(x) = EV_t C_m \exp[jk_0(x - x_m)]$. If d is the transducer separation, the two coordinates x and x' are related by $x = -x' - d$, so for transducer B this wave becomes an incident wave $\psi_i^{(m)}(x') = \psi_{i0}^{(m)} \exp(-jk_0 x')$, with

$$\psi_{i0}^{(m)} = EV_t C_m \exp[-jk_0(x_m + d)]. \tag{4.12}$$

The output current due to this wave is given by equation (4.9), where the beam width w must be equated with the source length a_m in transducer A. Thus the current is

$$I_{sc}^{(m)} = \psi_{i0}^{(m)} a_m E \sum_{n=1}^{M'} C_n' \exp(-jk_0 x_n'), \tag{4.13}$$

where transducer B is taken to have M' receiving elements, with polarities C_n' and locations x_n'. Equation (4.13) is the output current due to source m of transducer A. The total current I_{sc} is obtained by summing the contributions due to the M sources in this transducer, giving

$$I_{sc} = WV_t \exp(-jk_0 d) \left[E \sum_{n=1}^{M'} C_n' \exp(-jk_0 x_n') \right] H_a(\omega), \tag{4.14}$$

where

$$H_a(\omega) = E \sum_{m=1}^{M} \frac{a_m}{W} C_m \exp(-jk_0 x_m). \tag{4.15}$$

In equation (4.14), the term in square brackets is, by comparison with equation (4.2), the frequency response of transducer B. The term $H_a(\omega)$ can be taken as the frequency response of transducer A, which is apodised. Thus the device response is essentially the product of the two transducer responses. Equation (4.15) is consistent with the frequency response defined for an unapodised transducer, as can be seen by setting $a_m = W$. It should be noted that this formulation is valid only if transducer B is unapodised. It has also been assumed that the transducers do not reflect incident surface waves; this is found to be a reasonable assumption for many devices, provided the transducers are connected to zero electrical impedances, as they are in Figure 4.3.

The influence of the geometry of the apodised transducer is shown more clearly by considering its impulse response $h_a(t)$, which can be defined as the inverse Fourier transform of $H_a(\omega)$. Noting that $k_0 = \omega/v_0$, where v_0 is the free-surface velocity, independent of ω, equation (4.15) gives

$$h_a(t) = E \sum_{m=1}^{M} \frac{a_m C_m}{W} \delta(t - x_m/v_0) \qquad (4.16)$$

assuming E to be constant. Thus the impulse response is a sequence of delta functions at times corresponding to the element locations, with amplitudes given by the electrode overlaps a_m, as shown in the lower part of Figure 4.3. Physically this form for the impulse response is unrealistic because E is in fact a function of ω; however, if equation (4.16) is transformed to the frequency domain, taking E to be constant, the result is a good approximation in the region of the fundamental frequency response.

In view of this, it is convenient to consider an alternative definition of the impulse response, denoted $\tilde{h}_a(t)$, such that the harmonic responses in the frequency domain are eliminated. Thus the Fourier transform of $\tilde{h}_a(t)$ is the same as the transform of $h_a(t)$ for $\omega < 2\omega_c$, but is zero for $\omega > 2\omega_c$, with $\omega_c = \pi v_0/p$ as before. The form of $\tilde{h}_a(t)$ can be obtained from the equations of sampling theory, summarised later in Section 8.1.2 of Chapter 8. It is assumed that the element lengths a_m do not change rapidly, so that we can define a smooth function $a(x)$ such that $a_m = a(x_m)$, with the element locations given by $x_m = mp$. The smooth nature of $a(x)$ is expressed by requiring its Fourier transform $\bar{a}(\beta)$ to be zero for $|\beta| > \omega_c/v_0$. This is found to give

$$\tilde{h}_a(t) = \frac{\omega_c E}{\pi W} a(v_0 t) \cos(\omega_c t). \qquad (4.17)$$

This is an amplitude-modulated sinusoid, with peaks and troughs at the locations $t = x_m/v_0$ of the delta functions comprising $h_a(t)$. At these locations, $\tilde{h}_a(t)$ is proportional to the amplitudes of the delta-functions. The broken line in the lower part of Figure 4.3 shows the function $\pi \tilde{h}_a(t)/\omega_c$. It is often convenient to regard $\tilde{h}_a(t)$ as the effective impulse response of the transducer [81], because the harmonics are relatively insignificant.

This formulation leads to a simple design prescription, asuming that some required frequency response for the transducer has been given. The frequency response is Fourier transformed to the time domain. Assuming that this yields an amplitude-modulated carrier, the transducer elements are placed at locations corresponding to the peaks and troughs of this waveform, and their lengths (the electrode overlaps) are made proportional to the amplitude at these points. For example, if the frequency response is to be constant within a specified band and zero outside this band, the apodisation function $a(x)$ is required to have the form $(\sin Ax)/(Ax)$, where A is a constant, and the geometry of the transducer is illustrated by Figure 1.4. The design procedure is thus very versatile and, in principle, straightforward. Further details will be given in Chapter 8, where it will be shown that the impulse response may have phase modulation as well as amplitude modulation.

4.2. DISCUSSION OF SECOND-ORDER EFFECTS AND METHODS OF ANALYSIS

Although the delta-function model described above gives a very useful first-order method of transducer analysis, it fails to account for a number of second-order effects that are often significant in practice. In addition some important transducer properties, notably the element factor and the transducer admittance, are not given by this model. A more detailed analysis will be given below, but it is convenient to consider first the variety of transducer types and the relevant second-order effects, in order to clarify the requirements for transducer analysis. We also discuss some of the theoretical models that are used.

An important second-order effect is the reflection of incident surface waves by a transducer. In a device using two transducers, the output transducer will in general produce a reflected wave, which is then reflected a second time by the input transducer. Thus, a reflected wave reaches the output transducer after traversing the substrate three times, giving an unwanted output signal known as the *triple-transit* signal. There are also spurious signals due to additional reflections, but these are usually insignificant. For C.W. excitation of the input transducer, the triple-transit signal causes the device output, as a function of frequency, to exhibit a ripple. If a short pulse is applied to the input transducer, the triple-transit signal takes the form of an unwanted output pulse following the main pulse. The two pulses often overlap, but are resolved if the transducer separation is large enough.

In practice there are two particular considerations that affect the reflection coefficient: electrode interactions, and the value of the electrical load connected to the transducer. Suppose initially that the transducer is shorted, or connected to a source with zero impedance. In this situation, the transducer can reflect surface waves because the electrodes cause mechanical and electrical perturbations of the surface. This effect is here called *electrode interactions*, though the term "mechanical and electrical loading" (MEL) is also used. Each electrode can be considered to reflect the surface waves. Although the perturbation is small, the reflected waves add coherently if the electrode spacing equals half the wavelength, and the total reflection coefficient can therefore be quite large. It is found that this also distorts the frequency response of the transducer. The transducers of Section 4.1 have two electrodes per period, as shown in Figure 4.4(a), and consequently the interactions can be strong at the fundamental centre frequency ω_c. For a strongly piezoelectric substrate material, such

single-electrode double-electrode

(a) (b) (c)

FIGURE 4.4. Types of uniform transducer.

as lithium niobate, significant interactions occur if the number of electrodes, N, is such that $N\Delta v/v > 1$, where $\Delta v/v$ is the fractional velocity change due to a continuous metal film. To overcome this, *multi-electrode* transducers, such as those in Figure 4.4(b) and (c), are often used [82, 83]. Here there are more than two electrodes per period, so that strong reflections can occur only at frequencies well removed from the centre frequency ω_c. The simpler transducer of Figure 4.4(a) is often called a *single-electrode* transducer, while the transducer of Figure 4.4(b) is a *double-electrode* transducer.

Practical devices are usually designed such that electrode interactions are weak, since this enables a relatively straightforward design procedure to be used. Thus, for practical purposes, an analysis that neglects electrode interactions is usually adequate. This is the case for all of the analysis in this chapter, including the delta-function analysis of Section 4.1. In the sections to follow, the "quasi-static approximation" is used, with the consequence that the analysis neglects electrode interactions. However, Appendix E gives a second-order analysis which does not use this approximation, and describes electrode interactions in more detail.

In the above discussion the transducer was taken to be connected to an electrical impedance of zero, and, assuming electrode interactions are weak, its reflection coefficient is small in this situation. However, significant reflectivity can still arise when the load impedance is finite. For example, if the transducer is electrically matched to a load, thus optimising its conversion efficiency, the power reflection coefficient is theoretically $\frac{1}{4}$, giving a triple-transit signal too large for most applications. This large reflection coefficient is a consequence of the bidirectional nature of the transducer. The usual remedy is to avoid a close match of the impedances, accepting the consequent loss of efficiency. However, there are several special types of transducer that are essentially uni-directional, giving simultaneously a low reflection coefficient and good conversion efficiency. These will be described in Chapters 5 and 7.

In addition to its influence on the reflection coefficient, the use of a finite external impedance also affects the frequency response of a device to some extent. This is known as the "circuit effect", and is often significant for devices whose responses are accurately specified.

Some further second-order effects are associated with the electrostatic charge density on the transducer, that is, the charge density for unit applied voltage, calculated with the piezoelectric effect ignored. This function can be regarded, to a good approximation, as a distributed source of surface waves, thus giving the surface-wave amplitude generated by the transducer. With some restrictions, this approach enables the transducer response to be expressed in terms of an array factor and an element factor, as in Section 4.1, and yields an expression for the element factor, E. This procedure is necessary in order to evaluate the levels of the harmonic responses. The charge density on any one electrode can be regarded as a surface wave source associated with that electrode; however the charge density is affected by the configuration of several neighbouring electrodes on either side, which thus affect the associated surface wave amplitude. This is known as the *neighbour effect*. Fortunately, the neighbour effect can be allowed for implicitly by re-defining the surface wave component associated with each electrode, and so need not be considered explicitly. The transducers in Figure 4.4 are also affected by "end effects", that is, the electrodes near the ends have charge densities somewhat different to those in the centre. However, it is shown in Section 4.5.1 that end effects can be virtually eliminated by adding "guard" electrodes at each end of the transducer. Finally, it is noted that some "withdrawal-weighted" transducers have structures too irregular to allow the use of an array factor and element factor, but the response may nevertheless be deduced from the electrostatic solution.

Another complication is the excitation of bulk waves in addition to the intended surface waves. The bulk waves usually propagate away from the surface and therefore do not excite a receiving transducer very strongly, though it is often necessary to ensure this by roughening the rear surface of the substrate, suppressing specular reflections. Bulk waves travelling almost parallel to the surface reach the output transducer directly, and can be of some practical concern. In many devices, a multi-strip coupler is used to discriminate against them.

It will be seen that there is a considerable variety of transducer types and the behaviour is affected by a variety of second-order effects. In consequence, a variety of methods have been used for transducer analysis. The delta-function model described above is widely used because of its simplicity, despite its limitations — notably, the inability to calculate the transducer admittance and reflection coefficient, and, for a two-transducer device, the insertion loss. These limitations were largely overcome by the "crossed-field" network model introduced by Smith *et al.* [84–87], drawing an analogy between the surface-wave transducer and an array of bulk wave transducers. Each electrode (or, sometimes, pair of electrodes) in the surface wave transducer was replaced by an equivalent network, so that the response of a complex transducer could be found by analysing an array of networks, using conventional network analysis [79, 84]. The original model excluded electrode interactions and bulk wave excitation, and did not correctly allow for electrostatic effects. However, it was shown later that interactions could be modelled by including a repetitive impedance mis-match [85, 88, 89]. Also, electrostatic effects could be included by finding the electrostatic solution first and then incorporating it into the network model [90].

More rigorously, a number of theoretical approaches have been explored, including perturbation theory which has been applied to the analysis of both transducers [91] and multi-strip couplers. Another approach, more closely related to the analysis in this chapter, is based on the effective permittivity described in

Chapter 3. For transducer analysis, this concept was used in the Green's function method of Milsom *et al.* [92–94], where the Green's function was derived from the effective permittivity, as described in Section 3.4. The definition of the effective permittivity is such that, if applied rigorously, it implicitly allows for electrode interactions, bulk waves and electrostatic effects, though mechanical loading due to the electrodes is excluded. This method is therefore very versatile, though its complexity makes it somewhat inconvenient. The effective permittivity has also been applied to the analysis of multi-strip couplers, making use of Ingebrigtsen's approximation.

The analysis given below in this chapter is an approximate form of the Green's function method [95]. An approximate Green's function described in Section 3.4 is used, assuming bulk wave excitation to be negligible. The analysis also uses the quasi-static approximation, and thus neglects electrode interactions. The resulting method is relatively straightforward, giving simple results valid for a wide range of transducer types, and correctly allows for electrostatic effects. In most practical devices, electrode interactions and bulk wave excitation are relatively insignificant, and these topics are described in Appendices E and F, respectively. Diffraction of the surface waves is ignored in this chapter, but will be described in Chapter 6. Some fundamental relationships, derived from Auld's reciprocity relation, are given in Appendix B.

4.3. ANALYSIS FOR A GENERAL ARRAY OF ELECTRODES

In this section we consider an array of electrodes on the surface of a piezoelectric half-space, as illustrated in Figure 4.5(a). The analysis is based on the Green's function described in Section 3.4, and therefore excludes mechanical loading due to the electrodes. Section 3.4 also assumes the potential and charge density to be invariant in one direction in the surface, and to comply with this the aperture W in Figure 4.5(a) is assumed to be large, so that distortions near the ends of the electrodes can be neglected. Surface-wave diffraction is also neglected. It is assumed that the only acoustic wave present is a non-leaky piezoelectric Rayleigh wave, thus excluding bulk wave excitation. The electrodes are taken to have negligible resistivity.

The electrode voltages are all taken to have the same frequency ω, but are otherwise arbitrary. This enables the results to be used for analysis of multi-strip couplers, as well as transducers. The transducer analysis is given in the following sections of this chapter, while the multi-strip coupler analysis is given in Chapter 5.

4.3.1. The Quasi-Static Approximation

In Section 3.4 it was shown that the potential $\phi(x, \omega)$ and charge density $\sigma(x, \omega)$ at the surface of a piezoelectric half-space are related by a Green's function $G(x, \omega)$. Here $\phi(x, \omega)$ and $\sigma(x, \omega)$ are both proportional to $\exp(j\omega t)$, with this factor implicit. Since the only acoustic wave present is assumed to be a Rayleigh wave, $G(x, \omega)$ can be approximated as the sum of an electrostatic term $G_e(x)$ and a surface wave term $G_s(x, \omega)$, as in equation (3.48), so that

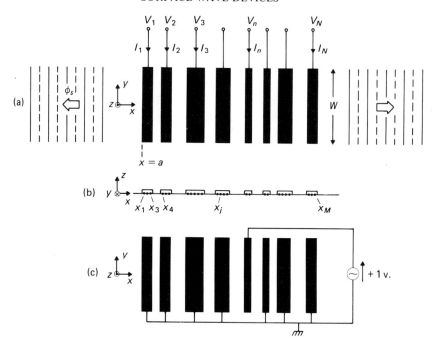

FIGURE 4.5. Generalised array of electrodes.

$$\phi(x, \omega) = [G_e(x) + G_s(x, \omega)] * \sigma(x, \omega), \qquad (4.18)$$

where the asterisk indicates convolution with respect to x. For convenience, the axis notation has been changed here, so that new coordinates x, y, z replace the earlier coordinates x_1, x_2, x_3. Thus the piezoelectric now occupies the region $z < 0$, and the surface waves propagate in the $\pm x$ directions. The two components of the Green's function are

$$G_e(x) = -\frac{\ln |x|}{\pi(\varepsilon_0 + \varepsilon_p^T)}, \qquad (4.19)$$

and

$$G_s(x, \omega) = j\Gamma_s \exp(-jk_0|x|), \qquad (4.20)$$

where $k_0 > 0$ is the wavenumber for surface wave propagation on a free surface at frequency ω. The constant ε_p^T is given by equation (3.37) with the elements ε_{ij} defined at constant stress, and the constant Γ_s is, from equation (3.39),

$$\Gamma_s \approx \frac{1}{\varepsilon_0 + \varepsilon_p^T} \frac{v_0 - v_m}{v_0}, \qquad (4.21)$$

where v_0 and v_m are respectively the surface wave velocities for a free surface and for a metallised surface.

Since the electrode resistivity is assumed to be negligible, each electrode will be an equipotential, and $\phi(x, \omega)$ must be equal to the electrode voltage at all points on the electrode. The charge density $\sigma(x, \omega)$ must be zero on all unmetallised regions. Using these boundary conditions and the equations given above, the surface potential and charge density are determined everywhere if the electrode voltages are specified.

At points remote from the electrodes, the surface potential is primarily the potential $\phi_s(x, \omega)$ associated with the surface waves generated by the structure. For example, consider the unmetallised region $x < a$, assuming that the electrodes are present only for $x \geq a$. Since $\sigma(x, \omega)$ is zero for $x < a$, the electrostatic term $G_e(x)$ in equation (4.18) gives a potential decaying with distance in this region. The surface-wave potential $\phi_s(x, \omega)$ can therefore be identified with $G_s(x, \omega) * \sigma(x, \omega)$, with $G_s(x, \omega)$ given by equation (4.20). Defining $\bar{\sigma}(\beta, \omega)$ as the Fourier transform of $\sigma(x, \omega)$, with ω held constant in the Fourier integral, the surface-wave potential is found to be

$$\phi_s(x, \omega) = j\Gamma_s\bar{\sigma}(k_0, \omega) \exp(jk_0 x), \quad \text{for } x < a \quad (4.22)$$

where the modulus sign in equation (4.20) has disappeared because $\sigma(x, \omega)$ is zero for $x < a$. Thus the surface-wave potential is given by the Fourier transform of the charge density. Although bulk wave excitation has been neglected here, equation (4.22) is still valid when bulk waves are excited, as shown in Appendix B, equation (B.30). In fact, equation (4.22) is derived indirectly from equation (B.30), since the latter was used in Section 3.4 to derive the surface wave Green's function $G_s(x, \omega)$.

It should be noted that $\bar{\sigma}(k_0, \omega)$ is not the Fourier transform of $\sigma(x, \omega)$ in the usual sense, because k_0 is not an independent parameter. If ω is changed, $\sigma(x, \omega)$ will in general change, so the Fourier integral must be calculated separately for each ω. For clarity, β is used here as the independent variable in the transform, while k, with various subscripts, refers to wavenumbers of propagating waves.

Although the above equations can be solved to obtain the surface potential and charge density, the calculation is generally very complex. Here the *quasi-static approximation* [95] is introduced in order to simplify the solution. The charge density $\sigma(x, \omega)$ is assumed to be dominated by an electrostatic term $\sigma_e(x, \omega)$, defined as the charge density obtained when acoustic wave excitation is ignored. Thus, from equation (4.18), $\sigma_e(x, \omega)$ is the solution of

$$\phi_e(x, \omega) = G_e(x) * \sigma_e(x, \omega). \quad (4.23)$$

Here $\phi_e(x, \omega)$ is a new surface potential, equal to the specified electrode voltage whenever x is on an electrode, and $\sigma_e(x, \omega) = 0$ on all unmetallised regions. Note that $\sigma_e(x, \omega)$ is independent of ω if the electrode voltages are independent of ω, and $\sigma_e(x, \omega)$ is zero if the electrode voltages are all the same. The total charge density is now taken to be

$$\sigma(x, \omega) \approx \sigma_e(x, \omega) + \sigma_a(x, \omega), \quad (4.24)$$

where $\sigma_a(x, \omega)$ is a relatively small contribution due to the piezoelectric effect; it can be regarded as a regenerated term due to the acoustic waves present. The surface potential $\phi(x, \omega)$ is given by equation (4.18), using equation (4.24) for $\sigma(x, \omega)$. On the right there are now four terms, but a term $G_s(x, \omega) * \sigma_a(x, \omega)$ is small, and can be omitted; both $G_s(x, \omega)$ and $\sigma_a(x, \omega)$ become zero if there is no piezoelectric coupling. We thus have, in the quasi-static approximation,

$$\phi(x, \omega) = [G_e(x) + G_s(x, \omega)] * \sigma_e(x, \omega) + G_e(x) * \sigma_a(x, \omega). \quad (4.25)$$

Given the function $\sigma_e(x, \omega)$, this equation, together with the boundary conditions, determines the acoustic charge density $\sigma_a(x, \omega)$, as discussed further in Section 4.3.3 below. Also, $\sigma_e(x, \omega)$ is determined by equation (4.23). Hence the total charge density, equation (4.24), and the surface potential, equation (4.25), are determined. The solution is approximate because one term was omitted in deriving equation (4.25). However, we define $\sigma_a(x, \omega)$ such that equation (4.25) is exactly true; it then follows that the charge density is not exactly the sum of $\sigma_e(x, \omega)$ and $\sigma_a(x, \omega)$, so that equation (4.24) is approximate.

As explained earlier, the potential $\phi_s(x, \omega)$ of a surface wave generated by the electrodes can be obtained by omitting terms arising from the electrostatic Green's function $G_e(x)$. Using equation (4.25) the surface-wave potential for $x < a$, to the left of the structure, is found to be

$$\phi_s(x, \omega) = j\Gamma_s \bar{\sigma}_e(k_0, \omega) \exp(jk_0 x), \quad \text{for } x < a, \quad (4.26)$$

where we define $\bar{\sigma}_e(\beta, \omega)$ as the Fourier transform of $\sigma_e(x, \omega)$. This is the same as the more accurate result of equation (4.22), except that $\bar{\sigma}(k_0, \omega)$ is replaced by $\bar{\sigma}_e(k_0, \omega)$. This change simplifies the analysis considerably, as can be seen for example by considering excitation by a two-terminal transducer. For this case it is sufficient to take the electrode voltages to be ± 1, and thus to be independent of the frequency ω (a term $\exp(j\omega t)$ is implicit). This implies that the electrostatic charge density $\sigma_e(x, \omega)$ is independent of ω, and hence the surface wave amplitude for all ω can be deduced by evaluating $\sigma_e(x, \omega)$ at $\omega = 0$. For this reason, the approximation is described as "quasi-static".

4.3.2. Electrostatic Equations and Charge Superposition

The electrostatic charge density $\sigma_e(x, \omega)$ is determined by equation (4.23), with the boundary conditions that the potential $\phi_e(x, \omega)$ is equal to the electrode voltage whenever x is on an electrode, and $\sigma_e(x, \omega)$ is zero on all unmetallised regions. An elegant technique for finding $\sigma_e(x, \omega)$ has been given by Milsom et al. [93], who in fact used it for the piezoelectric case. The application to the electrostatic problem is discussed here. We define a set of M points x_j, as shown in Figure 4.5(b). The points exist only on the electrodes, where they have a small spacing Δx. The electrostatic equation (4.23) is written as

$$\phi_e(x_i, \omega) = \sum_{j=1}^{M} A_{ij} \sigma_e(x_j, \omega), \quad (4.27)$$

which will be identical to equation (4.23) in the limit when Δx vanishes and M becomes infinite. The obvious choice for the coefficients A_{ij} is $\Delta x G_e(x_i - x_j)$, but for finite Δx this is infinite when $i = j$, as can be seen from equation (4.19). A suitable choice for A_{ij} is therefore

$$A_{ij} = \tfrac{1}{2}\Delta x[G_e(x_i - x_j + \tfrac{1}{2}\Delta x) + G_e(x_i - x_j - \tfrac{1}{2}\Delta x)]. \tag{4.28}$$

This matrix takes account of the electrode geometry, since the points x_j exist only on the electrodes. Owing to the symmetry of $G_e(x)$, A_{ij} is symmetrical, so that $A_{ji} = A_{ij}$. We may now invert equation (4.27) to express $\sigma_e(x, \omega)$ in terms of $\phi_e(x, \omega)$:

$$\sigma_e(x_i, \omega) = \sum_{j=1}^{M} B_{ij}\phi_e(x_j, \omega), \tag{4.29}$$

where B_{ij} is the reciprocal of the matrix A_{ij}. Since A_{ij} is symmetrical, it follows that B_{ij} is also symmetrical:

$$B_{ji} = B_{ij}. \tag{4.30}$$

In equation (4.29) all the $\phi_e(x_j, \omega)$ are equal to known electrode voltages, since the points x_j exist only on the electrodes. Thus the potential in the inter-electrode gaps, which is not known *a priori*, is not required. The condition that $\sigma_e(x, \omega)$ should be zero in the gaps is implied by equation (4.27), because the summation excludes points not on the electrodes.

Charge Superposition. A useful principle for simplifying the determination of $\sigma_e(x, \omega)$ can be found by considering the arrangement shown in Figure 4.5(c). Here the electrodes are as in Figure 4.5(a), but now electrode n has unit voltage while all the other electrodes are grounded. The electrostatic charge density for this case is denoted $\varrho_{en}(x)$, and is real and independent of frequency. In general there will be charges present on all the electrodes, not just on electrode n. To find $\varrho_{en}(x)$, we define an electrode polarity function $\hat{p}_n(x)$ by

$$\hat{p}_n(x) = 1 \quad \text{if } x \text{ is on electrode } n,$$
$$= 0 \quad \text{for other } x. \tag{4.31}$$

For a point x_j on any electrode, the potential in Figure 4.5(c) is $\hat{p}_n(x_j)$, and hence from equation (4.29) the electrostatic charge density is

$$\varrho_{en}(x_i) = \sum_{j=1}^{M} B_{ij}\hat{p}_n(x_j) \tag{4.32}$$

with $\varrho_{en}(x) = 0$ on the unmetallised regions.

The charge superposition principle states that, when some arbitrary set of voltages is applied to the electrodes, the charge density is given by a linear combination of the functions $\varrho_{en}(x)$ due to individual electrodes. Thus, for Figure 4.5(a), where the electrode voltages are V_n, the electrostatic charge density is

$$\sigma_e(x, \omega) = \sum_{n=1}^{N} V_n \varrho_{en}(x). \tag{4.33}$$

This equation follows directly from equation (4.29). It is of considerable practical value, since $\sigma_e(x, \omega)$ can be evaluated for any set of electrode voltages V_n, once the functions $\varrho_{en}(x)$ are known. The superposition principle is well known in electrostatics, and is discussed in more detail by, for example, Ramo et al. [96]. It should be noted that the charge density must be zero if the surface potential is independent of x, since for this case there is no electric field. It follows that $\sigma_e(x, \omega)$ is unaffected if a constant is added to all the V_n in equation (4.33).

Evaluation of the Electrostatic Charge Density. The electrostatic charge density $\sigma_e(x, \omega)$ can be expressed analytically if the electrodes are regular, that is, if they have the same width and have uniform spacing. This solution is described in Section 4.5 below, where it is used to deduce transducer behaviour. Analytic solutions are also known for some simple cases, where the structure has either 2 or 3 electrodes [97]. In general, however, numerical techniques must be used. For several reasons, notably analysis of withdrawal-weighted transducers, these techniques have been investigated quite extensively. The literature refers to two-terminal transducers, where each electrode voltage takes one of two values; however, the methods can be applied directly when the voltages are all different.

Numerical techniques were first used by Hartmann and Secrest [98] to investigate end effects. The method made use of the relation between charge density and potential in the β-domain, given by equation (3.12) of Chapter 3. The boundary conditions (specified voltages on the electrodes and zero charge density in the gaps) are given in the x-domain, so an iterative technique involving Fourier transformation was used. Another method is based on the relationship shown in equation (4.29), which expresses the charge density in terms of the known electrode voltages; the matrix A_{ij} is obtained directly from the electrostatic Green's function [equation (4.28)], and is inverted to give the coefficients B_{ij} required in equation (4.29). A particular feature that can cause difficulty is that the charge density is infinite at the electrode edges. In view of this Milsom et al. [93] refined the method by using unequal spacing for the points x_j.

Several authors have used a method in which the charge density on each electrode is written as a polynomial with unknown coefficients. Substituting into equation (4.27) gives a set of simultaneous equations for the electrode potentials, and these can be solved for the polynomial coefficients. Quite good accuracy can be obtained with only three coefficients per electrode [99], and excellent results can be obtained using Chebychev polynomials [90, 100]. Tabulated results [90, 100] can be used to analyse a variety of cases, making use of the superposition principle mentioned above.

Another method replaces the actual electrodes by a fine grid of regular electrodes, for which the charge density is known analytically, applying boundary conditions of zero charge or specified potential as appropriate [101].

4.3.3. Current Entering One Electrode

The current entering one electrode can be found from the charge density on it, and we first consider the charge density in more detail, showing that $\sigma_a(x, \omega)$ can be expressed in terms of $\sigma_e(x, \omega)$. The electrostatic term $\sigma_e(x, \omega)$ is defined as the solution of equation (4.23), which involves a potential $\phi_e(x, \omega)$. The actual potential $\phi(x, \omega)$ is given by equation (4.25), which can be seen to include a term equal to

$\phi_e(x, \omega)$. Since $\phi(x, \omega)$ and $\phi_e(x, \omega)$ must be equal when x is on an electrode, the remaining terms in equation (4.25) must be zero at such points, so that

$$[G_s(x, \omega) * \sigma_e(x, \omega) + G_e(x) * \sigma_a(x, \omega)]_{x_j} = 0. \tag{4.34}$$

This equation relates $\sigma_a(x, \omega)$ to $\sigma_e(x, \omega)$. It is convenient to define an acoustic potential $\phi_a(x, \omega)$ by the expression

$$\phi_a(x, \omega) = G_s(x, \omega) * \sigma_e(x, \omega). \tag{4.35}$$

This potential is associated with acoustic wave excitation, and is zero if the material is not piezoelectric. Equation (4.34) can thus be written

$$[G_e(x) * \sigma_a(x, \omega)]_{x_j} = -\phi_a(x_j, \omega). \tag{4.36}$$

This has the same form as equation (4.23), and may therefore be expressed in a form similar to equation (4.27). The charge density $\sigma_a(x, \omega)$ is therefore, by analogy with equation (4.29),

$$\sigma_a(x_i, \omega) = -\sum_{j=1}^{M} B_{ij} \phi_a(x_j, \omega), \tag{4.37}$$

where B_{ij} is the inverse of the matrix A_{ij} defined in equation (4.28). For values of x on the electrodes, equation (4.37) gives $\sigma_a(x, \omega)$ exactly, in the limit $\Delta x \to 0$. For other x, $\sigma_a(x, \omega)$ is of course zero. Thus, using equation (4.35) for $\phi_a(x, \omega)$, $\sigma_a(x, \omega)$ can be deduced from $\sigma_e(x, \omega)$.

The current I_n flowing into electrode n is found by integrating the total charge density over the surface of the electrode, and then differentiating with respect to time. Thus

$$I_n = j\omega W \int_n [\sigma_e(x, \omega) + \sigma_a(x, \omega)] \, dx, \tag{4.38}$$

where the integral is taken over electrode n. The current is taken as the sum of two terms,

$$I_n = I_{en} + I_{an}, \tag{4.39}$$

where I_{en} and I_{an} are respectively the contributions due to $\sigma_e(x, \omega)$ and $\sigma_a(x, \omega)$. Thus

$$I_{en} = j\omega W \int_n \sigma_e(x, \omega) \, dx. \tag{4.40}$$

It is assumed that $\sigma_e(x, \omega)$ can be evaluated, so this contribution will not be considered further here. The acoustic contribution can be written

$$I_{an} = j\omega W \int_{-\infty}^{\infty} \hat{p}_n(x) \sigma_a(x, \omega) \, dx, \tag{4.41}$$

where $\hat{p}_n(x)$, defined by equation (4.31), is unity on electrode n and zero elsewhere. Since the integrand here is zero in the unmetallised regions, the integral can be expressed as discrete sum using the points x_j, so that

$$I_{an} = j\omega W \sum_{j=1}^{M} \hat{p}_n(x_j) \sigma_a(x_j, \omega) \Delta x$$

$$= -j\omega W \Delta x \sum_{i=1}^{M} \phi_a(x_i, \omega) \sum_{j=1}^{M} \hat{p}_n(x_j) B_{ji}, \tag{4.42}$$

where equation (4.37) has been used for $\sigma_a(x_j, \omega)$ and the summations have been re-ordered. Now, since B_{ij} is symmetrical, comparison with equation (4.32) shows that the sum over j in equation (4.42) can be identified as $\varrho_{en}(x_i)$. Hence

$$I_{an} = -j\omega W \sum_{i=1}^{M} \phi_a(x_i, \omega) \varrho_{en}(x_i) \Delta x$$

or, taking the limit as $\Delta x \to 0$,

$$I_{an} = -j\omega W \int_{-\infty}^{\infty} \varrho_{en}(x) \phi_a(x, \omega) dx \qquad (4.43)$$

since $\varrho_{en}(x)$ is zero in the unmetallised regions.

4.3.4. Evaluation of the Acoustic Potential

The acoustic potential $\phi_a(x, \omega)$ in equation (4.43) is defined in terms of $\sigma_e(x, \omega)$ by equation (4.35). For the analysis later we will need an expression involving $\bar{\sigma}_e(\beta, \omega)$, the Fourier transform of $\sigma_e(x, \omega)$, and this is derived here. From equation (4.35),

$$\phi_a(x, \omega) = G_s(x, \omega) * \sigma_e(x, \omega),$$

where $G_s(x, \omega)$ is given by equation (4.20). Thus,

$$\phi_a(x, \omega) = j\Gamma_s \int_{-\infty}^{x} \sigma_e(x', \omega) e^{-jk_0(x-x')} dx'$$

$$+ j\Gamma_s \int_{x}^{\infty} \sigma_e(x', \omega) e^{jk_0(x-x')} dx'. \qquad (4.44)$$

If the electrodes are present only for $x \geq a$, as in Figure 4.5(a), this equation gives $\phi_a(x, \omega) = j\Gamma_s \bar{\sigma}_e(k_0, \omega) \exp(jk_0 x)$ for the region $x < a$ to the left of the structure. Comparison with equation (4.26) shows that $\phi_a(x, \omega)$ is equal to the surface wave potential $\phi_s(x, \omega)$ for $x < a$, and this is also found to be true for the unmetallised region to the right of the structure.

Equation (4.44) may be re-arranged using the step function $U(x)$, which is equal to 1 for $x > 0$ and to zero for $x < 0$. Thus,

$$\phi_a(x, \omega) = j\Gamma_s e^{-jk_0 x} \int_{-\infty}^{\infty} \sigma_e(x', \omega) U(x - x') e^{jk_0 x'} dx'$$

$$+ j\Gamma_s e^{jk_0 x} \int_{-\infty}^{\infty} \sigma_e(x', \omega) U(x' - x) e^{-jk_0 x'} dx', \qquad (4.45)$$

where the limits are now $\pm \infty$. These integrals may be evaluated by Fourier methods, taking x as a constant. The first integral is the transform of $[\sigma_e(x', \omega) U(x - x')]$, from the x'-domain to the β-domain, with the result evaluated at $\beta = -k_0$. A similar method is used for the second integral. The relationships required are given in Appendix A. The transform of $U(x')$ is, from equation (A.38),

$$U(x') \leftrightarrow \pi\delta(\beta) - j/\beta. \qquad (4.46)$$

The shifting and scaling theorems, equations (A.10) and (A.9), are used to obtain the transforms of $U(x - x')$ and $U(x' - x)$, and the products in equation (4.45) are transformed using the convolution theorem, equation (A.20). With $\bar{\sigma}_e(\beta, \omega)$ defined as the Fourier transform of $\sigma_e(x, \omega)$, the result obtained is

ANALYSIS OF INTERDIGITAL TRANSDUCERS

$$\phi_a(x, \omega) = \tfrac{1}{2}j\Gamma_s e^{-jk_0 x}[\bar{\sigma}_e(-k_0, \omega) + jF(-k_0)/\pi]$$
$$+ \tfrac{1}{2}j\Gamma_s^- e^{jk_0 x}[\bar{\sigma}_e(k_0, \omega) - jF(k_0)/\pi], \quad (4.47)$$

where the function $F(\beta)$ is defined by

$$F(\beta) = \bar{\sigma}_e(\beta, \omega) * \frac{\exp(-j\beta x)}{\beta}. \quad (4.48)$$

Here the terms $\bar{\sigma}_e(\pm k_0, \omega)$ arise because of the delta-function in equation (4.46). It will be found later that these give the parallel conductance, $G_a(\omega)$, of a transducer. The terms $F(\pm k_0)$, involving a convolution, are due to the j/β term in equation (4.46) and will be found to give the acoustic susceptance $B_a(\omega)$.

4.4. QUASI-STATIC ANALYSIS OF TRANSDUCERS

We now apply the results of Section 4.3 to a two-terminal unapodised transducer such as that shown in Figure 4.6, where each of the electrodes is connected to one of the two bus-bars. The assumptions mentioned at the beginning of Section 4.3 apply here also, and in addition we assume that the bus bars have no effect other than providing electrical connections to the electrodes, with no resistance. The analysis here uses the quasi-static approximation, since this was used for the derivations in Section 4.3. Consequently, electrode interactions are neglected, as discussed in Section 4.2; for example, the analysis predicts that a shorted transducer will not reflect incident surface waves. Another consequence of the approximation is that surface waves are predicted to travel through a shorted transducer with a velocity equal to the free-surface velocity, v_0. In practice, the electrodes cause a small reduction of the

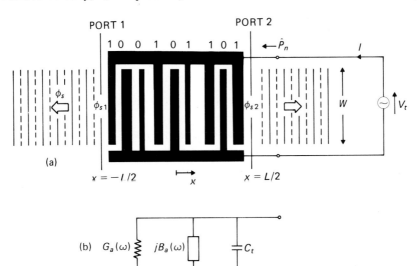

FIGURE 4.6. Two-terminal transducer.

velocity, even when electrode interactions are weak. For regular electrodes, with uniform width and spacing, the velocity change is derived in Section D.2. To allow for this, the results of this section may be modified by replacing the free-surface wavenumber k_0 by k_{sc}, given by equation (D.26).

4.4.1. Launching Transducer

We first consider the transducer of Figure 4.6(a), which is taken to be isolated, that is, there are no incident acoustic waves and the charge density is not affected by the presence of any other electrodes on the surface. A voltage V_t is applied, and the electrostatic part of the charge density is $\sigma_e(x, \omega)$, with Fourier transform $\bar{\sigma}_e(\beta, \omega)$. In the quasi-static approximation, the potential $\phi_s(x, \omega)$ of the surface-wave radiated in the $-x$ direction is, from equation (4.26),

$$\phi_s(x, \omega) = j\Gamma_s \bar{\sigma}_e(k_0, \omega) \exp(jk_0 x).$$

We define $\varrho_e(x) = \sigma_e(x, \omega)/V_t$ as the electrostatic charge density for unit applied voltage. This function is real and independent of frequency. If $\bar{\varrho}_e(\beta)$ is the Fourier transform of $\varrho_e(x)$, we then have

$$\phi_s(x, \omega) = j\Gamma_s V_t \bar{\varrho}_e(k_0) \exp(jk_0 x). \tag{4.49}$$

The transducer has two acoustic ports, and these are taken to be at $x = \pm L/2$. These points are taken to be close to the ends of the transducer, though their precise locations are immaterial. Defining $\phi_{s1}(\omega)$ as the potential of the wave launched at port 1, that is, at $x = -L/2$, we have

$$\phi_{s1}(\omega) = \phi_s(-\tfrac{1}{2}L, \omega) = j\Gamma_s V_t \bar{\varrho}_e(k_0) \exp(-\tfrac{1}{2}jk_0 L). \tag{4.50}$$

The power P_s carried by this wave is, from equation (3.34),

$$P_s = \tfrac{1}{4}\omega W |\phi_{s1}(\omega)|^2/\Gamma_s. \tag{4.51}$$

Similarly, the potential $\phi_{s2}(\omega)$ of the wave launched in the $+x$ direction, measured at port 2 ($x = L/2$), is found to be

$$\phi_{s2}(\omega) = j\Gamma_s V_t \bar{\varrho}_e(-k_0) \exp(-\tfrac{1}{2}jk_0 L). \tag{4.52}$$

Since $\varrho_e(x)$ is real, $\bar{\varrho}_e(-k_0) = \bar{\varrho}_e^*(k_0)$, so that $\phi_{s2}(\omega)$ is essentially the conjugate of $\phi_{s1}(\omega)$. Clearly, the two waves have the same power.

The electrostatic charge density $\varrho_e(x)$ can be evaluated by methods discussed in Section 4.3.2. For a two-terminal transducer, we define an electrode polarity vector $\hat{P}_n$ by

$$\hat{P}_n = 1 \quad \text{if electrode } n \text{ is connected to the upper bus.}$$
$$= 0 \quad \text{if electrode } n \text{ is connected to the lower bus.} \tag{4.53}$$

as shown in Figure 4.6(a). For unit voltage across the transducer, the charge density may be found by taking the voltage of electrode n to be $\hat{P}_n$, and hence, using equation (4.33),

$$\varrho_e(x) = \sum_{n=1}^{N} \hat{P}_n \varrho_{en}(x), \tag{4.54}$$

where N is the number of electrodes and $\varrho_{en}(x)$ is the electrostatic charge density associated with unit voltage on electrode n, defined in Section 4.3.2.

4.4.2. Transducer Admittance

When a voltage V_t is applied to an isolated transducer, as in Figure 4.6, the transducer draws a current I, and the ratio I/V_t is the transducer admittance, Y_t. A major part of the current is due to the electrostatic charge density $\sigma_e(x, \omega)$, which is in phase with V_t and gives a capacitive contribution to Y_t. This contribution is usually written explicitly, denoting the capacitance by C_t, so that

$$Y_t(\omega) = G_a(\omega) + jB_a(\omega) + j\omega C_t. \tag{4.55}$$

Here $G_a(\omega)$ and $B_a(\omega)$ are the real and imaginary contributions due to $\sigma_a(x, \omega)$, the acoustic charge density. $G_a(\omega)$ is the conductance, and $B_a(\omega)$ is the acoustic susceptance. The admittance may be represented as an electrical equivalent circuit with these three contributions in parallel, as in Figure 4.6(b).

If I_n is the current entering electrode n from the bus-bar, the transducer current I is given by

$$I = \sum_n \hat{P}_n I_n = j\omega W \sum_n \hat{P}_n \int_n [\sigma_e(x, \omega) + \sigma_a(x, \omega)] dx, \tag{4.56}$$

where equation (4.38) has been used for I_n, and the integral is over electrode n. The electrostatic contribution to I is due to the term $\sigma_e(x, \omega)$ and is equal to $j\omega C_t V_t$. Since $\sigma_e(x, \omega) = V_t \varrho_e(x)$, we have

$$C_t = W \sum_n \hat{P}_n \int_n \varrho_e(x) dx, \tag{4.57}$$

which is simply the sum of the electrostatic charges on the electrodes connected to one bus, for unit applied voltage.

The acoustic part of the current, due to the $\sigma_a(x, \omega)$ term in equation (4.56), is denoted I_a. The contribution due to electrode n is denoted I_{an}, and is given by equation (4.43). We thus have

$$I_a = \sum_n \hat{P}_n I_{an} = -j\omega W \int_{-\infty}^{\infty} \sum_n \hat{P}_n \varrho_{en}(x) \phi_a(x, \omega) dx. \tag{4.58}$$

Using equation (4.54) this can be expressed as

$$I_a = -j\omega W \int_{-\infty}^{\infty} \varrho_e(x) \phi_a(x, \omega) dx. \tag{4.59}$$

This expression is, by definition, equal to $V_t[G_a(\omega) + jB_a(\omega)]$. The acoustic potential $\phi_a(x, \omega)$ is given by equation (4.47). The two terms in equation (4.47) involving $\bar{\sigma}_e(\pm k_0, \omega)$ are found to give the real part of I_a, equal to $V_t G_a(\omega)$. The terms involving $F(\pm k_0)$ give the imaginary part, $jV_t B_a(\omega)$. Noting that $\bar{\sigma}_e(k_0, \omega) = V_t \bar{\varrho}_e(k_0)$ and $\bar{\varrho}_e(-k_0) = \bar{\varrho}_e^*(k_0)$, the conductance is found to be

$$G_a(\omega) = \omega W \Gamma_s |\bar{\varrho}_e(k_0)|^2. \tag{4.60}$$

More directly, the conductance can be found by evaluating the surface-wave power generated, using equations (4.50)–(4.52). Since the electrodes are taken to have zero resistivity, this power is equal to the power extracted from the voltage source, which is $V_t^2 G_a(\omega)/2$, and this gives equation (4.60).

The susceptance $B_a(\omega)$ is found by substituting the $F(\pm k_0)$ terms of equation (4.47) into equation (4.59). After some manipulation, the result obtained is

$$B_a(\omega) = -\frac{\omega W \Gamma_s}{\pi} |\bar{\varrho}_e(k_0)|^2 * \frac{1}{k_0}, \qquad (4.61)$$

This is clearly related to $G_a(\omega)$, equation (4.60). Using the fact that $k_0 = \omega/v_0$, where the free-surface velocity v_0 is independent of ω, $B_a(\omega)$ can be related to $G_a(\omega)$ without using the function $\bar{\varrho}_e(k_0)$. The result is [95]

$$B_a(\omega) = -G_a(\omega) * \frac{1}{\pi\omega}. \qquad (4.62)$$

This relation is the Hilbert transform of $G_a(\omega)$. Although it has been derived here using the quasi-static approximation, it is in fact much more general than this. The general proof uses the fact that the relation between I and V_t must be causal, that is, if the voltage is zero for $t < 0$, say, then the current must also be zero for $t < 0$. It follows from this that the real and imaginary parts of the admittance are related by the Hilbert transform [102, p. 198]. The proof assumes the admittance to be zero at infinite frequencies, so the capacitive term must be excluded. Thus the acoustic susceptance $B_a(\omega)$ can be obtained from the conductance without further analysis of the transducer itself [103].

4.4.3. Receiving Transducer

We now consider a transducer with a surface wave incident on it. As shown in Figure 4.7(a), the transducer geometry is the same as before, but now the two bus-bars are shorted, and a current I_{sc} flows between them. The incident surface wave has a potential $\phi_i(x, \omega)$, which is taken to be equal to $\phi_{i1}(\omega)$ at port 1 of the transducer, where $x = -\frac{1}{2}L$. Thus, in the absence of the transducer, the potential of the wave would be

$$\phi_i(x, \omega) = \phi_{i1}(\omega) \exp(-jk_0 x) \exp(-\tfrac{1}{2}jk_0 L). \qquad (4.63)$$

The transducer electrodes can be taken to be at zero potential, so there must be a distribution of charges present, such that the potential $\phi_i(x, \omega)$ is cancelled at the electrode locations.

To find this charge distribution, we return to equation (4.25), which is the quasi-static relation between potential and charge density. The charge density is $\sigma_e(x, \omega) + \sigma_a(x, \omega)$, but here the electrostatic term $\sigma_e(x, \omega)$ is zero because the electrode voltages are zero. Thus the charge density is $\sigma_a(x, \omega)$, and is related to the surface potential $\phi(x, \omega)$ by

$$\phi(x, \omega) = G_e(x) * \sigma_a(x, \omega). \qquad (4.64)$$

The potential of the incident wave, $\phi_i(x, \omega)$, is an additional term which can be

ANALYSIS OF INTERDIGITAL TRANSDUCERS

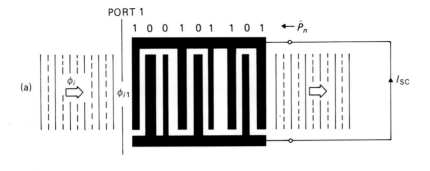

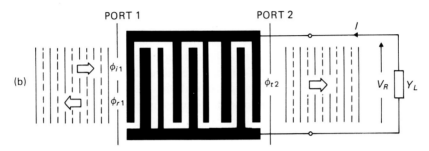

FIGURE 4.7. Reception by a two-terminal transducer.

considered to be due to some remote source whose charge density is not included in equation (4.64). To obtain zero potential on the electrodes we therefore have

$$\phi(x_j, \omega) = [G_e(x) * \sigma_a(x, \omega)]_{x_j} = -\phi_i(x_j, \omega), \quad (4.65)$$

where the points x_j exist only on the electrodes, as before.

A similar equation was found in Section 4.3 above when evaluating the current taken by individual electrodes. Equation (4.65) is the same as equation (4.36), except that $\phi_i(x, \omega)$ has replaced $\phi_a(x, \omega)$. It follows that the current I_{an} entering electrode n is given by equation (4.43), as before, but with $\phi_a(x, \omega)$ replaced by $\phi_i(x, \omega)$, so that

$$I_{an} = -j\omega W \int_{-\infty}^{\infty} \varrho_{en}(x) \phi_i(x, \omega) \, dx. \quad (4.66)$$

The transducer short-circuit current, I_{sc}, is therefore

$$I_{sc} = \sum_n \hat{P}_n I_{an} = -j\omega W \int_{-\infty}^{\infty} \varrho_e(x) \phi_i(x, \omega) \, dx \quad (4.67)$$

where $\varrho_e(x)$ has been introduced by using equation (4.54). Finally, substituting equation (4.63) for $\phi_i(x, \omega)$, we have

$$I_{sc} = -j\omega W \phi_{i1}(\omega) \bar{\varrho}_e(k_0) \exp\left(-\tfrac{1}{2} j k_0 L\right). \quad (4.68)$$

Comparing this with equation (4.50) gives a reciprocity relation for the processes of launching and receiving waves at port 1. We have

$$\left[\frac{I_{sc}}{\phi_{i1}(\omega)}\right]_{\text{receive}} = -\frac{\omega W}{\Gamma_s} \left[\frac{\phi_{s1}(\omega)}{V_t}\right]_{\text{launch}}. \quad (4.69)$$

80   SURFACE-WAVE DEVICES

A more general derivation of this equation is given in Appendix B, Section B.5, showing that it is valid when electrode interactions are present, and even when the transducer couples to bulk waves.

4.4.4. Scattering Coefficients

When a surface wave is incident on a transducer there will in general be a reflected wave, and the wave emerging on the other side of the transducer will have an amplitude different from the incident wave. Here we consider the reflection and transmission coefficients, and the conversion of incident power into the power dissipated in an electrical load

For a shorted receiving transducer, the potential associated with the charge density on the electrodes is, in the quasi-static approximation, given by equation (4.64) above. This involves the electrostatic Green's function, and therefore gives a potential localised on and near the transducer. The incident surface wave is therefore not affected by the transducer; there is no reflected wave, and the incident wave passes through the transducer with a velocity v_0, the free-surface velocity, and with no attenuation. Reflections therefore arise only if the transducer is not shorted. In practice, a shorted transducer can in some circumstances give significant reflections owing to electrode interactions as discussed in Section 4.2, but these are neglected here.

We consider a receiving transducer connected to an arbitrary electrical load with admittance Y_L, as in Figure 4.7(b). Since the transducer admittance is Y_t the voltage between the bus-bars, denoted V_R, is found to be

$$V_R = -I_{sc}/(Y_t + Y_L)$$
$$= j\omega W \phi_{i1}(\omega)\bar{\varrho}_e(k_0) \exp(-\tfrac{1}{2}jk_0 L)/(Y_t + Y_L), \quad (4.70)$$

where equation (4.68) has been used for the short-circuit current I_{sc}. Generally, Y_L and Y_t are both complex functions of frequency. The charge density on the electrodes now has two contributions — firstly, the charge density obtained for a shorted transducer and, secondly, a charge density arising from the voltage V_R. As discussed above, the first of these does not cause any surface wave excitation. The waves associated with V_R can be calculated as if a voltage V_R were applied to the transducer, with no surface wave incident. The wave radiated to the left, with potential denoted by $\phi_{r1}(\omega)$ at port 1, is therefore given by equation (4.50), with V_t replaced by V_R. Thus

$$\phi_{r1}(\omega) = j\Gamma_s V_R \bar{\varrho}_e(k_0) \exp(-\tfrac{1}{2}jk_0 L). \quad (4.71)$$

Using equation (4.70) for V_R, the reflection coefficient at port 1 is

$$r_1(\omega) \equiv \frac{\phi_{r1}(\omega)}{\phi_{i1}(\omega)} = -\frac{\omega W \Gamma_s}{Y_t + Y_L} [\bar{\varrho}_e(k_0)]^2 \exp(-jk_0 L). \quad (4.72)$$

The voltage V_R also causes generation of a wave to the right, with potential $\phi_{s2}(\omega)$, say, at port 2. This is given by equation (4.52), with V_R replacing V_t. Using equation (4.70) for V_R, we have

ANALYSIS OF INTERDIGITAL TRANSDUCERS

$$\phi_{s2}(\omega) = -\frac{\omega W \Gamma_s}{Y_t + Y_L} \phi_{i1}(\omega)|\bar{\varrho}_e(k_0)|^2 \exp(-jk_0 L), \qquad (4.73)$$

where we have used the relation $\bar{\varrho}_e(-k_0) = \bar{\varrho}_e^*(k_0)$, since $\varrho_e(x)$ is real. The total potential of the wave to the right, denoted by $\phi_{t2}(\omega)$ at port 2, also includes a contribution due to the incident wave, given by $\phi_{i1}(\omega) \exp(-jk_0 L)$. The total is divided by the incident wave potential $\phi_{i1}(\omega)$ to give the transmission coefficient $t_{12}(\omega)$. It is convenient to introduce $G_a(\omega)$ by using equation (4.60), giving

$$t_{12}(\omega) \equiv \frac{\phi_{t2}(\omega)}{\phi_{i1}(\omega)} = \left[1 - \frac{G_a(\omega)}{Y_t + Y_L}\right] \exp(-jk_0 L). \qquad (4.74)$$

The power ratios for reflection and transmission of waves incident on port 1 are, from equations (4.72) and (4.74),

$$|r_1(\omega)|^2 = \frac{G_a^2}{|Y_t + Y_L|^2} \qquad (4.75)$$

and

$$|t_{12}(\omega)|^2 = 1 - \frac{G_a^2 + 2G_a G_L}{|Y_t + Y_L|^2}, \qquad (4.76)$$

where G_L is the real part of the load admittance Y_L. There is also some power delivered to the load, given by $P_L = \frac{1}{2}|V_R|^2 G_L$. The power of the incident surface wave is, from equation (4.51), $P_s = \frac{1}{4}\omega W|\phi_{i1}|^2/\Gamma_s$. The ratio of these powers is the power conversion coefficient, $C_1(\omega)$. Using equation (4.70), we have

$$C_1(\omega) \equiv \frac{P_L}{P_s} = \frac{2G_a G_L}{|Y_t + Y_L|^2}. \qquad (4.77)$$

The sum of the three power coefficients, equations (4.75), (4.76) and (4.77), is unity, so that all the power incident on the transducer is accounted for. The power conversion coefficient, equation (4.77), also applies for a launching transducer driven by a source with internal admittance Y_L, giving the ratio of surface wave power generated (in one direction) divided by the available electric power. This can be shown by simple network analysis, and is of course expected by reciprocity. Some useful deductions from the above equations are:

(a) If the transducer is shorted, so that $Y_L = \infty$, the reflection and conversion coefficients are both zero, and the transmission coefficient is unity.
(b) If a loss-less reactance is connected across the terminals, cancelling the transducer susceptance, then $Y_t + Y_L = G_a$. The reflection coefficient is then unity, and no power is converted or transmitted. This applies for any frequency. The reactance required will normally be inductive.
(c) If the transducer is electrically matched, $Y_L = Y_t^*$ so that $Y_t + Y_L = 2G_a$. In this case the power reflection and transmission coefficients are both $\frac{1}{4}$, and the conversion coefficient is $\frac{1}{2}$.

Equations (4.75) and (4.77) also show that there is a simple relationship between the power reflection and conversion coefficients:

$$|r_1(\omega)|^2 = \frac{G_a}{2G_L} C_1(\omega). \tag{4.78}$$

The scattering coefficients derived above are consistent with experimental results, and with theoretical results obtained from the crossed-field network model, for uniform transducers operated at the fundamental centre frequency [84, 85]. In particular, the prediction of a zero reflection coefficient for a shorted transducer agrees well with experiment, though not for a single-electrode transducer on a strongly piezoelectric substrate (such as lithium niobate) — in this case electrode interactions are strong, and the quasi-static approximation is not valid.

Scattering matrix. The above equations may be combined to give a more general relationship between the potentials of incident and transmitted surface waves at the two ports and the transducer voltage and current. This situation is shown in Figure 4.8. The potentials of incident waves are denoted by ϕ_i, and those of waves leaving the transducer by ϕ_t, with additional subscripts 1 or 2 to indicate that these potentials are evaluated at port 1 or port 2, respectively. The four surface waves are all taken to be beams of width W, aligned with the active region of the transducer. The voltage across the bus-bars is V, and the transducer draws a current I.

The scattering matrix S_{ij} is defined by the equation

$$\begin{bmatrix} \phi_{t1} \\ \phi_{t2} \\ I/(\omega W) \end{bmatrix} = \begin{bmatrix} 0 & S_{12} & S_{13} \\ S_{21} & 0 & S_{23} \\ S_{31} & S_{32} & S_{33} \end{bmatrix} \begin{bmatrix} \phi_{i1} \\ \phi_{i2} \\ V\Gamma_s \end{bmatrix}, \tag{4.79}$$

where the coefficients are found to be

$$S_{12} = S_{21} = \exp(-jk_0 L),$$

$$S_{13} = -S_{31} = j\bar{\varrho}_e(k_0) \exp(-\tfrac{1}{2}jk_0 L),$$

$$S_{23} = -S_{32} = j\bar{\varrho}_e^*(k_0) \exp(-\tfrac{1}{2}jk_0 L),$$

$$S_{33} = Y_t(\omega)/[\omega W \Gamma_s],$$

where $Y_t(\omega)$ is the transducer admittance, given by equation (4.55).

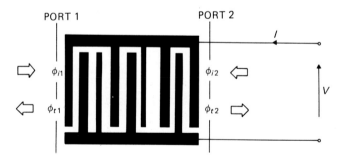

FIGURE 4.8. Parameters used in scattering matrix.

4.5. TRANSDUCERS WITH REGULAR ELECTRODES

In Section 4.4 the various properties of an unapodised transducer were derived, using the quasi-static approximation, and the results were expressed in terms of the electrostatic charge density $\varrho_e(x)$ and its Fourier transform $\bar{\varrho}_e(\beta)$. In general, these functions can only be evaluated numerically, and some methods were discussed in Section 4.3.2 above. However, for an important class of transducer structures the electrostatic charge density can be obtained analytically, and hence analytic results can be obtained for the transducer properties. This approach yields an element factor for transducer analysis, so that the analysis becomes very similar to the delta-function model of Section 4.1 above.

4.5.1. Electrostatic Charge Density and Element Factor

We consider the transducer shown in Figure 4.9(a). The electrodes here are assumed to be *regular*, that is, they all have the same width, a, and the centre-to-centre spacing, p, is constant throughout the array. Within the transducer, the fraction of the area covered by the electrodes is a/p, and this is called the *metallisation ratio*.

For unit voltage applied across the bus-bars, the electrostatic charge density $\varrho_e(x)$ is obtained by charge superposition. As in Section 4.4, equation (4.54), we have

$$\varrho_e(x) = \sum_{n=1}^{N} \hat{P}_n \varrho_{en}(x), \qquad (4.80)$$

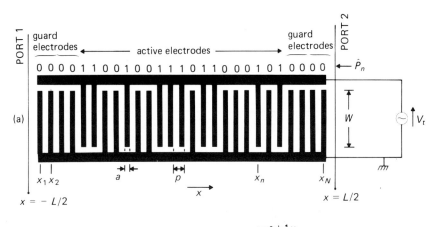

FIGURE 4.9. Transducer with regular electrodes.

where $\hat{P}_n = 1$ or 0 is the polarity of electrode n, as shown in Figure 4.9(a). The function $\varrho_{en}(x)$ is defined as the electrostatic charge density produced on the electrodes when unit voltage is applied to electrode n, with all the other electrodes grounded. This situation is illustrated in Figure 4.9(b). Now, the charge density produced here is largely determined by the geometry of electrode n and of a few neighbouring electrodes on either side. Here the electrodes all have the same geometry. It follows that, provided electrode n is not near either end of the array, the function $\varrho_{en}(x)$ is much the same for all electrodes, though it will of course have a lateral displacement corresponding to the electrode position [104, 105].

To exploit this feature, we define a more basic function $\varrho_f(x)$. We consider an infinite array of regular electrodes, as shown in Figure 4.10. One electode is centred at the origin $x = 0$ and has unit voltage applied, while all the other electrodes are grounded. The electrostatic charge density for this case is defined to be $\varrho_f(x)$, and will be called the *elemental charge density*.

For the finite array of Figure 4.9, electrode n is taken to be centred at $x = x_n$, and it follows that

$$\varrho_{en}(x) \approx \varrho_f(x - x_n) \tag{4.81}$$

provided electrode n is not near either end of the array. For the electrodes near the ends this relation is not valid. However, the contributions that the electrodes make to the total charge density $\varrho_e(x)$ depend also on the polarities $\hat{P}_n$, as shown by equation (4.80). It is assumed here that $\hat{P}_n = 0$ for the electrodes near the ends, as in Figure 4.9, so that the functions $\varrho_{en}(x)$ for these electrodes will not contribute to $\varrho_e(x)$. We thus have

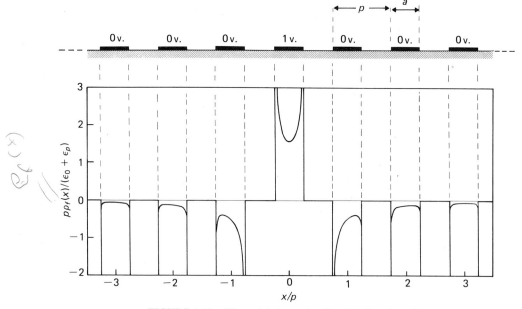

FIGURE 4.10. Elemental charge density $\varrho_f(x)$, for $a/p = \tfrac{1}{2}$.

$$\varrho_e(x) \approx \sum_{n=1}^{N} \hat{P}_n \varrho_f(x - x_n). \tag{4.82}$$

The electrodes at the two ends, with $\hat{P}_n = 0$, are described as *guard electrodes*, and are introduced in order to minimise end effects [104, 105]. Equation (4.82) becomes exact if an infinite number of guard electrodes is used at each end. However, the equation is found to be a good approximation if only a few are used, and for most practical purposes six guard electrodes at each end are sufficient. In the following analysis it is assumed that the number of guard electrodes is sufficient to make equation (4.82) valid to the required degree of accuracy. The electrodes in between the two groups of guard electrodes are called "active electrodes".

Equation (4.82) may be used to express the transducer response in terms of an element factor and an array factor. We first define an array factor in the x-domain by the relation

$$A_f(x) = \sum_{n=1}^{N} \hat{P}_n \delta(x - x_n) \tag{4.83}$$

so that equation (4.82) can be written

$$\varrho_e(x) = \int_{-\infty}^{\infty} A_f(x') \varrho_f(x - x') dx' = A_f(x) * \varrho_f(x), \tag{4.84}$$

where the asterisk indicates convolution. In the β-domain, the Fourier transform $\bar{\varrho}_e(\beta)$ of $\varrho_e(x)$ is obtained by using the convolution theorem; $A_f(x)$ and $\varrho_f(x)$ are transformed to give $\bar{A}_f(\beta)$ and $\bar{\varrho}_f(\beta)$ respectively, and $\bar{\varrho}_e(\beta)$ is then given by the product

$$\bar{\varrho}_e(\beta) = \bar{A}_f(\beta) \bar{\varrho}_f(\beta), \tag{4.85}$$

where, from equation (4.83) the array factor $\bar{A}_f(\beta)$ is give by

$$\bar{A}_f(\beta) = \sum_{n=1}^{N} \hat{P}_n \exp(-j\beta x_n). \tag{4.86}$$

As shown in Section 4.4, nearly all the transducer properties are directly related to $\bar{\varrho}_e(\beta)$, the exception being the capacitance C_t which is considered below. For example, when a voltage V_t is applied, the potential $\phi_{s1}(\omega)$ of the surface wave emerging at port 1 is given by equation (4.50), and with the aid of equation (4.85) this becomes

$$\phi_{s1}(\omega) = j\Gamma_s V_t \bar{A}_f(k_0) \bar{\varrho}_f(k_0) \exp(-\tfrac{1}{2} jk_0 L). \tag{4.87}$$

Here the function $\bar{A}_f(k_0)$ is an array factor, while the elemental charge density $\bar{\varrho}_f(k_0)$ has the role of an element factor. Thus the response is expressed in a simple form, similar to the delta-function model of Section 4.1. For most transducers it is found that the array factor varies with frequency much more rapidly than the elemental charge density $\bar{\varrho}_f(k_0)$, so that the array factor alone gives a good approximation to the shape of the frequency response.

An analytic solution for the elemental charge density has been given by Peach [97] and by Datta and Hunsinger [106]. A derivation is given in Appendix C, making use of the electrostatic analysis in Section 3.1. The function $\varrho_f(x)$, giving the charge density in the x-domain, is given by equation (C.20), and is shown in Figure 4.10 for a metallisation ratio $a/p = \frac{1}{2}$. It should be noted that the charge density is quantitatively significant only for the electrode centred at $x = 0$ and for a few electrodes on either side, though it is infinite at the edges of each electrode.

For most calculations the transformed function $\bar{\varrho}_f(\beta)$ is required, and Appendix C shows that this is given by

$$\bar{\varrho}_f(\beta) = (\varepsilon_0 + \varepsilon_p^T) \frac{2 \sin \pi s}{P_{-s}(-\cos \Delta)} P_m(\cos \Delta), \quad \text{for } m \leqslant \frac{\beta p}{2\pi} \leqslant m + 1, \quad (4.88)$$

where

$$\Delta = \pi a/p$$
$$s = \beta p/(2\pi) - m$$
$P_{-s}(-\cos \Delta)$ is a Legendre function

$P_m(\cos \Delta)$ is a Legendre polynomial.

This equation gives $\bar{\varrho}_f(\beta)$ for all β; the integer m must be chosen according to the value of β, and the parameter s is always in the range $0 \leqslant s \leqslant 1$. Appendix C gives some properties of Legendre functions. Since $\varrho_f(x)$ is clearly real and even, it follows from Fourier analysis that $\bar{\varrho}_f(\beta)$ is also real and even, so that $\bar{\varrho}_f(-\beta) = \bar{\varrho}_f(\beta)$. The variable β occurs only in the normalised form of the product βp. The function is shown in Figure 4.11, for several values of the metallisation ratio a/p. It consists of a sequence of lobes, all with the same shape, with relative amplitudes determined by the

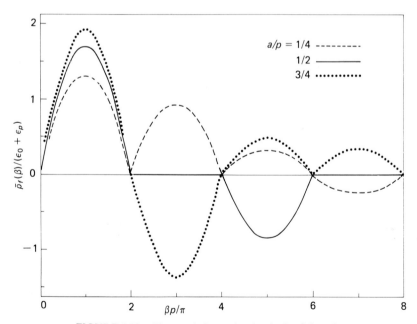

FIGURE 4.11. Elemental charge density, in the β-domain.

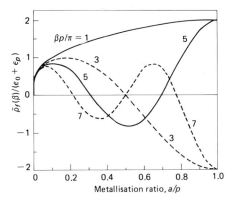

FIGURE 4.12. Elemental charge density: values at the peaks of the lobes, as functions of metallisation ratio.

polynomials $P_m(\cos \Delta)$ in equation (4.88). The relative amplitudes of the lobes are strongly influenced by the ratio a/p. This is clarified by Figure 4.12, where the values of $\bar{\varrho}_f(\beta)$ at the peaks of the lobes are plotted against a/p. The peaks of the lobes occur at $\beta = \pi/p$ and odd multiples of this value.

For a uniform single-electrode transducer, the peaks of the lobes occur at the fundamental centre frequency ω_c and at the odd harmonics, and the curves shown in Figure 4.12 give directly the relative harmonic strengths, as first pointed out by Engan [107].

The transducer capacitance C_t may be found by summing the electrostatic charges on all the electrodes connected to one bus-bar, as shown by equation (4.57). For transducers with regular electrodes, it is convenient to evaluate the net electrostatic charges on the electrodes when the charge density is $\varrho_f(x)$. These are denoted by Q_m, where $m = 0$ for the electrode at $x = 0$, $m = \pm 1$ for the electrodes on either side, and so on. Thus

$$Q_m = \int_{mp-a/2}^{mp+a/2} \varrho_f(x)\,dx, \qquad (4.89)$$

which gives the charge per unit width in the y-direction. When unit voltage is applied across a transducer, the charge on electrode n due to the voltage on electrode m is $W\hat{P}_m Q_{m-n}$. The total charge on electrode n is denoted q_n, and is thus given by

$$q_n = W \sum_{m=1}^{N} \hat{P}_m Q_{m-n}. \qquad (4.90)$$

The total electrostatic charge on all the electrodes connected to the upper bus is equal to the capacitance C_t, and is given by

$$C_t = \sum_{n=1}^{N} \hat{P}_n q_n. \qquad (4.91)$$

The capacitance is therefore readily found from the net charges Q_m, defined by equation (4.89). The charges are functions of the metallisation ratio a/p, and are given by equation (C.27) of Appendix C. In the particular case of $a/p = \frac{1}{2}$, the charges are given by the simple formula

$$Q_m = \frac{4(\varepsilon_0 + \varepsilon_p^T)}{\pi(1 - 4m^2)}, \quad \text{for } a/p = \tfrac{1}{2}. \tag{4.92}$$

4.5.2. End Effects

In practice, interdigital transducers often do not have the guard electrodes shown in Figure 4.9, and consequently the charge densities on the electrodes at and near the ends are distorted. These distortions are known as *end effects*. They affect the frequency response of the transducer, though the perturbation is small if the transducer has many electrodes. For example, in a transducer with more than, say, twenty electrodes, end effects generally have little effect on the frequency response, and are significant only if a very accurate response is required.

To analyse end effects, it is necessary to solve for the electrostatic charge distribution numerically, using methods discussed in Section 4.3.2 above [98, 108]. In addition, some of the tabulated results for withdrawal-weighted transducers [90, 100] give information on end effects. The effects cannot be characterised straightforwardly; they depend on the frequency and on the electrode structure. To give a quantitative example, in a uniform single-electrode transducer operated at the centre frequency ω_c of the fundamental pass-band, the coupling due to the electrode at the end is reduced by 29%, while that due to the adjacent electrode is increased by 5% [98].

The charge superposition principle can be used to simplify the analysis a little if the functions $\varrho_{en}(x)$ are first calculated numerically [104]. As mentioned in Section 4.5.1, most of the $\varrho_{en}(x)$ can be approximated by displaced versions of $\varrho_f(x)$; numerical calculations are needed only for the electrodes near the ends. The total electrostatic charge density on the transducer can then be obtained by using equation (4.80), and this is valid for any sequence of electrode polarities. Some examples showing net electrode charges corresponding to the functions $\varrho_{en}(x)$ are given in Ref. [104].

4.5.3. Transducer Response in Terms of Gap Elements

The above analysis gives the transducer response in terms of an array of elements, each associated with one electrode. An alternative formulation gives the response in terms of elements associated with the inter-electrode gaps [104, 109]. From equation (4.85), the electrostatic charge density, in the β-domain, is $\bar{\varrho}_c(\beta) = \bar{A}_f(\beta)\bar{\varrho}_f(\beta)$, with the array factor given by

$$\bar{A}_f(\beta) = \sum_{n=1}^{N} \hat{P}_n \exp(-j\beta x_n) = \sum_{n=1}^{N-1} \hat{P}_{n+1} \exp[-j\beta(x_n + p)]. \tag{4.93}$$

Here the second form follows because $x_{n+1} = x_n + p$ and $\hat{P}_1 = 0$. If the second form is multiplied by $\exp(j\beta p)$ and then subtracted from the first form, it is found that $\bar{A}_f(\beta)$ can be written as a sum of terms involving $(\hat{P}_{n+1} - \hat{P}_n)$. This enables the charge density $\bar{\varrho}_c(\beta)$ to be written as

$$\bar{\varrho}_c(\beta) = \bar{A}_g(\beta)\bar{\varrho}_g(\beta), \tag{4.94}$$

where the new array factor is

$$\bar{A}_g(\beta) = \sum_{n=1}^{N-1} (\hat{P}_{n+1} - \hat{P}_n) \exp[-j\beta(x_n + p/2)], \qquad (4.95)$$

and the new elemental charge density is

$$\bar{\varrho}_g(\beta) = -\tfrac{1}{2}j\bar{\varrho}_f(\beta)/\sin(\beta p/2). \qquad (4.96)$$

Here $\bar{\varrho}_f(\beta)$ is given by equation (4.88), and the term $\sin(\pi s)$ is cancelled by the term $\sin(\beta p/2)$ above. In consequence, $\bar{\varrho}_g(\beta)$ varies less rapidly with β than $\bar{\varrho}_f(\beta)$. In the x-domain, the inverse transform $\varrho_g(x)$ can be shown to be the charge density on an infinite array of regular electrodes, when those to the right of the origin have unit voltage and those to the left are grounded [109]. The new array factor, equation (4.95), corresponds to an array of elements located at $x = x_n + p/2$, the centres of the inter-electrode gaps, the strength of each element being zero if the adjacent electrodes have the same polarity; if the polarities are different, the strength is ± 1.

This formulation is often convenient for transducer analysis. It also corresponds to the delta-function analysis of Section 4.1, giving a formal justification for the delta-function approach. The element factor E in Section 4.1 can now be identified as

$$E = (\omega \Gamma_s)^{1/2} \bar{\varrho}_g(k_0). \qquad (4.97)$$

4.6. ADMITTANCE OF UNIFORM TRANSDUCERS

A uniform transducer is defined here as an unapodised transducer with regular electrodes, in which the electrode polarities $\hat{P}_n$ have a repetitive pattern instead of the arbitrary choice assumed in Section 4.5. Nearly all surface-wave devices have at least one uniform transducer, and the repetitive nature enables some simple and useful expressions for the main properties to be obtained. In this section the capacitance and the acoustic conductance and susceptance are considered, though other properties, such as the conversion coefficient, are readily obtained from the analysis given here.

The commonest types of uniform transducer are shown in Figure 4.13. Defining S_e as the number of electrodes per period, the types with $S_e = 2$ or 4 are the conventional "single-electrode" or "double-electrode" transducers, respectively, and a type with $S_e = 3$ is also shown. Engan [83] has derived the transducer properties from equations describing multi-strip couplers. The method used here is applicable for any repetitive polarity sequence, with any value of S_e. The presence of guard electrodes is assumed, so that end effects can be ignored. However, as discussed in Section 4.5, the analysis gives a good approximation if there are no guards (as in Figure 4.4, for example), provided the number of periods is not too small.

4.6.1. Acoustic Conductance and Susceptance
From Section 4.5, equation (4.83), the array factor can be written

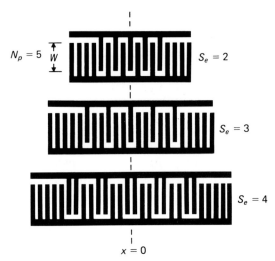

FIGURE 4.13. Uniform transducers, with guard electrodes.

$$A_f(x) = \sum_{n=1}^{N} \hat{P}_n \delta(x - x_n). \qquad (4.98)$$

For uniform transducers, this can be expressed as the convolution

$$A_f(x) = A_N(x) * A_1(x), \qquad (4.99)$$

where $A_1(x)$ is an array factor for one period of the transducer, and $A_N(x)$ is a sequence of delta functions with spacing equal to the transducer period, pS_e. For convenience these are defined such that $A_f(x)$ is symmetrical, so that $\bar{\varrho}_e(k_0)$ will be real. From Figure 4.13, $A_1(x)$ is given by

$$A_1(x) = \delta(x) \qquad \text{for } S_e = 2 \text{ or } 3$$
$$= \delta(x + p/2) + \delta(x - p/2), \quad \text{for } S_e = 4. \qquad (4.100)$$

Defining N_p as the number of periods in the transducer, the factor $A_N(x)$ is

$$A_N(x) = \sum_{n=1}^{N_p} \delta[x - (2n - N_p - 1)pS_e/2]. \qquad (4.101)$$

This factor has the same form for all uniform transducers. Its Fourier transform $\bar{A}_N(\beta)$ is a geometric progression which is readily summed to give

$$\bar{A}_N(k_0) = \frac{\sin(N_p k_0 p S_e/2)}{\sin(k_0 p S_e/2)}. \qquad (4.102)$$

From equation (4.100), the transform of $A_1(x)$ gives

$$\bar{A}_1(k_0) = 1, \qquad \text{for } S_e = 2 \text{ or } 3$$
$$= 2\cos(k_0 p/2), \quad \text{for } S_e = 4 \qquad (4.103)$$

The transform of $A_f(x)$ is, from equation (4.99),

$$\bar{A}_f(k_0) = \bar{A}_N(k_0)\bar{A}_1(k_0). \qquad (4.104)$$

This enables the parallel conductance to be calculated, since from equations (4.60) and (4.85),

$$G_a(\omega) = \omega W \Gamma_s |\bar{A}_f(k_0)\bar{\varrho}_f(k_0)|^2, \qquad (4.105)$$

where $\bar{\varrho}_f(\beta)$ is the elemental charge density [equation (4.88)], W is the transducer aperture, and $k_0 = \omega/v_0$ is the free-surface wavenumber at frequency ω.

For a single-electrode transducer ($S_e = 2$) we have $\bar{A}_f(k_0) = \bar{A}_N(k_0)$, given by equation (4.102). This function has maxima with magnitude N_p at $k_0 = 2\pi M/(pS_e)$, where $M = 0, 1, 2, \ldots$. The frequencies of the maxima are $\omega = M\omega_c$, where $\omega_c = \pi v_0/p$ is the centre frequency of the fundamental response. The function is similar to Figure 4.2, but here the even harmonics are present as well as the odd ones. The harmonic strengths are strongly affected by the $\bar{\varrho}_f(k_0)$ term in equation (4.105). The even-numbered harmonics are eliminated because $\bar{\varrho}_f(k_0)$ is zero at these points, and in addition the $\bar{\varrho}_f(k_0)$ term suppresses the 3rd, 7th, 11th, ... harmonics when $a/p = \frac{1}{2}$, as can be seen from Figure 4.11.

Similar remarks apply for the transducers with $S_e = 3$ and 4. In each case the fundamental response is centred at $\omega_c = 2\pi v_0/(pS_e)$. Harmonic responses are centred at frequencies $M\omega_c$, but are absent when M is a multiple of S_e because $\bar{\varrho}_f(k_0)$ is zero at these points. Thus for $S_e = 3$ the responses occur for $M = 1, 2, 4, 5, 7, \ldots$. For $S_e = 4$, the harmonics with $M = 2, 6, 10, \ldots$ are eliminated by the array factor for one period, equation (4.103), so for this case the responses occur only for odd values of M. In addition, harmonics in the region $2\pi < k_0 p < 4\pi$ disappear when $a/p = \frac{1}{2}$. For $S_e = 3$, this applies to the 4th and 5th harmonics, and for $S_e = 4$ it applies to the 5th and 7th harmonics.

If the number of periods, N_p, is not too small, a useful approximation can be obtained for the conductance in the region near the fundamental frequency ω_c. We define

$$X = N_p(\tfrac{1}{2}k_0 pS_e - \pi) = \pi N_p(\omega - \omega_c)/\omega_c, \qquad (4.106)$$

which is proportional to the fractional deviation of ω from ω_c. When ω is close to ω_c, the array factor of equation (4.102) is approximately

$$\bar{A}_N(k_0) \approx -N_p \frac{\sin X}{X}(-1)^{N_p}. \qquad (4.107)$$

If N_p is large, this function varies rapidly with ω. The conductance, equation (4.105), includes other frequency-dependent terms — the ω and $\bar{\varrho}_f(k_0)$ terms and, for $S_e = 4$, the array factor of equation (4.103) — but these vary slowly with ω. Thus, for frequencies near ω_c the conductance may be approximated by

$$G_a(\omega) \approx G_a(\omega_c)\left[\frac{\sin X}{X}\right]^2. \tag{4.108}$$

The constant $G_a(\omega_c)$ depends on the value of S_e, and is given below.

An approximate form for the acoustic susceptance $B_a(\omega)$ may be obtained from equation (4.108). As shown in Section 4.4.2, the susceptance is the Hilbert transform of the conductance. With X given by equations (4.106), the required Hilbert transform is

$$\left[\frac{\sin X}{X}\right]^2 * \frac{-1}{\pi\omega} = \frac{\sin(2X) - 2X}{2X^2}. \tag{4.109}$$

This may be demonstrated by transforming to the time domain, multiplying by the transform of $(-1/\pi\omega)$, and then transforming back to the frequency domain. The relationships needed are given in Appendix A. Thus, for frequencies near ω_c the susceptance is given by

$$B_a(\omega) \approx G_a(\omega_c)\frac{\sin(2X) - 2X}{2X^2}. \tag{4.110}$$

The approximate expressions for $G_a(\omega)$ and $B_a(\omega)$, equations (4.108) and (4.110), are plotted in Figure 4.14. The total susceptance of the transducer is $\omega C_t + B_a(\omega)$, and the capacitive term ωC_t is often much larger than $G_a(\omega_c)$; in addition $B_a(\omega)$ is zero at the centre frequency ω_c. Thus, in many cases the acoustic susceptance $B_a(\omega)$ is of little practical consequence.

Experimental measurements of $G_a(\omega)$ and $B_a(\omega)$ generally agree well with

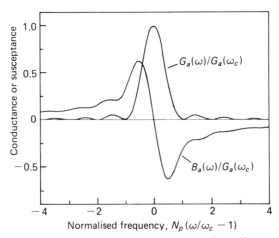

FIGURE 4.14. Acoustic conductance and susceptance for uniform transducers.

equations (4.108) and (4.110), as shown for example by Smith et al. [84] and Engan [83]. Usually there are also contributions due to bulk wave excitation, excluded in the above analysis, but provided N_p is not too small these are mostly confined to frequencies where the surface-wave terms are small. However, considerable distortion occurs if electrode interactions are strong.

For $M \leqslant S_e$, the conductance at the fundamental centre frequency ω_c and at the harmonics $M\omega_c$ can be written as

$$G_a(M\omega_c) = \alpha M\omega_c N_p^2 W \Gamma_s \left[\frac{2(\varepsilon_0 + \varepsilon_p^T) \sin(\pi s)}{P_{-s}(-\cos \Delta)} \right]^2, \quad (4.111)$$

where $\alpha = 1$ for $S_e = 2$ or 3. For $S_e = 4$, $\alpha = 2$ if $M = 1$ or 3 and $\alpha = 0$ if $M = 2$. Also, $s = M/S_e$ and $\Delta = \pi a/p$. Equation (4.111) follows from equation (4.105), with the array factor given by equations (4.102)–(4.104) and the elemental charge density $\bar{\varrho}_f(k_0)$ given by equation (4.88).

4.6.2. Capacitance

As shown by equation (4.91), the static capacitance of the transducer is

$$C_t = \sum_{n=1}^{N} \hat{P}_n q_n, \quad (4.112)$$

where N is the number of electrodes and q_n is the net electrostatic charge on electrode n when unit voltage is applied across the transducer. For a uniform transducer the electrode charges q_n in each period of the transducer are similar to those in other periods, and so the capacitance C_t is approximately proportional to the number of periods, N_p. This is not exactly true because the electrode charges are somewhat different near the ends of the transducer. For convenience an approximate formula is given here, assuming N_p to be large; the charges q_n are evaluated for each period assuming the transducer to be infinitely long, and C_t is then found by summing over N_p periods. Thus

$$C_t \approx N_p \sum_{n=1}^{S_e} \hat{P}_n q_n, \quad (4.113)$$

where $\hat{P}_n$ and q_n are evaluated for one period of a transducer of infinite length. The charges q_n are given by

$$q_n = W \sum_{r=-\infty}^{\infty} \sum_{i=1}^{S_n} \hat{P}_i Q_{i+rS_e-n}, \quad (4.114)$$

where Q_m is the charge on electrode n associated with the voltage on electrode $(n - m)$, as defined in equation (4.89). The summation over r in equation (4.114) may be done with the aid of equation (C.30) of Appendix C, and the remaining summations required are straightforward. For the transducers under consideration here, the result is

$$C_t = \gamma W N_p(\varepsilon_0 + \varepsilon_p^T) \frac{\sin(\pi/S_e)}{P_v(-\cos\Delta)} P_v(\cos\Delta), \qquad (4.115)$$

where $\gamma = 1$, $4/3$ or 2 for $S_e = 2$, 3 or 4, respectively, and $v = -1/S_e$.

4.6.3. Comparative Performance

In comparing the behaviour of the transducers, an important quantity is the electrical Q-factor, here denoted by Q_t. At frequency ω_c and at the multiples of this frequency, the acoustic susceptance $B_a(\omega)$ is zero. We define Q_t as the ratio of the susceptance to the conductance at these frequencies, so that

$$Q_t = M\omega_c C_t / G_a(M\omega_c). \qquad (4.116)$$

The value of Q_t will in general depend on the harmonic number, M. The reciprocal of Q_t is a measure of the surface-wave coupling strength, taking account of the transducer geometry and the parameters of the substrate material. It will be shown in Chapter 7 that this parameter has a strong influence on the bandwidth obtainable when the transducer is tuned in order to minimise its conversion loss.

The basic properties of the three types of transducer are summarised in Figure 4.15 and Table 4.1. For convenience, a normalised capacitance $\tilde{C}_t$, conductance $\tilde{G}_{aM}$ and Q-factor $\tilde{Q}_t$ are introduced, defined such that

$$C_t = W N_p(\varepsilon_0 + \varepsilon_p^T)\tilde{C}_t, \qquad (4.117)$$

$$G_a(M\omega_c) = M\omega_c(\varepsilon_0 + \varepsilon_p^T)^2 N_p^2 W \Gamma_s \tilde{G}_{aM}, \qquad (4.118)$$

$$Q_t = \tilde{Q}_t / [N_p(\varepsilon_0 + \varepsilon_p^T)\Gamma_s], \qquad (4.119)$$

where C_t, $G_a(M\omega_c)$ and Q_t are given by equations (4.115), (4.111) and (4.116) above. The normalised parameters are independent of the substrate properties ε_p^T and Γ_s, and of the number of periods N_p and the aperture W. The capacitance and Q-factor are

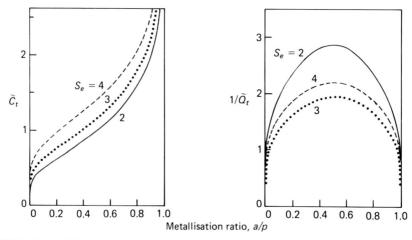

FIGURE 4.15. Uniform transducers: normalised capacitance and Q-factor, as functions of metallisation ratio.

TABLE 4.1
Data for uniform transducers, $a/p = \frac{1}{2}$

S_e	Normalised capacitance $\tilde{C}_t$	Harmonic number M	Normalised conductance $\tilde{G}_{aM}$	Normalised inverse Q-factor $1/\tilde{Q}_t$
2	1	1	2.871	2.871
3	1.155	1	2.231	1.932
		2	2.231	1.932
4	1.414	1	3.111	2.200
		2	0	0
		3	3.111	2.200

shown in Figure 4.15, plotted as functions of the metallisation ratio a/p. It can be seen that the metallisation ratio is not a critical parameter, and that for all three transducers the coupling strength is maximised for $a/p = \frac{1}{2}$. Numerical data are given in Table 4.1, for $a/p = \frac{1}{2}$. The coupling strength is strongest for the single-electrode transducer ($S_e = 2$). For $S_e = 3$ or 4, there is a harmonic with coupling strength equal to that of the fundamental. The capacitance is smallest for $S_e = 2$. Engan [83] has confirmed experimentally the values of $\tilde{C}_t$ and $\tilde{G}_{aM}$ in this table.

4.7. TWO-TRANSDUCER DEVICES

This section considers the analysis of a device comprising two interdigital transducers, each transducer having two terminals. One transducer is taken to be energised by a source with zero impedance, while the other is shorted. It is assumed that the transducers do not reflect incident surface waves in this situation, so that the quasi-static results of Section 4.4 can be used. It is also assumed that waves due to reflections from the edges of the substrate can be ignored, as is usually ensured by the use of acoustic absorbers. Reflections do however arise when the transducers are connected to finite electrical impedances, and this will be considered in Section 4.8. The analysis here is not valid if the transducers are of the unidirectional type using multi-strip couplers, described in Section 5.4, since these give appreciable reflections when they are shorted. Apodised transducers are introduced in this section, because the results of interest are meaningful only when a combination of transducers is considered.

Initially, a generalised transducer geometry is considered. For apodised transducers the more specialised case of regular electrodes is also considered, using the results of Section 4.5. This enables the response to be expressed in terms of an array factor and an element factor, as in the delta-function analysis of Section 4.1.

It is assumed that the transducer separation is sufficient to make electrostatic coupling between the transducers negligible, so that the current produced by the output transducer is entirely due to the surface waves incident on it.

4.7.1. Devices Using Unapodised Transducers

To analyse devices, it is convenient to define the frequency response, $H_t(\omega)$, of an unapodised transducer by the expression

$$H_t(\omega) = (\omega W \Gamma_s)^{1/2} \bar{\varrho}_e(k_0) \exp(-\tfrac{1}{2} j k_0 L). \tag{4.120}$$

The processes of launching and reception of surface waves can be written in terms of this function. For a launching transducer, with a voltage V_t applied, the surface wave potential at port 1 is, from equation (4.50),

$$\phi_{s1}(\omega) = jV_t H_t(\omega) \left[\frac{\Gamma_s}{\omega W}\right]^{1/2}. \tag{4.121}$$

For a receiving transducer, if the incident surface wave has a potential $\phi_{s1}(\omega)$ at port 1, the output current produced when the transducer is shorted is given by

$$I_{sc} = -j\phi_{i1}(\omega) H_t(\omega)(\omega W/\Gamma_s)^{1/2} \tag{4.122}$$

from equation (4.68).

These equations can be applied to the two-transducer device shown in Figure 4.16. For transducers A and B, the electrostatic charge density, for unit voltage applied across the bus-bars, is denoted by $\varrho_e^a(x)$ and $\varrho_e^b(x)$, respectively. Each of these functions is to be determined assuming the other transducer to be absent. To preserve the symmetry, different x-axes must be used for the two transducers; in each case the x-axis is directed away from the other transducer, and its origin is midway between the two acoustic ports. For each transducer, port 1 is defined as the port closest to the other transducer. The lengths of transducers A and B are respectively L_a and L_b. The frequency responses are $H_t^a(\omega)$ and $H_t^b(\omega)$, given by equation (4.120) with $\bar{\varrho}_e(k_0)$ given the superscript a or b and L given the corresponding subscript.

As shown in Figure 4.16, transducer A is connected to a source with zero impedance and voltage V_t, while transducer B is shorted. It is assumed that neither transducer reflects incident waves under these conditions, and that diffraction and attenuation are negligible. The wave incident on transducer B is therefore $\phi_{i1}(\omega) = \phi_{s1}(\omega) \exp(-jk_0 d)$, where d is the transducer separation. The ratio I_{sc}/V_t can then be found from equations (4.121) and (4.122). This ratio is denoted by $H_{sc}(\omega)$, and is given

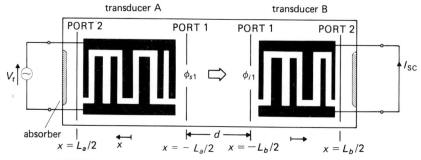

FIGURE 4.16. Two-transducer device, with unapodised transducers.

by

$$H_{sc}(\omega) \equiv I_{sc}/V_t = H_t^a(\omega) H_t^b(\omega) \exp(-jk_0 d). \quad (4.123)$$

This is called the *short-circuit response* of the device, since both transducers are connected to zero impedances.

Comparing equations (4.120) and (4.60) shows that the parallel conductance of an unapodised transducer is directly related to its frequency response:

$$G_a(\omega) = |H_t(\omega)|^2. \quad (4.124)$$

Thus, if the conductances of the two transducers are known, the magnitude of the device short-circuit response can be obtained directly.

4.7.2. Device Using One Apodised Transducer

We now consider the short-circuit response of the device in the upper part of Figure 4.17, where one transducer is apodised. The apodised transducer is assumed to have "dummy" electrodes in the acoustically inactive regions, so that the electrodes extend from both bus-bars and the breaks between them are small. This feature is commonly used in practice because the electrodes perturb the surface wave velocity; the addition of dummy electrodes gives a more uniform perturbation, and substantially reduces the consequent phase distortion of the surface wave [110].

Following Tancrell and Holland [79], the device is analysed by imagining it to be divided into a number of parallel "channels", whose edges correspond to the locations of the electrode breaks in the apodised transducer. Since diffraction is neglected, the launching and reception of surface waves in individual channels can be analysed independently. Thus, in each channel the electrodes of the apodised transducer constitute an imaginary unapodised transducer, and the apodised transducer may be replaced by an array of unapodised transducers, as in the lower part of Figure 4.17. The imaginary unapodised transducers, which can be analysed by methods given above, are electrically connected in parallel. Since the electrode resistivity is assumed to be negligible, the admittance of the apodised transducer is simply the sum of the admittances of the imaginary unapodised transducers. This approach is valid even if electrode interactions or bulk wave excitation are present, though the analysis here excludes these effects.

Consider first the response of transducer B, which is taken to be an unapodised transducer receiving surface waves and is assumed to be shorted. In channel j, the wave incident on port 1 is denoted $\phi_{i1}^j(\omega)$, and the width of channel j is denoted W_j. Considering the electrodes in this channel as a separate transducer, the output current produced when the transducer is shorted is I_{sc}^j, say, given by

$$I_{sc}^j = -j\omega W_j \phi_{i1}^j(\omega) \bar{\varrho}_e^b(k_0) \exp(-\tfrac{1}{2} jk_0 L_b) \quad (4.125)$$

from equation (4.68). The function $\bar{\varrho}_e^b(k_0)$ is the same in all channels because the transducer is unapodised. The total current I_{sc} is simply the sum of the I_{sc}^j. Using the frequency response defined in equation (4.120), this is

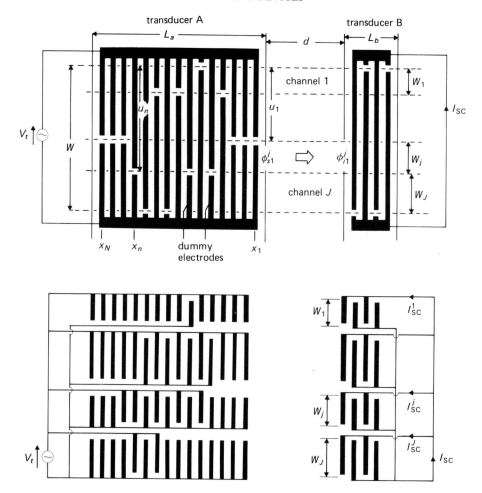

FIGURE 4.17. *Upper*: Two-transducer device with one transducer apodised. *Lower*: Equivalent representation as a network of unapodised transducers.

$$I_{sc} = -jH_t^b(\omega)(\omega W/\Gamma_s)^{1/2}\left[\sum_j W_j \phi_{i1}^j(\omega)/W\right]. \quad (4.126)$$

Here the term in square brackets is the average surface wave potential at the transducer input; the current is therefore the same as for a uniform input beam (equation 4.122), except that the average surface-wave potential is used.

For the apodised transducer, transducer A, we define $\varrho_{ej}^a(x)$ as the electrostatic charge density on the electrodes in channel j, when unit voltage is applied across the transducer, with Fourier transform $\bar{\varrho}_{ej}^a(\beta)$. Using this function, the potential of the wave generated in channel j, denoted by $\phi_{s1}^j(\omega)$, is given by equation (4.50). In addition, since there are no reflected waves, and diffraction and attenuation are ignored, we have $\phi^j_{i1}(\omega) = \phi^j_{s1}(\omega) \exp(-jk_0 d)$. The output current is then given by equation (4.126), and is thus related to the input voltage V_t. We define the response $H^a_t(\omega)$ of transducer A such that the device short-circuit response is essentially the product of the transducer responses:

$$H_{sc}(\omega) \equiv I_{sc}/V_t = H^a_t(\omega) H^b_t(\omega) \exp(-jk_0 d) \qquad (4.127)$$

as in the unapodised case. To satisfy this, the required definition for $H^a_t(\omega)$ is

$$H^a_t(\omega) = (\omega W \Gamma_s)^{1/2} \exp(-\tfrac{1}{2} jk_0 L_a) \sum_j W_j \bar{\varrho}^a_{ej}(k_0)/W. \qquad (4.128)$$

For an unapodised transducer, this reduces to the form already given in equation (4.120). The definition can also be expressed in terms of the surface wave potentials at port 1 when a voltage V_t is applied across the transducer; equation (4.121) is thus valid for an apodised transducer if $\phi_{s1}(\omega)$ is replaced by the average of the $\phi^j_{s1}(\omega)$. For an apodised transducer receiving a uniform surface-wave beam, the short-circuit current is found to be given by equation (4.122).

The response defined in equation (4.128) is valid only if at least one of the two transducers is unapodised, with its active region intercepting all of the surface waves generated (in one direction) by the other transudcer. It should also be noted that the simple relationship between $G_a(\omega)$ and $H_t(\omega)$ for an unapodised transducer, equation (4.124), is not valid for an apodised transducer.

4.7.3. Apodised Transducer with Regular Electrodes

If the electrodes are regular, so that they have the same width a and constant pitch p, the frequency response of an apodised transducer can be expressed in a convenient form, using the elemental charge distribution described in Section 4.5. As shown in Figure 4.17, it is assumed here that guard electrodes are included at both ends, thus minimising end effects.

The response is readily obtained from the definition of equation (4.128) and the analysis of Section 4.5.1, so the results are quoted here without the derivation. We define a parameter u_n as the location of the break between the electrodes centred at $x = x_n$, relative to the upper boundary of the active region, as in Figure 4.17. The transducer response is then given by

$$H_t(\omega) = (\omega W \Gamma_s)^{1/2} \bar{A}_f(k_0) \bar{\varrho}_f(k_0) \exp(-\tfrac{1}{2} jk_0 L), \qquad (4.129)$$

where $\bar{\varrho}_f(k_0)$ is the elemental charge density defined in equation (4.88) and $\bar{A}_f(k_0)$ is an array factor defined by

$$\bar{A}_f(k_0) = \sum_{n=1}^{N} (u_n/W) \exp(-jk_0 x_n). \qquad (4.130)$$

Thus the array factor is directly related to the transducer geometry. The formula can be applied to an unapodised transducer, in which case $\bar{A}_f(k_0)$ becomes the same as the definition given previously, equation (4.86).

An alternative formula can be given in terms of gap elements, using the equations of Section 4.5.3. The result is

$$H_t(\omega) = (\omega W \Gamma_s)^{1/2} \bar{A}_g(k_0) \bar{\varrho}_g(k_0) \exp(-\tfrac{1}{2} jk_0 L), \quad (4.131)$$

where the elemental charge density $\bar{\varrho}_g(k_0)$ is given by equation (4.96) and the array factor $\bar{A}_g(k_0)$ is defined by

$$\bar{A}_g(k_0) = \sum_{n=1}^{N-1} \frac{u_{n+1} - u_n}{W} \exp[-jk_0(x_n + p/2)]. \quad (4.132)$$

This shows that, in x-domain, the array factor can be taken as a set of delta functions located at the centres of the inter-electrode gaps, with the strength of each delta function proportional to the distance over which the adjacent electrodes with different polarities overlap. The transducer can therefore be regarded as an array of elements associated with the gaps, as in the delta-function analysis of Section 4.1. The elements can be taken to be located in the regions where the polarities of adjacent electrodes are different, as in Figure 4.3. The analysis of Section 4.1 is thus justified, with the element factor E given by equation (4.97).

Transverse End Effect. In the above analysis it has been assumed that the charge density on each electrode is uniform across the width of each channel. There is however some distortion of the charge density in the region near the break, where the electrodes connected to the two bus-bars almost meet. To find the charge density in this region, a two-dimensional analysis is needed. Some results obtained by Wagers [111] show that the distortion extends a distance approximately equal to the width of the break.

For practical purposes, this "transverse end effect" can be allowed for by adjusting the source strengths. Thus, if the response is expressed in terms of gap elements, a small constant is added to the terms $(u_{n+1} - u_n)$ in equation (4.132). The value of the constant depends on the width of the break.

4.8. DEVICE RESPONSE ALLOWING FOR TERMINATING CIRCUITS

Up to this point, we have only considered the response of a two-transducer device for the case where the transducers are connected to zero electrical impedances. In this section the electrical impedances are allowed to be finite, as they must be in any practical situation. This analysis is needed to evaluate the insertion loss of a device and to assess the circuit effect, that is, the distortion of the device frequency response due to the use of finite impedances. In addition the analysis gives the triple-transit spurious signal, discussed in Section 4.2 above.

It is quite common to connect each transducer to the electrical source or load via a simple lumped-element circuit, in order to improve the electrical matching. The analysis here therefore allows for unspecified circuits. The circuits in the analysis may include stray components, such as the capacitance between the "live" bus and the package and the inductance of the transducer bond wires. They may also include some

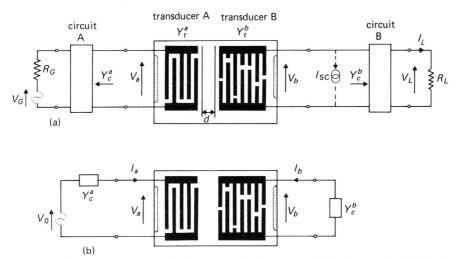

FIGURE 4.18. (a) Device analysis allowing for terminating circuits. (b) Equivalent circuit for analysis of multiple-transit terms.

resistance in order to model the resistivity of the transducer electrodes. Chapter 7 considers some specific circuits, and their effects on device performance.

It is assumed that the transducers do not reflect surface waves when they are shorted, as in previous sections. For finite terminating impedances the transducers reflect, and a triple-transit output signal is produced. The relative level of this signal is easily deduced if the transducers are unapodised. Consider the device shown in Figure 4.18(a), where the voltage on the output transducer, transducer B, is V_b. The main component of this voltage, denoted V_{b1}, is due to the wave generated by transducer A reaching transducer B directly. In addition there is a triple-transit component V_{b3}, due to reflection at transducer B and then again at transducer A, so that a second wave arrives at transducer B. The ratio of these components is therefore, for unapodised transducers,

$$V_{b3}/V_{b1} = r_1^a(\omega) r_1^b(\omega) \exp(-2jk_0 d), \qquad (4.133)$$

where d is the distance between the adjacent transducer ports, and $r_1^a(\omega)$ and $r_1^b(\omega)$ are the reflection coefficients referred to these ports. The reflection coefficients are given by equation (4.72). The total voltage across transducer B is the sum of V_{b1} and V_{b3}, plus contributions due to additional transits; thus $V_b = V_{b1} + V_{b3} + V_{b5} + \ldots$.

In Figure 4.18(a), the load R_L develops a voltage V_L proportional to V_b. Thus V_L includes components proportional to V_{b1} and V_{b3}, with the ratio given by equation (4.133). If V_G is the open circuit of the generator, we define the ratio V_L/V_G to be the device frequency response $H(\omega)$. It follows that the response can be written as

$$H(\omega) \equiv V_L/V_G = H_1(\omega) + H_3(\omega) + H_5(\omega) + \ldots, \qquad (4.134)$$

where $H_1(\omega)$ gives the main signal, due to the wave reaching transducer B directly, and

$H_3(\omega)$ gives the triple-transit signal. The ratio $H_3(\omega)/H_1(\omega)$ is equal to V_{b3}/V_{b1}, equation (4.133).

Equation (4.133) was derived assuming both transducers to be unapodised, and neglecting diffraction and attenuation. It will however be shown below that the response can be expressed as in equation (4.134), even when these assumptions are not valid. The results below involve the short-circuit response $H_{sc}(\omega)$. If the transducers are apodised, and diffraction and attenuation are significant, the analysis is still valid provided these factors are allowed for in the evaluation of $H_{sc}(\omega)$. However, the derivation of H_{sc} in Section 4.7 above assumes that diffraction and attenuation are negligible, and that at least one of the two transducers is unapodised.

Usually, the main response $H_1(\omega)$ in equation (4.134) is of most interest, the other terms being small unwanted components. It is therefore convenient to consider $H_1(\omega)$ first.

4.8.1. Main Response

The main response $H_1(\omega)$ is defined as the response obtained if reflections from the transducers are ignored. This is conveniently expressed by defining circuit factors $F_c^a(\omega)$ and $F_c^b(\omega)$. Considering circuit A and transducer A, at the input of the device in Figure 4.18(a), the circuit factor is defined as

$$F_c^a(\omega) = [V_a/V_G]_{V_b=0}, \quad (4.135)$$

where V_a and V_b are respectively the voltages across transducers A and B. The definition assumes that no acoustic waves are incident on the transducer, and it is therefore necessary to set $V_b = 0$ so that transducer B does not reflect. The circuit factor depends on the transducer admittance $Y_t^a(\omega)$ and the generator impedance R_G, as well as on the details of the circuit. For transducer B and circuit B, at the output, an imaginary current generator I_{sc} is connected between the terminals of the transducer and, with transducer A shorted so that it does not reflect, we define

$$F_c^b(\omega) = [I_L/I_{sc}]_{V_a=0}, \quad (4.136)$$

where I_L is the current in the load. The two definitions given by equations (4.135) and (4.136) are in fact equivalent, as can be shown using the reciprocity theorem of network analysis.

To find $H_1(\omega)$, the reflected surface waves are ignored. Thus V_a is given by equation (4.135) and I_L by equation (4.136), and we also have $I_{sc} = H_{sc}(\omega) V_a$, where $H_{sc}(\omega)$ is the short-circuit response. Thus

$$H_1(\omega) = V_L/V_G = H_{sc}(\omega) F_c^a(\omega) F_c^b(\omega) R_L, \quad (4.137)$$

which is essentially the product of the short-circuit response and the two circuit factors.

Since the multiple-transit terms $H_3(\omega)$, $H_5(\omega)$, ... are usually small in comparison, the insertion loss of the device may be obtained approximately from $H_1(\omega)$. If P_G is the power available from the generator, and P_L the power delivered to the load, these are in the ratio

$$P_L/P_G = 4|H_1(\omega)|^2 R_G/R_L \quad (4.138)$$

ANALYSIS OF INTERDIGITAL TRANSDUCERS

and the insertion loss, in decibels, is

$$IL = -10 \log (P_L/P_G) \quad (\text{dB}). \tag{4.139}$$

It is instructive to consider the theoretical minimum insertion loss predicted by these equations. Minimum loss is obtained if the circuits dissipate no power and the transducers are matched, so that all the power available from the generator is dissipated in transducer A, and all the power available from transducer B is dissipated in the load. These considerations determine the magnitudes of $F_c^a(\omega)$ and $F_c^b(\omega)$. For this calculation, diffraction and attenuation are assumed negligible and the short-circuit response $H_{sc}(\omega)$ of Section 4.7.2, equation (4.127), is used. Using standard network analysis, it is found that for this case the power ratio P_L/P_G is

$$[P_L/P_G]_{\max} = \tfrac{1}{4}|H_t^a(\omega)H_t^b(\omega)|^2/(G_a^a G_a^b), \tag{4.140}$$

where $H_t^a(\omega)$ and $H_t^b(\omega)$ are the transducer responses and G_a^a and G_a^b are their parallel conductances. Now, for an unapodised transducer we have, from equation (4.124), $G_a(\omega) = |H_t(\omega)|^2$. Thus, if both transducers are unapodised the maximum value of P_L/P_G is $\tfrac{1}{4}$, and hence the minimum insertion loss is 6 dB, as expected from the bidirectional nature of the transducers. If one of the transducers, say transducer B, is apodised, there is an additional contribution to the loss associated with the factor $|H_t^b(\omega)|^2/G_a^b$, which will be less than unity. This factor is sometimes known as the "apodisation loss".

In the above equations the multiple-transit terms have been excluded. The insertion loss including these terms is given by equation (4.138), with $H_1(\omega)$ replaced by the complete response $H(\omega)$, which is calculated below.

4.8.2. Multiple-transit Responses

To find the complete device response including the multiple-transit terms, it is convenient to consider the simpler circuit of Figure 4.18(b), which is equivalent to Figure 4.18(a). Here the generator and input circuit are replaced by a Thévenin equivalent, having a voltage generator V_0 in series with an admittance Y_c^a, where Y_c^a is the admittance "seen" by transducer A looking into the circuit in Figure 4.18(a). Using conventional network analysis, the equivalent generator voltage V_0 can be related to V_G using the circuit factor, giving

$$V_0/V_G = F_c^a(Y_t^a + Y_c^a)/Y_c^a. \tag{4.141}$$

At the output, transducer B is connected to an admittance Y_c^b, the admittance "seen" by transducer B looking into the circuit in Figure 4.18(a). The voltage on transducer B is denoted V_b, and this can be related to the load voltage V_L by means of the circuit factor:

$$V_L/V_b = -F_c^b R_L(Y_t^b + Y_c^b). \tag{4.142}$$

The device response $H(\omega)$ is defined as the ratio V_L/V_G, and in view of the above equations this can be obtained from the ratio V_b/V_0. For the main response, ignoring multiple transits, the ratio V_L/V_G is given by equation (4.137). The ratio of V_b/V_0 for the main response is denoted by $R_1(\omega)$, and can be obtained from equations (4.137), (4.141) and (4.142), giving

$$R_1(\omega) = \frac{-H_{sc} \cdot Y_c^a}{(Y_t^a + Y_c^a)(Y_t^b + Y_c^b)}. \tag{4.143}$$

To find the complete response, the behaviour of the surface wave device itself is expressed in terms of an admittance matrix [112]. We define V_a and V_b as the transducer voltages, and I_a and I_b as the transducer currents, as in Figure 4.18(b). For $V_b = 0$ is assumed that transducer B does not reflect surface waves, so that $I_a/V_a = Y_t^a$ and $I_b/V_a = H_{sc}(\omega)$. Similarly, for $V_a = 0$ we have $I_b/V_b = Y_t^b$ and, by reciprocity, $I_a/V_b = H_{sc}(\omega)$. The device therefore has an admittance matrix given by

$$\begin{bmatrix} I_a \\ I_b \end{bmatrix} = \begin{bmatrix} Y_t^a & H_{sc} \\ H_{sc} & Y_t^b \end{bmatrix} \cdot \begin{bmatrix} V_a \\ V_b \end{bmatrix}. \tag{4.144}$$

This enables the ratio V_b/V_0 to be evaluated. From Figure 4.18(b) we have $I_a = (V_0 - V_a)Y_c^a$ and $I_b = -Y_c^b V_b$. Using equation (4.144), and eliminating V_a, I_a and I_b, gives

$$V_b/V_0 = [R_1^{-1} + H_{sc}/Y_c^a]^{-1} \tag{4.145}$$

where R_1 is the function defined in equation (4.143). Using the binomial theorem, this is written as the series

$$V_b/V_0 = R_1[1 - R_1 H_{sc}/Y_c^a + (R_1 H_{sc}/Y_c^a)^2 + \ldots] \tag{4.146}$$

In this equation, the short-circuit response H_{sc} includes a factor giving the phase change due to the transducer separation. This can be seen, for example, in equation (4.127), where a term $\exp(-jk_0 d)$ is present. The function R_1, equation (4.143), is proportional to $H_{sc}(\omega)$, and therefore includes the same factor. Hence the first term of equation (4.146), which includes this factor, is the main response; the second term includes a factor with a phase change three times as large and is therefore the triple-transit term, and so on. The overall response of the device, $H(\omega)$, is defined as the ratio V_L/V_G, and can be obtained from equations (4.146), (4.141) and (4.142). However, since the contribution due to the main response, $H_1(\omega)$, has already been given in equation (4.137), it is sufficient to consider the ratios of the terms here. From equation (4.146), the ratio of the triple-transit response to the main response is

$$H_3(\omega)/H_1(\omega) = -R_1 H_{sc}/Y_c^a$$

$$= \frac{H_{sc}^2}{(Y_t^a + Y_c^a)(Y_t^b + Y_c^b)}, \tag{4.147}$$

where equation (4.143) has been used for R_1. The corresponding ratios for additional transits of the device are readily obtained from equation (4.146) and the overall response $H(\omega)$ can be obtained by summing these terms, as in equation (4.134).

It is emphasised here that the result of equation (4.147) is valid if both transducers are apodised, and if diffraction and attenuation are significant, provided these factors are allowed for in evaluating $H_{sc}(\omega)$. The main assumption is that reflections are absent when the transducers are connected to zero impedances; this is required in order to establish the admittance matrix of equation (4.144). In the particular case of unapodised transducers, with negligible diffraction and propagation loss, the triple-transit term can be expressed in terms of the transducer reflection coefficients, as shown by equation (4.133). This can be shown to be consistent with equation (4.147) by using the transducer reflection coefficient given in equation (4.72), together with the short-circuit response given by equations (4.123) and (4.120). However, for an apodised transducer a reflection coefficient cannot be defined straightforwardly because the amplitude of the reflected wave will vary across the aperture. Thus a more general approach, as given above, is called for.

Chapter 5

The Multi-Strip Coupler and Its Applications

The multi-strip coupler, first demonstrated by Marshall and Paige [113], consists of an array of identical electrodes oriented parallel to the surface-wave wavefront. The electrodes are electrically disconnected. As already mentioned in Chapter 1 the coupler may be used to transfer the surface wave power laterally, so that the output wave occupies a track displaced relative to that of the input wave, as in Figure 1.2. In many devices the coupler is used in this way to couple two interdigital transducers whose active regions do not overlap. This arrangement has the practical merit that spurious output signals due to excitation of bulk waves are much reduced, and the device response is easily calculated even if both transducers are apodised, thus introducing additional design flexibility. There are also many modified forms of the coupler with diverse applications, giving in particular several methods of reducing the level of spurious triple-transit signals in surface-wave devices. The number of electrodes required in the coupler depends inversely on the piezoelectric coupling strength of the substrate material, and for this reason the coupler is practicable only for strongly piezoelectric materials such as lithium niobate.

The basic mechanism of the coupler is a straightforward application of the piezoelectric effect. A surface wave incident in one track causes voltages to be induced on the electrodes, which therefore apply voltages in the second track, generating a second surface wave there. At any frequency, the induced voltages have relative phases corresponding to the propagating surface wave, so that waves generated by individual electrodes in the second track will be in phase. It is thus clear that the coupler will have a wide bandwidth, much wider than that of a transducer with a comparable length.

A quantitative description can be obtained by considering the propagating modes of the structure, that is, solutions in which all the field variables, such as the surface potential $\phi(x)$, have the property $\phi(x + p) = \phi(x) \exp(-jkp)$, where k is some wavenumber and p is the pitch. A solution such as this is found to exist for a uniform surface-wave beam, with the electrodes extending over the width of the beam. When two tracks are coupled by a set of electrodes, it is found that the basic solution splits into two modes with slightly different wavenumbers. A simple coupler, with identical geometries in the two tracks, gives a symmetric mode and an anti-symmetric mode;

the former has the surface-wave amplitudes the same in both tracks, while the latter has the amplitudes the same except for being opposite in sign. For a wave incident in one track, the waves in the coupler are readily found by superimposing the two modes, as will be shown in Section 5.2. It is found that the surface-wave energy is completely coupled from one track to the other in a distance related to the velocity difference of the symmetric and anti-symmetric modes, and that the waves in the two tracks have a phase difference of $\pi/2$. This behaviour is similar to many other systems involving weakly-coupled waves, for example waves in weakly-coupled waveguides. Another example is surface-wave propagation on a substrate of finite thickness, considered in Section 2.2.5.

To find the basic modes of the coupler, it is first necessary to consider propagation in an infinite array of electrodes, taking the surface-wave amplitude to be uniform in the transverse direction. Here the structure is periodic, and the wave properties show features in common with many other waves in periodic structures, for example electron waves in crystals and electromagnetic waves in a helix. A comprehensive review is given by Elachi [114]. The wave motion is evanescent if the frequency is within one of a set of bands known as stop bands. The stop bands are centred approximately at the frequencies where the pitch of the structure equals a multiple of the half-wavelength. This is readily appreciated if the structure is modelled as an array of weakly-reflecting elements, and Appendix E gives further details of this model, showing that the width of the stop bands is related to the reflection coefficient. The same conclusions can be obtained from coupled wave analysis [114]. Rigorous anlaysis is often facilitated by Floquet's theorem, which states that the amplitude of a propagating wave can be expressed as an infinite series with terms of the form $\exp[-j(k + 2\pi n/p)x]$, where p is the pitch of the structure. The individual terms are called space harmonics.

In most practical couplers the number of electrodes is large, and so end effects are not significant. Consequently, the results for a periodic structure may be applied directly to a coupler of finite length. An analysis using perturbation theory has been given by Maerfeld et al. and applied to a wide variety of cases [115–118]. An alternative, and very general, method given by Bløtekjaer et al. [119, 120] uses Floquet's theorem and the effective permittivity $\varepsilon_s(\beta)$ discussed in Chapter 3. For coupling to Rayleigh waves, Ingebrigtsen's approximation for $\varepsilon_s(\beta)$ is used, and the results are equivalent to the perturbation analysis. This is described in Appendix D.

The analysis given here is based on the quasi-static Green's function method, using equations given in Section 4.3. This method is relatively straightforward. The quasi-static approximation neglects electrode interactions, and this has the consequence that the stop bands are not predicted, and there are small errors in the predicted surface-wave velocities. However, for most practical devices, designed to operate outside the coupler stop bands, these errors are of little consequence. The analysis is subject to the assumptions made in deriving the equations of Section 4.3, so that the only acoustic wave present is assumed to be a piezoelectric Rayleigh wave. Mechanical loading and the resistivity of the electrodes are assumed to be negligible.

Section 5.1 gives the analysis for an infinite array of electrodes, with the surface-wave amplitude uniform in the transverse direction. This analysis is needed to establish the basic modes of a coupler. In particular, the solutions for shorted and for open-circuited electrodes give the symmetric and anti-symmetric modes for the basic form of coupler. The coupler itself is first considered in Section 5.2. Section 5.3 considers the use of the coupler to couple two interdigital transducers, the commonest application. The remaining sections describe a variety of other applications, involving modified forms of the coupler.

5.1. ANALYSIS FOR AN INFINITE ARRAY OF ELECTRODES

We first consider an infinite array of regular electrodes with pitch p and width a, as shown in Figure 5.1. The electrode length, W, in the y-direction is assumed to be large, so that all the field quantities can be taken to be independent of y over the electrodes. The origin for the x-axis is at the centre of one electrode, and the voltage on the electrode centred at $x = np$ is V_n, while the current entering this electrode is I_n. We consider specifically solutions in which the voltages and currents can be written

$$V_n = V_0 \exp(-j\kappa np) \qquad (5.1a)$$

and

$$I_n = I_0 \exp(-j\kappa np), \qquad (5.1b)$$

where a factor $\exp(j\omega t)$ is implicit. The constant κ is arbitrary at this stage, though it should be noted that the solution is unaffected if a multiple of $2\pi/p$ is added to κ. In some particular cases, notably when $V_0 = 0$ or $I_0 = 0$, it will be found that κ can be interpreted as the wavenumber of a perturbed surface wave, with a value close to the free-surface wavenumber k_0. The requirement here is to find a relationship between V_n and I_n and the amplitude of any accompanying surface wave.

The solution is obtained using the quasi-static analysis of Section 4.3. The charge density on the electrodes is dominated by the electrostatic term $\sigma_e(x)$, and by equation (4.33) this can be written in terms of the functions $\varrho_{en}(x)$, defined as the electrostatic charge density when unit voltage is applied to electrode n with other electrodes grounded. In the present case the electrodes are regular, so $\varrho_{en}(x)$ is equal to $\varrho_f(x - np)$, where $\varrho_f(x)$ is the elemental charge density introduced in Section 4.5.1. We thus have

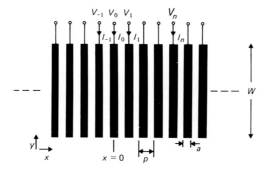

FIGURE 5.1 Infinite array of regular electrodes.

$$\sigma_e(x) = \sum_{n=-\infty}^{\infty} V_n \varrho_{en}(x) = V_0 \sum_{n=-\infty}^{\infty} \varrho_f(x - np) e^{-j\kappa np}. \quad (5.2)$$

The Fourier transform of $\sigma_e(x)$ is denoted $\bar{\sigma}_e(\beta)$. Using the shifting theorem, equation (A.10), this is seen to be a sum of complex exponentials, and equation (A.43) shows that this can be written as a sum of delta-functions. With $\bar{\varrho}_f(\beta)$ defined as the Fourier transform of $\varrho_f(x)$, it is found that

$$\bar{\sigma}_e(\beta) = \frac{2\pi V_0}{p} \sum_{m=-\infty}^{\infty} \bar{\varrho}_f(\kappa + 2\pi m/p)\delta(\beta + \kappa + 2\pi m/p), \quad (5.3)$$

where equation (A.24) has been used, and also the symmetry of $\bar{\varrho}_f(\beta)$. The function $\bar{\varrho}_f(\beta)$ is given by equation (4.88).

To find the current I_n entering electrode n, it is first necessary to evaluate the acoustic potential $\phi_a(x, \omega)$, defined as the convolution of the surface-wave Green's function $G_s(x, \omega)$ with $\sigma_e(x)$. It is shown in Section 4.3.4 that $\phi_a(x, \omega)$ may be written in terms of $\bar{\sigma}_e(\beta)$, as in equation (4.47). This equation involves terms proportional to $\bar{\sigma}_e(\pm k_0)$, and terms proportional to $F(\pm k_0)$, which are derived from $\bar{\sigma}_e(\beta)$. In the present case, $\bar{\sigma}_e(\beta)$ is given by equation (5.3), and it can be seen that $\bar{\sigma}_e(\pm k_0)$ is zero unless $\kappa + 2\pi m/p = \pm k_0$ for some m. For the moment this case is excluded; it will be seen later that it occurs when the electrode voltages are zero. Thus, $\phi_a(x, \omega)$ is given by equation (4.47) with the $\bar{\sigma}_e(\pm k_0)$ terms omitted, and with the aid of equation (A.23) this gives

$$\phi_a(x, \omega) = \frac{2\Gamma_s k_0 V_0}{p} \sum_{m=-\infty}^{\infty} \frac{\bar{\varrho}_f(\beta_m) \exp[-j\beta_m x]}{k_0^2 - \beta_m^2}, \quad (5.4)$$

where

$$\beta_m = \kappa + 2\pi m/p. \quad (5.5)$$

Thus $\phi_a(x, \omega)$ has the form of a Floquet expansion.

As in Section 4.3.3, the electrode current I_n can be written

$$I_n = I_{en} + I_{an}, \quad (5.6)$$

where I_{en} and I_{an} are respectively the electrostatic and acoustic contributions. The latter is given by equation (4.43), and since $\varrho_{en}(x) = \varrho_f(x - np)$ we have

$$I_{an} = -j\omega W \int_{-\infty}^{\infty} \varrho_f(x - np) \phi_a(x, \omega) \, dx. \quad (5.7)$$

Substituting equation (5.4) for $\phi_a(x, \omega)$ we have

$$I_{an} = -2j\omega W V_n \frac{\Gamma_s k_0}{p} \sum_{m=-\infty}^{\infty} \frac{[\bar{\varrho}_f(\beta_m)]^2}{k_0^2 - \beta_m^2}. \quad (5.8)$$

The electrostatic contribution, I_{en}, is obtained by integrating $\sigma_e(x)$ over the area of electrode n and differentiating with respect to time. This can be expressed in terms of Q_m, the net electrode charges corresponding to the elemental charge density $\varrho_f(x)$, defined in equation (4.89). For voltages V_n on the electrodes, the net charge on electrode n is q_n, given by

$$q_n = W \sum_{m=-\infty}^{\infty} Q_{m-n} V_m = W V_0 \sum_{m=-\infty}^{\infty} Q_{m-n} e^{-j\kappa mp}. \quad (5.9)$$

It is convenient to define a function $C(\kappa)$ by

$$C(\kappa) = \sum_{m=-\infty}^{\infty} Q_m e^{-j\kappa mp} \quad (5.10)$$

so that the electrostatic term is

$$I_{en} = j\omega q_n = j\omega W V_n C(\kappa). \quad (5.11)$$

The function $C(\kappa)$ is derived in Appendix C, equation (C.31), giving

$$C(\kappa) = 2(\varepsilon_0 + \varepsilon_p^T) \frac{\sin(\pi s)}{P_{-s}(-\cos \Delta)} P_{-s}(\cos \Delta), \quad (5.12)$$

where $\Delta = \pi a/p$, and s is defined such that $\kappa p = 2\pi(s + i)$, where i is an integer and $0 \leq s \leq 1$. The total current is the sum of equations (5.8) and (5.11), so that

$$\frac{I_n}{V_n} = j\omega W \left[C(\kappa) - \frac{2\Gamma_s k_0}{p} \sum_{m=-\infty}^{\infty} \frac{[\bar{\varrho}_f(\beta_m)]^2}{k_0^2 - \beta_m^2} \right]. \quad (5.13)$$

It is convenient to define also ϕ_n as the value of the acoustic potential $\phi_a(x, \omega)$ at $x = np$, and from equation (5.4),

$$\frac{\phi_n}{V_n} = \frac{2\Gamma_s k_0}{p} \sum_{m=-\infty}^{\infty} \frac{\bar{\varrho}_f(\beta_m)}{k_0^2 - \beta_m^2}, \quad (5.14)$$

where $\beta_m = \kappa + 2\pi m/p$.

If $\kappa + 2\pi M/p = \pm k_0$ for some M, we have $\beta_M = \pm k_0$, so that equations (5.13) and (5.14) are both infinite. This case was excluded earlier, but a solution can be obtained by a limiting process, taking the voltages V_n to be zero. For β_M close to $\pm k_0$ the summations are well approximated by omitting all terms except for $m = M$. The ratio of the two equations thus gives the limit

$$I_n = -j\omega W \bar{\varrho}_f(\pm k_0)\phi_n, \quad \text{for } V_n = 0, \quad (5.15)$$

when $\kappa + 2\pi M/p = \pm k_0$. We thus have a solution for all κ. For $V_n = 0$ the solution is consistent with the analysis for a shorted interdigital transducer receiving surface waves, and may be obtained from equation (4.66) of Section 4.4.

For subsequent analysis it is convenient to use an approximate form of the solution. Since the solution is unaffected when $2\pi/p$ is added to κ, we may restrict κ such that the integer parts of $\kappa p/(2\pi)$ and $k_0 p/(2\pi)$ are equal. Thus, for integer i,

$$2\pi i/p \leq k_0 \leq 2\pi(i + 1)/p, \quad (5.16)$$

and

$$2\pi i/p \leq \kappa \leq 2\pi(i + 1)/p. \quad (5.17)$$

With this restriction, the summation in equation (5.13) is significant only when κ is close to k_0, and when this is the case the term with $m = 0$ dominates. We may therefore omit the other terms, so that

$$\frac{I_n}{V_n} \approx j\omega W \left[C(\kappa) - \frac{2\Gamma_s k_0}{p} \frac{[\bar{\varrho}_f(\kappa)]^2}{k_0^2 - \kappa^2} \right]. \tag{5.18}$$

Similarly, equation (5.14) becomes

$$\frac{\phi_n}{V_n} \approx \frac{2\Gamma_s k_0}{p} \frac{\bar{\varrho}_f(\kappa)}{k_0^2 - \kappa^2}. \tag{5.19}$$

The potential ϕ_n can be interpreted as the surface wave amplitude at $x = np$. This is justified by the observation that $\phi_a(x, \omega)$ is proportional to $\exp(\pm jk_0 x)$ on all unmetallised regions, and therefore gives the surface wave potential in such regions. In the cases of interest here, $\phi_a(x, \omega)$ varies slowly across the width of any one electrode, and hence $\phi_n = \phi_a(np, \omega)$ can be taken as the surface wave amplitude at $x = np$.

Propagation in Shorted or Open-circuited Electrodes The wavenumbers for these two cases are particularly important in the analysis of multi-strip couplers. For shorted electrodes the voltages V_n are zero, and hence equation (5.18) gives $\kappa = k_0$, the wavenumber for propagation on a free surface. This solution was found previously, giving the currents shown by equation (5.15). It is convenient to denote the wavenumber for shorted electrodes by k_{sc}, so that

$$k_{sc} = k_0. \tag{5.20}$$

For open-circuited electrodes the currents I_n are zero. In this case the solution for κ is denoted k_{oc}, and from equation (5.18) we have

$$k_{sc}^2 - k_{oc}^2 = \frac{2\Gamma_s k_0}{p} \frac{[\bar{\varrho}_f(k_{oc})]^2}{C(k_{oc})}. \tag{5.21}$$

Here $\bar{\varrho}_f(k_{oc})$ is given by equation (4.88) of Section 4.5.1, and $C(k_{oc})$ is given by equation (5.12). These functions both vary slowly with k_{oc}. Since k_{oc} must be close to k_0 it is a good approximation to replace k_{oc} by k_0 in these functions. We also have, from equation (3.39) of Chapter 3,

$$\Gamma_s = \frac{1}{\varepsilon_0 + \varepsilon_p^T} \frac{\Delta v}{v} = \frac{1}{\varepsilon_0 + \varepsilon_p^T} \frac{v_0 - v_m}{v_0}, \tag{5.22}$$

where v_0 and v_m are respectively the surface wave velocities for a free surface and for a metallised surface. Equation (5.21) thus gives

$$k_{sc}^2 - k_{oc}^2 = \frac{4k_0}{p} \frac{\Delta v}{v} \frac{\sin(\pi s)}{P_{-s}(-\cos \Delta) P_{-s}(\cos \Delta)} [P_i(\cos \Delta)]^2 \tag{5.23}$$

where i is defined by equation (5.16) and $s = k_0 p/(2\pi) - i$.

Equation (5.23) is in agreement with the more rigorous analysis using Floquet's theorem, as shown by equation (D.31) of Appendix D, and also with the results of perturbation theory [118]. However, the theory here predicts that $k_{sc} = k_0$, and does not therefore allow for the perturbation of the surface wave velocity by an array of shorted electrodes. This is one of the consequences of the quasi-static approximation used in the analysis. In practical devices the precise value of k_{sc} is of little consequence, so the above relationships are adequate. A more accurate value for k_{sc} is given in Appendix D, equation (D.26). The above analysis also fails to predict the presence of stop bands. These occur when k_0 is close to multiples of π/p, as shown in Section D.3.

5.2. BASIC COUPLER BEHAVIOUR

In its simplest form the multi-strip coupler is an array of disconnected regular electrodes spanning two identical tracks. In each track the surface wave amplitude is independent of y, but the amplitudes in the two tracks will generally be different. This is illustrated in Figure 5.2. The structure is first analysed as if the number of strips were infinite, and then the results are applied to a finite structure. The pitch p and width a of the electrodes is the same in both tracks, and the two tracks have the same aperture, W.

The electrode voltages and currents are denoted by V_n^a and I_n^a for track A, and by V_n^b and I_n^b for track B. The voltages are the same in the two tracks, and we consider a solution in which they take the form

$$V_n^a = V_n^b = V_0 \exp(-j\kappa np), \tag{5.24}$$

where the value of κ is taken to be the same for both tracks in view of the restriction imposed by equation (5.17). In addition, continuity of the currents requires

$$I_n^a = -I_n^b. \tag{5.25}$$

These equations give two solutions for propagating waves. The first solution, known as the *symmetric mode*, has the acoustic potentials ϕ_n the same in both tracks. This is the same as the solution for open-circuit electrodes already given in Section 5.1, and therefore gives $\kappa = k_{oc}$. This solution can also be obtained from equations (5.18) and (5.25), which give $I_n^a = I_n^b = 0$.

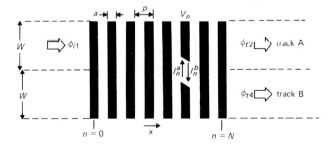

FIGURE 5.2. Simple multi-strip coupler.

The second solution, known as the *antisymmetric mode*, is obtained when the electrode voltages V_n^a and V_n^b are zero. Physically, this is to be expected when the acoustic potentials in the two tracks are equal and opposite. Since the voltages are zero each track will support propagation as if its electrodes were all connected, so that the wavenumber must be $\kappa = k_{sc}$. This solution also follows from equation (5.18), which shows that V_n/I_n is zero when $\kappa = k_0$, which in turn equals k_{sc}. In this case the acoustic potentials are related to the currents by equation (5.15). Denoting the potentials in tracks A and B by ϕ_n^a and ϕ_n^b, we have $\phi_n^a = -\phi_n^b$, since $I_n^a = -I_n^b$.

In order to evaluate the transmission of incident waves by a coupler, it is necessary to consider a more general solution obtained by adding the symmetric and antisymmetric modes. It is sufficient here to take the two modes to have equal amplitudes, denoted by A. The acoustic potentials for the symmetric mode are written

$$\phi_n^a = \phi_n^b = A \exp(-jk_{oc}np)$$

and, for the antisymmetric mode,

$$\phi_n^a = -\phi_n^b = A \exp(-jk_{sc}np).$$

Adding the two modes, the total acoustic potentials are, in track A,

$$\phi_n^a = 2A \cos(\tfrac{1}{2}\pi n/N_c) \exp(-jknp), \tag{5.26}$$

and in track B,

$$\phi_n^b = 2jA \sin(\tfrac{1}{2}\pi n/N_c) \exp(-jknp), \tag{5.27}$$

where $k = (k_{sc} + k_{oc})/2$. The parameter N_c is a function of ω, given by

$$pN_c(\omega) = \pi/(k_{sc} - k_{oc}).$$

Equations (5.26) and (5.27) show that for $n = 0$ the surface wave power is entirely in track A, while for $n = N_c$ the power is entirely in track B. Thus N_c is the number of electrodes required to transfer the surface wave power from one track to the other. N_c may be evaluated using equation (5.23), noting that k_{sc} and k_{oc} are both close to k_0, so that $k_{sc} + k_{oc} \approx 2k_0$. This gives

$$[N_c(\omega)]^{-1} = \frac{\Delta v}{v} \frac{2 \sin(\pi s)}{\pi P_{-s}(-\cos \Delta) \cdot P_{-s}(\cos \Delta)} [P_i(\cos \Delta)]^2, \tag{5.28}$$

where $\Delta v/v$ is the piezoelectric coupling parameter [equation (5.22)], $\Delta = \pi a/p$, i is defined by equation (5.16) and $s = k_0 p/(2\pi) - i$. Thus N_c is inversely proportional to $\Delta v/v$.

Figure 5.3 shows that $N_c(\omega)$ varies slowly with frequency, and has a minimum when $k_0 = \pi/p$, that is, when $\omega = \pi v_0/p$. However, this frequency corresponds to one of the stop bands, and is not therefore used in practice. A typical operating frequency is $\omega = 3\pi v_0/(4p)$. For a metallisation ratio $a/p = \tfrac{1}{2}$, equation (5.28) gives $N_c = 2.32/(\Delta v/v)$ at this frequency. Thus, for Y, Z lithium niobate, with $\Delta v/v = 2.15\%$, we have $N_c = 108$. At the operating frequency the electrode width is $3\lambda/16$, where λ is the wavelength, and this is somewhat less than the electrode width of $\lambda/4$ for a single-electrode transducer at the same frequency.

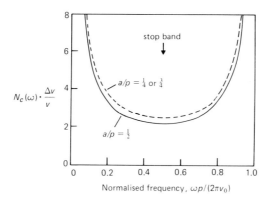

FIGURE 5.3. Number of electrodes for full transfer, as a function of frequency.

For a coupler of finite length, as in Figure 5.2, it is assumed that end effects are negligible, so that the above equations apply directly. The input and output wave amplitudes are measured at the centres of the first and last electrodes. It is assumed that a surface wave is incident in track A, and it is convenient to take $n = 0$ for the first electrode, since equation (5.27) gives $\phi_n^b = 0$ for $n = 0$, as required. The input wave amplitude, denoted ϕ_{i1}, is given by equation (5.26) with $n = 0$. The output wave amplitudes are denoted by ϕ_{t2} for track A and ϕ_{t4} for track B. Taking the number of electrodes to be $N + 1$, these are respectively given by equations (5.26) and (5.27) with $n = N$. It is convenient to relate the output amplitudes to the input amplitude by defining scattering coefficients S_{12} and S_{14}, so that

$$S_{12} \equiv \phi_{t2}/\phi_{i1} = \cos\left(\tfrac{1}{2}\pi N/N_c\right) \exp\left(-jkNp\right), \qquad (5.29)$$

$$S_{14} \equiv \phi_{t4}/\phi_{i1} = j \sin\left(\tfrac{1}{2}\pi N/N_c\right) \exp\left(-jkNp\right). \qquad (5.30)$$

The magnitudes of these functions are shown in Figure 5.4, for $a/p = \tfrac{1}{2}$. The solid

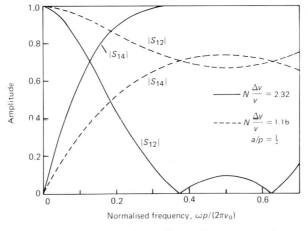

FIGURE 5.4 Scattering coefficients for simple couplers.

lines are drawn up for $N = 2.32/(\Delta v/v)$. This is equal to N_c when $\omega = 3\pi v_0/(4p)$, so all of the power emerges on track B at this frequency. In fact, the power transferred to track B varies little over a wide frequency range. Thus the coupler has a wide bandwidth, in contrast to a transducer with a comparable number of electrodes. The broken lines on Figure 5.4 refer to a coupler with half the number of electrodes, that is, with $N = 1.16/(\Delta v/v)$. In this case half of the power is transferred to track B at frequency $\omega = 3\pi v_0/(4p)$, so at this frequency the output amplitudes on the two tracks are the same. Such a coupler has a number of applications, discussed below, and is called a 3 dB coupler.

Coupler with Dissimilar Track Widths. The above analysis shows that a simple coupler can transfer all the surface-wave power from one track to the other if the appropriate number of electrodes is used. However, this is not true if the tracks have unequal widths. To illustrate this, we consider the coupler shown in Figure 5.5, where there are now three tracks. The electrodes in the outer tracks, tracks A and B, have the same pitch p and width a. Track C, in the centre, has angled electrodes so that the pitch, measured perpendicular to the electrodes, is different. This suppresses coupling to surface waves in track C, and is often used in practical devices in order to separate the active tracks and thus reduce cross-coupling due to diffraction spreading.

As before the analysis first assumes an infinite number of electrodes, and the results are then applied to a finite device. The electrode voltages in all three tracks are $V_n = V_0 \exp(-j\kappa np)$, and continuity requires that the sum of the currents entering electrode n in the three tracks is zero. For tracks A and B the currents are given by equation (5.18), with W replaced by the track widths W_a or W_b, as appropriate. For track C there is no acoustic coupling, so the electrode currents I_n^c are given by

$$I_n^c/V_n = j\omega W_c C(\kappa), \qquad (5.31)$$

where W_c is the electrode length in track C, as in Figure 5.5. The function $C(\kappa)$, given by equation (5.12), involves the constant ε_p^T, and this may have a value somewhat different to its value for the other tracks because of the anisotropy of the substrate.

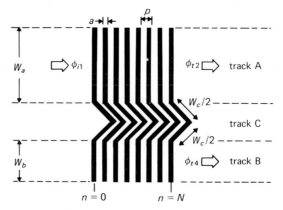

FIGURE 5.5. Coupler with unequal track widths.

However this may be allowed for simply by modifying the value of W_c in the equations.

As for the simple coupler described previously, it is found that there are two solutions for wave propagation in the infinite structure. One solution is obtained when the voltages V_n are zero, and from equation (5.18) this gives $\kappa = k_0 = k_{sc}$. The surface-wave potentials in tracks A and B are found using equation (5.15), and are related by $\phi_n^a W_a = -\phi_n^b W_b$. The other solution follows from equations (5.18) and (5.31), and gives a new solution for κ, here denoted by k_1. For this solution, $\phi_n^a = \phi_n^b$. For the finite structure of Figure 5.5 the two solutions are added, with one multiplied by a coefficient such that the total ϕ_n^b is zero for $n = 0$. Taking the number of electrodes to be $N + 1$, the output amplitude on track B is $\phi_{t4} = \phi_N^b$, and with $\phi_{i1} = \phi_0^a$ this is found to be given by

$$S_{14} \equiv \phi_{t4}/\phi_{i1} = \frac{2jW_a}{W_a + W_b} \sin(\tfrac{1}{2}\pi N/N_c') \exp(-jkNp), \qquad (5.32)$$

where $k = (k_1 + k_{sc})/2$ and N_c' is given by

$$N_c'(\omega) = \frac{\pi}{p(k_{sc} - k_1)} = N_c(\omega) \frac{W_a + W_b + W_c}{W_a + W_b}, \qquad (5.33)$$

where N_c is the number of electrodes for a full power transfer in a simpler coupler, given by equation (5.28). The surface-wave power in track B is thus maximised if the number of electrodes is equal to N_c'. However, not all the incident power is transferred to track B. From equation (3.34) of Chapter 3, the ratio of the output power on track B to the input power is $(W_b/W_a)|\phi_{t4}/\phi_{i1}|^2$, and equation (5.32) shows that this can be equal to unity only if $W_a = W_b$.

The amplitude of the output wave on track A is a more complicated expression, but for the particular case $W_a = W_b$ is given by

$$S_{12} \equiv \phi_{t2}/\phi_{i1} = \cos(\tfrac{1}{2}\pi N/N_c') \exp(-jkNp). \qquad (5.34)$$

Thus, if tracks A and B have equal widths the coupler behaves in the same way as the simple coupler of Figure 5.2. A complete transfer of power from track A to track B is obtained if $N = N_c'$. In most practical cases W_c is smaller than W_a or W_b, so that N_c', given by equation (5.33), is not substantially larger than N_c.

Second-order Effects. Provided the frequency is not in or near any of the coupler stop bands, the above equations are found to agree well with experimental measurements [116, 121]. Some loss of power due to the resistivity of the electrodes is observed, and this was studied in detail by Maerfeld et al. [117, 118]. For a simple coupler (Figure 5.2), designed to transfer all the power from one track to the other, it was found that the loss is about 1 dB if the sheet resistivity of the electrodes (in ohms per square) is equal to $2500 \lambda^2/W^2$, where λ is the wavelength. This formula assumes a Y, Z lithium niobate substrate, with $a/p = \tfrac{1}{2}$ and a frequency $\omega = 0.8\pi v_0/p$. A typical resistivity is 0.25 ohms per square, giving a 1 dB loss when the track width W is equal to 100 wavelengths. This is often the main cause of loss in practice. In addition to the loss, the resistivity causes the output beam to have a variation of amplitude across the width of the track, though this is usually of little consequence if the loss is 1 dB or less.

Another possible cause of loss is the generation of bulk waves. From considerations of phase matching, it can be deduced that significant excitation of bulk waves can only occur for radian frequencies higher than $\pi v_0/p$. In practice the coupler is usually designed for operation at frequencies below this value, so bulk wave excitation is not significant. Further discussion is given in Appendix F.

Diffraction of the surface waves becomes relevant if the width of either track is only a few wavelengths. In this case the amplitude may vary with y in each track. The velocities of all the modes described above are less than the free-surface velocity v_0, and consequently the coupler has some characteristics in common with a waveguide. Experimental work [122] shows that the wave amplitude, as a function of y, has irregularities, and these can be predicted theoretically [123]. Diffraction can generally be ignored if the track widths are greater than 10 wavelengths.

For an input on track A, the coupler also produces reflected waves on both tracks. These will be ignored in this chapter. Some discussion is given in Appendix E, which shows that the reflections are generally negligible if the frequency is not close to one of the stop bands.

5.3. INTERDIGITAL DEVICES USING COUPLERS

In this section we consider the response of a device comprising two interdigital transducers and a multi-strip coupler, as in Figure 5.6. It will be shown that the device response is essentially the product of the two transducer responses, even if both transducers are apodised [124–126]. This feature is of considerable value in the design of sophisticated devices. Another important advantage is that the coupler reduces unwanted output signals due to the excitation of bulk waves by the input transducer; this is because any bulk wave incident in one track is coupled to the other track much less efficiently than a surface wave. Although this is clearly demonstrated

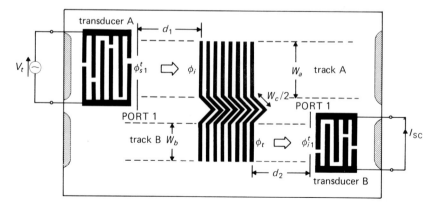

FIGURE 5.6. Interdigital device using a multi-strip coupler.

experimentally [126] it appears that a quantitative analysis has not been given. However, since bulk waves, observed at the surface, generally exhibit attenuation due to radiation into the bulk, it is clearly unlikely that the coupler could transfer bulk wave power efficiently from one track to another; furthermore, in practical devices the coupler is usually designed to optimise the coupling efficiency for surface waves.

For the analysis here, bulk wave excitation is ignored. As shown in Figure 5.6, the active region of each transducer is assumed to be confined within the appropriate track of the coupler, and the transducer apertures are taken to be equal to the coupler track widths W_a and W_b. It is assumed initially that transducer A is connected to a source with zero impedance, and transducer B is shorted. In this situation, the potential ϕ_i of the wave incident on the coupler in track A is a function of y, because transducer A is apodised. We show here that the amplitude of the output wave in track B can be found straightforwardly using superposition. Suppose first that a narrow beam of width Δy is incident, with amplitude $\phi_i(y_1)$ at the coupler input, where y_1 is the location of the centre of the beam. Assuming diffraction and electrode resistance to be negligible, the corresponding output wave must have an amplitude uniform across track B and its potential ϕ_t at the coupler output must be proportional to $\phi_i(y_1)$, so that

$$\phi_t = K\phi_i(y_1),$$

where the constant K is to be determined. Since electrode resistance is negligible the output wave amplitude is independent of the location y_1 of the input beam. Thus for a second input beam, centred at y_2 and with amplitude $\phi_i(y_2)$, the output amplitude is $\phi_t = K\phi_i(y_2)$. The total wave incident on the coupler is represented by J elementary beams, each of width Δy, and the total output wave is therefore, by superposition,

$$\phi_t = \sum_{j=1}^{J} K\phi_i(y_j).$$

This equation is still valid if all the $\phi_i(y_j)$ are equal, in which case $\phi_t = S_{14}\phi_i$, and since $J = W_a/\Delta y$ we conclude that $K = S_{14}\Delta y/W_a$. Taking the limit as $\Delta y \to 0$ we have

$$\phi_t = S_{14} \int_a \frac{\phi_i(y)}{W_a} dy, \qquad (5.35)$$

where S_{14} is given by equation (5.32) and the integral is taken over the width of track A. Thus the amplitude of the output beam is proportional to the average amplitude of the input beam.

The response of the overall device is now found with the aid of the transducer analysis of Chapter 4. The frequency responses of transducers A and B are respectively $H_t^a(\omega)$ and $H_t^b(\omega)$, both given by equation (4.128). In both cases, port 1 of the transducer is taken as the port nearest to the coupler. When a voltage V_t is applied to transducer A, the surface wave generated has potential $\phi'_{s1}(y)$ at port 1, where the superscript t indicates a potential at one of the transducer ports. From the analysis of Section 4.7.2 we have

$$\int_a \phi_{s1}^t(y)\,dy = jV_tW_aH_t^a(\omega)\left[\frac{\Gamma_s}{\omega W_a}\right]^{1/2}. \tag{5.36}$$

For transducer B, if a uniform surface wave beam of width W_b is incident, with potential ϕ_{i1}^t at port 1, the short-circuit output current is

$$I_{sc} = -j\phi_{i1}^t[\omega W_b/\Gamma_s]^{1/2}H_t^b(\omega). \tag{5.37}$$

Taking d_1 as the separation between transducer A and the coupler, and d_2 as the separation between transducer B and the coupler, the above equations give the overall device response as

$$H_{sc}(\omega) \equiv I_{sc}/V_t = H_t^a(\omega)H_t^b(\omega)S_{14}(W_b/W_a)^{1/2}\exp[-jk_0(d_1+d_2)], \tag{5.38}$$

which is essentially the product of the two transducer responses and the coupler response S_{14}, which varies slowly with frequency. This approach is valid only if reflected waves are insignificant, that is, if electrode interactions in the transducers are negligible, and if the frequency is not close to one of the coupler stop bands.

For finite source and load impedances the frequency response is modified somewhat by the circuit effect and the transducers reflect the waves, giving a triple-transit spurious signal. These features can be calculated using the analysis of Section 4.8, which is sufficiently general to allow for the presence of the coupler. Thus the main response of the device is given by equation (4.137), and the triple-transit response is given by equation (4.147); in both cases, equation (5.38) is used for the short-circuit response $H_{sc}(\omega)$.

5.4. UNIDIRECTIONAL TRANSDUCER

A 3 dB coupler may be used to modify a conventional bidirectional transducer in such a way that it becomes essentially unidirectional. As shown in Figure 5.7(a), the

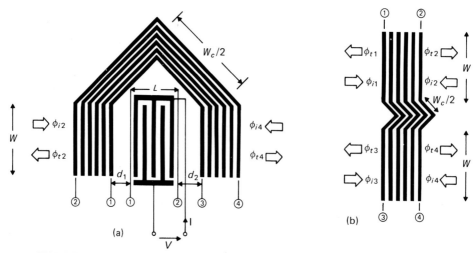

FIGURE 5.7. (a) Unidirectional transducer. (b) Coupler parameters for scattering matrix.

coupler is folded into a "U"-shape, and a symmetrical unapodised transducer is located between the parallel arms of the coupler. When a voltage is applied, the surface wave leaving the coupler at the right is the sum of two components, arising from the waves launched to left and right by the transducer. These two components cancel if the coupler provides 3 dB coupling and if the transducer is displaced slightly to the left, so that the waves arrive at the coupler inputs in phase quadrature. Thus, surface waves emerge only at the left. By reciprocity, a wave incident from the left will not be reflected provided the transducer is matched, and this feature is useful for minimising the triple-transit spurious signal in surface wave devices. However, the behaviour is strictly unidirectional only at one particular frequency.

For the analysis it is convenient to define a scattering matrix for the coupler. As shown in Figure 5.7(b), we consider a coupler whose acoustically-coupled tracks have equal apertures W, and denote the four inputs as ports 1 to 4. The potentials of incident surface waves are denoted by ϕ_i and those of surface waves leaving the coupler by ϕ_t. In each case these are also subscripted 1, 2, 3 or 4 to identify the port. For a wave incident from the left in one track, the output waves are given by equations (5.32) and (5.34), while for other incident waves the outputs are obtained by symmetry. The scattering matrix may therefore be written as

$$\begin{bmatrix} \phi_{t1} \\ \phi_{t2} \\ \phi_{t3} \\ \phi_{t4} \end{bmatrix} = \begin{bmatrix} 0 & S_{12} & 0 & S_{14} \\ S_{12} & 0 & S_{14} & 0 \\ 0 & S_{14} & 0 & S_{12} \\ S_{14} & 0 & S_{12} & 0 \end{bmatrix} \begin{bmatrix} \phi_{i1} \\ \phi_{i2} \\ \phi_{i3} \\ \phi_{i4} \end{bmatrix}, \quad (5.39)$$

where the coefficients are

$$S_{12} = \cos(\tfrac{1}{2}\pi N/N_c') \exp(-jkNp), \quad (5.40)$$

$$S_{14} = j \sin(\tfrac{1}{2}\pi N/N_c') \exp(-jkNp). \quad (5.41)$$

Here N is the number of electrodes, N_c' is given by equation (5.33), and $k = (k_1 + k_{sc})/2$. In equation (5.39) it is assumed that reflected waves can be ignored so that, for example $S_{11} = S_{31} = 0$. This is a reasonable assumption provided the frequency is not in or near one of the coupler stop bands. Strictly speaking, the analysis of the coupler in the unidirectional transducer, Figure 5.7(a), should allow for the fact that the acoustically inactive parts of the electrodes vary in length. However this is not found to change its behaviour appreciably, so the above scattering matrix can be applied, taking W_c as the average of the electrode lengths in the inactive region.

The transducer behaviour is described by the scattering matrix already given in Section 4.4, equation (4.79). To distinguish this from the coupler scattering matrix, the elements of the transducer matrix are here written as S_{ij}^t. Since the transducer is symmetrical, its electrostatic charge density $\varrho_e(x)$ is also symmetrical, and it follows that $\bar{\varrho}_e(k_0)$ is real. We thus have $S_{13}^t = S_{23}^t = -S_{31}^t = -S_{32}^t$. The transducer scattering matrix is combined with the coupler scattering matrix, equation (5.39), to give a scattering matrix for the unidirectional transducer. This is denoted by S_{ij}', and defined by

$$\begin{bmatrix} \phi_{t2} \\ \phi_{t4} \\ I/(\omega W) \end{bmatrix} = \begin{bmatrix} S'_{11} & S'_{12} & S'_{13} \\ S'_{21} & S'_{22} & S'_{23} \\ S'_{31} & S'_{32} & S'_{33} \end{bmatrix} \begin{bmatrix} \phi_{i2} \\ \phi_{i4} \\ V\Gamma_s \end{bmatrix}, \quad (5.42)$$

where I and V are the transducer current and voltage, as shown on Figure 5.7(a). Defining d_1 as the distance between the coupler port 1 and the transducer port 1, and d_2 as the distance between the coupler port 3 and the transducer port 2, the matrix elements in equation (5.42) are found to be

$$S'_{11} = S'_{22} = 2S'_{12} S_{12} S_{14} \exp\left[-jk_0(d_1 + d_2)\right]$$
$$S'_{12} = S'_{21} = S'_{12}(S_{12}^2 + S_{14}^2) \exp\left[-jk_0(d_1 + d_2)\right],$$
$$S'_{13} = -S'_{31} = [S_{12} \exp(-jk_0 d_1) + S_{14} \exp(-jk_0 d_2)]S'_{13},$$
$$S'_{23} = -S'_{32} = [S_{14} \exp(-jk_0 d_1) + S_{12} \exp(-jk_0 d_2)]S'_{13},$$
$$S'_{33} = S'_{33} = Y_t(\omega)/(\omega W \Gamma_s), \quad (5.43)$$

where $Y_t(\omega)$ is the admittance of the transducer in the absence of the coupler.

The coupler is designed to give 3 dB coupling at a frequency ω_c, which is the transducer centre frequency. At this frequency $N = \tfrac{1}{2} N'_c$, and $S_{14} = jS_{12}$. The transducer is displaced to the left by a small precise amount, such that $(d_2 - d_1)$ equals one quarter-wavelength at frequency ω_c, and hence $\exp[-jk_0(d_1 - d_2)] = j$. It follows that at this frequency S'_{23} is zero, and hence when a voltage is applied to the transducer there is no wave emerging at the right. It is also found that for a wave of frequency ω_c incident on port 2 there is no reflected wave if the transducer is electrically matched, that is, if it is connected to a load whose admittance is equal to $Y_t^*(\omega_c)$, the conjugate of the transducer admittance.

Figure 5.8 shows the reflection coefficient for surface waves incident on port 2 of the coupler, and the conversion coefficient for waves leaving port 2. The latter is the

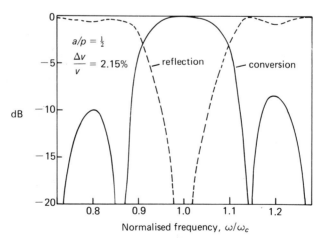

FIGURE 5.8. Reflection and conversion coefficients for a unidirectional transducer.

ratio of the surface wave power to the power available from a generator connected to the transducer. The transducer is taken to be of the uniform single-electrode ($S_e = 2$) type, with $N_p = 7$ periods, and its admittance is given by equations in Section 4.6. The electrical load is, as in common practice, a series inductor followed by a fixed resistance, with values chosen to match the transducer impedance at the centre frequency ω_c. The coupler is designed to give 3 dB coupling at frequency ω_c, and its electrodes have pitch p such that $\omega_c = 3\pi v_0/(4p)$. It can be seen that at the transducer centre frequency ω_c the conversion coefficient is 0 dB, and incident waves are not reflected. In contrast, a matched bidirectional transducer theoretically gives a conversion coefficient of 3 dB and a reflection coefficient of 6 dB, as shown in Section 4.4.4. Thus the unidirectional transducer is much superior at the centre frequency. On the other hand, the reflection coefficient is quite large at the edges of the transducer pass-band, for reasons discussed below. Thus the unidirectional transducer is best suited for applications where the signal bandwidth is less than the transducer bandwidth.

Experimental measurements give results similar to Figure 5.8, though generally with a few dB's of loss due mainly to the resistance of the electrodes in the transducer and in the coupler. For example [126] a transducer centred at 61 MHz gave a minimum conversion loss of 1.5 dB, and a delay line comprising two such transducers had an insertion loss of 3 dB.

5.5. OTHER APPLICATIONS OF 3 dB COUPLERS

Two further applications of 3 dB couplers are shown in Figure 5.9. On the left is a multi-strip *mirror*, an efficient reflector of surface waves. This comprises a 3 dB coupler folded into a "U"-shape, and is thus similar to a unidirectional transducer [Figure 5.7(a)] except that the transducer has been omitted. The reflection coefficient at port 2 can be obtained from the coupler scattering matrix of equation (5.39); alternatively, it can be obtained from the scattering matrix of a unidirectional transducer, equation

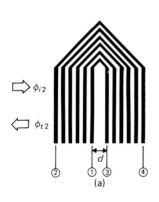
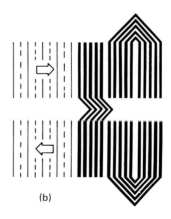

FIGURE 5.9. (a) Multi-strip mirror, (b) reflecting track-changer.

(5.42), setting $V = 0$. Thus, if there are no waves incident on port 4 the mirror gives

$$\phi_{t2}/\phi_{i2} = S'_{11} = j \sin(\pi N/N'_c) \exp[-j(2kNp + k_0 d)], \quad (5.44)$$

where d is the distance between ports 1 and 3, as in Figure 5.9(a), N is the number of electrodes and N'_c is given by equation (5.33). For a 3 dB coupler we have $N = \frac{1}{2}N'_c$ so that the magnitude of the reflection coefficient is unity. Further, N'_c is almost independent of frequency, so that the reflection coefficient is close to unity over a wide band. Experimental results [126] centred at 80 MHz demonstrate a reflection loss constant within about $\frac{1}{4}$ dB over a bandwidth exceeding 60%; however the loss is about 2 dB, which can be attributed to electrode resistance [118]. The operation of the mirror explains why the reflection coefficient of a unidirectional transducer is close to unity except near the centre frequency, as seen on Figure 5.8. For frequencies remote from the centre frequency the transducer has little effect on any incident surface wave, and the unidirectional transducer thus behaves like a mirror, enhancing the reflection coefficient instead of reducing it.

Figure 5.9(b) shows a *reflecting track-changer* which, in response to an input surface wave beam, generates an oppositely directed beam on an adjacent track. This device has a straight 3 dB coupler followed by two mirrors. The straight coupler partitions the input wave equally between the two mirrors, and the reflected waves pass through the straight coupler for a second time. Apart from the change of direction this process is equivalent to passage through two 3 dB couplers in cascade, and hence ideally all the power emerges in the output track. As in the case of the mirror alone, the output power varies very slowly with frequency. Experimentally [126] a loss of 2 to 3 dB was obtained, over a bandwidth exceeding 50%. The track-changer may be used in a surface-wave delay line to increase the delay obtainable for a given length of substrate. In fact, by using track-changers at both ends of the substrate the surface wave may be made to traverse the substrate length several times. A device with eight track-changers gave 130 μsec delay and 23 dB insertion loss [127].

Several other multistrip devices are described by Marshall *et al.* [126], including an "echo trap". This reduces triple-transit signals more effectively than a unidirectional transducer, but gives somewhat greater losses. Maerfeld *et al.* [117] investigated the effect of electrode resistance in a variety of devices, including novel types of trackchanger. Chapman and Bristol [128] have shown that the performance of some of these devices may be improved if slightly different pitches are used for the electrodes in the two tracks.

5.6. BANDPASS FILTERING USING MULTI-STRIP COUPLERS

The multi-strip devices described so far have wide bandwidths. There are however several methods of obtaining a narrower bandwidth, with potential application to bandpass filtering. In a technique introduced by Solie [129] the electrodes in one track are displaced in groups in the direction of surface wave propagation, as shown in Figure 5.10(a). The displacements introduce phase shifts in the waves launched in the output track by individual groups of electrodes. Denoting the displacements by d_i, as in Figure 5.10(a), the overall response can be written approximately as

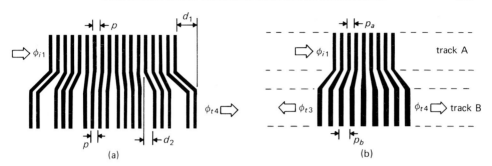

FIGURE 5.10. Narrow-band multi-strip couplers for bandpass filtering.

$$\phi_{t4}/\phi_{i1} \approx \sum_i A_i \exp[jk_0(d_i - Np)], \qquad (5.45)$$

where the coefficients A_i give the amplitudes of the waves generated by individual electrode groups. It is a reasonable approximation to take the A_i to be independent of frequency, in which case equation (5.45) has the form of the response of a transversal filter. Standard techniques, described in Chapter 8, may then be used to deduce appropriate values for the A_i and d_i, such that the response approximates some required response. In practice, this method is somewhat inflexible because the experimental A_i-values are controlled only by varying the numbers of electrodes in individual groups, and the accuracy is affected by end effects. However this disadvantage is offset by the fact that the coupler can be combined with weighted transducers, and in addition the loss can be very low. Solie [129] instances a coupler with a 3 dB bandwidth of 9 MHz centred at 213 MHz, with 3 dB loss at the centre frequency and 30 dB out-of-band rejection.

Another technique for narrow bandwidths is the use of different electrode pitches in the two tracks, as shown in Figure 5.10(b). In this structure, strong coupling between propagating surface waves in the two tracks can occur only if their phases, observed at the electrode locations, are equal or differ by a multiple of 2π. Taking p_a and p_b as the electrode pitches in tracks A and B respectively, and noting that the wavenumber must be close to k_0, the phase matching condition is

$$p_b \pm p_a \approx 2\pi m/k_0, \qquad (5.46)$$

where m is an integer. The upper sign applies for waves propagating in opposite directions in the two tracks, and the lower sign for waves propagating in the same direction. For a wave incident in track A, the response observed in track B is a narrow-band function, with bandwidth related to the number of electrodes and with centre frequency given by equation (5.46). For either output port, the amplitude, as a function of frequency, is observed [129] to have the form of the function $(\sin x)/x$. Several methods of introducing weighting are described by Feldmann and Henaff [130]. In particular, the electrodes in track B may be fanned out so that the pitch varies across the track, increasing the bandwidth over which efficient coupling occurs. In this case waves are generated in track B at locations dependent on the frequency. An array of output transducers may be used, so that different frequency components emerge from different output transducers, and the device is then a filter bank. A device with five outputs, each with about 2 MHz bandwidth, was demonstrated. Another type of filter bank, using 3 dB couplers, is described in Chapter 8.

5.7. BEAM COMPRESSION

Another modification of the basic coupler geometry can be introduced in order to couple efficiently two tracks with different widths. As shown in Section 5.2, some power loss occurs if the electrodes are identical in the two tracks; however efficient coupling can be obtained if the pitches are slightly different [131], as shown in Figure 5.11(a). This technique has been used in surface-wave convolvers, where the enhancement of power density due to the compression of the beam leads to an improvement in device efficiency. It is also of interest for efficient coupling of transducers with different apertures.

The device can be analysed using equations in Section 5.1 above. The metallisation ratio is taken to be the same in all three tracks. Taking p_a and p_b as the pitches in tracks A and B, the electrode voltages can be written as $V_n = V_0 \exp(-jn\kappa_a p_a)$ for track A and $V_n = V_0 \exp(-jn\kappa_b p_b)$ for track B. These must of course be equal, so that we can write

$$\kappa_a p_a = \kappa_b p_b = \gamma.$$

The voltages are also equal if a multiple of $2\pi/p_b$ is added to κ_b, but this is not of interest here. The currents entering the electrodes in the three tracks are given by equation (5.18), which involves the slowly-varying functions $C(\kappa)$ and $\bar{\varrho}_f(\kappa)$, dependent on κ and p. The solutions of interest here have κ close to k_0, and the pitches in the three tracks are closely similar, so these functions can be taken as $C(k_0)$ and $\bar{\varrho}_f(k_0)$ in all tracks, using an average value of the pitch. Using equation (5.18) and setting the sum of the currents entering the three tracks to zero shows that there are two positive solutions for γ, denoted by γ_+ and γ_-, and these are related to the pitches p_a and p_b and the track widths W_a, W_b and W_c. The surface wave amplitudes in tracks A and B for these two modes are given by equation (5.19). The two modes are added, with an amplitude ratio such that the total surface wave potential in track B is zero for $n = 0$, the left end of the coupler. If the coupler is to transfer all the power from

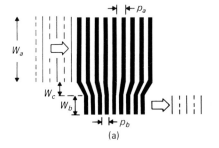
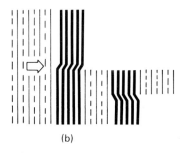

FIGURE 5.11. Beam compression.

track A to track B, it is also necessary that the wave amplitude in track A should be zero for some $n \neq 0$. This is found to give a further relationship between $\gamma_\pm$ and the pitches p_a and p_b. After some manipulation, it is found that the required relationship between the pitches is

$$p_a - p_b = \frac{\pi}{k_0 N_c} \frac{W_a - W_b}{W_a + W_b + W_c} \qquad (5.47)$$

and, if this relation is satisfied, a complete transfer of power from one track to the other is obtained if the number of electrodes is N'_c, given by

$$N'_c = N_c \frac{W_a + W_b + W_c}{2\sqrt{W_a W_b}}. \qquad (5.48)$$

In these equations, N_c is the number of electrodes required for a complete transfer in a simple coupler, given by equation (5.28). Equation (5.48) shows that the number of electrodes required is quite practicable, even for quite large ratios of track widths; for example, a 10:1 ratio of track widths requires about 190 electrodes on a Y, Z lithium niobate substrate. The required difference between the pitches, equation (5.47), is typically 1%. Experimental results [131] generally show good agreement for track width ratios up to 15:1, though a loss of up to about 1 dB is observed, due to the resistance of the electrodes. However, larger losses, due to surface wave diffraction, are found if the width of the narrower track is less than about five wavelengths.

An alternative type of multi-strip beam compressor makes use of 3 dB couplers [126], as illustrated in Figure 5.11(b). It was seen in Section 5.4 that for a 3 dB coupler the scattering coefficients S_{12} and S_{14} are of equal magnitude and are in phase quadrature. In the beam compressor an additional $\pi/2$ phase change is introduced by displacing the electrodes in one track by a quarter-wavelength relative to the other track. For incident waves of equal amplitude and phase on both tracks, this causes the output on one track to be cancelled, and thus an input beam spanning both tracks has its width reduced by a factor of two. Additional couplers can be used to provide further width reduction. Using three couplers a width compression of 8:1 was demonstrated [126], with only 0.5 dB of loss. In comparison with the device of Figure 5.11(a), this method generally requires a larger number of electrodes for a given compression ratio and is also rather more frequency-dependent, since the electrode displacement gives the required phase change only at one particular frequency.

Chapter 6

Propagation Effects and Materials

In previous chapters the behaviour of interdigital transducers and multi-strip couplers were considered, assuming idealised surface-wave propagation conditions. However, in practical devices several propagation effects, such as diffraction, propagation loss and temperature effects, are often significant, and these are considered in this chapter. Materials for surface-wave devices are also considered here because their comparative merits are strongly influenced by propagation effects, in addition to other parameters such as the strength of the piezoelectric coupling. It is convenient to consider first some experimental methods for investigating propagating surface waves.

6.1. SURFACE-WAVE PROBING

Some surface-wave propagation effects can be investigated using simple interdigital devices. Propagation loss is one example [132], while temperature effects can be measured using an interdigital delay-line oscillator. However, for many purposes the distribution of the surface-wave amplitude is required, over a specified area, with fine spatial resolution. This is the function of surface-wave probes, which are used both for measurements of propagation effects and for diagnostic investigation of operational devices.

Two types of probe have been developed: electrostatic and optical. The electrostatic probe [133, 134], suitable only for piezoelectric materials, senses the surface potential accompanying the wave. A sharp tungsten needle is used, and since the potential decays rapidly above the surface the needle is held in contact, using a balance arm so that the pressure is too light to cause surface damage. This technique gives a reproducibility of about 1 dB, though it is ineffective in or near metal structures such as transducers. Both the amplitude and phase of the surface wave can be obtained. In common with other types of probe, the substrate is usually mounted on a motor-driven stage with potentiometers sensing the position, so that a two-dimensional plot can be obtained on an X–Y recorder. The output signal is often mixed with the surface-wave input frequency, so that the wave motion appears "frozen" stroboscopically.

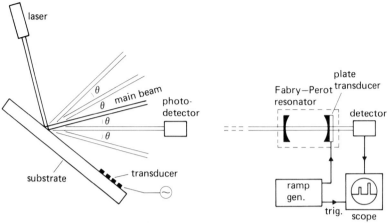

FIGURE 6.1. Left: diffraction probe. Right: optical spectrum analyser, used to distinguish waves propagating in diffraction directions.

Optical probes have been developed in a variety of forms reviewed by Stegeman [135], and here the main techniques are considered briefly. If an optical beam illuminates a surface on which a surface wave is propagating, the corrugations due to the wave act as an optical diffraction grating. Thus, in addition to the specularly-reflected main beam there are diffracted beams with approximately regular spacing θ, as shown in Figure 6.1, where θ is typically 1°. The intensity of the first-order diffracted beam, which is proportional to the surface-wave power density, is measured using a photomultiplier tube or photo-diode [132, 136, 137]. Usually the surface-wave signal is modulated and a lock-in amplifier is used to improve the sensitivity. Owing to the motion of the surface wave the first-order diffracted beam has a doppler shift, with magnitude equal to the surface-wave frequency and sign dependent on the propagation direction. This feature can be used to distinguish waves propagating in opposite directions, using an optical spectrum analyser [135, 138]. As

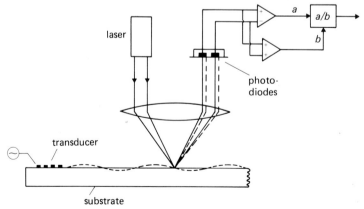

FIGURE 6.2. "Knife-edge" probe. After Engan [140], copyright © 1978 IEEE.

shown on the right in Figure 6.1, the beam is passed through a Fabry–Perot resonator in which one of the mirrors is vibrated by a piezoelectric plate transducer, thus sweeping the frequencies of the transmission peaks. If two oppositely-directed surface waves are present, two corresponding peaks are observed on the oscilloscope, giving the individual power densities. This method has been used to study surface wave generation within a transducer [138].

Figure 6.2 shows a method giving the surface-wave phase as well as amplitude. A laser beam is focussed by a lens, giving a spot on the surface smaller than the surface-wave wavelength, and the reflected wave is collimated by the same lens. A tilt of the surface, due to the presence of a surface wave, causes the output beam to be displaced to one side. This motion can be sensed by using a knife-edge to obstruct part of the beam, followed by a photo-detector [139]. Subsequently, Engan [140] used two photodiodes, as in Figure 6.2, aligned to give equal outputs when the surface wave is absent. The difference of the two diode signals gives the surface-wave amplitude and phase, though it also depends on the optical reflectivity of the surface, and therefore on whether the surface is metallised. However, an output almost independent of the reflectivity is obtained by dividing the difference of the diode signals by their sum. The laser beam can be scanned over the area beneath the lens by means of rotating mirros [139], though it is more usual to mount the substrate on a movable stage. The system has been used to examine very short surface-wave pulses, as well as C.W. signals, and can give excellent spatial resolution.

A third method, the optical heterodyne probe [141, 142], is shown in Figure 6.3. The light first passes through a water-filled Bragg cell, in which a longitudinal bulk acoustic wave of frequency f_b propagates. Most of the light is passed directly, but some is diffracted through a small angle and doppler-shifted by an amount f_b; the latter part is reflected by a mirror, and used as a reference beam. The direct beam, with frequency f_0, is focussed on to the substrate supporting the propagating surface wave, which has frequency f_s. The reflected light includes a zero-order specular component of frequency f_0, and two doppler-shifted components, with frequency $f_0 \pm f_s$, due to the presence of the surface wave. A further doppler shift of f_b occurs when the reflected beam is diffracted again by the Bragg cell. Thus the light incident on the photodiode

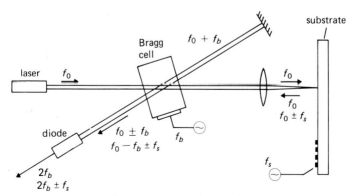

FIGURE 6.3. Heterodyne probe.

has components with frequencies $f_0 \pm f_b$ and $f_0 - f_b \pm f_s$. The system makes use of two components of the diode output, separated by bandpass filtering: one, at frequency $2f_b + f_s$, gives the surface-wave amplitude and phase, while another, at frequency $2f_b$, is used as a reference for synchronous demodulation, and also enables the final output to be made independent of the surface reflectivity.

Heterodyne probes are usable for surface-wave frequencies up to several hundred MHz, and can detect surface displacements of about 10^{-3} Å; for lithium niobate at 100 MHz, this corresponds to a surface wave power density of about 10^{-9} watt/mm. The other optical probes give a sensitivity of about 0.1 Å, but can operate above 1 GHz.

6.2. DIFFRACTION AND BEAM STEERING

In this section the diffraction of surface waves on an unmetallised surface is considered, assuming that the only wave motion present is a non-leaky surface wave. The primary aim is to develop relationships for analysis of diffraction effects in surface-wave devices. The surface-wave field generated by an unapodised transducer has many similarities to the diffracted light field produced by a slit aperture. In particular, there is a "near-field", or Fresnel, region in which the beam maintains a width roughly the same as the transducer aperture. In the "far-field", or Fraunhofer, region the beam diverges with an angle determined by the transducer aperture. In many interdigital devices the receiving transducer is in the near field of the launching transducer, so that diffraction effects are relatively small; nevertheless, they are often of practical significance when an accurate device response is required.

The main distinction between surface-wave diffraction and conventional optical diffraction arises from the anisotropy of the substrate material. This implies that a straight-crested wave, that is, a wave with a straight wavefront, has a phase velocity depending on the propagation direction. Consequently, the diffracted field depends markedly on the substrate material and the crystallographic orientation. Most analyses of diffraction in anisotropic media have relied on one of two basic techniques. This section uses the "angular spectrum of plane waves" method in which the diffracted field is represented as an infinite sum of straight-crested waves, using Fourier synthesis [143–146]. Alternatively, a Green's function can be used to express the field at some point (x, y) due to a point source elsewhere, and the total field at (x, y) is then found by integrating over the finite source region [147–151]. This method is a generalisation of Huygen's principle for anisotropic media.

In both methods, the surface-wave amplitude at some point (x, y) is represented by a scalar, here denoted $\psi(x, y)$. This may be taken to be the surface potential $\phi_s(x, y)$ (for a piezoelectric material), or one of the components of the elastic displacement $\mathbf{u}(x, y)$ at the surface. The use of scalar is in fact an approximation since, for a straight-crested wave with a given power density, the magnitudes of $\phi_s(x, y)$ and $\mathbf{u}(x, y)$ vary with the propagation direction. However, a scalar analysis is found to be adequate for practical purposes. A more rigorous approach is given by Milsom [152].

6.2.1. Formulation Using Angular Spectrum of Plane Waves

Taking the surface of the material to be the x–y plane, the amplitude of a straight-crested wave is proportional to $\exp[-j(xk_x + yk_y)]$, where a factor $\exp(j\omega t)$ is omitted and the frequency ω is taken to be constant. It is assumed that there is no propagation loss. As shown in Figure 6.4, k_x and k_y are the x and y components of the wave vector $\mathbf{k}(\phi)$, which makes an angle ϕ with the x-axis. Thus $k_x = k(\phi)\cos\phi$ and $k_y = k(\phi)\sin\phi$, where $k(\phi)$ is the magnitude of $\mathbf{k}(\phi)$. In addition, $k(\phi) = \omega/v(\phi)$, where $v(\phi)$ is the phase velocity for straight-crested waves with propagation direction ϕ, and may be calculated by method described in Chapter 2. The quantity $1/v(\phi)$ is termed the *slowness*, and a polar plot of $k(\phi)$ as a function of ϕ is called the *slowness curve*, since the constant ω is immaterial to the present argument. Figure 6.4 shows schematically a slowness curve and its relation to the wave vector. Note that the form of the slowness curve will depend on the substrate material and the orientation of the surface normal. For an isotropic material $k(\phi)$ and $v(\phi)$ are independent of ϕ, and the slowness curve is a circle.

The x-axis will later be taken to be normal to the electrodes of a transducer launching surface waves. The angle θ in Figure 6.4 relates this direction to some reference direction defined by the crystal lattice. Thus for constant θ the surface-wave velocity v is a function of ϕ, but it can also be expressed as a function of θ by taking $\phi = 0$. For the present, θ is taken to be constant.

Since the system is linear, a general solution can be obtained by summing straight-crested waves of the form $\exp[-j(xk_x + yk_y)]$, allowing the propagation direction to vary [144, 146]. It is assumed that k_x is determined by k_y, so that the total disturbance can be represented by a scalar field $\psi(x, y)$ given by

$$\psi(x, y) = \int_{-\infty}^{\infty} \Psi(k_y) \exp[-j\{xk_x(k_y) + yk_y\}]\,dk_y \quad \text{for } x \geq 0, \quad (6.1)$$

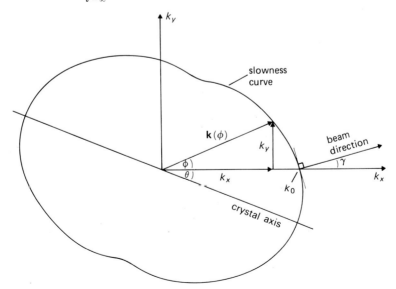

FIGURE 6.4. Diffraction analysis using the slowness curve.

where $\Psi(k_y)$ is the amplitude distribution of the contributing straight-crested waves. Setting $x = 0$ in this equation shows that $\psi(0, y)$ is the Fourier transform of $\Psi(k_y)$, and hence $\Psi(k_y)$ is the inverse transform of $\psi(0, y)$:

$$\Psi(k_y) = \frac{1}{2\pi} \int_{-\infty}^{\infty} \psi(0, y') \exp(jy'k_y) \, dy'. \tag{6.2}$$

Thus the disturbance at any point (x, y) can be obtained from the disturbance $\psi(0, y)$ on the line $x = 0$ by transforming to give $\Psi(k_y)$, and then using equation (6.1). The function $k_x(k_y)$ is given by

$$[k_x(k_y)]^2 = [k(\phi)]^2 - k_y^2. \tag{6.3}$$

It can be assumed that, for $x \geq 0$, only positive solutions for k_x are required, and this usually determines k_x uniquely from k_y. This condition is necessary if the field $\psi(x, y)$ is to be obtained from equation (6.1). In practice this is not always true; if the slowness curve has a minimum near $\phi = \pm\pi/2$, some values of k_y can give more than one solution for k_x. However, $\Psi(k_y)$ is nearly always very small in such regions, so the ambiguity in k_x does not invalidate the method. For large k_y, when $k_y > k$, k_x is imaginary and is taken to be negative, so that the contribution to equation (6.1) decays with x.

6.2.2. Beam Steering in the Near Field

In the near field, the waves due to an aperture with uniform illumination exhibit little diffraction spreading, so that diffraction need not be taken into account. This is generally true for both isotropic and anisotropic materials. However, anisotropic materials can exhibit *beam steering*, which causes the beam to propagate in a direction that is not normal to the wavefronts. In surface-wave devices, this phenomenon can be significant even when the receiving transducer is in the near field of the launching transducer, so that diffraction spreading is negligible.

To analyse this, suppose that $\psi(0, y)$ represents a line source at $x = 0$ extending many wavelengths in the y-direction, and with phase independent of y. In this case, the transform $\Psi(k_y)$ will be significant only for k_y close to zero, so that only a small part of the slowness curve is relevant. The phase xk_x in equation (6.1) can therefore be approximated by using a Taylor expansion for $k_x(k_y)$, provided x is not too large. Defining $k_0 = k_x(0)$, equal to the value of $k(\phi)$ for $\phi = 0$, we have for $k_y \ll k_x$

$$k_x(k_y) \approx k_0 - k_y \tan \gamma, \tag{6.4}$$

where quadratic and higher order terms are ignored, and γ is defined by

$$\tan \gamma = -[dk_x/dk_y]_{\phi = 0}. \tag{6.5}$$

This will generally depend on θ. Substituting equation (6.4) into equation (6.1) we find

$$\psi(x, y) \approx \exp(-jxk_0) \psi(0, y - x \tan \gamma). \tag{6.6}$$

Thus in the near field the disturbance propagates with no distortion, but the propagation direction of the beam makes an angle γ with the x-axis, that is, with the direction normal to the wavefronts. This phenomenon, known as beam steering, can occur only if the medium is anisotropic. The beam direction is normal to the slowness curve, as shown on Figure 6.4. The angle γ is called the beam steering angle, and generally depends on the orientation angle θ. From equation (6.5) we have

$$\tan \gamma = \left[\frac{1}{v}\frac{dv}{d\phi}\right]_{\phi=0} = \frac{1}{v}\frac{dv}{d\theta}. \qquad (6.7)$$

Usually γ is small, so that $\tan \gamma \approx \gamma$.

Figure 6.5 illustrates beam steering for a wave launched by an unapodised interdigital transducer, where $\psi(0, y)$ is taken as the surface-wave amplitude at the edge of the transducer. The wavefronts are parallel to the transducer electrodes. Owing to beam steering, only part of the beam is intercepted by a receiving transducer aligned with the launching transducer, so that the available output power is reduced [153]. The output power can be found from equation (4.126). The beam steering has been confirmed experimentally by probing [132, 136].

The orientation θ is usually chosen such that the velocity $v(\phi)$ is symmetrical about $\phi = 0$ for small angles, that is, a pure mode direction (Section 2.3.3) is chosen. The beam steering angle γ is then zero. However, beam steering can still arise in practice because of an error in the angle θ due to misalignment. For small errors, γ can be found from the differential $d\gamma/d\theta$, given by

$$\frac{d\gamma}{d\theta} = \frac{1}{v}\frac{d^2v}{d\theta^2}\cos^2\gamma - \sin^2\gamma \qquad (6.8)$$

$$= \frac{1}{v}\frac{d^2v}{d\theta^2}, \quad \text{if } \gamma = 0.$$

In the limit when the transducer aperture becomes infinite the disturbance becomes an infinite straight-crested wave, with wavenumber **k** directed along the x-axis. It is reasonable to conclude that the energy of this wave propagates in the direction normal to the slowness curve, making an angle γ with **k**, and hence that the energy velocity is $v/\cos \gamma$. A more rigorous derivation is obtained by defining a Poynting vector for

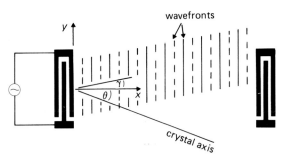

FIGURE 6.5. Beam steering, schematic.

an elastic piezoelectric medium [154, p. 307], analogous to the eletromagnetic Poynting vector. Auld [154, p. 225] shows that for plane bulk waves this gives an energy flow direction normal to the slowness curve, as deduced above. Lighthill [155] gives an alternative derivation for plane waves. For a surface wave the Poynting vector generally has a direction varying with depth, and must be integrated to give the overall energy flow direction [156]. The result is found to agree with the result obtained directly from the slowness curve, though there does not appear to be any proof that the two methods are equivalent.

6.2.3. Minimal-diffraction Orientations

In the above analysis it was assumed that only very small values of k_y give significant contributions to the integral of equation (6.1), so that, for small x, k_x could be approximated by equation (6.4). However, if we consider a hypothetical case in which k_x is a constant, independent of k_y, then the field $\psi(x, y)$ is given exactly by equation (6.6) for all x, with $\gamma = 0$. In this case the surface-wave beam propagates with no diffraction spreading. Physically, this occurs because the energy flow direction is parallel to the x-axis, irrespective of the direction of the wave vector $\mathbf{k}$. For k_x to be constant the phase velocity v must be proportional to $\cos \phi$. Noting that $dv/d\theta$ is equal to $dv/d\phi$ at $\phi = 0$, and using equation (6.8), we have in this case

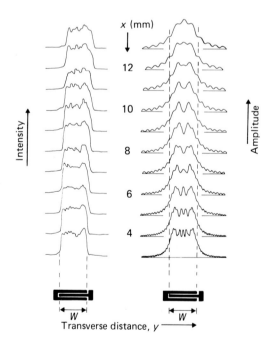

FIGURE 6.6. Diffraction of beams launched by unapodised transducers with aperture $W = 40$ wavelengths at 100 MHz. *Left*: Experimental intensity profiles for Y, Z lithium niobate, after Crabb *et al.* [150] with permission (British Crown Copyright reserved). *Right*: Theoretical amplitude profiles for isotropic material. In both cases, plots refer to distances $x = 3, 4, 5, \ldots, 13$ mm from launching transducer, and are displaced vertically.

$$\frac{d\gamma}{d\theta} = -1. \tag{6.9}$$

In practice this is unrealistic because $k_y = k_x \tan \phi$ is infinite at $\phi = \pm \pi/2$ if k_x is constant. However, for several materials there are orientations such that k_x is almost constant for small ϕ so that, for the chosen value of θ, $d\gamma/d\theta \approx -1$. These *minimal-diffraction orientations* give substantially less diffraction spreading than the isotropic case. An example is Y, Z lithium niobate, which gives $d\gamma/d\theta = -1.08$. However, for such cases the orientation of the transducers relative to the crystal axes can be quite critical, since the beam steering angle γ varies rapidly with the orientation θ.

Figure 6.6 shows a diffraction pattern for Y, Z lithium niobate, measured using an electrostatic probe. The wave was launched by an unapodised transducer with aperture equal to 40 wavelengths at the 100 MHz measurement frequency. Also shown is the diffraction pattern for an isotropic material, calculated from the analysis of Section 6.2.4 below. The anisotropy of lithium niobate clearly causes a very marked reduction of the diffraction spreading.

6.2.4. Diffracted Field in the Parabolic Approximation: Scaling

Up to this point, we have considered the surface-wave distribution only in the near field. In this section we consider the distribution for all x, including the far field. This can be obtained from the angular-spectrum-of-plane-waves formulation, equations (6.1) and (6.2), in which $k_x(k_y)$ is determined by the phase velocity $v(\phi)$ and thus depends on the material and orientation. However, a more convenient formulation can be obtained by using the *parabolic approximation*, in which k_x is taken to be a quadratic function of k_y. This is often valid since, for large apertures, $\Psi(k_y)$ in equation (6.1) is small except for the region near $k_y = 0$. We show here that the parabolic approximation gives a formulation which is more convenient to compute than the angular-spectrum-of-plane-waves formula. It also leads to the very useful conclusion that the diffraction pattern for an anisotropic material is related to the isotropic case simply by scaling in the x-direction. However, it must be emphasised that the validity of the approximation depends on the nature of the slowness curve, and for some materials, notably Y, Z lithium niobate, the parabolic approximation cannot be used. A discussion of the validity is given later.

Assuming the parabolic approximation to be valid, the function $k_x(k_y)$ is taken to have the form

$$k_x(k_y) \approx k_0 - ak_y - \tfrac{1}{2}bk_y^2/k_0, \tag{6.10}$$

where a, b and k_0 are constants for a given orientation though they will generally vary with θ. Using equations (6.5) and (6.8), a and b are given by

$$a = \tan \gamma, \tag{6.11a}$$

$$b = 1 + \frac{1}{v}\frac{d^2v}{d\theta^2} = \left(1 + \frac{d\gamma}{d\theta}\right)\sec^2\gamma. \tag{6.11b}$$

Equation (6.11b) is found by evaluating d^2k_x/dk_y^2 as a function of ϕ, at $\phi = 0$. For an isotropic material we have

$$k_x = [k_0^2 - k_y^2]^{1/2} \approx k_0 - \tfrac{1}{2}k_y^2/k_0,$$

where the approximate form applies for $k_y \ll k_0$. Thus the isotropic case is given by equation (6.10), with $a = 0$ and $b = 1$.

The field $\psi(x, y)$ is found by substituting equation (6.10) into equation (6.1), with $\Psi(k_y)$ determined by the source distribution. This shows that the effect of the parameter a can be expressed straightforwardly. If $\psi_0(x, y)$ is the field calculated with $a = 0$, it is found that

$$\psi(x, y) = \psi_0(x, y - ax), \qquad (6.12)$$

that is, the diffraction pattern can be calculated for $a = 0$, and then $(y - ax)$ is substituted for y. This is valid irrespective of the form of $\Psi(k_y)$. Thus the linear term in equation (6.10) skews the entire diffraction pattern, as found for the near-field case in equation (6.6); however, equation (6.12) is valid for all x, including the far-field.

Aperture with Uniform Illumination. We now consider a uniformly illuminated aperture, of width W, located at $x = 0$. This case can be taken to refer to an unapodised launching transducer with aperture W and with few electrodes, so that diffraction within it can be ignored. For simplicity it is assumed that $a = 0$, since the field for $a \neq 0$ is readily obtained from equation (6.12). The field $\psi(x, y)$ is obtained by substituting equation (6.10) into equation (6.1), using equation (6.2) for $\Psi(k_y)$. The field at the aperture is taken as $\psi(0, y) = 1$ for $|y| \leq W/2$ and $\psi(0, y) = 0$ for $|y| > W/2$. We thus have, for $a = 0$,

$$\psi(x, y) = \frac{\exp(-jxk_0)}{\pi} \int_{-W/2}^{W/2} dy' \int_{-\infty}^{\infty} dk_y \exp[j\{(y' - y)k_y + \tfrac{1}{2}xbk_y^2/k_0\}]. \qquad (6.13)$$

This can be re-arranged using the standard integral [157, p. 301]

$$\int_{-\infty}^{\infty} \exp(jKt^2) \, dt = \sqrt{\frac{\pi}{|K|}} \, e^{\pm j\pi/4}, \qquad (6.14)$$

where $K \neq 0$ is a real constant and the sign in the exponential on the right is the same as the sign of K. Using equation (6.14), equation (6.13) can be written

$$\psi(x, y) = \frac{1}{\sqrt{2}} \exp(-jxk_0 \pm j\pi/4) \int_{A_-}^{A_+} \exp(\mp \tfrac{1}{2}j\pi u^2) \, du, \qquad (6.15)$$

where the upper signs are used for $b > 0$ and the lower signs for $b < 0$, and the limits for the integral are

$$A_\pm = (y \pm W/2) \left[\frac{k_0}{\pi x |b|}\right]^{1/2}. \qquad (6.16)$$

Equation (6.15) is readily computed using Fresnel integrals, and an example is shown on the right in Figure 6.6. The anisotropy of the material is expressed by the constant b, which occurs in the product $x|b|$ in equation (6.16). Thus, apart from the phase xk_0 in equation (6.15), the diffraction pattern is the same as that for an *isotropic* material ($b = 1$), except for scaling in the x-direction by a factor $|b|$. This useful result has also been established using the Green's function theory of diffraction [147–150]. In the limit $b \to 0$, k_x becomes a constant and there is then no diffraction spreading, as in equation (6.6).

In most practical cases the orientation is chosen such that the phase velocity $v(\phi)$ is symmetrical about $\phi = 0$, so that $a = \gamma = 0$ and, from equation (6.11b), $b = 1 + d\gamma/d\theta$. The parameter $d\gamma/d\theta$ thus determines the diffraction scaling. In addition, $d\gamma/d\theta$ gives the alignment accuracy required in order to keep the near-field beam steering within specified limits, as noted in Section 6.2.2, and minimal-diffraction orientations occur when $d\gamma/d\theta \approx -1$. Thus, $d\gamma/d\theta$ is a very significant parameter for any particular orientation, and its numerical values for a number of cases are given in Section 6.5 below.

In the far-field, where $4\pi x|b| \gg k_0 W^2$, the limits $A_\pm$ given by equation (6.16) are close together, and equation (6.15) gives

$$|\psi(x, y)|^2 \approx \frac{\chi W}{\pi x} \left[\frac{\sin (\chi y/x)}{\chi y/x}\right]^2, \tag{6.17}$$

where $\chi = \frac{1}{2}Wk_0/|b|$. Thus the diffraction pattern spreads out radially in this region. On the other hand, in the near-field the beam propagates with little distortion or spreading, as shown in Section 6.2.2. It is useful to define a *Fresnel distance* x_f which gives approximately the demarkation between the two regions, so that diffraction effects are negligible when $x \ll x_f$. For a substrate with parabolic anisotropy, equation (6.16) shows that x_f must be proportional to $W^2 k_0/|b|$. A convenient definition is the distance at which the central peak, given for the far-field by equation (6.17), first appears. Crabb *et al.* [150] have shown that this distance is given by

$$x_f \approx \frac{W^2 k_0}{10\pi|b|}. \tag{6.18}$$

Measurements of diffraction patterns on Y-cut lithium niobate, using a variety of transducer orientations, gave good agreement with this formula [150], thus confirming the scaling of the diffraction pattern according to the value of $|b|$. An electrostatic probe was used for these measurements.

Validity of the Parabolic Approximation. The above analysis is valid provided the slowness curve can be taken to be a parabola, as in equation (6.10). This approximation will generally break down when x is very large, or when the aperture is very small. Moreover, for any particular aperture and particular observation point, the accuracy depends on how well the slowness curve can be approximated by a quadratic, and this in turn depends on the material and orientation. For many cases, it is found that the parabolic approximation gives the diffraction pattern with good accuracy well into the far-field region, provided a realistic aperture is assumed, and the approximation is then acceptable for practical purposes. This was established by Szabo and Slobodnik [158], who compared the predictions of the parabolic approximation with measurements obtained by probing, and with the predictions of the more accurate angular-spectrum-of-plane-waves formulation. The validity was established for a variety of materials and orientations, some of which are indicated in Table 6.1 of Section 6.5 below.

For Y, Z lithium niobate it is found that the parabolic approximation is not valid, so the less convenient angular-spectrum-of-plane-waves formulation must be used. However, an additional difficulty arises here because the diffraction pattern is found to be very sensitive to the details of the velocity anisotropy. The velocity may be calculated by the method of Section 2.3.2, but this makes use of elastic and piezoelectric constants that are obtained from bulk measurements, and are therefore subject to experimental error. The resulting velocity errors are found to cause theoretical diffraction patterns to disagree with experiment [158]. However, the sensitivity of the diffraction pattern to velocity errors can be exploited in order to obtain more accurate velocity data [159]. Using a narrow-aperture transducer to generate waves spreading over a relatively wide angle, the amplitude and phase are scanned as functions of y at two values of x, using a heterodyne probe. The two scans are related by equations (6.1) and (6.2); by Fourier transforming them, k_x can be found as a function of k_y, thus giving the slowness curve. This method improved the velocity accuracy to some 3 parts in 10^5, enabling the diffraction analysis to give results in quite good agreement with experiment [160].

6.2.5. Two-transducer Devices

The analysis of the previous section gives the field $\psi(x, y)$ at all points, due to an unapodised launching transducer of aperture W. Further development is necessary to find the effect of diffraction on a surface-wave device. We consider here a device comprising two transducers, though the methods are readily adapted for other cases such as devices including multi-strip couplers.

We consider initially the case in which both transducers are unapodised, and both have relatively few electrodes so that diffraction within them can be neglected. For this case, the effect of diffraction can be found by considering the simple geometry of Figure 6.7, where the transducers are replaced by a line source and a line receiver, both parallel to the y-axis. More complicated cases, involving apodised transducers and diffraction within transducers, are considered below.

In Figure 6.7, the source and receiver are taken to have apertures W_a and W_b, respectively. The field $\psi(x, y)$ due to the source can be found by methods given above. The response of the receiver is obtained by noting that, for an unapodised transducer, the short-circuit output current is proportional to the average surface-wave potential at the transducer input, as shown by equation (4.126) of Chapter 4. The effect of diffraction can therefore be obtained by integrating $\psi(x, y)$ over the receiver aperture, and comparing with diffractionless analysis [151, 161]. The integral required is therefore

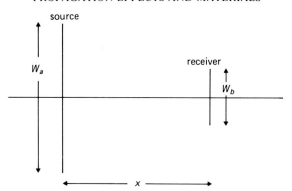

FIGURE 6.7. Co-axial line source and receiver.

$$R = \int_{-W_b/2}^{W_b/2} \psi(x, y) \, dy, \qquad (6.19)$$

where x gives the location of the receiver relative to the source. For brevity it is assumed here that both transducers are centred on the line $y = 0$. For a receiver displaced in the y-direction the results are more complex [161] though they do not show any essentially new features.

For the general case, $\psi(x, y)$ is given by equation (6.1), but here the parabolic approximation is assumed to be valid as this gives some convenient results. Thus $\psi(x, y)$ is given by equations (6.15) and (6.16), using W_a for W. It is convenient to define the functions

$$F_\pm(t) = \int_0^t \exp(\pm \tfrac{1}{2} j\pi u^2) \, du = C(t) \pm jS(t), \qquad (6.20)$$

where $C(t)$ and $S(t)$ are the Fresnel integrals. We also define the functions $X_\pm(s)$ as integrals of $F_\pm(t)$, so that

$$X_\pm(s) \equiv \int_0^s F_\pm(t) \, dt = sF_\pm(s) \pm \frac{j}{\pi}[\exp(\pm \tfrac{1}{2} j\pi s^2) - 1], \qquad (6.21)$$

where the final result is obtained using integration by parts [151]. Using these formulae, and equation (6.15) for $\psi(x, y)$, the function R defined in equation (6.19) is found to be

$$R = \frac{\sqrt{2} \exp(-jxk_0 + j\pi/4)}{\eta k_0} [X_\mp(B_+) - X_\mp(B_-)], \qquad (6.22)$$

where the upper signs are used for $b > 0$ and the lower signs for $b < 0$, and we define

$$\eta = [\pi x k_0 |b|]^{-1/2} \qquad (6.23)$$

and

$$B_\pm = \tfrac{1}{2} \eta k_0 [W_a \pm W_b]. \qquad (6.24)$$

Equation (6.22) gives the transducer output, allowing for diffraction. The output obtained in the absence of diffraction can be found by taking the limit $b \to 0$. Defining R_0 as the value of R for this case, equation (6.22) gives

$$R_0 = W_b \exp(-jxk_0), \qquad (6.25)$$

where it is assumed that $W_b \leqslant W_a$. The receiver output R can therefore be expressed in the form

$$R = DW_b \exp(-jxk_0), \qquad (6.26)$$

where $D = R/R_0$ is a function expressing the effect of diffraction, equal to unity when diffraction is absent. The function D depends on the transducer apertures W_a and W_b, the separation x, the material anisotropy and the frequency, which enters through the wavenumber k_0. However the above equations show that D can be regarded as a function of only two normalised variables, the aperture ratio W_b/W_a and a normalised transducer separation

$$\hat{x} = \frac{|b|(x/\lambda)}{(W_a/\lambda)^2}, \qquad (6.27)$$

where $\lambda = 2\pi/k_0$ is the wavelength. Figure 6.8 shows the amplitude and phase of D, as functions of $\hat{x}$, for several values of W_b/W_a. For $W_b = 0$, D is simply the field $\psi(x, y)$ due to the source, evaluated on the centre line $y = 0$, with the term $\exp(-jxk_0)$ omitted.

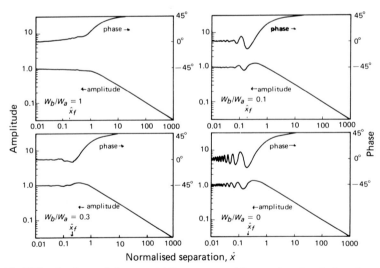

FIGURE 6.8. Diffraction factor D for co-axial line source and line receiver. Parabolic anisotropy, with $a = 0$ and $b > 0$.

Diffraction for Long Transducers and Apodised Transducers. It was assumed above that the two transducers had relatively few electrodes and were unapodised, so that the effect of diffraction could be found by replacing them by a line source and a line receiver, as in Figure 6.7. However, in a transducer with many electrodes the effect of diffraction can vary significantly along the length, and it becomes necessary to consider the diffraction within the transducer itself. A similar complication arises if the transducer is apodised, since the effect of diffraction depends on the extent of the electrode overlaps.

For such cases, the effect of diffraction can be found by regarding each of the two transducers as an *array* of line elements, as in Figure 6.9. Here is it assumed that only one of the transducers is apodised, as is usually the case in practice. To a good approximation, the elements may be identified with the inter-electrode gaps, as illustrated in Figure 4.3. Element m in the apodised receiving transducer has aperture w_m and a polarity $C_m = \pm 1$ determined by which of the adjacent electrodes is connected to the upper bus-bar. In the launching transducer, element n has polarity C_n. The output due to these two elements has the form of equation (6.26) and thus, using superposition, the output current of the overall device is given by

$$I_{sc} \propto \sum_n \sum_m D_{nm} C_n C_m w_m \exp[-j(x_m - x_n)k_0], \qquad (6.28)$$

where, for source n and receiver m, the separation is $x_m - x_n$ and D_{nm} is the diffraction factor. It is assumed here that each source overlaps the entire region of y occupied by the receivers, as is usually true in practice. If diffraction is negligible all the D_{nm} are unity, and the result is then as given by the delta-function analysis of Section 4.1. Thus equation (6.28) can be used to deduce the diffraction factor, as a function of frequency, for the overall device. A more accurate device response can then be calculated by combining the overall diffraction factor with the diffractionless analysis of Section 4.7.3, which allows for the element factor.

The diffraction factors D_{nm} in equation (6.28) are usually evaluated as if the waves were everywhere propagating on a free surface, ignoring perturbations due to the transducer electrodes. This is generally a reasonable assumption, since diffraction has a relatively small effect in most practical devices. Milsom [152] has given a formulation allowing for the presence of the electrodes. If free-surface diffraction is

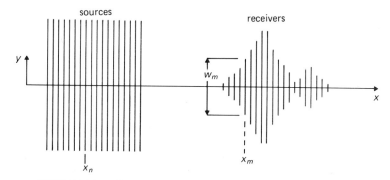

FIGURE 6.9. Diffraction analysis for extended or apodised transducers.

assumed, the factors D_{nm} can be obtained from the analysis given earlier. Thus, if the anisotropy is parabolic and the sources and receivers are co-axial, D_{nm} is equal to R/R_0, equations (6.22) and (6.25), with w_m replacing W_b.

In practice, the computation of equation (6.28) is in some cases very time-consuming. The two transducers may well have several hundred electrodes, so that for equation (6.28) it is necessary to calculate some 10^5 values of D_{nm} for each frequency. However, if the transducer apodisation varies slowly, as in many chirp filters, D_{nm} generally varies slowly with n and m; it is then feasible to calculate D_{nm} for only a few values of n and m, interpolating to obtain intermediate values. It is also beneficial to pre-calculate the function $X_\pm(s)$ of equation (6.21) and interpolate as required. Several authors have given other techniques for reducing the computation time [160, 162–165], and these are applicable when the apodisation varies rapidly.

In practical devices, diffraction effects are often compensated for by modifying the transducer design. Methods for doing this are discussed later, in Section 8.4.3.

6.3. PROPAGATION LOSS AND NON-LINEAR EFFECTS

(a) *Propagation Loss.* In addition to diffraction the surface wave amplitude is attenuated because of propagation loss, and this is of practical significance in many devices, particularly at high frequencies.

Consider a crystalline material with a free surface, assuming the surface wave amplitude to be small so that non-linear effects can be ignored. The beam is also assumed to be wide, so that diffraction is negligible. At room temperature, it is found that the propagation loss is primarily due to two effects. Firstly, the surface waves interact with thermal lattice waves (phonons), and from analysis using a viscosity model [166, 167] the consequent attenuation coefficient, in dB per unit length, is proportional to f^2. Using measured attenuation coefficients for bulk waves, this model enables the surface-wave attenuation to be deduced. This applies when the crystal is in a vacuum; generally however the crystal is in air, or in an inert gas, and then the second effect, known as "air loading" is present. Analysis shows [167, 168] that, in the absence of other loss mechanisms, this gives an attenuation coefficient proportional to f, due to excitation of acoustic waves in the gas.

Measurements on a variety of materials confirm the form of these predictions [132, 137, 169]. For example [137], if f is the frequency in GHz, the measured loss of Y, Z lithium niobate is typically

$$\alpha \approx 0.19f + 0.88f^{1.9} \quad (\text{dB}/\mu\text{sec}) \tag{6.29}$$

and for ST, X quartz, an orientation discussed below,

$$\alpha \approx 0.47f + 2.62f^2 \quad (\text{dB}/\mu\text{sec}), \tag{6.30}$$

where α is the attenuation per μsec of propagation path. These relations are plotted in Figure 6.10. The linear terms account for air loading, and are omitted if the crystal is in a vacuum. Theoretical predictions of the attenuation in a vacuum have been

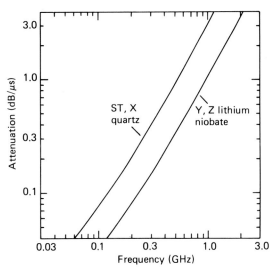

FIGURE 6.10. Free-surface attenuation, in air at room temperature, for quartz and lithium niobate. From data given by Slobodnik [137].

found to agree well with measurements on Y, X quartz [169]. The theory for gas loading gives fair agreement with experiment [137].

Surface imperfections can also contribute to the loss. The attenuation due to an uneven surface has been analysed [170], and the loss due to the presence of cracks or pits has been investigated experimentally [137]. However, for good quality polished crystals these factors are not usually very significant. Foreign material such as grease can cause substantial attenuation, but this is not significant if good cleanliness is maintained. At very low temperatures the attenuation can be much smaller [137], but this is not relevant to practical surface-wave devices.

(b). Non-linear Effects. If high surface-wave power levels are used, non-linear effects can become significant. Thus, in addition to the intended fundamental wave, with frequency ω, there will also be harmonic waves with frequencies 2ω, 3ω, 4ω, ... Consequently, there is additional attenuation of the fundamental, and, for a surface-wave device, unwanted frequency components may emerge at the output. The amplitudes of the fundamental and harmonic waves have been measured, as functions of position, using the optical probe of Figure 6.1 [136, 137, 169, 171]. In addition to generation of harmonics the non-linear process can also regenerate the fundamental by, for example, mixing the second and third harmonics, with frequencies 2ω and 3ω. Thus the energy is continually exchanged between the various components, and the amplitudes are complicated functions of position.

In practice, these effects are often affected by small amounts of dispersion due to, for example, the presence of a metal film. Dispersion reduces the severity of non-linear effects, as can be seen from simple phase-matching considerations. If the fundamental wave has frequency ω and wavenumber k, the non-linearity gives a 2ω term with "wavenumber" $2k$; however, if dispersion is present the wavenumber for surface waves with frequency 2ω is not $2k$ but rather $2k + \Delta k$, say. Thus the phase is not accurately matched, and non-linear effects become progressively weaker as $|\Delta k|$ increases.

These effects can be analysed using a coupled-wave approach [169]. If the fundamental power level is low enough, so that the harmonics have powers much less than the fundamental, the power $P_2(x)$ of the second harmonic is found to be given by [171, 172]

$$P_2(x) = \frac{Ck^4[P_1(0)]^2}{\omega W[(\Delta k)^2 + (2\alpha_1 - \alpha_2)^2]} \{\exp(-4\alpha_1 x)$$
$$+ \exp(-2\alpha_2 x) - 2\exp(-2\alpha_1 x - \alpha_2 x)\cos(x\Delta k)\}, \quad (6.31)$$

where the fundamental is taken to be generated at $x = 0$, so that the harmonic powers are zero at this point. Here $P_1(0)$ is the power of the fundamental at $x = 0$, Δk is defined above, W is the beam width and C is a constant depending only on the material and orientation. The factors α_1 and α_2 are attenuation constants defined so that, for low power levels, waves of frequency ω and 2ω have amplitudes proportional to $\exp(-\alpha_1 x)$ and $\exp(-\alpha_2 x)$ respectively.

Experimental results agree well with equation (6.31), though it is found that non-zero values for Δk must be used, even for a nominally free surface. There is therefore some dispersion, even though the analysis of Section 2.3 predicts that the free-surface velocity is independent of frequency. The dispersion is presumed to arise from surface imperfections induced by the crystal preparation (cutting and polishing). A summary of experimental measurements on Y, Z lithium niobate [173] shows that the dispersion varies substantially from sample to sample. Typically, the fractional change of velocity with frequency is 0.7×10^{-3} per GHz. This significantly affects the non-linear phenomena, though for most other purposes it is negligible.

For Y, Z lithium niobate, the constant C is found experimentally to be $1.1 \times 10^{-11} \text{m}^3 \text{watt}^{-1} \text{s}^{-1}$. With this value, Lean and Powell [172] found good experimental agreement with equation (6.31), for propagation on a "free" surface with an input frequency of 1.09 GHz, taking $\alpha_1 = 30 \text{m}^{-1}$, $\alpha_2 = 180 \text{m}^{-1}$ and $\Delta k = 2740 \text{m}^{-1}$. Figure 6.11 shows the second-harmonic power density, from equation (6.31), for an input power density $P_1(0)/W = 16.8$ watt/m, corresponding to Lean's experiment. In the absence of non-linear effects, the fundamental power density would be proportional to $\exp(-2\alpha_1 x)$. The second-harmonic power density has its first maximum at $x \approx \pi/\Delta k$, and at this point the ratio of the second-harmonic power to the ideal fundamental power is 0.17. Assuming that the fundamental power is depleted such that the second-harmonic power is accounted for, it can be concluded that the non-linear effect causes 0.8 dB attenuation of the fundamental at this point.

Although the non-linear phenomena have rather complex behaviour, Williamson [173] has shown that results for a variety of devices give quite good consistency when the depletion of the fundamental is considered. For devices using "free" surface propagation on Y, Z lithium niobate, a 1 dB depletion of the fundamental was

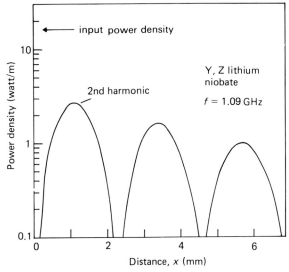

FIGURE 6.11. Second-harmonic power density, as a function of distance from launching point.

obtained for input power densities given by

$$\frac{P_1(0)}{W\lambda} \approx 4 \text{ to } 8 \text{ watt/mm}^2,$$

where λ is the wavelength of the fundamental. This relation gives a practical estimate of the power limitation necessary if non-linear effects are to be insignificant.

For a surface with aluminium metallisation, equation (6.31) is found to agree with experimental results, using the same value of C, but the non-linear effects are much weaker because of the increased dispersion [171, 172].

Equation (6.31) has also been confirmed experimentally for unmetallised Y, X quartz [169], taking $C = 1.5 \times 10^{-13} \text{m}^3 \text{watt}^{-1} \text{s}^{-1}$. For a fundamental frequency of 281 MHz, Δk was found to be 50 m^{-1}.

Although non-linear effects are readily avoided in most cases by restricting the input power level, they are in fact exploited in the surface-wave convolver, described in Chapter 10. There have also been a number of studies on the mixing of two non-collinear surface waves to produce a third surface wave at the sum or difference frequency [174].

6.4. TEMPERATURE EFFECTS AND VELOCITY ERRORS

Many surface-wave devices, particularly chirp filters and oscillators, are required to meet exacting specifications in which the temperature stability and velocity accuracy of the substrate are significant issues. Here, we first consider temperature changes and their effects on the response of a surface-wave device. Velocity errors, which give effects similar in form, are considered later.

Temperature Coefficient of Delay. Temperature effects can be characterised by considering a line source and a line receiver on a free surface, as in Figure 6.7, representing two transducers fabricated on the surface. Diffraction, propagation loss and dispersion are ignored here. If the source and receiver are separated by a distance l, the delay is $T = l/v_0$. The velocity v_0 depends on the density and the elastic and piezoelectric constants of the material, and if the temperature variations of these constants are known the variation of v_0 may be found using the method of Section 2.3.2. In addition, l varies with temperature because of the expansion of the material. We can thus define a temperature coefficient of delay, given by

$$\alpha_T \equiv \frac{1}{T}\frac{dT}{d\Theta} = \frac{1}{l}\frac{dl}{d\Theta} - \frac{1}{v_0}\frac{dv_0}{d\Theta}, \qquad (6.32)$$

where Θ is the temperature. Note that α_T is independent of the transducer separation l. Experimentally, α_T can be found from the frequency variation of a surface-wave delay line oscillator [175, 176], which is a sensitive measure of T.

In practice the fractional change of T is small, and T is often linear with Θ so that α_T is practically constant. Slobodnik [137] lists, for a variety of materials, the values of α_T and the temperature coefficient of velocity. For Y, Z lithium niobate, $\alpha_T = 94 \times 10^{-6}/°C$, though most practical materials give smaller values than this. For quartz the two terms on the right of equation (6.32) are generally both positive, and for some particular orientations they are equal at room temperature [177], so that $\alpha_T = 0$. One of these orientations is often used for devices requiring good temperature stability, and, by analogy with the AT- and BT-cuts for bulk crystals, this cut is called *ST-cut quartz*. Propagation is in the X-direction, so the orientation is called "ST, X quartz". As shown in Figure 6.12, this is a rotated Y-cut. The angle μ is 42.75°. The delay T is an approximately parabolic function of temperature, given experimentally by [176]

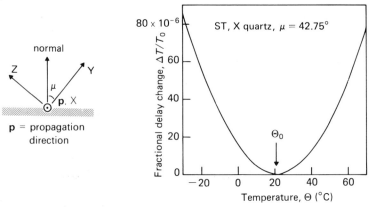

FIGURE 6.12. Temperature variation of delay for ST, X quartz.

PROPAGATION EFFECTS AND MATERIALS

$$T(\Theta) \approx T(\Theta_0) \cdot [1 + c(\Theta - \Theta_0)^2], \tag{6.33}$$

where $c = 32.3 \times 10^{-9} (^\circ C)^{-2}$ and $\Theta_0 = 21.1^\circ C$. The delay is a minimum at temperature Θ_0, which is known as the *turn-over temperature*. The delay variation is shown in Figure 6.12, where $\Delta T = T(\Theta) - T(\Theta_0)$ and $T_0 = T(\Theta_0)$. From equation (6.33), the temperature coefficient for this case is

$$\alpha_T = 2c(\Theta - \Theta_0) \tag{6.34}$$

and is therefore zero at the turn-over temperature.

Experiments on other orientations, using different values of the cut angle μ, show similar behaviour but with the turn-over temperature Θ_0 taking values from $-6^\circ C$ to $114^\circ C$ [176]. This feature can be exploited to optimise the stability over a specified temperature range.

Temperature Effects in Device Responses. In a surface-wave device, temperature effects cause the response to vary with temperature. Consider a two-transducer device, such as that in Figure 4.3, at temperatures Θ_1 and Θ_1', where Θ_1 is a temperature at which the response can be taken to be ideal. If T and T' give the delay between two points at temperatures Θ_1 and Θ_1' respectively, we can define a small quantity ε such that

$$T' = T(1 + \varepsilon). \tag{6.35}$$

Thus if T varies linearly with Θ we have $\varepsilon = \alpha_T(\Theta_1' - \Theta_1)$; for ST, X quartz, ε can be found from equation (6.33), noting that Θ_1 is not necessarily equal to Θ_0. It is assumed that equation (6.35) is valid for any two points in the propagation path. Since dispersion of the wave can be neglected in this context, the temperature change causes a change of time-scale in the impulse response, which changes from $h(t)$ at temperature Θ_1 to $h'(t)$ at temperature Θ_1', and hence

$$h'(t) = h\left(\frac{t}{1 + \varepsilon}\right). \tag{6.36}$$

These responses are taken to refer to the short-circuit case, where the transducers are connected to zero electrical impedances, thus excluding temperature effects in terminating circuits. There is also a small amplitude change, but this is insignificant and is neglected here. The frequency responses at temperatures Θ_1 and Θ_1' are $H_{sc}(\omega)$ and $H_{sc}'(\omega)$ given respectively by the Fourier transforms of $h(t)$ and $h'(t)$. Using the scaling theorem [equation (A.8)] we have

$$H_{sc}'(\omega) \approx H_{sc}(\omega[1 + \varepsilon]), \tag{6.37}$$

where, again, an insignificant amplitude change has been neglected. If we write $H_{sc}(\omega) = A(\omega) \exp[j\phi(\omega)]$ and $H_{sc}'(\omega) = A'(\omega) \exp[j\phi'(\omega)]$, we have

$$A'(\omega) = A(\omega[1 + \varepsilon])$$

and

$$\phi'(\omega) = \phi(\omega[1 + \varepsilon]). \tag{6.38}$$

Thus the amplitude and phase of the frequency response are simply scaled in frequency by the factor $(1 + \varepsilon)$.

For practical purposes the above equations are usually adequate for assessing temperature effects. However, special considerations apply when high temperature stability is required and ST, X quartz is used. For this case, a continuous aluminium film has been shown to depress the turn-over temperature Θ_0 by typically 20 to 50°C; for an interdigital device, in which the surface is partially metallised, temperature effects can be estimated by assuming a somewhat smaller reduction of Θ_0 [178]. A similar effect is found when the transducers are tuned using temperature-dependent inductors [176]. Also, adhesive mounting of the substrate can cause substantial changes of the temperature coefficient due to differential expansion, even when a flexible elastomer is used [179, p. 326]; to obtain the stability indicated by equation (6.33), the substrate must be free to expand.

Special considerations apply also to Reflective Array Compressors, which use surface wave propagation in two different directions, and these will be described in Section 9.6.1.

Velocity Accuracy. If a surface-wave device is designed assuming the velocity to be v, and the actual velocity v' differs slightly from v because of some error, this modifies the response in a manner similar to a temperature change. If T is the intended delay between two points and T' is the actual delay, we have $T' = T(1 + \varepsilon)$ as in equation (6.35), but now ε is given by

$$\varepsilon \approx -\frac{v' - v}{v} \tag{6.39}$$

With this value of ε, the effect of a velocity change is given by equations (6.36)–(6.38) above. This is valid provided the dispersion of the wave is negligible and the velocity is uniform.

Velocity errors can arise for several reasons. Errors in the crystal orientation are particularly important. Figure 6.13 defines the angular errors; $\delta\mu$ and $\delta\nu$ refer to the orientation of the surface normal, and $\delta\theta$ to the propagation direction. All three angles are affected by the accuracy of the crystal preparation, and $\delta\theta$ is also affected by the accuracy of the alignment during photolithography. For small angles, the velocity error for Y, Z lithium niobate is

$$(v' - v)/v = \{13.8\,(\delta v)^2 + 618\delta\mu + 138(\delta\theta)^2\} \times 10^{-6} \tag{6.40}$$

and, for ST, X quartz,

$$(v' - v)/v = \{343\,\delta v + 99(\delta\mu)^2 + 58(\delta\theta)^2\} \times 10^{-6}, \tag{6.41}$$

where the angles are in degrees [179]. In consequence, it is found that for practical devices it is often necessary to orient the crystal to an accuracy of 1° or better, in order to obtain adequate velocity accuracy.

Lithium niobate crystals are found to show sample-to-sample velocity variations

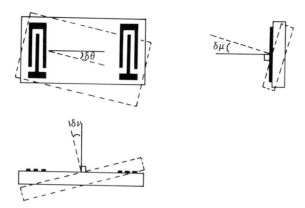

FIGURE 6.13. Definitions of misorientation angles. Broken lines indicate the ideal orientation.

of approximately 1 part in 10^3 [134, 180]. Such errors are unacceptable for some devices. However, if the device temperature is regulated the relatively large temperature sensitivity, usually a disadvantage, can sometimes be exploited to compensate for the velocity error.

Velocity changes are also caused by the presence of electrodes (Section D.2), groove arrays or other structures on the surface. The effects of these can often be estimated using equations (6.36)–(6.38), though in some cases dispersion and non-uniformity of the velocity are significant, and a more detailed analysis is then necessary.

6.5. MATERIALS FOR SURFACE-WAVE DEVICES

Table 6.1 summarises the relevant properties of some crystalline materials of practical interest, and the orientations of the rotated cuts are shown in Figures 6.12 and 6.14. In all cases the wave involved is a non-leaky piezoelectric Rayleigh wave and the propagation direction is a pure mode direction, so that the beam steering angle γ is zero. The parameters $\Delta v/v$ and ε_p^T are needed for analysis of transducers and multi-strip couplers. The sensitivity of beam steering to misorientation is given by $d\gamma/d\theta$ (Section 6.2.2). The table also indicates orientations for which the parabolic approximation is valid [137]. In these cases the diffraction pattern is a scaled version of the isotropic case with the scaling given approximately by $d\gamma/d\theta$, as in Section 6.2.4. The room-temperature attenuation is shown for a frequency of 1 GHz, and this includes the loss due to air loading, as in equations (6.29) and (6.30). Data for a wide range of other materials are given by Auld [170] and, in more detail, by Slobodnik and Conway [181].

The parameter $\Delta v/v$ is a measure of the piezoelectric coupling for surface waves. Schulz and Matsinger [182] have determined this parameter experimentally by measuring the change of phase of the surface wave as a metal film is evaporated on

TABLE 6.1
Crystalline materials for surface-wave devices

Material	Orientation	Velocity v_0 m/s	$\Delta v/v$ %	$\varepsilon_p^T/\varepsilon_0$	$\frac{d\gamma}{d\theta}$	Temp. coeff., $\alpha_T \times 10^6$ (°C)$^{-1}$	Attenuation at 1 GHz dB/μs	Ref.
	(a)	(e)	(b)		(c, e)	(d, e)	(e)	
LiNbO$_3$	Y, Z	3488	2.41 (2.15)	50.2	$-1.083^\dagger$	94	1.07	137
	128° rot.	3992	2.72	—	—	75	—	183
Quartz	Y, X	3159	0.09 (0.095)	4.52	0.65*	-24	2.6	137
	ST, X	3158	0.058 (0.067)	4.55	0.38*	0	3.1	137
LiTaO$_3$	Y, Z	3230	0.33 (0.36)	47.9	-0.211^*	35	1.14	137
	167° rot.	3394	0.75	47.9	-0.95	64	—	185
Bi$_{12}$GeO$_{20}$	001, 110	1681	0.68 (0.72)	43.6	-0.304^*	120	1.64	137, 186
	40.04° rot.	1830	0.31	43.6	-0.99^*	—	—	137, 187
Li$_2$B$_4$O$_7$	X, Z	3542	0.51	—	-0.11	0	—	189
AlPO$_4$	80.4° rot.	2741	0.245	—	0.901	0	—	190
GaAs	001, 110	2868	0.036	—	—	52	—	191

(a) Surface normal, followed by propagation direction. For rotated orientations, see Figure 6.14.
(b) $100(v_0 - v_m)/v_0$. Experimental values from Schulz and Matsinger [182] in brackets.
(c) $\gamma = 0$ in all cases, * = Parabolic approximation for diffraction is valid. $\dagger$ = Parabolic approximation is invalid [137].
(d) Zero entries apply only at one temperature, see text. For non-zero entries α_T is approximately constant.
(e) For free-surface propagation.

PROPAGATION EFFECTS AND MATERIALS 153

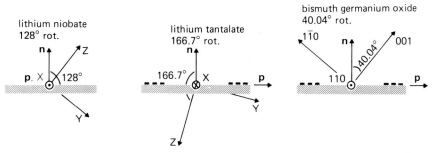

FIGURE 6.14. Orientations for rotated cuts shown in Table 6.1 **n** is surface normal, **p** is surface-wave propagation direction.

to the substrate. Their experimental values are shown in brackets on Table 6.1. Theoretical values are obtained by the method of Section 2.3.2, using piezoelectric and stiffness constants obtained from bulk measurements. There are some significant discrepancies, notably for Y, Z lithium niobate, attributed to small errors in the measured values of the constants. The experimental values in brackets are therefore more reliable.

The piezoelectric coupling for surface waves is often expressed by a parameter k^2. This is usually defined [80, 81, 84] such that the Q-factor for a single-electrode uniform transducer at its fundamental centre frequency ω_c is

$$Q_t = \omega_c C_t / G_a(\omega_c) = \pi/(4k^2 N_p)$$

provided electrode interactions are weak. With this definition, the quasi-static analysis of Section 4.6 gives $k^2 = 2.255 \Delta v/v$, and this result also follows from normal mode theory [91]. However, some authors define $k^2 = 2\Delta v/v$.

The most significant parameters for a material are the piezoelectric coupling $\Delta v/v$ and the temperature coefficient α_T. Large values of $\Delta v/v$ generally give lower insertion losses for devices, though they also give stronger electrode interactions and so are not always desirable. *Lithium niobate*, $LiNbO_3$ (trigonal, class 3m) has the popular Y, Z orientation giving exceptionally strong coupling and low attenuation, and this is also a minimal-diffraction orientation, since $d\gamma/d\theta$ is close to -1. However, the temperature coefficient is large. Also, transducers on this material can couple quite strongly to bulk waves, though since $\Delta v/v$ is large bulk-wave effects can be minimised by using a multi-strip coupler. Further information is given in Figures 2.8 and 2.9. The 128° rotated orientation gives much weaker coupling to bulk waves [183].

Quartz, SiO_2 (trigonal, class 32) is the other popular material. The *ST*-cut is used for devices requiring good temperature stability, such as oscillators and chirp filters, though $\Delta v/v$ is small. Searches for orientations giving even better temperature stability have been conducted, with some success. For example, the SST orientation [184] uses a *BT*-cut crystal and gives a delay variation similar to equation (6.33), with $c = 20 \times 10^{-9} (°C)^{-2}$, and with $\Delta v/v = 0.068\%$. *Lithium tantalate*, $LiTaO_3$ (trigonal, class 3m) has a Y, Z orientation with properties intermediate between lithium niobate and quartz, in that the coupling is stronger than quartz while the temperature stability is better than lithium niobate. Also, the rotated orientation shown in Table 6.1 gives minimal diffraction, less than Y, Z lithium niobate, and very little coupling to bulk waves [185]. *Bismuth germanium oxide*, $Bi_{12}GeO_{20}$ (cubic, class 23) is noteworthy for its unusually low surface-wave velocities, implying that for long delays there is the advantage that devices become more compact. The rotated orientation shown gives exceptionally low diffraction spreading [187].

Although quartz can be used for applications requiring good temperature stability, its weak coupling is a disadvantage and consequently other temperature-stable materials have been sought [188]. Two recent examples, *lithium tetraborate*, $Li_2B_4O_7$, and *berlinite*, $AlPO_4$, are shown in Table 6.1. Both give delays with the form of equation (6.33), with $c = 230 \times 10^{-9}(°C)^{-2}$ and $c = 220 \times 10^{-9}(°C)^{-2}$ respectively. However, neither of these has yet come into common use.

Gallium arsenide, GaAs, is included because recent investigations have explored a variety of possibilities for novel surface-wave devices, exploiting the semiconducting property [191].

Piezoelectric ceramics, such as PZT, have also been used experimentally. In the past the large attenuation and poor velocity repeatability have precluded the practical use of these materials. However they are relatively inexpensive, and recent technical developments [192] have enabled better acoustic properties to be obtained.

Effects Due to Metallic Films. Since metallic films are used in all practical surface-wave devices, their influence on the wave propagation characteristics is of considerable significance. In particular, the propagation becomes slightly dispersive. For a continuous film, also discussed in Section 2.3.6, results for a variety of materials are given by Penunuri and Lakin [193]. Provided the film is thin, the phase velocity v is given by

$$v \approx v_m(1 - Ah/\lambda_0) \qquad (6.42)$$

where $\lambda_0 = 2\pi v_0/\omega$ is the free-surface wavelength, h is the film thickness and A is a constant. For example, an aluminium film on ST, X quartz gives $A = 0.183$, while on Y, Z lithium niobate it gives $A = 0.287$. For these cases equation (6.42) is a good approximation provided $h/\lambda_0 \leq 0.01$, which is usually valid in practice.

A metal film also increases the surface-wave attenuation. For example, room-temperature measurements [194] on Y, Z lithium niobate samples with continuous aluminium films gave an attenuation coefficient $\alpha \approx Kf^{2.2}$ dB/μs, with $K = 3.0$ for a 500 Å film and $K = 5.2$ for a 2000 Å film, taking the frequency f to be in GHz. In addition, aluminium films on ST, X quartz modify the temperature stability, as mentioned in Section 6.4.

Dielectric Films. A *piezoelectric* film can be used to enable an interdigital transducer to generate surface waves on a non-piezoelectric substrate, thus extending the range of applicable substrate materials. The foremost example of this is zinc oxide (ZnO), already mentioned in Section 3.5. Zinc oxide films have been used on glass substrates, and on silicon substrates. In addition they are of interest for gallium arsenide substrates. The piezoelectric coupling for this material is very weak, and the use of a zinc oxide film enables much better transducer efficiency to be obtained [191].

Another notable piezoelectric film is aluminium nitride (AlN), which has been investigated mainly on sapphire (Al_2O_3) substrates [195]. This combination has been found to give low dispersion and an unusually high surface-wave velocity, about 6000 m/s. This is potentially attractive for high frequency devices since, for a given transducer design, the centre frequency is proportional to the velocity. However, the films produced to date have not generally been of good enough quality for practical devices.

A dielectric film can also be used to compensate for temperature effects in the substrate. The primary example of this is a silicon dioxide (SiO_2) film on a Y, Z lithium tantalate substrate [196]. In this case the film is polycrystalline, and so non-piezoelectric. For a film thickness equal to about half the surface-wave wavelength the temperature coefficient of the substrate is compensated, so that good temperature stability and strong piezoelectric coupling are both obtained. In fact, the temperature stability is found to be considerably better than that of ST, X quartz. This and several other examples are discussed further by Lewis [188].

Chapter 7

Delay Lines and Multi-phase Transducers

In previous chapters we have considered the analysis of interdigital transducers and multi-strip couplers, the essential components of many surface-wave devices, and the variety of propagation effects that can be of practical significance. Here, and in subsequent chapters, we are mainly concerned with the design and performance of practical devices, making use of the earlier analysis to deduce many of the important features.

The simplest form of surface-wave device is the elementary delay line shown in Figure 7.1, consisting essentially of two identical uniform transducers. The function of this device is to delay an applied input signal without appreciably distorting it. The delay, related directly to the transducer separation, is typically in the range 1 to 50 μsec. Section 7.1 discusses this device, and is mainly concerned with the practical performance of uniform transducers. It should be noted that much of this discussion is directly relevant to other types of device, since uniform transducers are widely used. In addition, many of the conclusions apply qualitatively to more complex transducers, including the apodised transducers often used in bandpass filters and chirp transducers used in chirp filters. It will be seen that a delay line cannot generally be electrically matched such that the insertion loss is minimised, because this gives an unacceptable triple-transit output signal due to the acoustic reflectivity of the transducers. There are several special techniques for overcoming this limitation, in particular the use of multi-phase unidirectional transducers which are considered in Section 7.2.

In this chapter, the transducer analysis makes use of the quasi-static results of Chapter 4. Thus, electrode interactions are neglected, and this is usually valid in practice as discussed in Section 4.2. The reader is reminded that electrode interactions cause a small reduction of the surface-wave velocity within a transducer, and this is not allowed for in the analysis (Section 4.4).

7.1. DELAY LINES

In this section we consider delay lines using uniform transducers, such as that shown in Figure 7.1. In order to minimise the distortion of the applied signal the device must

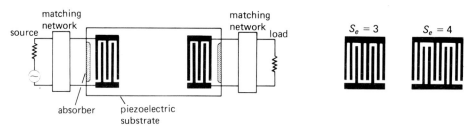

FIGURE 7.1. *Left*: Delay line using single-electrode transducers ($S_e = 2$). *Right*: Transducers with $S_e = 3$ or 4.

be non-dispersive, that is, its group delay must be almost constant over the band occupied by the signal spectrum. Chirp filters, which are very dispersive, are sometimes called dispersive delay lines, and are described in Chapter 9. We also exclude here non-dispersive delay lines using two chirp transducers, arranged such that the dispersion cancels.

Most of this section is concerned with the performance of transducers and their influence on the device performance, particularly on its bandwidth, insertion loss, and the level of the triple-transit output signal. These parameters can be found from the analysis of Chapter 4. The objective here is to illustrate some important practical features which arise when uniform transducers are used. For clarity some simplifying approximations are introduced, but the basic conclusions are nearly always valid in practice. Section 7.1.1 below considers the bandwidth and insertion loss of the device, including the effect of matching networks, and Section 7.1.2 discusses parasitics. The triple-transit signal is considered in Section 7.1.3, which also discusses methods of minimising it. Section 7.1.4 discusses the performance of practical delay lines, including special techniques of obtaining long delays or high frequencies.

7.1.1. Bandwidth and Conversion Loss of Uniform Transducers

Surface-wave devices often include simple matching networks, as indicated in Figure 7.1, to reduce the insertion loss. However, this is not always necessary, and sometimes it is advisable to omit the matching networks in order to minimise the triple-transit signal. Both cases will be considered here, and it will be seen that there is an interesting contrast in the role of the piezoelectric coupling parameter $\Delta v/v$. For unmatched transducers, $\Delta v/v$ has an important influence on the transducer conversion loss, which decreases as $\Delta v/v$ increases. When networks are included to reduce the loss, it is found that there is generally a trade-off between loss and bandwidth, and the trade-off is determined by the value of $\Delta v/v$. Thus the two cases have very significant practical differences.

We consider the common types of uniform transducer with $S_e = 2$, 3 or 4 electrodes per period, as illustrated in Figure 7.1. For the present, the matching networks will be excluded, so that each transducer in the device is connected directly to the source or load. For each transducer, an electrical equivalent circuit can be defined, such that its admittance equals the transducer admittance $Y_t(\omega)$. As shown in Figure 7.2(a), the circuit comprises a capacitance C_t, and acoustic conductance

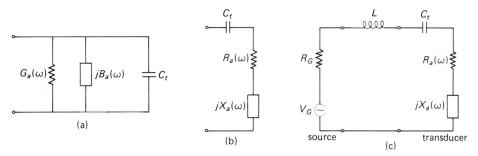

FIGURE 7.2. (a), (b): Equivalent circuits for transducer. (c): Tuned transducer connected to source.

$G_a(\omega)$ and susceptance $B_a(\omega)$, connected in parallel. Parasitic elements are ignored here, but will be considered in Section 7.1.2.

The fundamental centre frequency of the transducer is $\omega_c = 2\pi v_0/(pS_e)$, where p is the electrode pitch. For frequencies near ω_c, the conductance is, from equation (4.108),

$$G_a(\omega) \approx G_a(\omega_c)\left[\frac{\sin X}{X}\right]^2 \qquad (7.1)$$

where $X = \pi N_p(\omega - \omega_c)/\omega_c$ and N_p is the number of periods. Thus $G_a(\omega)$ is maximised at $\omega = \omega_c$, as shown in Figure 4.14. The acoustic susceptance $B_a(\omega)$ is ignored here since it generally has little effect on the response (Section 4.6). When a generator is applied to the transducer, the power dissipated in the conductance accounts for the power of the generated surface waves. Assuming initially that the voltage across the transducer is independent of frequency, the surface-wave power is proportional to $G_a(\omega)$, and this gives the transducer bandwidth. Here we consider the 1.5 dB bandwidth, between the points where $G_a(\omega) = G_a(\omega_c)/\sqrt{2}$, since this is the 3 dB bandwidth of the delay line. From equation (7.1), the 1.5 dB transducer bandwidth $\Delta\omega$ is given by

$$\Delta\omega/\omega_c = 0.638/N_p. \qquad (7.2)$$

This is called the *acoustic bandwidth*, since it does not allow for circuit effects.

The transducer Q-factor Q_t is, as in Section 4.6, given by

$$Q_t \equiv \frac{\omega_c C_t}{G_a(\omega_c)} = \frac{\tilde{Q}_t}{N_p \Delta v/v}, \qquad (7.3)$$

where $\tilde{Q}_t$ is a constant near unity, given in Table 4.1 and Figure 4.15; for a single-electrode transducer ($S_e = 2$) with metallisation ratio $a/p = \frac{1}{2}$, we have $1/\tilde{Q}_t = 2.871$. Note that the acoustic bandwidth is not determined by Q_t, which refers to the equivalent electrical circuit.

For many transducers $N_p \Delta v/v \ll 1$, so that $Q_t \gg 1$, and this is assumed to be true here. For single-electrode transducers this condition is nearly always used, in order to minimise electrode interaction effects. For $Q_t \gg 1$, equation (7.3) shows that the transducer admittance is dominated by the capacitive term $j\omega C_t$. Thus, quite large losses are generally incurred if the transducer is connected directly to a resistive source or load. The power conversion coefficient is given by equation (4.77).

Lower loss can be obtained by using an inductor to tune out the capacitance at frequency ω_c. If the inductor is connected across the transducer, the impedance seen by the source, at frequency is ω_c, is $1/G_a(\omega_c)$. For most transducers this is not close to 50 Ω, the usual source impedance. It is however often found that a good match can be obtained using a series inductor, and to appreciate this we consider an alternative equivalent circuit.

Series Equivalent Circuit and Matching. With an appropriate choice of component values, the transducer can be represented by the series equivalent circuit of Figure 7.2(b). It is convenient to include the transducer capacitance C_t explicitly, since this is usually the dominant term. The remaining components are the *radiation resistance* $R_a(\omega)$ and an acoustic reactance $X_a(\omega)$. When a voltage is applied to the transducer, the power dissipated in $R_a(\omega)$ accounts for the surface-wave power generated. Since the series and parallel circuits are equivalent, $R_a(\omega)$ and $X_a(\omega)$ can be found by equating the impedances of the two circuits of Figures 7.2(a) and (b). The reactance $X_a(\omega)$ generally has a form similar to $B_a(\omega)$, and is small at the centre frequency. The radiation resistance is found to be

$$R_a(\omega) = \frac{G_a(\omega)}{[G_a(\omega)]^2 + [B_a(\omega) + \omega C_t]^2}. \tag{7.4}$$

Since we assume $Q_t \gg 1$, the $G_a(\omega)$ and $B_a(\omega)$ terms in the denominator are insignificant, so that $R_a(\omega)$ is approximately proportional to $G_a(\omega)$, given by equation (7.1). Also, at the centre frequency, $R_a(\omega_c)$ is approximately equal to $G_a(\omega_c)/(\omega_c C_t)^2$. It is convenient to denote this quantity by $\hat{R}_a$, so that

$$R_a(\omega_c) \approx \hat{R}_a, \quad \text{for } N_p \frac{\Delta v}{v} \ll 1$$

and

$$\hat{R}_a \equiv G_a(\omega_c)/(\omega_c C_t)^2. \tag{7.5}$$

It was shown in Section 4.6 that $G_a(\omega_c)$ is proportional to $\omega_c N_p^2 W$, where W is the transducer aperture, and that C_t is proportional to WN_p. Thus $\hat{R}_a$ is proportional to $1/(\omega_c W)$, and is *independent* of the number of periods, N_p. It is also found that the aperture W can usually be chosen such that $\hat{R}_a$ is close to 50 Ω, and hence the transducer can be well matched to a 50 Ω source by means of a series inductor, as on Figure 7.2(c), where $L = 1/(\omega_c^2 C_t)$. When this is done, the centre-frequency conversion loss is ideally 3 dB, and the insertion loss for a delay line using two matched transducers is 6 dB. The value of $\hat{R}_a$ is therefore of considerable practical utility, as first noted by Smith et al. [197]. Transducers are usually designed with metallisation ratio $a/p = \frac{1}{2}$ since this maximises the efficiency, as shown by Figure 4.15. For a given value of a/p, $\hat{R}_a$ depends only on the value of S_e, the aperture W expressed in wavelengths, and the material parameters $\Delta v/v$, v_0 and ε_p^T.

DELAY LINES AND MULTI-PHASE TRANSDUCERS

TABLE 7.1
Apertures for uniform transducers with $\hat{R}_a = 50\,\Omega$.
Fundamental response, $a/p = \frac{1}{2}$

S_e	W/λ_c	
	Y, Z lithium niobate	ST, X quartz
2	111	42
3	65	25
4	60	23

Table 7.1 gives data for Y, Z lithium niobate and ST, X quartz. The entries are values of W/λ_c required to give $\hat{R}_a = 50\,\Omega$, where $\lambda_c = 2\pi v_0/\omega_c = pS_e$ is the wavelength at the centre frequency. The data are obtained from materials parameters given in Table 6.1 and the analysis of Section 4.6, equations (4.111) and (4.115), together with equation (7.5). Typical apertures required are thus in the range 25 to 100 wavelengths; these are convenient practically since they generally give little diffraction spreading and the physical size, typically a few mm, is convenient for fabrication.

In some cases the transducer cannot be designed such that $R_a(\omega_c)$ is close to $50\,\Omega$. However, it is still possible to match to a $50\,\Omega$ source if two reactive components are used. As shown in Figure 7.3(a), we consider a complex load $Z_0 = R_0 + jX_0$ representing the transducer, with a reactance X_1 added in series and a further parallel reactance X_2. Generally, R_0, X_0, X_1 and X_2 will all be functions of frequency. For any specified frequency it is possible to choose X_1 and X_2 such that the impedance seen at the terminals on the left is equal to a specified real value R_G. Using conventional network analysis, the required values of X_1 and X_2 are given by

$$(X_0 + X_1)^2 = R_0(R_G - R_0) \tag{7.6}$$

and

$$X_2 = -R_0 R_G/(X_0 + X_1). \tag{7.7}$$

Since the left side of equation (7.6) must be positive, this method is valid only if $R_0 \leq R_G$. If this is so, there are two solutions for X_1 and X_2. If X_1 is positive the required reactance can be obtained by using an inductor with $L = X_1/\omega$, while for negative X_1 a capacitor can be used, with $C = -1/(\omega X_1)$. The same remarks apply for X_2.

An alternative circuit, in which the first reactance X_1 is in parallel with the load, is shown in Figure 7.3(b). It is convenient here to define G_0 and B_0 such that $G_0 + jB_0 = 1/Z_0$. For a resistive impedance R_G at the terminals on the left, it is found that X_1 and X_2 are given by

$$(B_0 - 1/X_1)^2 = -G_0(G_0 - 1/R_G)$$

and

$$X_2 = (B_0 - 1/X_1) R_G/G_0. \tag{7.8}$$

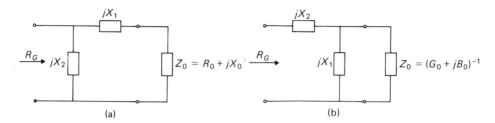

FIGURE 7.3. Two-component matching networks.

This method is valid if $G_0 \leq 1/R_G$. It is readily shown that at least one of the two circuits is valid for any specified values of Z_0 and R_G.

Bandwidth for Minimum loss. If the transducer is electrically matched to a source, at frequency ω_c, then all of the available power is converted into surface waves at this frequency. Thus the conversion loss is minimised, with a value of 3 dB due to the bidirectional nature of the transducer. One of the penalties paid for this low loss is that the matching circuit may substantially reduce the overall bandwidth, which is determined by the frequency variation of the surface-wave power generated. This circuit effect is significant when N_p is relatively small, and is associated with the electrical Q-factor of the circuit.

For the moment we consider only the series-tuned circuit of Figure 7.2(c), where the inductor tunes out the capacitance at frequency ω_c. Suppose first that N_p is small, so that the transducer acoustic bandwidth is large. In this case, for frequencies near ω_c, $R_a(\omega)$ can be taken as a constant approximately equal to $\hat{R}_a$, and $X_a(\omega)$ is small. Thus the circuit becomes a conventional series-tuned circuit. Since the transducer is matched we have $\hat{R}_a \approx R_G$, so that the Q-factor of the circuit is $1/(2\omega_c C_t \hat{R}_a)$. Using equation (7.5) we have $Q = \frac{1}{2}Q_t$, where Q_t is the transducer Q-factor of equation (7.3). We consider the bandwidth between the 1.5 dB points, where the power dissipated in the radiation resistance (that is, the surface-wave power) is equal to $1/\sqrt{2}$ times its maximum value. This is found to give a fractional bandwidth of $0.644/Q$, so that

$$\frac{\Delta\omega}{\omega_c} \approx \frac{1.29}{Q_t} = 1.29\frac{N_p \Delta v/v}{\tilde{Q}_t}, \quad \text{for small } N_p. \tag{7.9}$$

For large N_p (but still subject to the constraint $N_p \Delta v/v \ll 1$), the transducer acoustic bandwidth is small, so that $R_a(\omega)$ varies rapidly with ω and the electrical bandwidth of the circuit is not relevant. In this case the bandwidth is approximately given by the transducer acoustic bandwidth of equation (7.2). There is however an additional complication arising because the transducer is matched to the source, and this will be considered further in Section 8.4.2 below. For a matched transducer, it is found that the 1.5 dB bandwidth is

$$\Delta\omega/\omega_c \approx 1.14/N_p, \quad \text{for large } N_p. \tag{7.10}$$

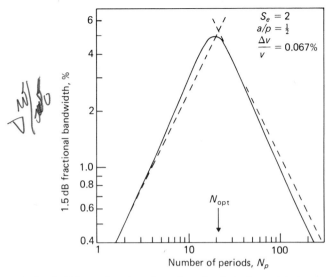

FIGURE 7.4 Bandwidth as a function of number of periods, for matched series-tuned single-electrode transducers on ST, X quartz.

Equations (7.9) and (7.10) are shown as broken lines on Figure 7.4, taking $\Delta v/v = 0.067\%$, appropriate for ST, X quartz. For equation (7.9), $\tilde{Q}_t$ is taken to be $1/2.871$, appropriate for a single-electrode ($S_e = 2$) transducer with $a/p = \frac{1}{2}$. The bandwidth is proportional to N_p when N_p is small, and is proportional to $1/N_p$ when N_p is large. There is therefore an intermediate value of N_p at which the bandwidth is maximised, as first noted by Smith et al. [197]. The solid curve on Figure 7.4 is the 1.5 dB bandwidth calculated by analysing the circuit of Figure 7.2(c), using equation (7.1) for $G_a(\omega)$ and equation (4.110) for $B_a(\omega)$. The broken lines are seen to give a reasonable estimate for the bandwidth, for most values of N_p. The value of N_p for maximum bandwidth is denoted N_{opt}, and can be estimated by equating the bandwidths of equations (7.9) and (7.10), giving

$$N_{opt} \approx 0.55/\sqrt{\Delta v/v} \qquad (7.11)$$

for $S_e = 2$ and $a/p = \frac{1}{2}$. Figure 7.4 also shows that the maximum bandwidth, obtained when $N_p \approx N_{opt}$ is given by $\Delta\omega/\omega_c \approx 1/N_{opt}$. Thus, for ST, X quartz ($\Delta v/v = 0.067\%$) we have $N_{opt} = 21$ and the maximum bandwidth is 4.8%. For Y, Z lithium niobate ($\Delta v/v = 2.15\%$), $N_{opt} = 4$ and the maximum bandwidth is 25%. Transducers with $S_e = 3$ or 4 give larger values of N_{opt}, and the maximum bandwidth is smaller, because $\tilde{Q}_t$ is larger (Table 4.1).

Trade-off Between Bandwidth and Loss. Clearly the overall bandwidth of a matched transducer cannot substantially exceed the electrical bandwidth or the acoustic bandwidth. If N_p is reduced, the acoustic bandwidth increases but the electrical bandwidth decreases. However, it is possible to increase the *electrical* bandwidth if the transducer does not need to be matched to the source. If the acoustic bandwidth is also increased, by reducing N_p, this enables a larger overall bandwidth to be obtained, at the expense of larger conversion loss [198].

In the case of the series-tuned circuit of Figure 7.2(c), where now R_G is not necessarily equal to $\hat{R}_a$, the electrical Q of the circuit is $1/[\omega_c C_t (\hat{R}_a + R_G)]$, giving $Q = Q_t/(1 + R_G/\hat{R}_a)$. Thus a larger electrical bandwidth is obtained if $R_G > \hat{R}_a$. As before, the 1.5 dB electrical bandwidth is $0.644/Q$, and is therefore proportional to N_p, while the acoustic bandwidth, equation (7.2), is proportional to $1/N_p$. The optimum value of N_p can be estimated by equating the bandwidths. The centre-frequency conversion loss, obtained by analysing Figure 7.2(c), is given by

$$\text{conversion loss} \approx 10 \log \left[\tfrac{1}{2}(1 + \hat{R}_a/R_G)^2 R_G/\hat{R}_a\right] \quad \text{(dB)}, \qquad (7.12)$$

which gives 3 dB when $R_G/\hat{R}_a = 1$, and greater losses for other values of $R_G/\hat{R}_a$. The maximum 1.5 dB bandwidth can therefore be evaluated as a function of conversion loss by increasing $R_G/\hat{R}_a$, and evaluating the optimum value of N_p for each value of $R_G/\hat{R}_a$. Figure 7.5 shows the result obtained by numerical analysis [199]. To a good approximation, a universal curve valid for all materials is obtained if the fractional bandwidth is divided by $\sqrt{\Delta v/v}$. The figure includes scales giving the required optimum value of N_p, for ST, X quartz and Y, Z lithium niobate. For conversion losses greater than about 7 dB, the optimum value of N_p is approximately $1/(2\Delta\omega/\omega_c)$.

The value of $R_G/\hat{R}_a$ can be increased by increasing the aperture W, since $\hat{R}_a$ is proportional to $1/W$. However, a sufficient increase cannot always be conveniently

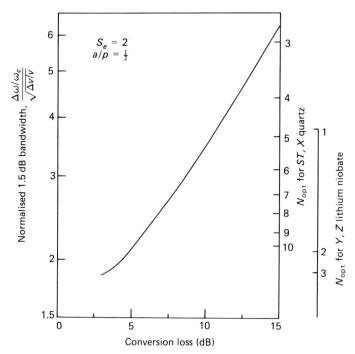

FIGURE 7.5. Maximum bandwidth as a function of conversion loss, for series tuned single-electrode transducers.

obtained by this method. An alternative method is to use a transformer, so that the transducer "sees" a source impedance greater than 50Ω. The two-component circuits of Figure 7.3 can be used for this purpose, though since they only act as ideal transformers at one frequency the above discussion is not strictly valid for this case. Reeder et al. [200] have developed this method using an inverter network, essentially a cascade of several two-component sections of the type shown in Figure 7.3. Using $N_p = 1$ in each transducer, they demonstrated a Y, Z lithium niobate delay line with a 3 dB bandwidth of 43%, centred at 105 MHz, and with 15 dB insertion loss.

A related problem is that of efficient matching into a series resonant circuit, given that some specified bandwidth is required. Reeder and Sperry [201] have investigated the use of a transmission line to give the required impedance transformation for this case.

An alternative method of obtaining wide fractional bandwidths is to use chirp transducers (Chapter 9), arranged such that the dispersion cancels. This is sometimes preferable as it gives transducer designs less sensitive to strays.

Standing-wave Ratio. In some applications the reflection of the electrical signal applied to the device is of some concern, since this causes standing-wave effects if the device is used with cables of length comparable with the electromagnetic wavelength. The requirement is usually specified in terms of the voltage standing wave ratio (VSWR), equal to $(1 + R)/(1 - R)$, where R is the magnitude of the voltage reflection coefficient. If there are no resistive losses in the transducer or matching network, the transducer conversion loss is $-10 \log (1 - R^2)/2$ (dB), and is thus directly related to the VSWR. In practice this requirement can be very restrictive. For example, if the VSWR is not to exceed 2, as is often the case, the conversion loss must not exceed 3.5 dB, and therefore must not vary by more than 0.5 dB over the specified band. Thus the VSWR requirement can usually be met only over a small part of the transducer acoustic bandwidth.

Better VSWR performance can be obtained by adding a series resistor, since this enables an impedance close to the characteristic impedance of the cable to be obtained, without requiring the transducer itself to be closely matched. Smith [202] has discussed this method in detail, showing the relevant trade-offs. There is however the disadvantage that the additional resistor dissipates power, thus increasing the insertion loss of the device. An alternative solution is to incorporate well-matched amplifiers, with good reverse isolation, into the same package as the surface-wave device. This improves the VSWR without significantly degrading the overall noise figure.

7.1.2. Parasitic Components

In this chapter it has been assumed so far that any parasitic components, such as stray capacitance or the resistivity of the electrodes, have negligible effects on the transducer performance. This is often true for transducers operating at relatively low frequencies, but at high frequencies strays can be very significant. We consider first the electrode resistivity, a particularly important effect which can often be predicted fairly accurately.

(a) Electrode Resistivity.

We assume here that the resistivity of the electrodes is small, and that it can be represented by adding a series resistance R_E to the ideal equivalent circuit, as shown in Figure 7.6(a). For the analysis here the transducer is taken to be unapodised and to have regular electrodes, though it is not necessary to constrain it to be uniform; thus the discussion here applies, for example, to the transducer of Figure 4.9(a), which was analysed in Section 4.5. To estimate the value of R_E it is assumed that piezoelectric coupling can be neglected, so that the transducer behaves as a capacitor with capacitance C_t.

We consider first the current in electrode n, when the resistivity is negligible. When unit voltage is applied across the transducer, the total charge on electrode n is q_n, given by equation (4.90), and the current entering the electrode is therefore $j\omega q_n$, at frequency ω. Since the charge density is uniform in the y-direction, along the length of the electrode, the current varies linearly in this direction. Thus the current at some point y is given by

$$I_n(y) = j\omega q_n(\tfrac{1}{2} \pm y/W), \tag{7.13}$$

where W is the transducer aperture and the origin for y is taken to be centrally located in the transducer. The sign in equation (7.13) depends on which bus-bar the electrode is connected to. The current is invariant across the width of the electrode.

A small amount of resistivity, such that $R_E \ll 1/(\omega C_t)$, has little effect on the voltage across the ideal transducer, or on the electrode currents given by equation (7.13). The main consequence is a loss of power, which can be accounted for by the resistance R_E. It is assumed that the bus-bars are wide enough for their resistance to be negligible. Defining r_E as the resistance of one electrode per unit length, the power dissipated in electrode n is

$$P^{(n)} = \tfrac{1}{2} r_E \int_{-W/2}^{W/2} |I_n(y)|^2 \, dy = r_E \omega^2 q_n^2 W/6. \tag{7.14}$$

The total power lost is the sum of the $P^{(n)}$, and is equated with the power dissipated in R_E. Since the current through R_E is approximately $j\omega C_t$, we have

$$\sum_{n=1}^{N} P^{(n)} = \tfrac{1}{2}\omega^2 C_t^2 R_E,$$

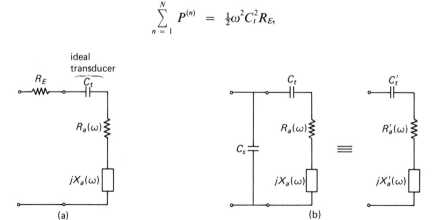

FIGURE 7.6. Representation of (a) electrode resistance, (b) stray capacitance.

where N is the number of electrodes. Using equation (4.91) for C_t, it is found that R_E is given by

$$R_E = \frac{r_E W}{3} \left[\sum_{n=1}^{N} q_n^2\right] \Bigg/ \left[\sum_{n=1}^{N} \hat{P}_n q_n\right]^2, \tag{7.15}$$

where $\hat{P}_n = 0$ or 1 is the polarity of electrode n. The values of the q_n are given by the analysis in Section 4.5.1.

For the uniform transducers of Figure 7.1 the relative values of the q_n can be readily reduced from symmetry arguments, assuming end effects to be negligible. We thus find

$$\begin{aligned} R_E &= 2r_E W/(3N_p), \quad \text{for } S_e = 2, \\ &= r_E W/(2N_p), \quad \text{for } S_e = 3, \\ &= r_E W/(3N_p), \quad \text{for } S_e = 4. \end{aligned} \tag{7.16}$$

The resistance of one electrode per unit length is $r_E = \varrho/a$, where a is the electrode width and ϱ is the resistivity of the metal film, in ohms per square. For the frequencies and film thicknesses of interest here the skin effect is insignificant and, except for very thin films, the resistivity is found to correspond quite well with the bulk resistivity of the metal. Measurements on aluminium films [203] give

$$\varrho \approx 0.04/h \quad \text{ohm/square} \tag{7.17}$$

for $h \geqslant 0.05$, where h is the film thickness in μm. For smaller values of h the resistivity increases more rapidly. Typical experimental film thicknesses are in the range 0.05 to 0.3 μm.

For practical transducer designs, R_E values of 20 Ω or greater are not unusual. The practical significance of this can be found by analysing the circuit of Figure 7.6(a), including the source and any matching network. It is often the case that the losses become significant when R_E is comparable to, or greater than, $\hat{R}_a$. Thus, it is generally important to consider the electrode resistivity when designing transducers. If the predicted value of R_E is too large it may be reduced by increasing h or by reducing the aperture W, since $R_E/\hat{R}_a$ is proportional to W^2.

A more detailed analysis of the effects of electrode resistivity has been given by Lakin [204] who showed that, in addition to the loss of power, there is a variation of voltage along the length of any one electrode. This implies that the power of the surface wave generated by the transducer will vary across its aperture. However, if the resistivity is small, Lakin's analysis shows that an external resistance R_E may be added, as above, to account for the power loss. Also, for $R_E \leqslant \hat{R}_a$ the voltage variations are usually insignificant. Thus the expressions given above generally give an adequate representation of electrode resistivity, for practical purposes.

(b) Other Parasitic Components. In many cases, stray capacitance between the transducer and the package can be significant. Thus, if R_E is negligible the transducer may be represented as in Figure 7.6(b), where a stray capacitance C_s is connected across the ideal transducer. It is convenient to consider the effect of this in

terms of an equivalent series circuit with components C_t', $R_a'(\omega)$ and $X_a'(\omega)$; these components can be evaluated by equating the impedances of the two circuits in Figure 7.6(b). For simplicity it is assumed that, as for most transducers, $R_a(\omega) \ll 1/(\omega C_t)$ and $X_a(\omega) \ll 1/(\omega C_t)$, and it is then found that

$$R_a'(\omega) \approx R_a(\omega) \frac{C_t^2}{(C_t + C_s)^2} \qquad (7.18)$$

and

$$C_t' \approx C_t + C_s \qquad (7.19)$$

Thus the stray capacitance causes an apparent reduction of the radiation resistance, affecting the matching, and an apparent increase of the transducer capacitance. These effects are usually of practical significance if C_s is in the region of $C_t/4$ or greater. In practice, C_s is typically $\frac{1}{2}$ to 1 pF, but transducer capacitances can sometimes be comparable with this, particularly for high-frequency operation.

Several other types of parasitic component can be significant, for example the resistance of tuning inductors, and stray capacitance associated with pins or connectors on the package. Resistance of the transducer bus-bars can be significant, though usually the bus-bar width can be made large enough for this resistance to be negligible. In long transducers the inductance of the bus-bars can be relevant, causing them to behave as a transmission line; this effect is discussed briefly in Section 9.5.4.

Slobodnik [205] has discussed in more detail the effects of stray components on device performance. It can be concluded that for high frequency devices, above about 300 MHz, a detailed assessment of the effects is needed as part of the design procedure, and considerable care is needed in package design. At lower frequencies, strays are generally less consequential.

7.1.3. Triple-transit Signal

In addition to the desired output signal, a surface-wave delay line also produces an unwanted triple-transit signal due to acoustic reflections, as discussed in Section 4.2. It is usually important to ensure that the triple-transit signal is adequately suppressed. Here we consider the triple-transit signal for a device with uniform transducers, and show that there is a trade-off between triple-transit suppression and insertion loss.

Considering the delay line of Figure 7.1, the transducers and matching networks are taken to be identical, and the load impedance is assumed to be equal to the source impedance. The reflection coefficients of the two transducers are therefore the same. As discussed in Section 4.8, the output voltage is the sum of the main signal V_1 and the triple-transit signal V_3. If the device input signal is C.W., with frequency ω, the magnitudes of these are in the ratio

$$|V_3/V_1| = |r(\omega)|^2, \qquad (7.20)$$

where $r(\omega)$ is the amplitude reflection coefficient of each transducer. Propagation loss and diffraction are assumed to be negligible here. For an unapodised transducer with negligible electrode interactions we have, from equation (4.78),

DELAY LINES AND MULTI-PHASE TRANSDUCERS 169

$$|r(\omega)|^2 = \frac{G_a(\omega)}{2G_L} C(\omega), \qquad (7.21)$$

where $C(\omega)$ is the power conversion coefficient, so that the conversion loss is $-10 \log [C(\omega)]$. Equation (7.21) is valid provided there are no strays and the matching network is purely reactive. The quantity G_L is the real part of the admittance $Y_L(\omega)$ "seen" by the transducer.

The transducers are taken to be uniform, and we consider the triple-transit signal at the centre frequency ω_c. Each transducer is assumed to be tuned with a series inductor, as in Figure 7.2(c), with $L = 1/(\omega_c^2 C_t)$. In this case G_L equals the real part of $(R_G + j\omega L)^{-1}$. Assuming that Q_t is large we have $\omega_c C_t \hat{R}_a \ll 1$ and $X_a(\omega_c)$ is negligible, and it is found that $G_a(\omega_c)/G_L \approx \hat{R}_a/R_G$. Equations (7.20) and (7.21) thus give

$$|V_3/V_1| = \tfrac{1}{2}C(\omega_c)\hat{R}_a/R_G. \qquad (7.22)$$

For a launching transducer, the power conversion coefficient $C(\omega)$ is defined as the ratio of the surface-wave power generated (in one direction) to the power available from the source, and from Figure 7.2(c) we find

$$C(\omega_c) = \frac{2\hat{R}_a/R_G}{(1 + \hat{R}_a/R_G)^2}. \qquad (7.23)$$

The insertion loss of the device is $-20 \log [C(\omega_c)]$, in dB. We define the *triple-transit suppression* as the power ratio, in dB, of the main and triple-transit signals, equal to $-20 \log |V_3/V_1|$. Both quantities are thus expressed as functions of the ratio $\hat{R}_a/R_G$, so the triple-transit suppression can be plotted as a function of insertion loss. This is shown as the solid curve in Figure 7.7. When $R_G = \hat{R}_a$ the transducers are matched

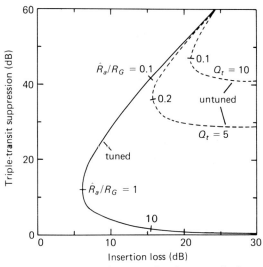

FIGURE 7.7. Triple-transit suppression for unapodised transducers.

and the triple-transit suppression is 12 dB. Better triple-transit suppression is obtained, at the expense of greater loss, by increasing $R_G/\hat{R}_a$. For large values of $R_G/\hat{R}_a$, the triple-transit suppression is equal to twice the insertion loss, plus 12 dB. In practice, the triple-transit suppression is often better than the predictions of Figure 7.7, due to the presence of strays and propagation loss.

The broken lines in Figure 7.7 refer to untuned transducers, that is, the same situation except that the inductors are omitted. The curves are obtained from equations (7.20) and (7.21), using circuit analysis to evaluate $C(\omega_c)$. In this case the trade-off between triple-transit and insertion loss depends on the transducer Q-factor, Q_t.

It should be noted that the above discussion refers to C.W. excitation of the device. If a rectangular pulse is applied, with carrier frequency ω_c, the device output gives a pulse due to the main response followed by a smaller pulse due to the triple-transit signal, and these are resolved if the transducer separation is sufficient. If the input pulse is long enough, so that its bandwidth is much less than the delay line bandwidth, the ratio of the output pulse powers gives the C.W. triple-transit suppression discussed above. For a short input pulse this is not so; to calculate the ratio of the output pulse amplitudes it is necessary to use the device frequency responses $H_1(\omega)$ and $H_3(\omega)$ of Section 4.8, allowing for the spectrum of the input pulse.

Reduction of Triple-transit Signal. Surface-wave devices are often required to give a triple-transit suppression of 40 dB or more, and Figure 7.7 shows that this requires the insertion loss to exceed 15 dB. Although this is often acceptable, there are many applications where good triple-transit suppression and low loss are required simultaneously. For this purpose, a number of special methods have been developed. Since the finite reflection coefficient of a matched transducer is associated with its bidirectional nature, one type of solution is to use a *unidirectional* transducer, designed such that it can generate surface waves in only one direction. If such a transducer is matched, the reflection coefficient for incident surface waves is, ideally, zero and the triple-transit signal vanishes. A type of unidirectional transducer using a multi-strip coupler is considered in Section 5.4 above. There are also several types of *multi-phase* transducers that are unidirectional, and these are considered in Section 7.2 below. In addition to reducing the triple-transit signal, unidirectional transducers also give very low insertion loss, ideally 0 dB. However, these transducers are strictly unidirectional only at one frequency, so the triple-transit signal is not entirely eliminated.

Other methods of reducing the triple-transit signal are shown in Figure 7.8. In the *three-transducer scheme* [206] of Figure 7.8(a), the two outer transducers are connected in parallel. All three transducers are unapodised, and the entire structure is symmetrical about the centre line. Thus the central transducer is symmetrical, and the outer transducers have the same geometry as each other. The waves generated by the outer transducers arrive at the central transducer with equal amplitude and phase, and it is found that no reflected waves are produced if the central transducer is electrically matched. This is to be expected by reciprocity, since a symmetrical transducer generates waves of equal amplitude and phase when a voltage is applied

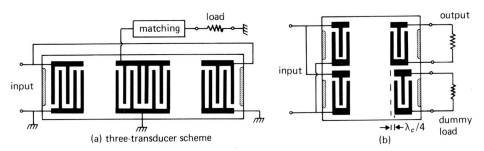

FIGURE 7.8. Methods of reducing triple-transit signals.

to it. The device may be analysed by the methods of Chapter 4. In particular, it can be shown from the scattering matrix of equation (4.79) that the central transducer does not reflect if it is matched; this follows because the symmetry implies that $\bar{\varrho}_e(k_0)$ is real. An additional advantage is that the central transducer converts all the incident surface-wave power into output power, so that the minimum insertion loss for the device is ideally 3 dB.

The three-transducer scheme is also effective if the central transducer is apodised, or if the outer transducers are apodised, provided the symmetry is preserved. In these cases, the triple-transit signal is eliminated if a suitable choice is made for the electrical admittance "seen" by the central transducer, though the admittance required does not match the transducer and so the insertion loss is not minimised. The required admittance can be deduced from the analysis of Chapter 4.

Another method, shown in Figure 7.8(b), essentially comprises two delay lines constructed on one substrate [206]. The transducers at one end are connected together, while those at the other end are connected separately to equal loads; one of the loads is the device output, while the other is a dummy. One of the output transducers is displaced relative to the other by an amount $\lambda_c/4$, where λ_c is the surface-wave wavelength at frequency ω_c. At this frequency, waves reflected by the output transducers are in anti-phase and so do not excite the input transducers; thus, no reflected waves reach the output transducers. However, this is true only at one frequency. In addition, because of the power dissipated in the dummy load the minimum insertion loss obtainable is 9 dB. There are however several variants of the method. Tanski and Van De Vaart [207] have shown that, if double-electrode transducers are used, the required phase relationship between the reflected waves can be obtained by modifying the transducer geometries instead of using a displacement. The triple-transit suppression obtained is then, in principle, independent of frequency. Gunton [208] has investigated several schemes in which the transducers at each end are connected to a hybrid coupler, showing that the loss due to the presence of the dummy load in Figure 7.8(b) can be eliminated. A related method is used in the multi-strip echo trap, mentioned in Section 5.5.

7.1.4. Delay Line Types and Performance

The simplest type of surface-wave delay line is the two-transducer device illustrated in Figure 7.1. The performance of this device has been discussed above, though diffraction and propagation loss have not been considered. These effects can contribute significantly to the insertion loss, especially at high frequencies, and are discussed in Sections 6.2 and 6.3 above.

In some cases the device is required to have a series of outputs, giving different delays of the input signal. Such a device is called a *tapped* delay line, and is readily implemented by using a separate transducer for each output; in this case the output transducers are often called "taps". A closely related device is the PSK filter designed to correlate a phase-coded waveform, using a sequence of taps connected together. This is considered in Chapter 10.

Long Delay Techniques. As for other surface-wave devices, the maximum delay of the basic type of delay line is limited to about 50 μsec because of the lengths of available substrate materials. Several special techniques for obtaining longer delays have been investigated. The use of multi-strip track-changers has already been mentioned in Section 5.5. Another method is the *wrap-around* technique in which the ends of the substrate are rounded into smooth cylindrical surfaces, so that a surface wave can circulate around the substrate several times [209]. For example, a device using a bismuth germanium oxide substrate gave 908 μsec delay at 85 MHz centre frequency, with 65 MHz bandwidth and 65 dB insertion loss. A similar principle is used in the *disk delay line* [210], where the substrate is in the form of a circular disk; this exploits a subtle topographic effect in which the rounded edge acts rather like a surface-wave lens, limiting the diffraction spreading. However, neither the disk nor the wrap-around method has been widely used, owing to fabrication and mounting complications.

Another approach for long delays is *cascading*, in which several devices are connected in sequence, alternating with amplifiers to compensate for the device losses. In this case the amplitude response of each device needs to be exceptionally flat in order to obtain adequate fidelity for the overall system. For example, a tapped 400 MHz delay module including amplifiers gave 40 μsec delay, and had a response flat within $\pm$ 0.5 dB over a 100 MHz bandwidth [211]. Such modules can be cascaded to give delays of several hundred microseconds. Another example is an 800 MHz module giving 7.5 μsec delay, with a response flat to $\pm$ 0.2 dB over a 230 MHz bandwidth [212]. In both cases lithium niobate substrates were used.

Several special techniques have been used to obtain the *memory* function, in which the delay required is considerably larger than the duration of the input signal. These have been investigated mainly in connection with correlation of coded signals, and are considered in Chapter 10.

High-frequency Techniques. Generally, the maximum centre frequency obtainable is about 1.5 GHz, since this requires 0.5 μm line widths when single-electrode transducers are used [213]. Narrower line widths can be obtained by electron beam lithography [214], though this is not convenient for commercial devices. However, some special techniques enable higher frequencies to be obtained without requiring narrower line widths. A simple approach is to use a harmonic response of a transducer, for example, the third harmonic of a double-electrode transducer. If the presence of the fundamental response is undesirable, it may be virtually eliminated in a two-transducer device by using transducers operated at different harmonics, as shown by Engan [215]. For example, the device may have an $S_e = 3$ transducer operated at the second harmonic and an $S_e = 4$ transducer operated at the third harmonic. In this case there is little overlap of the fundamental responses of the transducers. An additional advantage is that electrode interaction affects are minimised.

Generally, only the first few harmonics are useful for high-frequency operation, because for harmonic numbers greater than S_e the coupling efficiency of the transducer is found to be quite sensitive to the metallisation ratio a/p, which is difficult to control accurately for narrow electrodes. The sensitivity to a/p is due to rapid variations of the elemental charge density $\bar{\varrho}_f(\beta)$ with a/p, as shown in Figure 4.12. To overcome this difficulty, modified types of transducer have been proposed [216], using electrodes with different widths in each period. These transducers give harmonic responses less sensitive to errors in the electrode widths, though the coupling strength is rather weak.

Another method of obtaining higher centre frequencies is to use a wave with higher phase velocity. Appendix F discusses the use of bulk waves for this purpose, and another potential possibility is the use of aluminium nitride films, mentioned in Section 6.5.

7.2. MULTI-PHASE UNIDIRECTIONAL TRANSDUCERS

In Section 7.1.3 several methods for minimising triple-transit signals were discussed, and it was noted that unidirectional transducers offer both good triple-transit suppression and low losses. The multi-strip type of unidirectional transducer has been described in Section 5.4. In this section we describe the use of multi-phase unidirectional transducers. In contrast to the multi-strip type, the multi-phase method can be applied to transducers that are many wavelengths long, and also to transducers that are apodised. For these reasons, multi-phase transducers have been investigated extensively, and several types have been developed.

7.2.1. Transducer Types and Performance

The earliest type of multi-phase transducer was the *three-phase* type, introduced by Hartmann et al. [217] and shown in Figure 7.9(a). The electrodes are regular, and are connected sequentially to three bus-bars, with terminals A, B and C. Connections to one of the bus-bars must be made via insulating *cross-overs*, which are usually made with the aid of an insulating silicon oxide film, though air gaps have also been used [218]. The periodicity of the transducer is made equal to $\lambda_c = 2\pi v_0/\omega_c$, which is the surface-wave wavelength at the required centre frequency ω_c. Thus there are three electrodes per wavelength. The operation of the transducer is considered in detail later, but the essential features are readily appreciated if each electrode is regarded as a surface-wave source. For a surface wave of frequency ω_c travelling through the transducer, the phase angle corresponding to the electrode spacing is 120°. If voltages are applied to the three bus-bars, with the same amplitudes but with phases

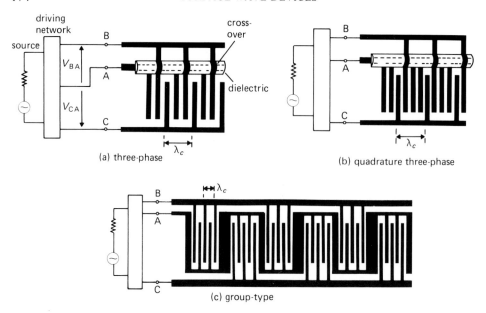

FIGURE 7.9. Multi-phase transducers.

incrementing by 120°, the waves generated in the "forward" direction have the same phase and thus reinforce. The waves launched in the "backward" direction, due to the three electrodes in any one period of the transducer, have relative phases of 0, 240° and 480°, and add vectorially to give a resultant of zero. Thus surface waves are generated only in one direction. It should be noted that only relative electrode voltages are significant, since a uniform voltage applied to all the electrodes gives no acoustic excitation. It is therefore sufficient to note the voltages relative to terminal A, say. The relative voltages, denoted V_{BA} and V_{CA}, have a phase difference of 60°.

The required voltages are produced by applying a source to a driving network comprising a few reactive components. This can be designed such that the transducer is matched, so that all of the power available from the source is converted into a surface wave propagating in one direction. In this case any surface wave incident on the transducer on the "forward" side is fully converted into electrical output power, with no acoustic reflection, as expected by reciprocity. However, a practical driving network can give the required phase relationship only at one frequency, so the transducer is unidirectional only at one frequency.

Figure 7.9(b) shows the *quadrature three-phase* transducer [219, 220] which operates in a very similar manner. There are now four electrodes per period and the period, as before, is equal to λ_c. For a propagating surface wave the electrode pitch corresponds to a phase change of 90°. Thus, to cancel the waves radiated in one direction the voltage on terminals B and C, relative to terminal A, should differ in phase by 90°. The structure is closely related to the earlier "hybrid-junction" transducer [221] in which the electrodes are connected sequentially to four bus-bars; this has been shown to be usable as a unidirectional transducer, though it does not have appreciable advantages over the quadrature three-phase type. It should be noted that the transducers in Figure 7.9(a) and (b) give negligible electrode interaction effects, because the electrode pitch is not a multiple of $\lambda_c/2$.

The *group-type* transducer introduced by Yamanouchi et al. [222] is shown in Figure 7.9(c). The transducer consists of two interleaved sets of "groups", each group being essentially a conventional single-electrode transducer, though other types of structure may be used for the individual groups. The centre-frequency wavelength λ_c is equal to twice the electrode pitch in any one group. One set of groups is displaced relative to the other by $(N + \frac{1}{4})\lambda_c$, where N is some integer. Thus, unidirectional operation is obtained at frequency ω_c by driving the two sets of groups in phase quadrature. Since the centre-to-centre spacing of adjacent groups is several wavelengths, the appropriate phase relationship is obtained, to an adequate accuracy, only over a relatively narrow bandwidth. Thus this type of transducer is best suited for narrow-band devices. However, the transducer is particularly suitable for high-frequency devices since the electrode width required, for a given frequency, is larger than that of other multi-phase types.

Another advantage of the group-type transducer is that one of the three connections required can be provided by a meander line, as shown in Figure 7.9(c). This avoids the need for cross-overs and thus considerably simplifies the fabrication. However, care is needed to ensure that the resistance of the meander line is low enough. In some cases this resistance would cause unacceptable losses, but cross-overs can if necessary be introduced to overcome this problem, using an additional bus-bar at the side.

Malocha and Hunsinger [223] have introduced another group-type transducer, shown in Figure 7.10. This again has the advantage that a meander line can

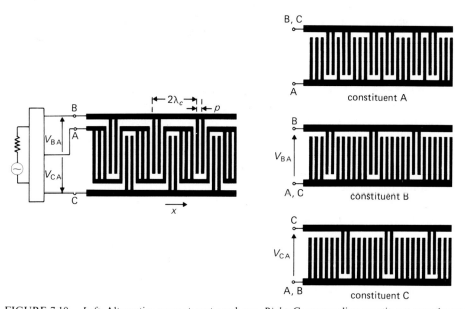

FIGURE 7.10. *Left*: Alternative group-type transducer. *Right*: Corresponding constituent transducers.

sometimes be used, simplifying the fabrication. The electrodes are regular here. The transducer period is of length $2\lambda_c$ and includes 8 electrodes, so that there are four electrodes per wavelength. The two-electrode groups connected to terminals B and C in Figure 7.10 have a relative displacement of $3\lambda_c/4$, so unidirectional operation is obtained at frequency ω_c if the relative voltages V_{BA} and V_{CA} are in phase quadrature with each other.

As for conventional two-terminal transducers, the structure of a multi-phase transducer may be modified in order to weight the response, for example, to produce a bandpass filter response with good sidelobe rejection. The commonest weighting technique is essentially the same as the apodisation technique applied to two-terminal transducers, that is, the electrode overlaps are varied along the length of the transducer. This method can be applied to all the transducer types mentioned above. Provided the electrode overlaps do not vary substantially in any one period, the transducer retains its unidirectional character, enabling low loss and good triple-transit suppression to be obtained. For narrow-band devices, withdrawal weighting is sometimes used.

Multi-phase transducers enable exceptionally low insertion losses to be obtained. For the best results, considerable care is taken to minimise losses due to imperfections such as electrode resistivity, surface-wave diffraction and resistance in the driving networks. In this way, Brown [224] demonstrated a 34 MHz three-phase device with only 0.6 dB insertion loss, and Yamanouchi *et al.* [225] obtained an insertion loss of 3 dB at 1.0 GHz using a modified form of group-type transducer. A variety of recent results show the performance achievable for bandpass filters using weighted transducers [226–228]. For example, a 650 MHz lithium niobate filter, using group-type transducers with apodisation, gave 7 dB insertion loss, 9 MHz bandwidth and 40 dB out-of-band rejection [227]. Withdrawal-weighted group-type transducers have been used effectively on *ST*, *X* quartz [228], giving for example, a 900 MHz filter with 5 MHz bandwidth, 8 dB insertion loss and 30 dB out-of-band rejection. In most of the devices the in-band ripple, due mainly to the triple-transit signal, was 0.2 dB peak-to-peak or less.

7.2.2. Analysis of Multi-phase Transducers

The analysis here is based on a charge superposition argument [229], which shows that the operation of a multi-phase transducer can be derived from the analysis of two-terminal transducers, given in Chapter 4. An alternative approach, obtained by generalising the crossed-field network model, is given by Farnell *et al.* [230]. For illustration we consider the group-type transducer, Figure 7.10, though the method is also applicable to the other types. It is convenient to assume that guard electrodes are used at each end of the transducer to eliminate end effects, though the analysis is approximately valid if the guards are absent.

(a) *Launching Transducer*. We first consider launching of surface waves. At the frequencies of interest the electrode pitch is not close to the half-wavelength, or to a multiple of this value, and therefore electrode interactions can be ignored. Thus the quasi-static analysis of Section 4.3, which allows for arbitrary electrode voltages, is

DELAY LINES AND MULTI-PHASE TRANSDUCERS

valid here. In particular, the potentials $\phi_s^\pm(x, \omega)$ of the surface waves generated in the $\pm x$ directions are, from equation (4.26),

$$\phi_s^\pm(x, \omega) = j\Gamma_s \bar{\sigma}_e(\mp k_0) \exp(\mp jk_0 x), \tag{7.24}$$

where $k_0 = \omega/v_0$ is the free-surface wavenumber at the frequency ω under consideration, and Γ_s is the piezoelectric coupling parameter given by equation (4.21). The function $\bar{\sigma}_e(\beta)$ is the Fourier transform of the electrostatic charge density of $\sigma_e(x)$. The charge superposition principle shows that this is given by equation (4.33), that is,

$$\sigma_e(x) = \sum_{n=1}^{N} V_n \varrho_{en}(x), \tag{7.25}$$

where N is the number of electrodes, V_n is the voltage on electrode n and $\varrho_{en}(x)$ is the electrostatic charge density produced when unit voltage is applied to electrode n with the other electrodes grounded.

In the present case there are only three distinct values of the electrode voltages, and since only relative voltages are significant it is sufficient to take terminal A to be at zero voltage, and to consider only the relative voltages V_{BA} and V_{CA} of terminals B and C. Each of the V_n in equation (7.25) can thus be equated with V_{BA}, V_{CA} or zero, and the charge density can therefore be written in the form

$$\sigma_e(x) = V_{BA} \varrho_e^b(x) + V_{CA} \varrho_e^c(x). \tag{7.26}$$

Here $\varrho_e^b(x)$ is the electrode charge density obtained when terminals A and C are grounded and unit voltage is applied to terminal B. Similarly, $\varrho_e^c(x)$ is the charge density when terminals A and B are grounded and unit voltage is applied to terminal C.

The method can be clarified by considering imaginary two-terminal transducers, called *constituent* transducers, shown on the right in Figure 7.10. Each of these is defined by imagining two terminals of the group-type transducer to be connected together. For constituent A terminals B and C are connected, for constituent B terminals A and C are connected, and for constituent C terminals A and B are connected. Thus, in equation (7.26), $\varrho_e^b(x)$ is equal to the electrostatic charge density on constituent B for unit applied voltage, and $\varrho_e^c(x)$ is the electrostatic charge density on constituent C for unit applied voltage. The constituent two-terminal transducers can be analysed by the methods of Chapter 4, and since the electrodes are regular here the methods of Section 4.5 can be applied.

From the geometries of constituents B and C, shown in Figure 7.10, it is clear that $\varrho_e^b(x)$ and $\varrho_e^c(x)$ are the same except for a displacement $3p$, where p is the electrode pitch. Thus $\varrho_e^c(x) = \varrho_e^b(x - 3p)$. The Fourier transforms of $\varrho_e^b(x)$ and $\varrho_e^c(x)$, denoted respectively by $\bar{\varrho}_e^b(\beta)$ and $\bar{\varrho}_e^c(\beta)$, are therefore related by

$$\bar{\varrho}_e^c(\beta) = \bar{\varrho}_e^b(\beta) \exp(-3j\beta p).$$

Taking the transform of equation (7.26), this enables $\bar{\sigma}_e(\beta)$, the transform of $\sigma_e(x)$, to be written in terms of $\bar{\varrho}_e^b(\beta)$. Substituting into equation (7.24), the potentials of the two surface waves launched by the transducer are

$$\phi_s^\pm(x, \omega) = j\Gamma_s \bar{\varrho}_e^b(\mp k_0)[V_{BA} + V_{CA} \exp(\pm 3jk_0 p)] \exp(\mp jk_0 x). \tag{7.27}$$

Now, at the centre frequency ω_c we have $\lambda_c = 4p$ and hence $k_0 p = \pi/2$, giving $\exp(\pm 3jk_0 p) = \mp j$. Thus if $V_{CA} = jV_{BA}$, so that the voltages are in phase quadrature, we have $\phi_s^-(x, \omega) = 0$, so that the transducer radiates surface waves only in the $+x$ direction. Alternatively, if $V_{CA} = -jV_{BA}$ then $\phi_s^+(x, \omega) = 0$ and the transducer radiates only in the $-x$ direction. The surface-wave powers are given by equation (4.51).

The superposition principle can still be applied if the multi-phase transducer is apodised. In this case the constituent transducers are also apodised. Assuming the electrode voltages to be known, the waves generated by the multi-phase transducer can be found by summing the waves that would be produced by the constituents B and C.

(b) Equivalent Circuit. For a two-terminal transducer the admittance seen between the terminals is $Y_t(\omega) = G_a(\omega) + jB_a(\omega) + j\omega C_t$. Here we consider the admittances seen between the three terminals of a multi-phase transducer. These are needed in order to evaluate the electrode voltages V_{BA} and V_{CA} as functions of frequency, for a specified driving network, and to analyse reception of surface waves by the transducer.

Assuming that no acoustic waves are incident on the transducer, the terminal currents and voltages are taken to be related by the equivalent circuit of Figure 7.11(a), where admittances Y_{AB}, Y_{BC} and Y_{AC} are connected between the three terminals. This arrangement is sufficiently general to represent any linear passive three-terminal device. The terminals A, B and C correspond to the transducer terminals in Figure 7.10. The admittances are straightforwardly related to the admittances of the constituent transducers in Figure 7.10, denoted by Y_t^a, Y_t^b and Y_t^c for transducers A, B and C, respectively. If terminals B and C of the multi-phase transducer are connected together, the admittance between them and terminal A is, from Figure 7.11(a), $Y_{AB} + Y_{AC}$. This must be equal to the admittance Y_t^a of constituent transducer A. Similarly, by imagining terminals A and B or A and C to be connected, two further relations are found. These can be solved for the admittances of the multi-phase transducer, giving

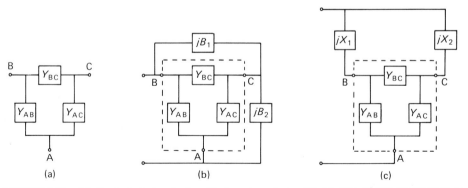

FIGURE 7.11. (a): Equivalent circuit of multi-phase transducer. (b), (c): Use of two components to provide required phase shift.

$$Y_{AB} = \tfrac{1}{2}(Y_t^a + Y_t^b - Y_t^c),$$

$$Y_{BC} = \tfrac{1}{2}(Y_t^b + Y_t^c - Y_t^a),$$

$$Y_{AC} = \tfrac{1}{2}(Y_t^a + Y_t^c - Y_t^b). \tag{7.28}$$

Since the admittances of the constituent transducers can be found by the methods of Chapter 4, these equations enable Y_{AB}, Y_{BC} and Y_{AC} to be evaluated. For the transducer of Figure 7.10 it can be seen that constituents B and C have essentially the same structure, so that $Y_t^b = Y_t^c$ and hence $Y_{AB} = Y_{AC}$. For the three-phase transducer of Figure 7.9(a), the three constituents are all the same except for a displacement, and hence $Y_{AB} = Y_{BC} = Y_{AC}$. In fact, the constituent transducers for this case are the same as the uniform $S_e = 3$ transducer, whose admittance has already been considered in Section 4.6.

(c) Driving Networks. Once the transducer admittances are known, either from the above analysis or by measurement, a driving network can be designed in order to provide the appropriate phase relationship between the terminal voltages, and to match the transducer to a source. The driving network usually consists of a few lumped reactive components. Two common methods of obtaining the required phase relationship are shown in Figures 7.11(b) and (c). In Figure 7.11(b), reactive elements with susceptances B_1 and B_2 are connected between terminals B and C and between terminals C and A, respectively. This method [224] is often used for the three-phase transducer of Figure 7.9(a). The susceptances B_1 and B_2 are to be chosen such that $V_{CA} = V_{BA} \exp(j\alpha)$ at the centre frequency ω_c, with $\alpha = \pi/3$ or $\pi/2$ for the transducers considered above. Writing $Y_{AB} = G_{AB} + jB_{AB}$, and corresponding expressions for Y_{BC} and Y_{AC}, analysis of the circuit of Figure 7.11(b) gives

$$B_1 = (G_{AC} + G_{BC})\operatorname{cosec}\alpha - G_{BC}\cot\alpha - B_{BC}$$

and

$$B_2 = -(G_{AC} + 2G_{BC})\tan\alpha/2 - B_{AC}. \tag{7.29}$$

Thus two reactive components are sufficient to obtain the required phase relationship, whatever the values of Y_{AB}, Y_{AC} and Y_{BC}. However since these admittances are frequency dependent, the phase difference between V_{CA} and V_{BA} will also be frequency dependent, so B_1 and B_2 are chosen to satisfy equation (7.29) at the centre frequency ω_c. The required value of B_1 is then obtained by using a capacitor, with $C = B_1/\omega_c$, or an inductor, with $L = -1/(\omega_c B_1)$, depending on the sign of B_1. The same remarks apply to B_2. A more general form of the solution is described by Farnell et al. [230].

In addition to providing the required phase relationship, the driving network must also match the transducer to a resistive source in order to minimise the loss and to suppress acoustic reflections. The impedance seen looking into the circuit of Figure 7.11(b) is usually complex, so some additional reactive elements are added to provide the matching. These can take the form of one of the two-component circuits of Figure 7.3.

An alternative circuit, using series reactances X_1 and X_2, is shown in Figure 7.11(c). This method [223] is often used for the group-type transducer of Figure 7.10, though it is also applicable to other types of transducer. As before, the required reactances can be deduced from circuit analysis, applying the condition that $V_{CA} = V_{BA} \exp(j\alpha)$. In this case the expressions obtained are more complicated and it is convenient to simplify them by assuming that $Y_{AB} = Y_{AC}$, which is valid for all of the transducer types considered above. With this assumption, the required values of X_1 and X_2 are found to be given by

$$X_1 + X_2 = 2[B_{AB} + 2B_{BC} + G_{AB}(G_{AB} + 2G_{BC})/B_{AB}]^{-1},$$

$$X_1 - X_2 = (\tan \alpha/2)(X_1 + X_2)(G_{AB} + 2G_{BC})/B_{AB}. \qquad (7.30)$$

As before, the required phase relationship can be obtained only at one frequency, and additional reactances are needed to match the transducer to a resistive source.

(d) Reception of Surface Waves. The above analysis can be used to deduce the conversion efficiency for a multi-phase transducer launching surface waves and, by reciprocity, this is equal to the conversion efficiency for the same transducer receiving surface waves. Thus for a device using two multi-phase transducers the insertion loss, as a function of frequency, can be obtained. However, it is necessary to consider the reception process explicitly if the acoustic reflection coefficient is to be obtained.

For a receiving transducer the equivalent circuit of Figure 7.11(a) must be modified to include sources due to the incident surface wave. The sources can be deduced by considering the terminal currents produced when the three terminals are shorted, as shown in Figure 7.12(a). For an incident wave with potential $\phi_i(x, \omega)$ uniform across the transducer aperture, we define I_B and I_C as the currents entering terminals B and C, respectively. Comparison with Figure 7.10 shows that I_B is equal to the short-circuit current produced by constituent transducer B when a wave with potential $\phi_i(x, \omega)$ is incident on it. Similarly, I_C is the current produced by constituent transducer C when the same wave is incident. The currents I_B and I_C can therefore be obtained by analysing the two-terminal constituent transducers, using the methods of Chapter 4; they are related to the electrostatic charge densities $\varrho_e^b(x)$ and $\varrho_e^c(x)$ by equation (4.67).

Figure 7.12(b) shows the equivalent circuit for this case, where the broken line encloses the transducer itself. To account for the currents produced in the external

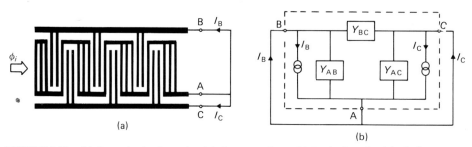

FIGURE 7.12. (a) Reception by shorted multi-phase transducer. (b) Equivalent circuit including current sources.

short circuit, a current generator I_B is added to the transducer equivalent circuit between terminals A and B, and a current generator I_C is added between terminals A and C.

The equivalent circuit of Figure 7.12(b), within the broken line, enables the performance of the receiving transducer to be calculated for any network connected to the terminals. Using conventional network analysis, the voltages V_{BA} and V_{CA} across the terminals can be found and the power delivered to the load can be calculated, thus giving the conversion loss. To calculate the reflected wave amplitude, it is first noted that electrode interactions can be assumed to be weak, so that the reflection coefficient is negligible if the transducer is shorted. The waves generated by the transducer can therefore be found by assuming terminal voltages V_{BA} and V_{CA}, and applying the analysis for a launching transducer as if the incident wave were absent. Thus the waves generated are given by equations (7.24) and (7.26), and the reflection coefficient can be obtained. For zero reflection coefficient it is found, as expected, that the transducer must be matched to the load, and the network must be such that it would provide the appropriate phases when launching surface waves.

Chapter 8

Bandpass Filters

In this chapter we consider the surface-wave bandpass filter, whose basic function is to pass signals with frequencies within a specified band, known as the pass-band, and to reject signals with frequencies outside this band. This is one of the commonest applications of surface waves, exemplified by the very large volume production of surface-wave filters for colour television receivers. A basic principle employed in most bandpass filters is that of apodisation, which enables an interdigital transducer to be designed such that its frequency reponse approximates a required response, as already discussed in Chapter 1. The earliest filters [231, 232] used this principle, and a variety of complementary techniques were developed later.

There are several important distinctions between the surface-wave filter and the more familiar L–C filter, comprising a network of inductors and capacitors. The response of an L–C filter is usually considered in terms of poles and zeros, in the complex frequency plane, and the design problem is expressed in terms of finding appropriate locations for these poles and zeros. For a surface-wave filter the response has no poles, and consequently quite different approaches must be used in design, similar to the methods used for digital finite-impulse-response filters. The number of zeros can be as large as the number of electrodes, typically several hundred; this is much larger than the number of poles or zeros for typical L–C filters. In addition, the contribution due to any one electrode is often accurate to 1% or better. These features enable surface-wave filters to achieve impressive performances with, for example, very flat pass-bands, sharp skirts and good stop-band attenuation. The phase of the frequency response may be either a linear or a non-linear function of frequency, and can be specified independently of the amplitude, a facility not available in the basic design method for L–C filters. Another distinction lies in the fact that surface-wave transducers give zero response at zero frequency. Thus, surface-wave devices cannot be used for low-pass filtering.

Suitable methods for analysis of interdigital bandpass filters are given in Chapter 4, and the relevant propagation effects are described in Chapter 6. This chapter is therefore concerned with the design and performance of the devices. The first two sections are concerned with apodised transducers. Section 8.1 considers the close analogy between an apodised transducer and a transversal filter, and Section 8.2 uses this analogy in describing transducer design techniques. Section 8.3 is mainly concerned with withdrawal weighting, a technique alternative to apodisation. In Section 8.4 the design and performance of interdigital filters are described, concluding with some remarks on other types of surface-wave bandpass filter. Finally, Section 8.5 discusses filter banks, which are essentially arrays of bandpass filters.

In common with many other devices, including most surface-wave devices and L–C filters, surface-wave bandpass filters can be described as *linear filters*. The meaning of this term is explained in Appendix A, Section A.2, which also defines the important terms "frequency response" and "impulse response". In practice some non-linear effects do occur, as decribed in Section 6.3. However, if it is assumed that any signal applied to the device has a small enough power level, the device will behave as a linear filter to a very good approximation. This is nearly always true in practice, and is assumed to be the case throughout this chapter.

8.1. APODISED TRANSDUCER AS A TRANSVERSAL FILTER

We consider here a filter comprising two transducers, with one apodised and the other unapodised, as illustrated in Figure 8.1. The transducers are assumed to have regular electrodes and are taken to be of the two-terminal type, that is, multi-phase transducers are excluded. It is assumed that the only acoustic wave present is a piezoelectric Rayleigh wave, and that electrode interactions, propagation loss and surface-wave diffraction are negligible.

It was shown in Chapter 4 that the short-circuit response $H_{sc}(\omega)$ of this device is essentially the product of the two transducer responses $H_t^a(\omega)$ and $H_t^b(\omega)$. From equation (4.127),

$$H_{sc}(\omega) \equiv I_{sc}/V_t = H_t^a(\omega) H_t^b(\omega) \exp(-jk_0 d), \qquad (8.1)$$

where d is the separation between the transducer acoustic ports and $k_0 = \omega/v_0$ is the free-surface wavenumber at frequency ω. V_t is the voltage applied to the input

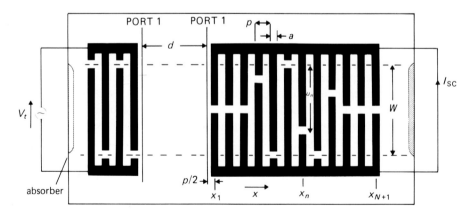

FIGURE 8.1. Bandpass filter using apodised transducer with regular electrodes.

BANDPASS FILTERS

transducer, and I_{sc} the current produced by the shorted output transducer. Distortions due to the circuit effect, which arise when the transducers are connected to finite impedances, are ignored for the present but will be considered in Section 8.4.2.

Here we consider the design of the transducers such that the device response $H_{sc}(\omega)$ meets some required specification. For convenience we assume initially that the unapodised transducer has already been designed, so that its response can be predicted. The problem is therefore reduced to that of designing the apodised transducer, whose required response is simply the required device response divided by the response of the unapodised transducer, as shown by equation (8.1).

In this section it is shown that the transducer geometry can be obtained by sampling a waveform $v(t)$, defined such that its Fourier transform $V(\omega)$ is closely related to the required frequency response of the transducer. The calculation of $v(t)$ itself is considered in Section 8.2.

8.1.1. Transversal Filter Analogy

The analysis of Section 4.7.3 shows that, for regular electrodes, the frequency response of an apodised transducer is given by equation (4.131). With minor changes of notation, this reads

$$H_t(\omega) = E(\omega) \sum_{n=1}^{N} v_n \exp[-jk_0(x_n - x_1 + p)], \qquad (8.2)$$

where

$$v_n = (u_{n+1} - u_n)/W. \qquad (8.3)$$

As shown in Figure 8.1, p is the electrode pitch, x_n is the centre location of electrode n, u_n is the location of the break in electrode n, N is the number of gaps and W is the aperture. The response can be regarded as a sum of contributions due to gap elements; the term $(x_n - x_1 + p)$ is the distance between element n, located at $x = x_n + p/2$, and the transducer acoustic port, which is here taken to be at $x = x_1 - p/2$. The term $E(\omega)$ is the gap element factor, given by

$$E(\omega) = (\omega W \Gamma_s)^{1/2} \bar{\varrho}_g(k_0), \qquad (8.4)$$

where Γ_s is the piezoelectric coupling factor of equation (4.21). The function $\bar{\varrho}_g(k_0)$ is the Fourier transform of the elemental charge density for gap elements, given by equation (4.96), and is plotted in Figure 8.2 for two values of the metallisation ratio a/p. Note that $\bar{\varrho}_g(k_0)$ is zero when $k_0 p$ is a multiple of 2π, that is, at frequencies such that p is a multiple of the surface-wave wavelength; however, for frequencies not close to these points it varies quite slowly with frequency. Also, $\bar{\varrho}_g(k_0)$, and therefore $E(\omega)$, is imaginary. It is assumed that a few guard electrodes (Section 4.5.1) are included at each end of the transducer in order to minimise end effects.

The above form for the frequency response is also given quite simply by the delta-function model of Section 4.1, though this does not give the element factor. An alternative formulation in terms of elements associated with electrodes rather than gaps is given in Section 4.7.3, equation (4.129); for this case v_n in equation (8.2) is

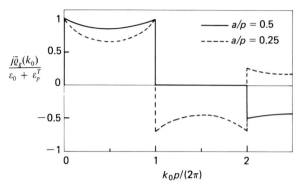

FIGURE 8.2. Elemental charge density for gaps, in the frequency domain.

taken as u_n/W and a different element factor is used. In this chapter the design is considered in terms of gap elements, though the use of electrode elements is equally valid and the design procedures are very similar.

It is convenient to define $\tau_s = p/v_0$ as the delay corresponding to the distance between successive gaps. Noting that $x_{n+1} = x_n + p$, equation (8.2) becomes

$$H_t(\omega) = E(\omega) \sum_{n=1}^{N} v_n \exp(-jn\omega\tau_s). \qquad (8.5)$$

This shows that the transducer behaves essentially as a *transversal filter*, a conceptual device shown in Figure 8.3. Here an ideal tapped delay line, with regularly spaced taps, produces delayed replicas of an input waveform. These replicas, with delays $n\tau_s$, are weighted using real amplitude coefficients v_n and then summed to give the output waveform. Since the frequency response of an ideal delay line with delay τ is $\exp(-j\omega\tau)$, the frequency response of the transversal filter is

$$H_s(\omega) = \sum_{n=1}^{N} v_n \exp(-jn\omega\tau_s). \qquad (8.6)$$

This is the same as the transducer response, equation (8.5), except for omission of the element factor $E(\omega)$. The significance of this comparison is that a variety of techniques have been developed for transversal filter design, determining the coefficients v_n required for the response $H_s(\omega)$ to meet a specification. These techniques can be

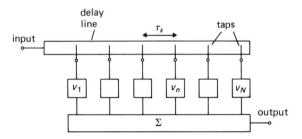

FIGURE 8.3. Transversal filter.

applied to transducer design if the required response is first divided by $E(\omega)$; the resulting values of v_n then give the transducer geometry, using equation (8.3).

The transversal filter concept was introduced by Kallman [233] as a versatile technique for obtaining accurate frequency responses. In addition to surface-wave devices, there are several other technologies using the same principle. For example, a transversal filter can be realised by means of a long cable with weakly-coupled taps, or by a charge-coupled device, in which charge packets representing the input signal are transferred sequentially along an array of electrodes in a metal-oxide-semiconductor structure [234]. Another realisation is a type of digital filter, which simply computes the output waveform in real time [235–237]. The digital realisation is called a "finite impulse response" (FIR) filter to distinguish it from the recursive digital filter, which employs feedback and gives an impulse response of infinite duration.

Although these devices are all based on the same concept, there are some important practical differences. Surface-wave devices give zero response at zero frequency, and so cannot be used for low-pass filtering. Also, the elements in a surface-wave device can have non-uniform spacing, though here uniform spacing is assumed except for some remarks at the end of Section 8.1.3. Another distinction is that a surface-wave device must have two transducers, and so is essentially *two* transversal filters, giving a useful degree of design flexibility. Charge-coupled devices and digital filters sample the input waveform before processing it, and so, unlike surface-wave devices, are not strictly linear filters.

8.1.2. Sampling and Surface-wave Transducers

Transforming equation (8.6) from the frequency domain to the time domain gives the impulse response $h_s(t)$ of the transversal filter, which is the delta-function sequence

$$h_s(t) = \sum_{n=1}^{N} v_n \delta(t - n\tau_s). \quad (8.7)$$

This equation also follows directly from Figure 8.3. To appreciate the design principle of a transversal filter, it is supposed that we can define a smooth real function $v(t)$, such that its values at times $n\tau_s$ are equal to the coefficients v_n. We also assume that $v(t) = 0$ for $t \leq 0$ and for $t > N\tau_s$. With these assumptions, equation (8.7) may be written in the form

$$h_s(t) = \sum_{n=-\infty}^{\infty} v(n\tau_s) \delta(t - n\tau_s)$$

$$= v(t) \sum_{n=-\infty}^{\infty} \delta(t - n\tau_s). \quad (8.8)$$

This is a sampled form of the waveform $v(t)$. The quantity τ_s is called the *sampling interval*, and since we take $\tau_s = p/v_0$ this is equal to the delay corresponding to the electrode spacing in a transducer. The corresponding frequency $\omega_s = 2\pi/\tau_s$ is the *sampling frequency*, and at this frequency the electrode pitch p is equal to the surface-wave wavelength.

The frequency response $H_s(\omega)$ of the transversal filter is the Fourier transform of $h_s(t)$. If $V(\omega)$ is the Fourier transform of $v(t)$, $H_s(\omega)$ can be expressed as

$$H_s(\omega) = \frac{\omega_s}{2\pi} \sum_{m=-\infty}^{\infty} V(\omega - m\omega_s). \tag{8.9}$$

This follows from standard theorems of Fourier analysis given in Appendix A, equations (A.20), (A.23) and (A.42). Figure 8.4 illustrates the relationship, showing the magnitudes of $V(\omega)$ and $H_s(\omega)$. The terms in the above equation with $m \neq 0$ are shown by broken lines. The magnitude of $V(\omega)$ is symmetrical about $\omega = 0$ because $v(t)$ is real. The figure assumes that $V(\omega)$ is a *bandpass* function, that is, for positive frequencies its magnitude is negligible except in the frequency range between two points ω_1 and ω_2. This condition is always valid for surface-wave transducer design.

In Figure 8.4 it has been assumed that the sampling frequency ω_s exceeds $2\omega_2$. For this case, the individual terms in equation (8.9) do not overlap, because $V(\omega)$ is negligible for $|\omega| > \omega_2$. In particular the "fundamental" component of $H_s(\omega)$, the term with $m = 0$, is essentially the same as $V(\omega)$. Thus,

$$H_s(\omega) = \omega_s V(\omega)/(2\pi), \quad \text{for } |\omega| \leq \tfrac{1}{2}\omega_s, \tag{8.10}$$

provided ω_s exceeds a minimum value $2\omega_2$, which is known as the *Nyquist frequency*. This result gives an important part of the design procedure for a transversal filter. It is assumed that the required response is specified for frequencies up to ω_2, at which point its magnitude has fallen to zero. We first generate a finite-length waveform $v(t)$ such that its transform $V(\omega)$ is a good approximation to the required response for $|\omega| \leq \omega_2$, and is negligible for $\omega > \omega_2$. We then sample $v(t)$ with some sampling frequency $\omega_s \geq 2\omega_2$, giving real weighting coefficients $v_n = v(n\tau_s)$. The frequency response $H_s(\omega)$ of the transversal filter is then a good approximation to the required response, for $|\omega| \leq \tfrac{1}{2}\omega_s$, apart from a multiplying constant. Since this procedure is valid for *any*

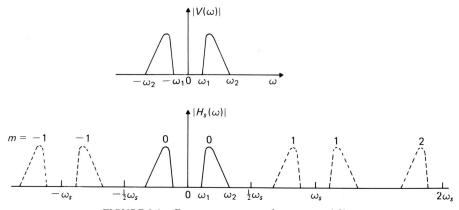

FIGURE 8.4. Frequency response of a transversal filter.

specified frequency response, the transversal filter is very flexible. In practice, some complications arise in the calculation of $v(t)$, and the methods used for this are considered in Section 8.2 below.

In addition to the required response, the transversal filter also gives "image" responses at frequencies above $\frac{1}{2}\omega_s$, as shown by the broken lines in Figure 8.4. It is usually necessary to suppress these by low-pass filtering. The sampling frequency ω_s is usually chosen to be somewhat larger than the Nyquist frequency $2\omega_2$, so that the low-pass filter does not need to have a sharp cut-off.

If the sampling frequency ω_s is less than the Nyquist frequency $2\omega_2$, the individual terms in equation (8.9) will in general overlap. This phenomenon is known as "aliasing". In this case, the original spectrum $V(\omega)$ cannot be recovered from $H_s(\omega)$ by low-pass filtering.

For a surface-wave transducer we have, from equations (8.5) and (8.6), $H_t(\omega) = E(\omega)H_s(\omega)$. Thus if the sampling frequency exceeds the Nyquist frequency, the transducer frequency response is given by

$$H_t(\omega) = \omega_s E(\omega) V(\omega)/(2\pi), \quad \text{for } |\omega| \leq \tfrac{1}{2}\omega_s. \tag{8.11}$$

The design procedure is therefore the same as for a transversal filter, except that the required frequency response is first divided by the element factor $E(\omega)$ in order to obtain $V(\omega)$. Although $E(\omega)$ has a sequence of zeros, these do not affect the method because they occur at zero frequency and at multiples of ω_s, where the transversal filter response $H_s(\omega)$ is zero.

It should be noted that, although the transversal filter has an impulse response with the simple form of equation (8.7), there is no corresponding expression for the transducer because of the frequency-domain distortion caused by the element factor, as shown by equation (8.5). However, the transducer response in the *fundamental* pass-band is given by equation (8.11), and if the bandwidth is not too large the slowly-varying element factor $E(\omega)$ has little effect, so that the response is approximately proportional to $V(\omega)$. Thus $v(t)$ can be regarded approximately as the transducer impulse response, provided the image responses are of no interest. This concept was mentioned earlier in Section 4.1, and was used in Hartmann's impulse model [198].

Note also that $V(\omega)$ was taken to be a bandpass function. This is not in fact valid, because the finite length of $v(t)$ implies, according to Fourier analysis, that $V(\omega)$ must have infinite extent in the frequency domain. However, it will be shown in Section 8.2 that $V(\omega)$ can be designed such that its amplitude outside a specified band is very small. Assuming that this is done, the above analysis can be taken to be valid to a good approximation, and can then provide the basis for a valid design technique.

8.1.3. Examples of Particular Cases

Here we give some illustrations of the relationship between the "impulse response" $v(t)$, the transducer geometry, and the transducer frequency response. It should be noted that $v(t)$ may be delayed without significantly affecting the transducer response – the delay of $v(t)$ causes a corresponding group delay, independent of frequency, in the transducer frequency response. For practical applications this change of response is usually insignificant, though the transducer geometry may be changed substantially.

Since $V(\omega)$ is a bandpass function, the waveform $v(t)$ will be oscillatory and is conveniently written in the form

$$v(t) = \hat{a}(t) \cos [\omega_r t + \hat{\theta}(t)], \qquad (8.12)$$

where ω_r is some reference frequency between ω_1 and ω_2, and $\hat{a}(t)$ is the envelope. If the phase $\hat{\theta}(t)$ is a non-linear function of t, the waveform is phase-modulated; it will also be amplitude-modulated if $\hat{a}(t)$ varies with t. The term "amplitude-modulated waveform" is used when there is no phase modulation, so that $\hat{\theta}(t)$ is a linear function of t. For this case ω_r may be chosen such that $\hat{\theta}(t)$ is a constant, and ω_r is then the carrier frequency, ω_c.

A simple example of an amplitude-modulated waveform is a pulse of carrier with a rectangular envelope, as shown in Figure 8.5(a). This figure shows samples taken at a sampling frequency $\omega_s = 4\omega_c$, so that there are four samples per period. With the waveform positioned such that the samples occur where the phase is a multiple of $\pi/2$, the weights v_n give the sequence 0, 1, 0, -1, 0, 1 . . . . When these are interpreted as interelectrode gaps $(u_{n+1} - u_n)/W$, as in equation (8.3), the transducer geometry is seen to be the conventional uniform double-electrode type, which is unapodised. A few zero-valued samples can be added beyond the ends of the waveform $v(t)$ to provide guard electrodes, minimizing end effects. Figure 8.5(b) shows the same case, but with the waveform $v(t)$ displaced slightly so that the samples occur where the phase is $\pi/4 + n\pi/2$. In this case the transducer is apodised, even though its frequency response is essentially the same. Similar observations apply if three samples are taken per wavelength, so that $\omega_s = 3\omega_c$. In this case a uniform $S_e = 3$ transducer is produced if the waveform is sampled at points where the phase is a multiple of $\pi/3$.

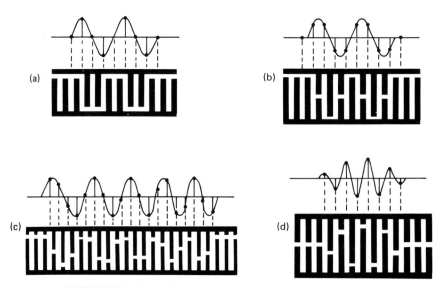

FIGURE 8.5. Examples of transducer design using regular sampling.

The transducer response remains essentially the same whatever value of the sampling frequency ω_s is used, provided it exceeds the Nyquist frequency $2\omega_2$. However, it is usually preferable to avoid apodised designs where possible, since they are more prone to errors due to diffraction. Unapodised designs can be obtained only when ω_s is a multiple of the carrier frequency ω_c. The image responses, shown by broken lines in Figure 8.4, are centred at $m\omega_s \pm \omega_c$, and become harmonics of the fundamental when ω_s is a multiple of ω_c. Similar considerations apply if the envelope of $v(t)$ is not flat; in this case rapid changes of apodisation are generally minimised by choosing ω_s to be a multiple of ω_c.

Figure 8.5(c) shows a transducer design obtained by sampling a phase-modulated waveform, taking a chirp waveform as an example. For a phase-modulated waveform an apodised transducer design is obtained if uniformly-spaced samples are used. However, chirp waveforms are usually sampled non-uniformly, and chirp transducers using this principle are discussed in Chapter 9.

Figure 8.5(d) shows an amplitude-modulated waveform, with carrier frequency ω_c, sampled with two samples per period, so that $\omega_s = 2\omega_c$. This gives a single-electrode apodised transducer, discussed using the delta-function model in Section 4.1. If the envelope of the waveform is flat, a uniform single-electrode transducer is produced. The sampling theory given above is *invalid* for this case because, since $\omega_2 > \omega_c$, the sampling frequency $\omega_s = 2\omega_c$ is less than the Nyquist frequency $2\omega_2$. However, it is shown in Section A.5 that, if $v(t)$ is an *amplitude-modulated* waveform, the sampling procedure remains valid if the particular sampling frequency $\omega_s = 2\omega_c$ is chosen. The transducer response $H_t(\omega)$ is then proportional to $E(\omega)V(\omega)$ for $|\omega| < \omega_s$. It is of course also necessary to ensure that the samples are not taken at the zeros of the waveform. This case is of practical importance because the low sampling frequency implies that the electrodes are relatively wide, making the fabrication easier.

Waveform Characteristics. Some relationships between a bandpass waveform $v(t)$ and is spectrum $V(\omega)$ are given in Appendix A, Section A.5. Bandpass filters are often required to have a frequency response whose phase is a linear function of frequency, since this phase variation causes a delay of an applied signal but does not distort it. This condition is satisfied if $V(\omega)$ has linear phase, since $H_t(\omega)$ is proportional to $E(\omega)V(\omega)$, by equation (8.11). Section A.5 shows that this implies that the waveform $v(t)$ of equation (8.12) must have an envelope $\hat{a}(t)$ which is symmetric or anti-symmetric about some time $t = t_0$. Also, the phase $\hat{\theta}(t)$ should be a constant plus a term anti-symmetric about t_0; it may for example be linear with t. The phase of $V(\omega)$ is then $-\omega t_0$ plus a constant, and the group delay is therefore t_0.

It is also shown in Section A.5 that, if $v(t)$ is an amplitude-modulated waveform, so that there is no phase modulation, its spectrum $V(\omega)$ has a magnitude symmetric about the carrier frequency ω_c, and the phase of $V(\omega)$ is a constant plus a function anti-symmetric about ω_c. For the particular cases when the envelope $\hat{a}(t)$ is symmetric or anti-symmetric about $t = t_0$, the phase of $V(\omega)$ varies linearly with frequency, and the group delay is t_0.

For many surface-wave filters the frequency response is required to have its magnitude symmetrical about a centre frequency ω_c, and its phase linear with ω. If

$V(\omega)$ is taken to have these characteristics, the waveform $v(t)$ will be an amplitude-modulated waveform, and can therefore be sampled with a sampling frequency $\omega_s = 2\omega_c$, giving a single-electrode transducer. However, the transducer response $H_t(\omega)$ is in fact proportional to $E(\omega)V(\omega)$ and, for accurate results, the distortion due to the element factor $E(\omega)$ must be compensated for, particularly if the bandwidth is large. If this is done, $V(\omega)$ will not have a symmetrical magnitude, and therefore $v(t)$ must have some phase modulation. It is then necessary to use a higher sampling frequency, above the Nyquist rate. Alternatively, the waveform $v(t)$ may be sampled non-uniformly, as discussed below.

Non-uniform Sampling. It has been assumed above that the waveform $v(t)$ is sampled uniformly, with sample spacing τ_s corresponding to the electrode spacing. However, for surface-wave transducers this is not a necessary constraint, as is clearly shown by considering the chirp waveform of Figure 8.5(c), for example. For this case it is usual to sample at the peaks and troughs of the waveform, so that the sample spacing varies and the electrode pitch p varies along the length; the result is a chirp transducer, considered further in Chapter 9.

This principle can also be applied to bandpass filter design, for cases where $v(t)$ has phase modulation. If this is done the analysis above is invalid, though the analysis in Chapter 9 can be applied provided the sample spacing does not vary rapidly. The non-uniform sampling causes some distortion in the response of the transducer. However, for bandpass filters it is often the case that the sample spacing is almost uniform and it is then found that the results of this section are approximately valid in the fundamental pass-band.

Non-uniform sampling has been used quite extensively in bandpass filters [238]. The main advantage is that for phase-modulated waveforms it is not necessary to sample above the Nyquist rate, so that a single-electrode transducer, with relatively wide electrodes, can be used. However, the method becomes invalid if the sample spacing varies rapidly, since the analysis breaks down in this case.

8.2 DESIGN OF APODISED TRANSDUCERS

It was shown above that an apodised transducer can be designed by sampling a continuous waveform $v(t)$, giving a transducer response proportional to $E(\omega)V(\omega)$, where $E(\omega)$ is the element factor. The sample values $v_n = v(n\tau_s)$ give the transducer geometry, as shown by equation (8.3). Methods of obtaining $v(t)$ are considered in Section 8.2.1 below. The main consideration is that $v(t)$ must have finite length. It is shown that the use of window functions enables $v(t)$ to be designed such that the transducer frequency response is a good approximation to some required response.

Several other design techniques are considered in Section 8.2.2. Most of these consider the problem in terms of designing the weights v_n, without considering the continuous waveform $v(t)$ as an intermediate step. Thus the sampling theory of Section 8.1 above need not be considered explicitly. Section 8.2.3 is concerned with the design of minimum-phase filters, where only the amplitude of the frequency response is specified, and the design is required to minimise the group delay.

Although this section is primarily concerned with apodised transducers, the basic methods are applicable to a wide variety of finite-impulse-response filters; in particular they can be applied to surface-wave transducers with other types of weighting, and to digital filters. Digital design techniques are reviewed in, for example references [235–237], while the methods for surface-wave filters are reviewed in references [238–241].

8.2.1. Use of Window Functions

Suppose that the response required of the transducer is some function $H_0(\omega)$. Since the actual response $H_t(\omega)$ is proportional to $E(\omega)V(\omega)$, it appears that we can divide $H_0(\omega)$ by $E(\omega)$ to obtain $V(\omega)$, and then take the inverse Fourier transform to obtain $v(t)$. However, this is not acceptable in practice because it gives a waveform $v(t)$ of infinite length, and hence the transducer response cannot be exactly equal to $H_0(\omega)$.

To allow for this, we define a function $V_0(\omega) = H_0(\omega)/E(\omega)$ with inverse Fourier transform $v_0(t)$, which will have infinite duration. The design problem is then to find a function $v(t)$, of finite length, such that its transform $V(\omega)$ is a good approximation to $V_0(\omega)$; the transducer response $H_t(\omega)$ will then be a good approximation to $H_0(\omega)$. The function $v(t)$ is sampled to give the transducer geometry, as explained in the previous section. The accuracy of the approximation will depend on the method used to evaluate $v(t)$, and it will be shown that the accuracy obtainable generally increases with the duration of $v(t)$. Note that $V_0(\omega)$ must be a bandpass function, as required for the sampling technique, but is otherwise arbitrary.

An obvious method of obtaining $v(t)$ is simply to truncate $v_0(t)$ to a finite length. For present purposes, it is convenient to represent this as a multiplication by a *window* $W(t)$, so that

$$v(t) = W(t)v_0(t). \tag{8.13}$$

This equation truncates $v_0(t)$ if $W(t)$ is unity for some finite time interval, and is zero for other t; however other forms of $W(t)$ will be considered later. In the frequency domain, the multiplication in equation (8.13) causes $V_0(\omega)$ to be convolved with $\bar{W}(\omega)$, the Fourier transform of $W(t)$, so that

$$V(\omega) = \frac{1}{2\pi}\int_{-\infty}^{\infty} V_0(\omega')\bar{W}(\omega - \omega')\,d\omega'. \tag{8.14}$$

Equation (8.13) truncates the waveform $v_0(t)$ to a length T if the window is taken as

$$W(t) = \text{rect}(t/T) \tag{8.15}$$

where $\text{rect}(x) = 1$ for $|x| \leq \frac{1}{2}$ and is zero for other x. Strictly speaking this form for $W(t)$ is not valid, because causality requires that $v(t)$ should be zero for $t < 0$. However causality can easily be satisfied at a later stage simply by delaying $v(t)$; in the frequency domain, this adds a phase term proportional to ω, as wown by the shifting theorem. The Fourier transform of equation (8.15) is

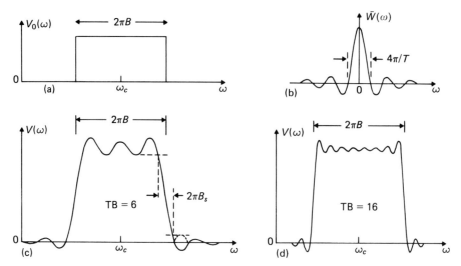

FIGURE 8.6. Effect of time-domain truncation, for a filter with an ideally rectangular frequency response.

$$\bar{W}(\omega) = T \operatorname{sinc}(\tfrac{1}{2}\omega T), \qquad (8.16)$$

where sinc $x = (\sin x)/x$.

To illustrate the effect of truncation, suppose that the required frequency response $V_0(\omega)$ is flat within a band of width $\Delta\omega = 2\pi B$ and zero outside this band, so that

$$V_0(\omega) = 1, \quad \text{for } |\omega - \omega_c| \leq \pi B,$$
$$= 0, \quad \text{for } |\omega - \omega_c| > \pi B. \qquad (8.17)$$

The inverse transform, $v_0(t)$, of this function is an amplitude-modulated waveform with carrier frequency ω_c and envelope proportional to sinc $(\pi B t)$, and is therefore infinite in length. Figure 8.6(a) shows $V_0(\omega)$, and Figure 8.6(b) shows the spectrum $\bar{W}(\omega)$ of the window function, equation (8.16). These are convolved in accordance with equation (8.14) to give $V(\omega)$, the spectrum of the finite-length waveform $v(t)$. This is shown in Figure 8.6(c) for $TB = 6$ and in Figure 8.6(d) for $TB = 16$. In comparison with the ideal response $V_0(\omega)$, the actual response $V(\omega)$ exhibits ripples in the pass-band and sidelobes in the stop bands. These are due to the sidelobes of the function $\bar{W}(\omega - \omega')$ in equation (8.14). At any ω, some of the sidelobes of $\bar{W}(\omega - \omega')$ are in the band occupied by $V_0(\omega')$ and therefore contribute to the integral; as ω changes, sidelobes enter the band at one side and leave at the other side, thus giving an oscillatory contribution. In addition the transitions at the band edges are no longer sharp, so that $V(\omega)$ has finite skirt widths; these are associated with the width of the main peak of the function $\bar{W}(\omega)$.

Although the above refers to a specific case, the pass-band ripple, stop-band sidelobes and skirt broadening are in fact quite general phenomena in finite-impulse-response filters, present to some extent irrespective of the design method used. The distortions arise basically because the finite length of $v(t)$ implies that $V(\omega)$ cannot be strictly confined to a band of finite width, hence the presence of the stop-band sidelobes.

The distortion due to truncation can be reduced by increasing the window length T, as can be seen by comparing the results for $TB = 6$ and $TB = 16$ shown in Figures 8.6(c) and (d). A larger value of T reduces the skirt width and the ripple at the band centre. However, the ripple near the band edge, and the sidelobes near the band edge, have magnitudes almost independent of T; the largest sidelobe is 21 dB below the pass-band level then TB is large. For practical purposes this is usually unacceptable.

The solution to this problem is to modify the weighting function $W(t)$. The basic requirement is for a function of finite length, whose Fourier transform has sidelobes substantially less than those of the sinc function. This requirement arises in several fields; in addition to surface-wave and digital filters, it arises in spectral estimation from finite-length records [242] and in the design of antennas whose polar diagrams need to have low sidelobes. Consequently, a variety of suitable window functions are available. An example is the *Kaiser* window [235], given by

$$W_K(t) = \frac{I_0[\alpha\sqrt{1 - 4t^2/T^2}]}{I_0(\alpha)}, \quad \text{for } |t| \leq T/2,$$

$$= 0, \quad \text{for } |t| > T/2, \quad (8.18)$$

where $I_0(x)$ is the modified Bessel function of the first kind and zero order. The parameter α is chosen to suit the application. Figure 8.7 shows $W_K(t)$ for $\alpha = 6$ and its transform $\bar{W}_K(\omega)$, which is given by a simple formula [241]. The figure includes for comparison the function sinc $(\frac{1}{2}\omega T)$, which is the transform of the rectangular window function, as in equation (8.16). These frequency-domain functions are plotted logarithmically. It can be seen that the Kaiser function $\bar{W}_K(\omega)$ gives much smaller sidelobes, though its main peak is about twice as wide as that of the sinc function. Thus, when used for filter design, the Kaiser window gives smaller ripples and sidelobes but, for a given value of T, the skirt width is larger. The skirt width can

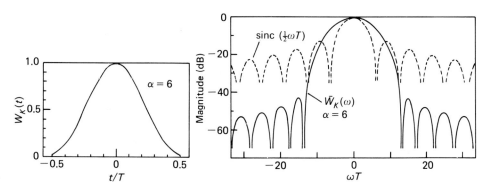

FIGURE 8.7. Kaiser window function $W_K(t)$ and its Fourier transform.

however be reduced by increasing T, since this simply reduces the width of $\bar{W}_K(\omega)$ in proportion.

The parameter α in equation (8.18) affects the sidelobe levels of $\bar{W}_K(\omega)$ and the width of the main peak. Larger values of α give smaller sidelobes and wider peaks, enabling these features to be traded against each other. When the Kaiser window is used to design a filter with an ideally rectangular frequency response [as in Figure 8.6(a)], α determines the pass-band ripple, stop-band attenuation and skirt width. Tancrell [239] gives these parameters as functions of α. For example, if $\alpha = 6$, the largest stop-band sidelobe is at -62 dB, the ripple at the band edge is 0.009 dB peak-to-peak, and the skirt width is $B_s = 4.0/T$, where B_s is in Hz and is defined as shown in Figure 8.6(c). To design the filter, the value of α is chosen first, such that the pass-band ripple and stop-band sidelobes are acceptable, and the value of T needed to give the required skirt width can then be deduced straightforwardly. Design data equivalent to Tancrell's are also given by Rabiner and Gold [237, p. 101].

A number of other window functions have been considered for surface-wave filters [239, 241], and Harris [242] has compared the performance of a large variety of windows for spectral estimation. In all cases, the window function in the time domain is real and symmetric about $t = 0$, so its transform in the frequency domain is real and symmetric about $\omega = 0$. None of these functions gives a performance appreciably better than the Kaiser window, in the context of filter design. A particular example is the *Dolph–Chebyshev* window [241, 243]. For a filter design using this window the stop-band sidelobes are all the same size, and the window is an optimum in the sense that the narrowest possible skirt width is obtained for a given sidelobe level. However, the Dolph–Chebyshev window is a complicated function, rather inconvenient for practical usage, and gives a performance little better than that of the Kaiser window [241]. Some other window functions are described in Section 9.2.3, in connection with chirp filters.

Having designed $v(t)$ with the aid of a window function, the final stage of the design procedure is to sample to obtain the weights $v_n = v(n\tau_s)$, as explained in Section 8.1. A complication arises here because the sampling theory of Section 8.1 assumes $V(\omega)$ to be a band-pass function. This is not in fact true here because $v(t)$ has finite length, hence the presence of the sidelobes of $V(\omega)$, extending indefinitely on either side of the pass-band. Thus, some aliasing occurs when $v(t)$ is sampled, so that the image responses generated by sampling contribute sidelobes in the required pass-band. However, the distortion due to this aliasing is usually small and acceptable, provided $V(\omega)$ is designed to have well-suppressed sidelobes. It is also necessary that the sidelobes should decrease in amplitude for frequencies more remote from the pass-band. This is the case for most window functions, including the Kaiser window. The Dolph–Chebyshev window is an exception, giving sidelobes of equal amplitude, but for this case a specially adapted form of the window function can be used to avoid the aliasing problem for sampled waveforms [243].

Note that, although the above discussion has illustrated the method for the case when the ideal response $V_0(\omega)$ is a rectangular function, the method is in fact quite general. Thus there are no constraints on either the amplitude or phase of $V_0(\omega)$, except that it must be a band-pass function.

8.2.2. Optimised Design Methods

Although the use of window functions described above is an effective and straightforward design technique, it does not give optimised designs. For example, the relative magnitude of the pass-band ripple and the stop-band sidelobes are determined by the weighting function, and the pass-band ripple is maximised at the band edges. For some cases a more sophisticated method is desirable, taking account of specified tolerances, so that an optimised design can be produced.

A quite simple approach is based on an iterative use of the Fourier transform. The required response $V_0(\omega)$ is first transformed to the time-domain, truncated to a finite length T, and then transformed back to the frequency domain. Due to the truncation, this new frequency response will generally fall outside the specified tolerances in some regions. The frequency response is modified to bring it within the tolerances and the procedure is the repeated, transforming to the time domain, truncating, and returning to the frequency domain. Provided T is large enough, iteration of this process generally gives progressive improvement, yielding finally a frequency response meeting the specification and a finite-length impulse response which gives the transducer geometry. This principle was used, with some modifications, by Moulding and Parker [244], and by De Vries [245]. Although the final design is not strictly optimal, the method has the advantages of simplicity and versatility; it can accommodate a non-linear phase in the frequency domain, and can allow for the tolerances being different in different frequency regions.

A variety of other optmisation techniques have been developed for design of digital finite-impulse-response filters [236, 237, 239], and these may be applied directly to the design of surface-wave transducers if the required response is first divided by $E(\omega)$ to give the required response $V_0(\omega)$ of the corresponding transversal filter. These methods give directly the weights v_n of the transversal filter, related to the transducer geometry by equation (8.3). The continuous waveform $v(t)$ is not considered explicitly, so the aliasing problem mentioned in Section 8.2.1 does not occur. A particular example, which has been used extensively for surface-wave filters, is that of McClellan and Parks [237, 246, 247]. Filters designed by this method have linear phase, so that the response $H_s(\omega)$ of the transversal filter has the form

$$H_s(\omega) = A(\omega) \exp[j(c - \omega t_0)], \tag{8.19}$$

for $\omega > 0$, where $A(\omega)$ is the amplitude and t_0 is a constant determined by the length of the response in the time domain. The constant c is either 0 or $\pi/2$. The method makes use of the Remez exchange algorithm to design the filter such that the amplitude response $A(\omega)$ is a good approximation to a required amplitude response $A_0(\omega)$. The design obtained minimises an error function E_A, defined by

$$E_A = \text{Max } \{e(\omega)|A(\omega) - A_0(\omega)|\}, \quad 0 < \omega < \tfrac{1}{2}\omega_s, \tag{8.20}$$

where $e(\omega)$ is an error weighting function chosen by the designer. The method thus allows the designer to specify that the response is required to be more accurate in some regions of the band than in others. For example, the tolerances on pass-band ripple and stop-band sidelobes can be specified independently. The accuracy obtained

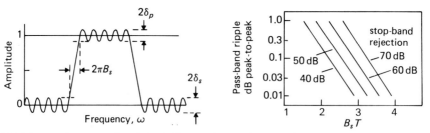

FIGURE 8.8. Response of equi-ripple filter, and approximate performance theoretically obtainable.

depends on the length of the impulse response, which is therefore increased if a better accuracy is needed.

If the required amplitude response $A_0(\omega)$ is rectangular, as in Figure 8.6(a), this method can give a design with an "equi-ripple" response, illustrated in Figure 8.8. This type of result is obtained if the tolerances in the pass-band and in the stop band are independent of frequency. The normalised amplitude in the pass-band oscillates between extrema at $1 \pm \delta_p$, and the sidelobes all have magnitude δ_s. Such a response is found to give the smallest possible skirt width consistent with a give maximum error in the pass-band and a given maximum sidelobe level. From studies of large numbers of designs of this type, it can be concluded that the skirt width B_s is approximately given by [248]

$$\log(\delta_p \delta_s) \approx -1.05 - 1.45 B_s T, \qquad (8.21)$$

where T is the duration of the impulse response. This equation is useful for estimating the duration T needed to meet a given specification. Figure 8.8 shows the relation between pass-band ripple, stop-band rejection and skirt width B_s. Typical values of $B_s T$ are in the range 2 to 4.

Although the basic method assumes the required frequency response to have linear phase, it can be adapted quite readily to design filters with non-linear phase [249]. The adaptation exploits the fact that the basic method gives a response with the form of equation (8.19), where the phase constant c is either 0 or $\pi/2$. The constant is zero if the design is specified to be symmetric in the time domain, so that $v_n = v_{N+1-n}$, and is $\pi/2$ if the design is anti-symmetric, so that $v_n = -v_{N+1-n}$. The response of a filter with non-linear phase can be written

$$H(\omega) = A(\omega) \exp[j\phi(\omega)] \exp(-j\omega t_0) \qquad (8.22)$$

This can be regarded as a sum of two linear-phase responses, an "in-phase" response with $c = 0$:

$$H^i(\omega) = [A(\omega) \cos \phi(\omega)] \exp(-j\omega t_0)$$

and a "quadrature" response with $c = \pi/2$:

$$H^q(\omega) = j[A(\omega) \sin \phi(\omega)] \exp(-j\omega t_0).$$

The two linear-phase responses can be designed by the basic method, taking the

BANDPASS FILTERS

required amplitude responses to be $A(\omega) \cos \phi(\omega)$ and $A(\omega) \sin \phi(\omega)$, with $c = 0$ and $\pi/2$ respectively. The number of time-domain samples, N, is taken to be the same for both responses, so that the value of t_0 is the same for both. The time-domain samples are at the same points, and are simply added to obtain the weights for the non-linear phase response. This method has been shown to be effective for design of a chirp type of response, where the phase $\phi(\omega)$ is a quadratic function of frequency [249].

A quite different design technique makes use of non-linear programming to optimise the response, and has recently been applied to surface-wave filter design [250].

8.2.3. Minimum-phase Filters

As already noted in Section 8.1.3, if the frequency response of a filter has a phase linear with ω, the group delay is equal to t_0, where t_0 is the centre point of the impulse response. Thus, if the impulse response has duration T, the group delay must be at least $T/2$. For some applications it is important to minimise the group delay, and we therefore consider whether the delay can be reduced without appreciably changing the magnitude of the frequency response. This implies that the frequency-domain phase will become a non-linear function of ω, causing some distortion of a signal applied to the device, but this will be acceptable if the phase non-linearity is small enough.

A particular case is the *minimum-phase* filter [251, 252]. If the response is $H(\omega) = A(\omega) \exp[j\phi(\omega)]$, the phase $\phi(\omega)$ is, for a minimum-phase filter, related to the amplitude by

$$\phi(\omega) = -\frac{1}{\pi} \int_{-\infty}^{\infty} \frac{\ln |A(\omega')|}{\omega - \omega'} d\omega' \qquad (8.23)$$

Thus $\phi(\omega)$ is the Hilbert transform of $\ln |A(\omega)|$, that is, the convolution with $-1/(\pi\omega)$. The significance of this result is that, for any specified amplitude response $A(\omega)$, the magnitude of the phase $\phi(\omega)$ is smaller for a minimum-phase filter than for any other feasible filter response. It is assumed that the responses under consideration are causal, that is, the corresponding impulse responses are zero for $t < 0$. Equation (8.23) ensures this for the minimum-phase case. In comparison with a linear-phase filter with the same amplitude response, the minimum-phase filter is generally found to give a smaller group delay. Moreover, the phase non-linearity is often found to be acceptable. This is true particularly when the amplitude $A(\omega)$ varies relatively slowly with frequency, because functions of this type have Hilbert transforms that are relatively small and smoothly varying.

Given some required amplitude response, the phase required for a minimum-phase filter can be obtained from equation (8.23), and the filter may then be designed by the methods described earlier. For a transversal filter, an alternative method can be used [253, 254]. If we define $z = \exp(-j\omega\tau_s)$, the response $H_s(\omega)$ of a transversal filter, equation (8.6), becomes a polynomial in z, and can be specified by its zeros. These are the values of z, generally complex, at which the polynomial is zero. A linear-phase transversal filter is first designed, and its zeros evaluated. Some of the zeros can then be deleted in a systematic way, such that the resulting new polynomial gives a minimum-phase filter with the same amplitude response.

For a surface-wave filter, the method is of course applied to the design of one of the two transducers [254]. In addition to giving a smaller delay, a minimum-phase design can also give a shorter transducer, making more economical use of the substrate area. It has also been found that the minimum-phase design is less affected by second-order effects [254].

8.3. THINNING AND WITHDRAWAL WEIGHTING

Withdrawal weighting is a technique in which selected sources are omitted from a transducer. This enables the transducer to be weighted without using apodisation, so that the apertures of all the remaining sources are the same. In narrow-band filters, which require transducers many wavelengths long, this is advantageous in reducing the perturbation due to diffraction. In addition, a withdrawal-weighted transducer can be used in conjunction with an apodised transducer, so that the flexibility obtained by using two weighted transducers becomes available for weakly piezoelectric substrates, such as quartz, on which a multi-strip coupler cannot be used.

It is convenient first to introduce a simple modification known as "thinning", which is often applied to narrow-band transducers. Figure 8.9(a) shows thinning as applied to a uniform unapodised transducer. The design is obtained simply by omitting some of the electrodes in a conventional transducer, such that the result obtained consists of a regular sequence of identical groups of electrodes. The nature of the response of this transducer is readily appreciated from sampling theory. Since the bandwidth of any one group of electrodes is much larger than the overall transducer bandwidth, the thinning process is approximately equivalent to sampling the impulse response with a sampling interval, τ'_s, corresponding to the spacing of the groups. The sampling process causes image responses to appear in the frequency response, with frequency spacing $2\pi/\tau'_s$, as shown by equation (8.9). These image responses may be attenuated by filtering elsewhere, for example in a second transducer or in a matching network. Thus, for practical purposes the response of a thinned transducer can often be regarded as essentially the same as that of a

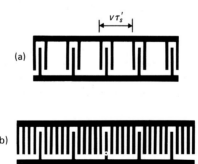

FIGURE 8.9. Two types of thinning, as applied to a uniform transducer.

conventional transducer, apart from a reduction of the strength of coupling to surface waves. An accurate analysis would of course need to account for the response of the individual groups of electrodes, and these are affected somewhat by end effects (Section 4.5.2).

The commonest reason for using thinning is to reduce electrode interaction effects in single-electrode transducers. For a shorted single-electrode transducer the acoustic reflection coefficient, which is due to electrode interactions, is approximately proportional to the number of electrodes and is therefore substantially reduced by thinning. Alternatively, interactions can be reduced by increasing the number of electrodes per period, as discussed in Section 4.2, but this requires narrower electrodes and so makes the fabrication more difficult. The scheme of Figure 8.9(a) is frequently used in delay-line oscillators, and also occurs in a modified form in PSK filters.

Thinning also has the effect of reducing the transducer admittance. The capacitance C_t and the centre-frequency conductance $G_a(\omega_c)$ are both reduced, while the transducer Q-factor $\omega_c C_t / G_a(\omega_c)$ is increased. These changes are sometimes beneficial in reducing the severity of circuit effects.

An alternative method of thinning, shown in Figure 8.9(b), is to change some of the electrode polarities so that selected gap elements are eliminated. This reduces the capacitance and conductance, though it does not substantially reduce electrode interaction effects. For this transducer the electrodes are regular and the individual groups are not affected by end effects. Either of the two types of thinning may also be applied to an apodised transducer.

Withdrawal weighting is illustrated in Figure 8.10. The technique is very similar to thinning, but here the thinning process is applied non-uniformly so that the remaining groups of sources have differing lengths. This enables the transducer to be designed such that its frequency response approximates a required response, without using apodisation. The technique was first demonstrated by Hartmann [255], who weighted

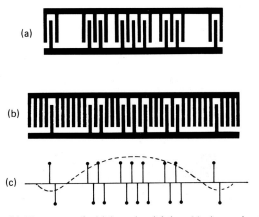

FIGURE 8.10. (a), (b): Two types of withdrawal weighting. (c): Approximate impulse response.

the transducer by omitting electrodes as in Figure 8.10(a). This is known as "electrode-withdrawal weighting". The alternative method of Figure 8.10(b), where selected gap sources are eliminated by changing the electrode polarities [256], is called "source-withdrawal weighting". Withdrawal weighting is suitable only for transducers with relatively narrow bandwidths. As in the case of a thinned transducer, additional responses arise at frequencies outside the main pass-band, so that the stop-band attenuation is poor. In practical devices, a withdrawal-weighted transducer is often used in conjunction with an apodised transducer, with the latter designed such that adequate stop-band attenuation is obtained.

Both types of withdrawal-weighted transducer may be analysed by the methods of Chapter 4, giving the frequency response, admittance and scattering parameters. For a source-withdrawal transducer the electrodes are regular and the response can be written in terms of an array factor and an element factor. This considerably simplifies the analysis, as explained in Section 4.5. Figure 8.10(c) shows the approximate impulse response, where each of the gap sources is represented by a delta function. The broken line indicates an approximate form for the envelope of the impulse response, obtained by smoothing out the actual envelope. The smoothed impulse response has a Fourier transform approximating the transducer frequency response, for frequencies in the main pass-band.

As for apodised transducers, withdrawal-weighted transducers can be designed by first generating an impulse response $v(t)$ of finite duration, such that its Fourier transform is a good approximation to the required frequency response. The envelope of $v(t)$ is exemplified by the broken line in Figure 8.10(c). Since the actual impulse response has sharp discontinuities and so is very different from $v(t)$, the design procedure seeks to synthesise the required frequency response only in the main pass-band, neglecting any stop-band requirements.

To design an electrode-withdrawal transducer, the responses of the possible types of electrode group are first evaluated, at the centre frequency. The ideal impulse response $v(t)$ is assumed to be an amplitude-modulated waveform, and its envelope is integrated to give a new function of time. The types and locations of the electrode groups are then chosen such that the integral of the designed impulse response envelope is as close as possible to the integral of the ideal impulse response envelope [255, 257]. The transducer design then gives a frequency response approximating the spectrum of $v(t)$, for frequencies close to the centre frequency. In practice, the design must allow for the fact that the responses of individual electrode groups are affected by end effects, as discussed in Section 4.5.2. In addition, the electrodes perturb the surface-wave velocity non-uniformly, owing to mechanical and electrical loading. This causes phase errors which must be compensated for by making small adjustments to the locations of individual electrode groups [257]. A sophisticated design procedure taking account of these complications is described by Laker *et al.* [258].

For a source-withdrawal transducer, Figure 8.10(b), the design procedure is less complicated. Because the electrodes are regular, end effects are virtually eliminated and the frequency response has the simple form of equation (8.2). Also, velocity perturbations due to the electrodes are uniform throughout the transducer, and so do not complicate the design.

8.4. FILTER DESIGN AND PERFORMANCE

8.4.1. Basic Types of Bandpass Filter

The simplest form of bandpass filter comprises one apodised transducer and one uniform transducer, as illustrated in Figure 8.1. The response of this device is essentially the product of the responses of the individual transducers, as shown by equation (8.1). To design the device, the length of the uniform transducer is chosen first, such that the variation of its frequency response over the band of interest is acceptable; if the specification calls for "traps", where the filter response is to have a particularly small amplitude, the uniform transducer is usually designed to give zeros at one or more of these locations. The response required for the filter is divided by the calculated response of the uniform transducer, giving the required response of the apodised transducer, and the latter is then designed by methods discussed in Section 8.2.

While this type of design is often acceptable, there are some significant limitations and hence a variety of other filter types are often used, the choice depending on the specific requirements. Nearly always, the design is chosen such that the filter response is essentially the product of two transducer responses, as this is very convenient for design purposes. For a strongly piezoelectric substrate material the two transducers can be located in different tracks and coupled via a multi-strip coupler, as illustrated in Figure 5.6. This reduces unwanted output signals due to bulk wave excitation, as discussed in Section 5.3, and is beneficial particularly for Y, Z lithium niobate substrates.

An additional advantage introduced by using a coupler is that the device response is still essentially the product of the two transducer responses, even if both transducers are apodised. This is not true for two apodised transducers located in the same track. The coupler thus enables two weighted transducers to be combined to synthesise the required response. This is often of considerable value when an exacting specification is to be met, because second-order effects, such as transverse end effects and diffraction, are often less deleterious if two weighted transducers are used. For example, if a rectangular frequency response is required, a combination of two similar apodised transducers enables smaller skirt widths and better stop-band rejection to be obtained [240]. The two transducers may have identical designs, in which case each is designed to approximate the square root of the required filter response. Strictly, the coupler response S_{14} should be allowed for, as shown by equation (5.38), though in fact S_{14} varies only slowly with frequency.

For narrow-band filters, long transducers are required. In this case withdrawal-weighting is often attractive since it gives transducers less susceptible to diffraction errors, as discussed in Section 8.3. In addition, since a withdrawal-weighted transducer is unapodised, it may be combined with an apodised transducer in the same track. The filter response is then the product of two weighted transducer responses, without requiring a multi-strip coupler. The advantages of using two weighted transducers are thus obtainable for weakly-piezoelectric substrates, such as quartz, on which a multi-strip coupler cannot be used. In view of the somewhat limited design methods for withdrawal weighting, it is usual to design the withdrawal-weighted transducer first, and then to design the apodised transducer taking account of the predicted response of the withdrawal-weighted transducer.

8.4.2. Circuit Effect

The design procedures described earlier refer to the short-circuit response $H_{sc}(\omega)$ of the device. This is defined as the short-circuit output current I_{sc} produced when a given voltage V_t is applied to the input transducer, as in equation (8.1). In practice the transducers must be connected to finite impedances and this causes a distortion, due to the circuit effect, which may be calculated by methods described in Section 4.8. If the transducers are connected directly to a resistive source and a resistive load, the distortion is generally small. However, it is often necessary to tune the transducers in order to reduce the insertion loss, and we show here that this generally degrades the stop-band attenuation.

For the present discussion it is sufficient to consider only the input transducer, and it is assumed that this is tuned by a series inductor which connects it to a source with open-circuit voltage V_G and impedance R_G, as in Figure 8.11. The transducer is represented by a series equivalent circuit with radiation resistance $R_a(\omega)$, as discussed in Section 7.1.1. The transducer voltage, V_t, varies with frequency because of the inductor and the frequency variation of the transducer impedance. Since the short-circuit response $H_{sc}(\omega)$ gives the device output for a given value of V_t, the response allowing for the circuit effect can be obtained by multiplying by V_t/V_G, which is the circuit factor $F_c(\omega)$ defined in Section 4.8. From the circuit of Figure 8.11, this is approximately given by

$$\frac{V_t}{V_G} \approx \frac{-j/(\omega C_t)}{R_G + R_a(\omega) + j\omega L - j/(\omega C_t)}, \tag{8.24}$$

where C_t is the transducer capacitance. Here a term $R_a(\omega)$ has been omitted from the numerator on the assumption that the transducer Q-factor is large, which is often the case. The acoustic reactance $X_a(\omega)$ has also been neglected.

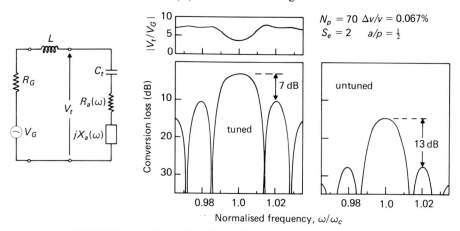

FIGURE 8.11. Circuit effect for a uniform single-electrode transducer.

Generally, $R_a(\omega)$ has a maximum at or near the centre frequency ω_c, and is small for frequencies outside the pass-band. It is assumed here that the transducer is matched at frequency ω_c, so that $R_a(\omega_c) = R_G$ and $\omega_c L - 1/(\omega_c C_t) = 0$. Thus, at frequency ω_c we have $V_t/V_G \approx -j/(2\omega_c C_t R_G)$. At other frequencies in the pass-band, the term $\omega L - 1/(\omega C_t)$ in equation (8.24) is still small – if this were not so, the response would be dominated by the electrical resonance of the circuit, and this condition is avoided in practical filters. Thus, for frequencies on either side of ω_c, where $R_a(\omega)$ is decreasing, the value of V_t/V_G increases because $R_a(\omega)$ is present in the denominator of equation (8.24). The circuit effect therefore tends to flatten the pass-band response. For frequencies in the stop-bands, but still close to the pass-band, $R_a(\omega)$ is small and hence $V_t/V_G \approx -j/(\omega C_t R_G)$. This is about twice the centre-frequency value, and hence the circuit effect increases the relative sidelobe levels by about 6 dB. For frequencies more remote from the centre frequency the electrical resonance is significant, so that $\omega L - 1/(\omega C_t)$ is no longer small, and the sidelobes are suppressed. Although a specific circuit has been considered here, distortions of this nature are generally observed when transducers are tuned; however they are generally less severe because accurate matching is avoided in order to ensure adequate triple-transit suppression, and because of the presence of parasitics.

The conversion loss curves in Figure 8.11 illustrate the circuit effect for a uniform transducer with $N_p = 70$ periods, calculated using the transducer admittance formulae in Section 4.6. For the tuned case the transducer is matched at the centre frequency, and the magnitude of V_t/V_G is also shown. The untuned conversion loss refers to the same case except that the inductor is omitted. The pass-band flattening and the increase of sidelobe level, due to the circuit effect, are clearly seen.

Because of the pass-band flattening, the circuit effect increases the bandwidth somewhat when N_p is large. From equation (8.24), V_t/V_G is approximately proportional to $1/(R_a + R_G)$, and hence the surface-wave power generated is proportional to $R_a/(R_a + R_G)^2$. For a uniform transducer, the radiation resistance $R_a(\omega)$ is approximately equal to $\hat{R}_a[(\sin X)/X]^2$, where $X = \pi N_p(\omega/\omega_c - 1)$, and here $\hat{R}_a = R_G$ because the transducer is matched. It follows that the 1.5 dB bandwidth $\Delta\omega \approx 1.14\, \omega_c/N_p$, when N_p is large. This formula is used in the discussion in Section 7.1.1.

8.4.3. Second-order Effects and Design

Many of the second-order effects that can affect the response of a bandpass filter are usually rendered negligible by an appropriate initial choice of the types of transducer and of the substrate material. For example, in a single-electrode transducer, electrode interactions are generally negligible if the number of electrodes, N, is such that $N\Delta v/v \ll 1$. If this is not the case, a multi-electrode transducer may be used, as described in Section 4.2, or, if the bandwidth is small, the transducer may be thinned as in Section 8.3. The triple-transit spurious signal, due to acoustic reflections, is normally suppressed adequately by ensuring that the transducers are not closely matched to the electrical source and load. As discussed for uniform transducers in Section 7.1.3, this implies a trade-off with the device insertion loss. Bulk-wave effects can be minimised by an appropriate choice of substrate material or, if a strongly piezoelectric material is used, by incorporating a multi-strip coupler. Temperature changes cause the amplitude and phase of the frequency response to scale with frequency, as discussed in Section 6.4. This is sometimes significant for filters with narrow bandwidths, or with narrow skirts, requiring the choice of a temperature-stable substrate material such as *ST, X* quartz. Electromagnetic breakthrough between the input and output of the device must be adequately suppressed, and this is achieved by careful package design and lay-out; this is particularly important at high frequencies.

Generally, the only remaining second-order effects of significance are diffraction, the circuit effect and the transverse end effect, though diffraction is usually insignificant if the substrate orientation is a minimal-diffraction orientation. These effects are, when necessary, compensated for by modifying the transducer design. The analysis of a device allowing for diffraction is discussed in Section 6.2.5, while the analysis including the circuit effect is described in Section 4.8, which allows for arbitrary terminating circuits. For accurate results, it is often important to include stray components, as discussed in Section 7.1.2. To design a device allowing for these effects, the simplest approach is an iterative process. The device is first designed ignoring the second-order effects. The frequency response of the design is then calculated allowing for the second-order effects, so that the distortion due to these effects can be evaluated. The original specification is then modified to compensate for the distortion, and the design stage is repeated. This process may be repeated several times, and is usually effective if the distortion due to second-order effects is small. It is usual to add at least one iteration involving an experimental device, repeating the design procedure to compensate for the distortion observed experimentally.

In some cases diffraction effects are too severe for the above approach to be effective, so a more basic method of compensation, based on the diffraction analysis of Section 6.2, is necessary. This is a problem of considerable complexity and hence a variety of techniques have been considered, as discussed for example in Refs. [259–261]. The difficulties arise because of the complexity of the basic analysis for diffraction, and because the effect is frequency-dependent. For a narrow-band filter it is sufficient to consider diffraction at the centre frequency, ignoring the frequency dependence [259]. Assuming that at least one of the two transducers is apodised, the effect of diffraction is calculated for each element of this transducer in turn, allowing for the geometry of the other transducer. The aperture of each element is then adjusted such that it gives the required contribution when diffraction is present; this requires an iterative procedure, since the equations involved are transcendental. For wider bandwidth devices this method is of limited value because of the frequency variation of the diffraction effect, but some sophisticated and effective methods have been developed for this case [260, 261].

8.4.4. Performance

Generally, surface-wave bandpass filters are suitable for centre frequencies in the range 10 MHz to 2 GHz, with bandwidths ranging from a maximum of 50% of the centre frequency to a minimum of about 100 kHz. The minimum bandwidth is related to the physical size, since narrower skirt widths generally require longer impulse responses. A stop-band rejection of 60 dB is often achievable. For a filter with a nominally rectangular frequency response a common figure of merit is the *shape factor*, defined as the ratio of the 40 dB bandwidth to the 3 dB bandwidth. Shape factors down to about 1.1 can be obtained, giving very sharp roll-off in the skirts of the response. Within the pass-band a relative amplitude accuracy of 0.1 dB is obtainable and, for a nominally linear-phase filter, the phase can be linear to within 1°. An important factor related to this fidelity is the triple-transit signal, which causes ripples in the amplitude and phase. For example, a triple-transit suppression of 40 dB corresponds to an amplitude ripple of 0.2 dB peak-to-peak, and a phase ripple of 1.1° peak-to-peak. It is thus important that the triple-transit signal should be adequately suppressed, and hence the insertion loss is not usually minimised. Typical insertion losses are in the range 15 to 30 dB. A more extensive discussion of the performance obtainable is given, with practical examples, by Hays and Hartmann [262].

Figure 8.12 shows the experimental response of a bandpass filter for a domestic television receiver [249]. This device had one uniform and one apodised transducer on a Y, Z lithium niobate substrate, with the transducers coupled by a multi-strip coupler. The hatched lines on the figure indicate the specification. The complexity of the specification illustrates well the versatility of surface-wave devices. Note that the phase $\phi(\omega)$ of the frequency response is required to be a non-linear function of frequency, so that the group delay $\tau_g(\omega) = -d\phi(\omega)/d\omega$ varies with frequency. The reason for this is that the surface-wave filter replaces an earlier *L–C* filter giving a non-linear phase, and consequently the broadcast signal is pre-distorted to compensate for this. The device is enclosed in a circular TO-8 package of 14 mm

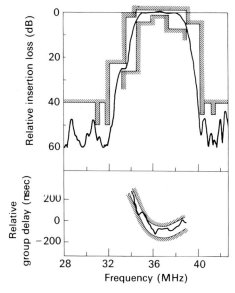

FIGURE 8.12. Response of a surface-wave bandpass filter for domestic television receivers. (Courtesy, Plessey Research)

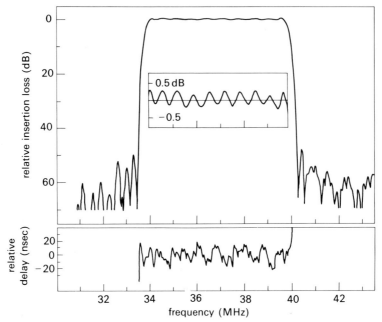

FIGURE 8.13. Response of a bandpass filter for television broadcast equipment (courtesy J. M. Deacon, Signal Technology Ltd.)

diameter. Such devices are now used almost universally in colour television receivers since, in comparison with the earlier L–C filters, they are smaller, give better performance, and do not require trimming after fabrication. In the interest of economy, the lithium niobate substrate may be replaced by a zinc oxide film on a glass substrate [263], a material mentioned previously in Section 3.5. The use of surface-wave filters in television receivers is discussed by Ash [264], who proposes an improved system configuration involving in addition a surface-wave oscillator.

An example of a filter response meeting very stringent requirements is shown in Figure 8.13. In this case the application is in television broadcasting equipment. The frequency response is required to have linear phase and a rectangular amplitude characteristic, with very narrow skirts. The upper part of the figure shows the insertion loss, with the pass-band detail shown in the inset, and the lower part shows the group delay. The device gave a shape factor (40 dB bandwidth divided by 3 dB bandwidth) of 1.2 and a pass-band ripple of 0.4 dB peak-to-peak. In this case two apodised transducers were used, coupled via a multi-strip coupler, on a Y, Z lithium niobate substrate. The mid-band insertion loss was 29 dB and the triple-transit suppression 63 dB. Kodama et al. [250] illustrate a similar result giving a pass-band ripple of 0.24 dB peak-to-peak, though with smaller out-of-band rejection.

Some other experimental results illustrating the performance of bandpass filters are given in, for example, references [262, 265–268], and in Sections 7.2.1 and 8.5.

8.4.5. Other Types of Bandpass Filter

Up to this point, this chapter has considered filters using transducers that are uniform, withdrawal-weighted, or apodised with regular (or almost regular) electrodes. Most filters rely on one or more of these types of transducer. However, a variety of other techniques have been investigated, and these are considered briefly here.

Chirp transducers, in which the electrode pitch varies with position, can be used effectively in filters requiring narrow skirts [249]. For this case, a transducer with regular electrodes has several time-domain sidelobes, and hence there are many small-aperture sources. A chirp transducer gives fewer small-aperture sources and is found to be less affected by diffraction and electrode interactions. This type of transducer is considered in Chapter 9. Its frequency response is dispersive, but for applications requiring linear phase two similar transducers can be used, arranged such that the dispersion cancels.

For applications requiring low insertion loss, some special techniques can be used to maintain good triple-transit suppression, as mentioned in Section 7.1.3. In particular, multi-phase unidirectional transducers have been found to be very effective, and the performance of bandpass filters using these is discussed in Section 7.2.1.

In addition to apodisation and withdrawal weighting, several other weighting techniques have been demonstrated. These include a capacitive weighting technique [269] and other methods reviewed by Engan [270]. There are several methods in which the transducer generates a uniform surface-wave beam, minimising diffraction effects and also avoiding the design limitations of withdrawal weighting. However, none of these has come into common usage.

Bandpass filters have also been made using reflective arrays of either grooves [271, 272] or metal dots [273]. The reflective array is used to reflect surface waves through 90°, in a manner similar to the operation of a reflective array compressor, and the frequency-dependent behaviour of the array provides the required frequency selectivity. Reflective arrays are also used in the surface-wave resonator, which has two arrays each reflecting the waves through 180°, forming a resonant cavity. This principle has been considered extensively as a method of obtaining narrow-band filters [274], the advantage being that the high Q-factor enables the device to be much shorter than an interdigital device. Resonators are considered further in Chapter 10.

Several methods for introducing frequency selectivity in a multi-strip coupler have been discussed in Section 5.6. These are somewhat inflexible, but have the advantages that the coupler can have low loss and can readily be combined with weighted transducers.

Filters in which interdigital transducers are used to generate surface-skimming bulk waves are described in Appendix F.

8.5. FILTER BANKS

The term "filter bank" refers to a device with one input port and several output ports. The response measured between the input and any one output is a relatively narrow band-pass function. The responses for the various outputs are arranged to be contiguous in frequency so that, taken together, they cover a relatively wide band. The

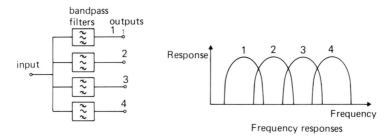

FIGURE 8.14. Principle of filter bank.

device functions as an array of bandpass filters with their inputs connected together, as shown in Figure 8.14. A narrow-band signal applied to the input will appear predominantly at one or two of the outputs, the choice of output depending on the centre frequency of the signal.

This device has two main applications. Firstly, it may be used for frequency measurement by detecting the output signals and comparing their amplitudes. Receivers which perform this function are known as channelised receivers. The second application is in frequency synthesis, where a comb spectrum is first generated and the filter bank is used to select the individual spectral components. An electronic switch is connected to each output of the filter bank, so that the required component can be selected.

A surface-wave filter bank may be made quite straightforwardly by connecting the inputs of several bandpass filters. For example, Slobodnik et al. [275] describe a filter bank which had 9 channels with 3 MHz spacing, covering an overall band of 321 to 345 MHz. Each channel had a 3 dB bandwidth of about 1 MHz, with an insertion loss of 20 dB and a relative stop band rejection of 61 dB or more. The individual filters were made on *ST*-cut quartz substrate, each filter using one withdrawal-weighted and one apodised transducer. A compact arrangement was obtained by fabricating three filters on each substrate, so that only three substrates were needed for the 9 channels.

Another type of filter bank, introduced by Solie, makes use of 3 dB multistrip couplers [276]. In this device a surface wave generated by an input transducer is routed to one of several output transducers, depending on its frequency. The principle is shown in Figure 8.15(a). A broad surface-wave beam is directed toward a 3 dB coupler in which the electrodes have an offset of $v\tau$, corresponding to a delay τ, where v is the velocity. If a beam is incident in *one* track of a 3 dB coupler, output beams are produced in both tracks with equal amplitude, as explained in Chapter 5. If there were no offset, the output beams would differ in phase by $\pi/2$. In Figure 8.15(a) the coupler outputs, at A and B, are the sums of signals due to inputs on both tracks, and the offset introduces an additional phase change of $\omega\tau$. If $v\tau = (n - \frac{1}{4})\lambda$, where λ is the surface-wave wavelength, all of the incident power emerges at A; if $v\tau = (n + \frac{1}{4})\lambda$ all of the power emerges at B. This behaviour is essentially the same as that of the beam compressor using 3 dB couplers, described in Section 5.7, though here a larger offset is used. If the coupler frequency response is ignored, a good approximation here, and

BANDPASS FILTERS

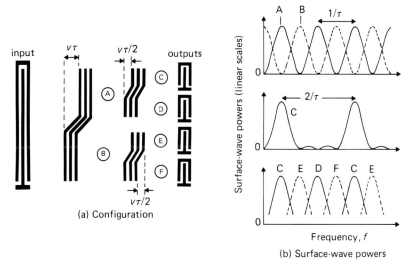

FIGURE 8.15. Filter bank using 3 dB multi-strip couplers.

if the input wave power is independent of frequency, the wave amplitudes at A and B are proportional to $\sin(\omega\tau/2 - \pi/4)$ and $\sin(\omega\tau/2 + \pi/4)$ respectively. The powers of the waves, which repeat with a frequency interval $\Delta f = 1/\tau$, are shown in the upper part of Figure 8.15(b).

The waves at A and B are applied to two further 3 dB couplers which act in a similar manner, except that the offset is approximately $v\tau/2$ instead of $v\tau$. Thus the responses give peaks with spacing $\Delta f = 2/\tau$ instead of $1/\tau$, and half of the peaks observed at A or B are eliminated. The response observed at C is shown in the centre of Figure 8.15(b), and the peaks of the responses for the four output channels C, D, E, F are shown below. The four output transducers, one in each channel, can be weighted in order to shape the response and to improve the stop-band rejection. Additional couplers, with smaller offsets, can be added if a larger number of channels is required.

An experimental device of this type [276] had 16 channels with 0.5 MHz channel spacing, covering the band 81.5 to 89 MHz. Each channel had a 3 dB bandwidth of 1 MHz, an insertion loss of 16 dB and a stop-band rejection of typically 30 dB. The relatively low loss is obtained because the signal is routed to an output dependent on its frequency, instead of being divided initially between a set of parallel filters. The authors describe an application in an FMCW radar system, where the filter bank is used for frequency measurement, determining the range to the target.

Another type of filter bank, using a fanned multi-strip coupler, is described in Section 5.6.

Chapter 9

Chirp Filters and Their Applications

A chirp filter is a linear device designed to give a delay varying substantially with frequency. An equivalent definition is that the impulse response is a frequency-modulated pulse, often known as a "chirp" pulse. We have seen in Chapter 1 that such devices are used in pulse-compression radar systems, where the transmitter generates a chirp pulse and a matching chirp filter is used in the receiver; the chirp filter compresses the pulse to a much shorter duration and also increases the signal-to-noise ratio, thus increasing the sensitivity of the receiver. The pulse compression principle was first put forward in the 1940's, but depended for its practical implementation on the development of a suitable technology for the chirp filters. A variety of technologies have been developed, including surface-wave devices which first emerged in the late 1960's. Today, pulse compression is commonly used in radar systems, and surface-wave technology is usually chosen for the chirp filters because of the exceptional performance capabilities and high precision that it offers. As already mentioned in Chapter 1, there are two main types of surface-wave chirp filter — the interdigital device, using an interdigital transducer with graded periodicity, and the Reflective Array Compressor (RAC), using arrays of grooves.

In addition to the use of chirp filters in pulse-compression systems, a number of other applications will be considered in this Chapter. Most of these make use of linear-chirp waveforms, in which the frequency varies linearly with time. The concept of a time-dependent "frequency" will be justified in Section 9.2; strictly, we should refer to the "instantaneous frequency". A basic property of a linear-chirp waveform is that if it is amplitude modulated its spectral amplitude, in the frequency domain, has the same shape as the time-domain envelope. This property is exploited in systems that use linear-chirp filters to perform Fourier transformation, giving an output waveform proportional to the spectrum of the input waveform. In particular the "compressive receiver", which uses two chirp filters, gives the magnitude of the spectrum of the input waveform, and gives this information much faster than a conventional spectrum analyser. There are also several other applications for linear-chirp filters, including for example a technique giving a variable delay.

This chapter commences with a summary of the principles of pulse-compression radar in Section 9.1. It will be seen in particular that there are several types of chirp waveform suitable for use as the transmitted pulse, and as the impulse response of the chirp filter in the receiver, and the properties and design of these waveforms are considered in Section 9.2. This material is not simply incidental to the subject; it is usually the case that the starting point for development of a surface-wave device is the specification of some basic performance requirements for the radar system, and hence the first task is the design of chirp waveforms satisfying these requirements. In addition, Section 9.2 introduces the very useful "stationary-phase approximation", which is also used in later sections in connection with device analysis. Interdigital devices are considered in the following two sections: the analysis and design of chirp transducers are described in Section 9.3, and Section 9.4 describes the design and performance of interdigital devices. Section 9.5 is concerned with a number of second-order effects, particularly the phase errors associated with velocity errors and temperature changes, and includes the consequent degradation of the radar system performance which must be considered when establishing tolerances. This topic is conveniently placed here since most of it is relevant not only to interdigital devices, considered in the previous sections, but also to RAC's which follow in Section 9.6. Finally, Section 9.7 describes a variety of system applications other than pulse-compression radar, including the compressive receiver and Fourier transformation.

9.1 PRINCIPLES OF PULSE-COMPRESSION RADAR

In a radar system, the transmitter emits a short electromagnetic pulse which is reflected by a target and then arrives at the receiver after a delay proportional to the target range. As in all electronic receivers the input signal will be accompanied by random noise, originating in the receiver itself or elsewhere, and the ability of the receiver to detect the signal is determined by the relative powers of the signal and noise. To optimise the performance of the system the linear portion of the receiver, prior to the detector, should be designed such that the signal-to-noise ratio at the detector input is maximised. A conventional radar system is illustrated in Figure 9.1(a), omitting amplifiers and frequency-conversion stages that are necessary

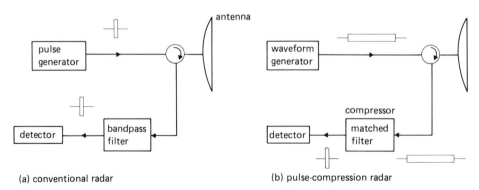

FIGURE 9.1. Comparison between conventional and pulse-compression radar systems. Output waveforms are shown as they appear for a point target.

in practice. The transmitter here generates a simple pulse of carrier with a rectangular envelope, and the system performance is close to optimal if the linear part of the receiver has a simple bandpass response, passing most of the signal spectral components and rejecting noise outside this band.

Suppose now that we wish to increase the sensitivity of such a radar system by increasing the signal-to-noise ratio at the detector input; this will enable the system to detect targets that are more distant, or have lower reflectivity. Clearly the power of the transmitter could be increased, but in practice this would generally be very expensive since microwave frequencies are involved. Alternatively, the length of the transmitted pulse could be increased, thus reducing its bandwidth; the receiver could then have a smaller bandwidth, thus reducing the output noise power. However, this would give a longer output pulse and would therefore degrade the spatial resolution of the system, that is, its ability to distinguish two targets close together.

The pulse-compression concept [277–281] resolves this dilemma, enabling the signal-to-noise ratio to be increased without increasing the transmitted power or degrading the system resolution. The principle is illustrated in Figure 9.1(b). Compared with a conventional system, a pulse-compression system transmits a pulse that is longer and also has a more complex form; the use of a complex waveform enables the pulse to be longer without reducing its bandwidth. In the linear part of the receiver the signal is applied to a special type of filter known as a *matched filter* [277–279, 282, 283], and this is usually a surface-wave device. The term "matched filter" refers to a linear filter designed such that, for a specified input waveform accompanied by noise, the signal-to-noise ratio at the filter output is maximised. The requirement is analysed in Appendix A, Section A.3, which shows that the required impulse response of the matched filter is essentially the time-reverse of the input waveform. It will be seen later that there are two key features associated with the use of a matched filter. Firstly, the output pulse produced has a width determined by the *bandwidth* of the input pulse, and not its duration; thus the latter can be increased without degrading the radar resolution. Secondly, the output signal-to-noise ratio is determined by the *energy* of the input pulse, and can therefore be enhanced by increasing the length of the transmitted pulse without changing its power level. Since the transmitter generates pulses repetitively, the length increase will in fact increase the average power, but this is generally acceptable because most radar transmitters are not limited in average power. On the other hand the *peak* power is often limited by breakdown considerations. Thus the pulse-compression principle enables the receiver signal-to-noise ratio to be increased without making radical changes to the transmitter, and without degrading the radar resolution.

For a pulse-compression system, the transmitted pulse is required to have a length much greater than the reciprocal of the bandwidth, that is, the time-bandwidth product, TB, must be large. This implies the use of some form of modulation. Most commonly, a *chirp* waveform is used, in which the frequency (strictly, the "instantaneous" frequency, Section 9.2) is swept monotonically with time. This is illustrated in Figure 9.2. The waveform $s(t)$ at the left represents the receiver input waveform due to a point target, and is therefore the same as the transmitted waveform except for a delay and a reduction of amplitude. In the receiver, the matched filter is

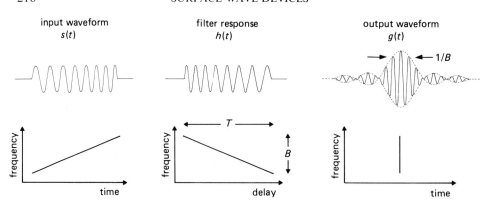

FIGURE 9.2. Pulse compression using linear-chirp waveforms.

matched to the transmitted waveform so that its impulse response is the time-reverse, a chirp waveform $h(t)$ with frequency sweeping in the opposite direction. A filter whose impulse response is a chirp waveform is known as a *chirp filter*. Here the chirp filter acts as a particular type of matched filter, though a number of other applications for chirp filters will be described in this chapter. The output waveform, shown in Figure 9.2, typically has a narrow peak called the *correlation peak*, with small *time-sidelobes* on either side. It will be shown later that the width of the correlation peak is approximately $1/B$, where B is the bandwidth (in Hz) of the input waveform. This is approximately equal to the range of frequencies swept by the waveform. Generally the width of the output peak is much less than the length T of the input waveform, and hence the process is often called *pulse compression*, and the matched filter is also called a *compressor*. The ratio of the input and output pulse widths is called the *compression ratio*. This is approximately equal to TB, the time-bandwidth product of the input waveform, and of the matched filter.

The lower part of Figure 9.2 shows a simplified interpretation of the pulse compression process. The matched filter, whose impulse response has a frequency decreasing with time, can be regarded as a dispersive delay line. Its group delay decreases with frequency, and has a slope opposite to that of the frequency–time characteristic of the input waveform. The frequency components of the input waveform are thus delayed such that they all arrive at the output at the same time, so that a large narrow correlation peak is produced. Although this view is over-simplified, some justification for it is provided by the stationary phase principle, considered in Section 9.2 below.

If there are two point targets, two similar chirp waveforms will arrive at the receiver input, with delays corresponding to the target ranges. Since the matched filter is linear, the output will be simply the sum of the outputs due to each target taken individually. This is true even if the two input chirps overlap in time; two output correlation peaks are produced, and these will be resolved if the delay difference is greater than the peak width $1/B$. The range resolution, expressed in terms of delay, is therefore $1/B$, which is quite different from the length T of the input waveform. In a conventional radar, using a simple rectangular pulse, the resolution is approximately equal to the pulse width, and this in turn is approximately the reciprocal of the bandwidth. Thus, in both cases the resolution is approximately $1/B$. A typical bandwidth is 20 MHz.

Quantitatively, the advantage to be gained from pulse compression can be assessed by considering the signal-to-noise ratio (SNR) at the output of the matched filter. From Section A.3, the output signal-to-noise power ratio is $2E_s/N_i$, where E_s is the signal energy, and N_i the noise spectral density, at the filter input. In a radar system the input signal has a flat envelope, so its energy can be written $E_s = P^s T$, where P^s is the average power during the pulse. The output SNR is therefore

$$\text{SNR}_0 = 2P^s T/N_i. \tag{9.1}$$

This is valid for any waveform with a flat envelope, applied to an appropriate matched filter. Consider now the output SNR for a pulse-compression radar and for a conventional radar, assuming both systems use a matched filter. If the pulse length is T for the pulse compression system, and T_0 for the conventional system, the output SNR's will be in the ratio T/T_0, assuming P^s and N_i to be the same for both. If the two radars have the same resolution, the bandwidth B of the pulse-compression system must be approximately $1/T_0$, and hence the ratio of the SNR's is

$$\frac{\text{SNR}_0(\text{chirp})}{\text{SNR}_0(\text{conventional})} \approx TB. \tag{9.2}$$

This ratio is called the *processing gain*, and is seen to be approximately equal to the compression ratio. In practice, conventional radars do not generally use matched filters, so the output SNR is somewhat less than the value given by equation (9.1); however, the difference is quite small because the time-bandwidth product is close to unity [278, p. 413]. The benefit obtained can be substantial: surface-wave devices give TB-products typically in the range 50 to 1000, so the use of pulse compression gives an SNR increase equivalent to increasing the peak power of a conventional radar by a factor of 50 to 1000.

In a practical system the linear part of the receiver includes components for amplification and frequency conversion in addition to the surface-wave matched filter, which is usually at an I.F. stage. Ideally, the entire pre-detection system should be matched to the input waveform. Usually, the surface-wave device is designed to match an I.F. version of the waveform, and the other components are specified to be adequately free of distortion. The tolerances required are considered in Section 9.5 below.

Waveform Considerations. Although only chirp waveforms have been considered so far, several other types of waveform are applicable to pulse compression radar. There are several basic requirements. Nearly always, the envelope of the transmitted waveform must be flat, because practical transmitters are most efficient when such waveforms are used. The TB-product should be large, so that significant processing gain can be obtained [equation (9.2)]. These requirements imply that some form of frequency or phase modulation is required. In addition, the waveform should be such that the matched filter output gives a narrow peak with relatively small time-sidelobes. The sidelobes appear as false targets, and can cause a weakly-reflecting target to be obscured by a nearby strongly-reflecting target. Thus it is usually important to ensure that the sidelobes are adequately suppressed.

The commonest waveform types satisfying these criteria are chirp waveforms and phase-shift-keyed (PSK) waveforms. The latter are often used in communication systems, and are considered in Chapter 10. In radar systems, chirp waveforms are usually preferred because there are some very effective techniques available for minimising the time-sidelobe levels. In addition, chirp waveforms are in general less affected by doppler shifts due to motion of the target relative to the radar, and this is often significant when high-velocity targets must be catered for. A variety of waveforms with large TB-products, including chirp and PSK waveforms, is discussed by Cook and Bernfeld [279].

In the simplest form of chirp radar the transmitted waveform is a *linear chirp*, that is, its instantaneous frequency (Section 9.2) varies linearly with time. When this waveform is applied to its matched filter, the output waveform has an envelope of the form $(\sin x)/x$, as illustrated in Figure 9.2. The time-sidelobes of this waveform are too large for most radar applications. Often, the sidelobes are reduced by incorporating amplitude weighting in the filter, giving sidelobes typically 35 dB below the main peak. This incurs an SNR penalty of about 1 dB, due to the fact that the receiver is no longer matched to the input waveform. Alternatively, a *non-linear chirp* waveform, in which the frequency is a prescribed non-linear function of time, enables low sidelobe levels to be obtained without the SNR penalty. However, non-linear chirps are not always applicable, because they are more sensitive to doppler shifts than linear chirps.

Generation of Chirp Waveforms. A variety of techniques have been developed for generation of chirp waveforms in the radar transmitter [279, 280]. Techniques involving the use of active circuits are generally called *active generation*. Conceptually, the simplest of these is a voltage-controlled oscillator, with the control voltage having the form of a ramp. This method is relatively straightforward, but since the frequency is required to change rapidly the required accuracy can be difficult to achieve; usually, it is necessary to distort the applied ramp to compensate for distortion in the response of the oscillator.

An alternative approach is to impulse a filter whose impulse response is essentially the waveform required. This is known as *passive generation*. Clearly, the device required is essentially the same as the matched filter in the receiver, except that its impulse response has the frequency sweeping in the opposite direction. A device used in this way is known as an *expander*. This method has the attraction that a highly accurate waveform can be obtained. Moreover some forms of error can be significantly reduced by this method. For example, if a surface-wave expander and compressor are used, errors due to temperature changes are much reduced if the devices are mounted close together, so that their temperatures are similar. Many systems therefore use a "matched pair" of devices. The main disadvantage of passive generation is that the expansion process reduces the power level of the signal, particularly if the *TB*-product is large. For this reason, the SNR of the expanded signal may be inadequate, and then passive generation cannot be used.

Technologies for Chirp Filtering. In this chapter we are mainly concerned with surface-wave chirp filters, which are widely used in modern radar systems. Interdigital surface-wave devices first emerged in the late 1960's [284, 285], and rapidly demonstrated the ability to give highly accurate and reproducible performance. Shortly afterwards, reflective array compressors (RAC's) using arrays of grooves were developed [286], and these enable very large time-bandwidth products, up to several thousand, to be used. The RAC is very similar in principle to an earlier device called the IMCON which also uses groove arrays, but in this case the wave motion involved is a bulk acoustic wave rather than a surface wave. In all three of these devices, the wave motion is non-dispersive, though the device response is dispersive. Interdigital devices are described in Sections 9.3 to 9.5, while RAC's and IMCON's are considered in Section 9.6.

A variety of other techniques have been developed [279, 280, 287], though most of these have been superseded by surface-wave devices and IMCON's. The required dispersion can be obtained by using an acoustic wave with dispersive propagation. For example, in a layered half-space Rayleigh waves and Love waves are both dispersive, as described in Section 2.2.4. Both have been considered for pulse compression, particularly Love waves [288]. Alternatively, a parallel-sided plate supports dispersive waves, as shown in Section 2.2.5. Chirp devices using these modes take the form of thin metal strips with a bulk-wave transducer at each end, and are known as "strip delay lines". Other acoustic devices use non-dispersive bulk waves, arranging for the path length to vary with frequency. This principle is used in perpendicular diffraction delay lines and wedge delay lines. Several optical techniques are also applicable, including acousto-optic devices in which a light beam is diffracted by an acoustic wave.

Digital processing [283] has considerable attractions because of the flexibility obtainable. In comparison with surface-wave techniques, digital methods are generally limited to lower bandwidths, and result in devices that are more bulky and consume more power. However current developments in integrated circuit technology, increasing the operating speed and the number of components per chip, can be expected to make digital methods increasingly attractive in future.

9.2. WAVEFORM CHARACTERISTICS AND DESIGN

In Section 9.1 we have seen that chirp waveforms are commonly used in pulse-compression radar systems, both for the transmitted pulse and as the required impulse response of the compressor in the receiver. Some of the basic performance criteria for the system – the range resolution, processing gain and the level of the time-sidelobes — are determined by the design of the chirp waveforms and by the accuracy with which they are implemented in practical devices. In the design of surface-wave chirp filters it is usually the case that the first task is to design chirp waveforms satisfying the basic system requirements, and this is the topic for this section.

The stationary-phase approximation, introduced in Section 9.2.1, is widely used in this context, and will also be used later in connection with analysis of surface-wave devices. Properties of linear-chirp waveforms are given in Section 9.2.2, and the amplitude weighting of linear chirps to reduce the time-sidelobes of the compressor output waveform is explained in Section 9.2.3. The use of non-linear chirp waveforms to reduce time-sidelobes is described in Section 9.2.4. The effects of a doppler shift, due to motion of the target relative to the radar, must usually be considered when designing the waveforms, and these effects are dependent on the waveform design. However, it is convenient to defer this topic to Section 9.5.3, because the analysis makes use of some results in Section 9.5.1 concerning the effects of phase errors.

We first define some basic parameters. A chirp waveform $v(t)$, representing either the transmitted radar waveform or the impulse response of the chirp filter in the receiver, can be written

$$v(t) = a(t) \cos [\theta(t)]. \qquad (9.3)$$

Here $\theta(t)$ is the time-domain phase, and will be a non-linear function of t. The envelope $a(t)$ is zero outside a time interval of length T, the length of the waveform. Clearly, $\theta(t)$ must be a monotonic function of time. We define $\theta(t)$ such that it *increases* with time; this can be done without loss of generality, because reversing the sign of $\theta(t)$ does not affect the waveform $v(t)$. The differential of $\theta(t)$ is defined to be the *instantaneous frequency* $\Omega_i(t)$, so that

$$\Omega_i(t) = \dot{\theta}(t). \qquad (9.4)$$

Thus, $\Omega_i(t)$ will always be positive. It will be shown later than $\Omega_i(t)$ can be regarded approximately as the "frequency" at time t. We also define the *chirp rate* $\mu(t)$ as the rate of change of instantaneous frequency. It is convenient to express this in units of Hz/sec, so that

$$2\pi\mu(t) = \dot{\Omega}_i(t) = \ddot{\theta}(t). \qquad (9.5)$$

For the waveforms considered in this chapter, the instantaneous frequency is a monotonic function, either increasing or decreasing with time. Thus $\mu(t)$ is either positive throughout the waveform, or negative throughout. Waveforms with $\mu(t) > 0$ are called *up-chirps*, while those with $\mu(t) < 0$ are called *down-chirps*. For a linear chirp $\mu(t)$ is a constant, and hence $\Omega_i(t)$ is a linear function of t and $\theta(t)$ is a quadratic function.

If $v(t)$ in equation (9.3) represents the trasmitted radar waveform, the required impulse response for the matched filter is $h(t) = v(-t)$, apart from an arbitrary delay and a multiplying constant. The phase of $h(t)$ is taken to be $-\theta(-t)$ rather than $\theta(-t)$, because here it is assumed to be increasing with time. In practice there is usually an unknown additional constant α, so that $h(t) = a(-t) \cos [\alpha - \theta(-t)]$. Although this is not in fact matched to the input waveform, the constant α has little

CHIRP FILTERS AND THEIR APPLICATIONS

effect on the output waveform: it leaves the envelope unchanged, but affects the phase of the carrier. This change is nearly always of no consequence.

9.2.1. Stationary-phase Approximation

This approximation [279, p. 34] gives a very convenient method for estimating the spectrum of a chirp waveform, and will be used later to solve a wide variety of problems. We consider the chirp waveform of equation (9.3), in the complex form

$$v(t) = \tfrac{1}{2}a(t) \exp[j\theta(t)] + \tfrac{1}{2}a(t) \exp[-j\theta(t)] \tag{9.6}$$

It can be assumed that $v(t)$ is a bandpass waveform, so that its spectrum is negligible for frequencies near zero. Since $\theta(t)$ is taken to be increasing with time, it follows that the Fourier transform of the first term in equation (9.6) gives the positive-frequency part of the spectrum, which is denoted $V_+(\omega)$. We thus have

$$V_+(\omega) = \frac{1}{2}\int_{-\infty}^{\infty} a(t) \exp[j\theta(t) - j\omega t] \, dt, \tag{9.7}$$

which is the spectrum of $v(t)$ for $\omega > 0$. The negative-frequency part can be deduced from this by noting that $v(t)$ is real. Since $\theta(t)$ is a non-linear function of t, the phase $\theta(t) - \omega t$ in the above equation generally varies rapidly with t, so that the exponential gives rapid oscillations which contribute little to the integral. However, the phase varies slowly for times close to the point where the differential is zero, that is, where $\dot\theta(t) = \omega$. The main contribution to the integral therefore arises from times near this point. The time satisfying $\dot\theta(t) = \omega$ is known as the *stationary phase point*, denoted by $T_s(\omega)$, so that

$$\dot\theta[T_s(\omega)] = \omega. \tag{9.8}$$

For any given ω, this equation has only one solution for $T_s(\omega)$, because $\dot\theta(t)$ is assumed to be monotonic.

Since the main contribution to the integral in equation (9.7) arises from times near $T_s(\omega)$, it is a good approximation to replace $\theta(t)$ by its Taylor expansion about this point, neglecting cubic and higher order terms. It is also assumed that the envelope $a(t)$ varies slowly with t, and may therefore be approximated by its value $a[T_s(\omega)]$ at the stationary-phase point. Using also equation (9.8), we find

$$V_+(\omega) \approx \tfrac{1}{2}a(T_s) \exp[-j\omega T_s + j\theta(T_s)] \int_{-\infty}^{\infty} \exp[\tfrac{1}{2}j(t - T_s)^2 \ddot\theta(T_s)] \, dt, \tag{9.9}$$

where the frequency argument of $T_s(\omega)$ is omitted for brevity. Here the integral is a form of the standard integral [289, p. 301]

$$\int_{-\infty}^{\infty} \exp(jKt^2) \, dt = \sqrt{\frac{\pi}{|K|}} \, e^{\pm j\pi/4}, \tag{9.10}$$

where $K \neq 0$ is a real constant and the sign in the exponential on the right is the same as the sign of K. We define $A(\omega)$ and $\phi(\omega)$ as the amplitude and phase of the spectrum $V_+(\omega)$, so that

$$V_+(\omega) = A(\omega) \exp[j\phi(\omega)]. \tag{9.11}$$

Using equation (9.10), equation (9.9) gives

$$A(\omega) \approx \frac{a(T_s)}{2} \sqrt{\frac{2\pi}{|\ddot{\theta}(T_s)|}} = \frac{1}{2} \frac{a(T_s)}{\sqrt{|\mu(T_s)|}} \qquad (9.12a)$$

and

$$\phi(\omega) \approx \theta(T_s) - \omega T_s \pm \pi/4. \qquad (9.12b)$$

These results assume that $\ddot{\theta}(T_s) \neq 0$, which is valid here because $\dot{\theta}(t)$ is monotonic. The sign in equation (9.12b) is the same as the sign of $\mu(T_s)$. Note that $a(t)$ is zero outside a specified interval of length T, and hence $A(\omega)$ is zero outside a corresponding frequency band. If $v(t)$ is the impulse response of a filter, the group delay at frequency ω is, from equations (9.12b) and (9.8),

$$\tau_g(\omega) \equiv -d\phi(\omega)/d\omega \approx T_s(\omega). \qquad (9.13)$$

The stationary-phase approximation shows that, for a particular frequency ω, the spectrum is determined mainly by the part of the waveform in the vicinity of the stationary phase point $T_s(\omega)$, defined as the solution of equation (9.8). Conversely, it is clear that the waveform at some time t is determined mainly by the part of the spectrum in the vicinity of the frequency $\omega = \dot{\theta}(t)$, which by definition is equal to the instantaneous frequency $\Omega_i(t)$. Thus $\Omega_i(t)$ can be regarded as the time-dependent "frequency" of the waveform. The stationary-phase approximation therefore gives a formal meaning to the statement that the frequency varies with time. This statement has no precise meaning otherwise, since the term "frequency" generally refers to periodic waveforms, or to the independent parameter in the Fourier transform. Although $\Omega_i(t)$ is precisely defined, its interpretation as the time-dependent "frequency" is meaningful only when the stationary-phase approximation is valid, and this depends on the nature of the waveform as shown in Section 9.2.2 below.

9.2.2. Linear Chirp Waveforms

For a linear chirp waveform, the instantaneous frequency varies linearly with time so that the chirp rate $\mu(t)$ is a constant. The phase $\theta(t)$ is a quadratic function of t. The waveform can be written as

$$v(t) = a(t) \cos [\theta(t)] = a(t) \cos (\omega_c t + \pi\mu t^2 + \phi_0) \qquad (9.14)$$

where ϕ_0 is a constant. We take the waveform to be centred at $t = 0$, so that $a(t) = 0$ for $|t| > T/2$, where T is the length. Thus, ω_c is the centre frequency. If the envelope is flat, as in Figure 9.2, then $a(t)$ will be a constant for $|t| \leq T/2$. The instantaneous frequency is

$$\Omega_i(t) = \omega_c + 2\pi\mu t$$

and the stationary-phase point is, from equation (9.8),

$$T_s(\omega) = (\omega - \omega_c)/(2\pi\mu). \qquad (9.15)$$

If $v(t)$ is the impulse response of a filter, this expression is approximately equal to the group delay, as shown by equation (9.13). The number of cycles in the waveform is readily found to be $T\omega_c/(2\pi)$, which is independent of the chirp rate μ.

In the stationary-phase approximation, the spectrum of the waveform is given by equations (9.12). Thus the spectral amplitude is

$$A(\omega) \approx \frac{a[T_s(\omega)]}{2\sqrt{|\mu|}}.\qquad(9.16a)$$

The spectral phase is

$$\phi(\omega) \approx -\frac{(\omega - \omega_c)^2}{4\pi\mu} + \phi_0 \pm \pi/4,\qquad(9.16b)$$

where the sign is the same as the sign of μ. Note that, since $T_s(\omega)$ is a linear function of ω, the spectral amplitude $A(\omega)$ has the same shape as the time-domain envelope $a(t)$.

Since the envelope $a(t)$ falls to zero at the ends of the waveform, where $t = \pm T/2$, $A(\omega)$ is finite only between the corresponding instantaneous frequencies $\dot{\theta}(\pm T/2) = \omega_c \pm \pi\mu T$. It is convenient to define the *bandwidth* B (in Hz) as the range of instantaneous frequencies, so that

$$B = |\mu|T\qquad(9.17)$$

and the band edges are at $\omega_c \pm \pi B$. Note that the 3 dB bandwidth of the spectrum is dependent on the form of $a(t)$, and can be quite different from B.

Linear Chirp with Flat Envelope. In a radar system the transmitted signal has a flat envelope, as illustrated in Figure 9.2. In this case the envelope $a(t)$ is a constant for the interval $|t| < T/2$ occupied by the pulse. The spectral amplitude of equation (9.16a) is therefore flat between the points $\omega = \omega_c \pm \pi B$, and is zero elsewhere, according to the stationary-phase approximation.

Some actual spectra of flat-envelope linear chirp waveforms are shown in Figure 9.3, for several values of the time-bandwidth product TB. These were obtained by taking the Fourier transform of the waveform, without approximation. For the phase curves, the quadratic term predicted by the stationary-phase approximation, equation (9.16b), has been subtracted from the data. For large TB-products, the amplitude is quite flat within a band of width $\Delta\omega = 2\pi B$ and falls off rapidly outside this band. However, ripples are present in both the amplitude and the phase, and they become more significant as TB is reduced. Thus the stationary-phase approximation is valid only for large TB. For most purposes the approximation is found to give adequate accuracy if $TB \geqslant 100$. Figure 9.3 also shows that B is approximately the 6 dB bandwidth, in Hz, for the flat-envelope case.

The amplitude curves in Figure 9.3 are closely related to the isotropic diffraction plots shown earlier on the right of Figure 6.6. Mathematically, both cases refer to the Fourier transform of a function with rectangular amplitude and quadratic phase. However, the appearance is rather different because the "TB-products" in Figure 6.6 are very small.

Suppose now that a flat-envelope linear chirp waveform is applied to its matched filter, as in Figure 9.2. The input waveform $s(t)$ is taken to be equal to $v(t)$, equation (9.14), with $a(t)$ set equal to unity for $|t| < T/2$. Taking the filter response

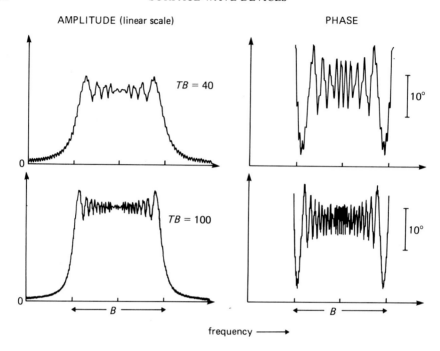

FIGURE 9.3. Spectra of flat-envelope linear chirp waveforms.

to be the time-reverse of this, the spectrum of the output waveform is $G(\omega) = |S(\omega)|^2$, where $S(\omega)$ is the spectrum of $s(t)$. This result follows from equation (A.66) of Appendix A. For positive frequencies the magnitude of $S(\omega)$ is equal to $A(\omega)$, given by equation (9.16a) in the stationary-phase approximation. Transforming $G(\omega)$ to the time-domain, the output waveform is found to be

$$g(t) \approx \tfrac{1}{2} T \cos(\omega_c t) \, \text{sinc}(\pi B t), \qquad (9.18)$$

where $\text{sinc}(x) = (\sin x)/x$. This is illustrated in Figure 9.2. The envelope has its maximum at $t = 0$, and the first zeros are at $t = \pm 1/B$. The 4 dB points are almost exactly at $t = \pm 1/(2B)$.

An exact analysis [279, p. 133] shows that equation (9.18) is a good approximation for times close to the main peak. The main differences are that the exact expression has a finite length $2T$, and, for times remote from the main peak, the locations of the zeros are different.

9.2.3. Weighting of Linear-chirp Filters

It was seen above that, for a flat-envelope linear chirp waveform applied to its matched filter, the output waveform has an envelope of the form $\text{sinc}(\pi B t)$. The largest time-sidelobes of this waveform are only 13 dB below the main peak, and this is unacceptable for most radar applications, which typically require a sidelobe rejection of 30 dB.

CHIRP FILTERS AND THEIR APPLICATIONS 225

A common method of reducing the sidelobes is to apply amplitude weighting in the filter [277, 279, 280], as illustrated in Figure 9.4. The filter impulse response is weighted to give a maximum amplitude at the centre, reducing towards the ends. In addition to reducing the sidelobe levels this also increases the width of the main peak somewhat, though this can be compensated for by increasing the bandwidth B. There is also a small reduction of output signal-to-noise ratio, since the filter is no longer exactly matched to the input waveform. The use of non-linear chirps to avoid this loss of SNR will be considered later, in Section 9.2.4.

In the analysis here it is assumed initially that $TB \geqslant 100$, so that the stationary-phase approximation can be used. The limitations imposed by this assumption are discussed later.

As before, we consider a waveform $s(t)$ applied to a filter with impulse response $h(t)$, giving an output waveform $g(t)$. The input waveform is a flat-envelope linear chirp

$$s(t) = \cos(\omega_c t - \pi\mu t^2 - \phi_0), \quad \text{for } |t| \leqslant T/2, \tag{9.19}$$

with $s(t) = 0$ for other t. The matched filter for this waveform has an impulse response $s(-t)$, but here the filter is taken to have amplitude weighting with an envelope $a(t)$, so that

$$h(t) = a(t) \cos(\omega_c t + \pi\mu t^2 + \phi_0), \quad \text{for } |t| \leqslant T/2. \tag{9.20}$$

The transforms of $s(t)$ and $h(t)$ are respectively denoted $S(\omega)$ and $H(\omega)$, and the spectrum of the output waveform is $G(\omega) = S(\omega)H(\omega)$. The waveforms $s(t)$ and $h(t)$ have the same form as equation (9.14), so their transforms are given by equations (9.16) in the stationary-phase approximation. The spectrum of the output waveform, for $\omega > 0$, is thus found to be the real function

$$G(\omega) \approx \frac{a[T_s(\omega)]}{4|\mu|}, \quad \text{for } |\omega - \omega_c| \leqslant \pi B,$$

$$\approx 0, \quad \text{for } |\omega - \omega_c| > \pi B, \tag{9.21}$$

where

$$T_s(\omega) = (\omega - \omega_c)/(2\pi\mu) \tag{9.22}$$

is the stationary-phase point for the filter.

It is convenient to express this result in terms of a real frequency-domain *weighting function* denoted by $\bar{W}(\omega)$, which is considered to be centred at zero frequency and

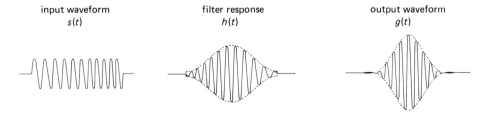

FIGURE 9.4. Pulse compression using linear chirps with amplitude weighting.

is zero for $|\omega| > \pi B$. The output spectrum $G(\omega)$ is thus proportional to $\bar{W}(\omega - \omega_c)$ for positive ω, and can be written

$$G(\omega) = K\bar{W}(\omega - \omega_c), \quad \text{for } \omega > 0, \tag{9.23}$$

where K is an arbitrary real constant. If $W(t)$ is the inverse Fourier transform of $\bar{W}(\omega)$, the output waveform $g(t)$ is given by the shifting theorem, equation (A.11), so that

$$g(t) = 2KW(t)\cos\omega_c t. \tag{9.24}$$

The envelope of the output waveform is therefore proportional to $W(t)$. We also have, from equations (9.21)–(9.23),

$$a(t) = 4K|\mu|\bar{W}(2\pi\mu t). \tag{9.25}$$

Thus the time-domain weighting $a(t)$ is, with a change of scale, proportional to the frequency-domain weighting function. Both sides of equation (9.25) are, by definition, zero for $|t| > \tfrac{1}{2}T$.

Types of Weighting Function. To design the filter impulse response, we need to find a weighting function $\bar{W}(\omega)$ which is zero for $|\omega| > \pi B$ and has an inverse Fourier transform $W(t)$ giving well-suppressed time-sidelobes. A similar problem was encountered in Section 8.2.1, which discussed the use of window functions for bandpass filter design. This is the dual of the chirp filter case; for a bandpass filter the window function has finite length in the time domain, while for a chirp filter the weighting function has finite length in the frequency domain. Thus a window function suitable for bandpass filters can also be applied to chirp filters if the time and frequency variables are interchanged, with a scaling factor such that the required bandwidth is obtained.

In practice, the weighting functions used for chirp filters have usually been functions developed earlier for antenna weighting, where the polar diagram is required to have well-suppressed sidelobes. While a variety of weighting functions have been considered [277, 279, 280], the usual choice for a chirp filter is either Hamming weighting or Taylor weighting. The Hamming weighting function $\bar{W}_H(\omega)$ is defined by

$$\bar{W}_H(\omega) = 0.54 + 0.46\cos\omega/B, \quad \text{for } |\omega| \leq \pi B,$$
$$= 0, \quad \text{for } |\omega| > \pi B. \tag{9.26}$$

This function, and its transform $W_H(t)$, are shown in Figure 9.5. The output envelope $W_H(t)$ gives a series of sidelobes, the largest of which is the fourth one, peaking at $Bt = 4.5$; this sidelobe is 42.8 dB below the main peak. The width of the main peak, at the 3 dB points, is about 50% larger than the width obtained for an unweighted filter, for the same bandwidth, B. There is therefore some loss of radar resolution, in exchange for the suppression of sidelobes. This weighting is also shown in Figure 9.4, where the envelope of the filter response is proportional to $\bar{W}_H(2\pi\mu t)$ and the envelope of the output waveform is proportional to $W_H(t)$.

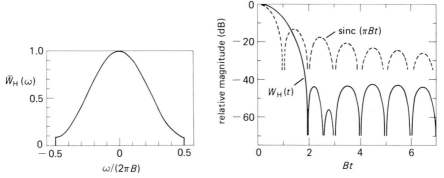

FIGURE 9.5. Hamming weighting function (left) and its transform. The transform is symmetrical about $t = 0$.

The Taylor weighting function [277, 279, 280, 290] enables the time-sidelobe level to be set by the designer, and is given by

$$\bar{W}_T(\omega) = 1 + 2 \sum_{m=1}^{\bar{n}-1} F_m \cos m\omega/B, \quad \text{for } |\omega| \leq \pi B,$$
$$= 0, \quad \text{for } |\omega| > \pi B. \quad (9.27)$$

For this function, the largest sidelobe is the one adjacent to the main peak. The coefficients F_m and the number of terms $\bar{n}$ are chosen in accordance with the performance required. To evaluate the function, the maximum sidelobe level is first chosen, and a value for $\bar{n}$ is selected. These values are then inserted into a formula to give the values of the F_m. For a given sidelobe level, increasing the value of $\bar{n}$ reduces the width of the main peak. For large $\bar{n}$ the peak width approaches that given by the Dolph–Chebyshev function, which gives the minimum possible peak width for a given sidelobe level. In practice, the value of $\bar{n}$ is not critical, and typical values are $\bar{n} = 5$ for 30 dB sidelobes or $\bar{n} = 10$ for 50 dB sidelobes. Generally, $\bar{W}_T(\omega)$ has an appearance similar to $\bar{W}_H(\omega)$, shown in Figure 9.5, though for large $\bar{n}$ it can increase at the band edges.

Figure 9.6 shows, on the left, the width of the main output peak as a function of the sidelobe level relative to the peak. The peak width is measured at the 3 dB points of the output waveform, and is normalised by multiplying by B. It can be seen that lower sidelobe levels give wider peaks. The point marked "unweighted" refers to a filter whose impulse response has a flat envelope, that is, the filter matched to the input waveform. In this case the output waveform has an envelope proportional to sinc (πBt), as shown by equation (9.18).

Figure 9.6 also shows the *mis-match loss* for a Taylor-weighted device. This is the reduction of output signal-to-noise power ratio (SNR) due to the fact that the filter is not matched to the input signal. The output SNR for the weighted filter is given by equations (A.55)–(A.57) in Appendix A. For a matched filter the output SNR is $2E_s/N_i$, as shown by equation (A.61). With these two SNR's expressed in dB, the difference is the mis-match loss plotted in Figure 9.6. This figure also shows the mis-match loss and output peak width for a Hamming weighted device.

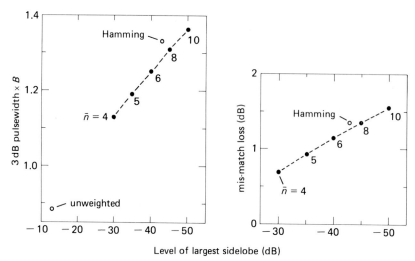

FIGURE 9.6. Output pulsewidth and mis-match loss for linear chirps with Taylor weighting (solid circles) and Hamming weighting. From data given by Farnett et al. [280].

Limitations of the Stationary-phase Approach. In the above description, the stationary-phase approximation was used both to establish the design of the filter response and to deduce the nature of the output waveform. In practice, the design method, giving the amplitude weighting of equation (9.25), is usually acceptable if TB exceeds 100. According to the stationary-phase analysis, the output waveform, equation (9.24), has an envelope proportional to $W(t)$. This gives a sequence of sidelobes which decrease progressively for times more remote from the main peak, except for the first three sidelobes in the case of Hamming weighting.

For a device designed as above, accurate calculations show that, for $TB > 100$, the output waveform predicted by the stationary-phase method is a good approximation for times close to the main peak. The main discrepancy is that the more remote sidelobes can be somewhat larger than those predicted by the stationary-phase method. The additional sidelobes are associated with the abrupt truncation of $s(t)$ and $h(t)$, and are known as "gating sidelobes". They are maximised at $t \approx \pm \frac{1}{2}T$. It is common practice to reduce the gating sidelobes by adding short "extensions" to $h(t)$, so that the envelope falls smoothly to zero instead of the abrupt truncation shown in Figure 9.4. When this is done, the maximum level of the gating sidelobes is typically about $20 \log (TB) + 3$ dB below the main peak, and this is usually acceptable.

For small TB, the stationary-phase design method generally gives unacceptable sidelobe levels. For this case a more sophisticated design method, called the "reciprocal-ripple" method, was introduced by Judd [291], and is also described by Armstrong [292, 293]. The spectrum of the output waveform is $G(\omega) = S(\omega)H(\omega)$, and is required to be proportional to the weighting function $\bar{W}(\omega - \omega_c)$. This requirement can be satisfied by taking the filter response to have the form

$$H(\omega) = \frac{\bar{W}(\omega - \omega_c)}{S(\omega)}. \qquad (9.28)$$

For large TB the amplitude of $S(\omega)$ is almost flat within the band, so that $1/S(\omega)$ is approximately proportional to $S^*(\omega)$, and hence the form of $H(\omega)$ given by equation (9.28) is almost the same as that given by the stationary-phase method. However, the ripples of $S(\omega)$, evident in Figure 9.3, are cancelled when $H(\omega)$ is given by equation (9.28). This enables low time-sidelobe levels to be obtained from waveforms with low TB-products, which give large ripples. In practice, the weighting function $\bar{W}(\omega - \omega_c)$ is extended beyond the band edges to avoid abrupt discontinuities. The resulting function $H(\omega)$ is transformed to the time domain, and since this gives a waveform of infinite duration it is necessary to truncate and add short extensions decaying smoothly to zero. It is usually found that the filter impulse response must be somewhat longer than the input waveform. An experimental device [291] gave a time-sidelobe rejection of 34 dB, with a TB-product as low as 8; for this case, the sidelobe rejection for a stationary-phase design is 20 dB.

The main limitation of the reciprocal-ripple method is that it increases the sensitivity to doppler shifts of the input waveform, which cause the sidelobe levels to increase. The doppler sensitivity becomes worse as the TB-product increases, and hence the reciprocal ripple method is usually applied only when $TB < 100$.

9.2.4. Weighting Using Non-linear Chirps

It was seen above that, for linear chirps, amplitude weighting in the filter causes a mis-match loss of typically 1 to 1.5 dB. If weighting could be incorporated in such a way that the filter were still matched to the input waveform, then low time-sidelobe levels would be obtainable without this loss of SNR. An obvious way of doing this is to use linear chirps, with amplitude weighting in both the input waveform and the filter impulse response, but this is usually precluded by the requirement that the waveform transmitted by the radar should have a flat envelope. However, a viable solution is to employ non-linear chirps, which use a form of phase weighting. These enable the time-sidelobes to be suppressed without significant loss of SNR, while still retaining a flat envelope for the input waveform. The method was introduced by Cook and Paolillo [294], and is also described by Cook and Bernfeld [279, p. 222] and by Newton [295].

The method is illustrated by Figure 9.7, which shows the frequency-time curves for the input waveform and for the matched filter. The frequency varies slowly with time at the centre, and more rapidly at the ends. This causes the spectrum of the waveform to have a power density decreasing toward the band edges, even though the time-domain envelope is flat. The input waveform and the filter impulse response appear much as in Figure 9.2, but the spectral weighting reduces the time-sidelobes so that the output waveform is typically as in Figure 9.4.

As in the previous section, the stationary-phase approximation is used in the analysis. We consider a finite-length waveform $s(t)$ applied to its matched filter, whose impulse response is $h(t)$. These waveforms are both flat-envelope chirps, and are given by the expressions

$$h(t) = \cos[\theta(t)], \quad \text{for } |t| \leq \tfrac{1}{2}T,$$
$$= 0, \quad \text{for } |t| > \tfrac{1}{2}T, \quad (9.29)$$

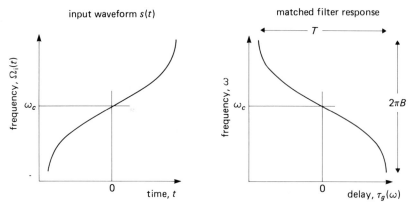

FIGURE 9.7. Dispersion curves for non-linear chirps with Hamming weighting, in the stationary-phase approximation.

and
$$s(t) = h(-t).$$

For convenience, the waveforms are both centred at $t = 0$, and have the same amplitude. According to the matched filter analysis of Section A.3, the waveform spectrum $S(\omega)$ and the filter frequency response $H(\omega)$ are related by $S(\omega) = H^*(\omega)$, and the output waveform $g(t)$ has a spectrum $G(\omega) = |H(\omega)|^2$. In the stationary-phase approximation the frequency response of the filter is given by equations (9.12), and we thus find

$$G(\omega) \approx \frac{\pi}{2|\ddot{\theta}(T_s)|}, \quad \text{for } |\omega - \omega_c| \leq \pi B,$$
$$\approx 0, \quad \text{for } |\omega - \omega_c| > \pi B, \quad (9.30)$$

where $T_s(\omega)$ is the stationary-phase point, defined as the solution of $\dot{\theta}(T_s) = \omega$. The band-edge frequencies $\omega = \omega_c \pm \pi B$ are defined as the values of the instantaneous frequency $\Omega_i(t) = \dot{\theta}(t)$ at the two ends of the waveform, that is, at $t = \pm \frac{1}{2}T$. The centre frequency ω_c is midway between these points.

If $\dot{\theta}(t)$ is a non-linear function of t, so that the chirps are non-linear, the factor $\ddot{\theta}(T_s)$ in equation (9.30) will vary with frequency, thus modifying the amplitude of the output spectrum. This enables the sidelobes of the output waveform to be suppressed by designing an appropriate time-domain phase $\theta(t)$. Since $G(\omega)$ is zero outside a band of width $2\pi B$, it is appropriate to design the device such that

$$G(\omega) = C\bar{W}(\omega - \omega_c), \quad \text{for } \omega > 0, \quad (9.31)$$

where C is a constant. Here $\bar{W}(\omega)$ is a weighting function which is zero for $|\omega| > \pi B$, and has an inverse Fourier transform $W(t)$ with well-suppressed sidelobes. The output waveform $g(t)$ has its envelope proportional to $W(t)$, as can be seen by comparison with equations (9.23) and (9.24). As discussed in the previous section, suitable choices for $\bar{W}(\omega)$ include the Hamming function $\bar{W}_H(\omega)$ and the Taylor function $\bar{W}_T(\omega)$.

From equations (9.30) and (9.31) we have

$$\ddot{\theta}[T_s(\omega)] \;\dot{=}\; \pm \frac{\pi}{2C\bar{W}(\omega - \omega_c)}. \tag{9.32}$$

Now, since $\dot{\theta}[T_s(\omega)] = \omega$ we have $\ddot{\theta}(T_s) \cdot dT_s/d\omega = 1$, and hence

$$dT_s/d\omega \;=\; \pm 2C\bar{W}(\omega - \omega_c)/\pi. \tag{9.33}$$

Thus, $\bar{W}(\omega - \omega_c)$ can be integrated to give $T_s(\omega)$, which in the stationary-phase approximation is equal to the group delay of the filter. The constant of integration determines the value of $T_s(\omega)$ at the centre frequency ω_c. The constant C determines the length T of the waveform, and is chosen such that the values of $T_s(\omega)$ at the band edges $\omega_c \pm \pi B$ differ by T.

Once $T_s(\omega)$ has been found, the instantaneous frequency $\Omega_i(t) = \dot{\theta}(t)$ follows by solving the equation $\dot{\theta}[T_s(\omega)] = \omega$, so that values of $T_s(\omega)$ give $\dot{\theta}(t)$ directly if T_s is interpreted as t and ω is interpreted as $\dot{\theta}(t)$. From $\dot{\theta}(t)$, the phase $\theta(t)$ is obtained by integration, and here the constant of integration is immaterial.

To illustrate the method, consider Hamming weighting, so that the function $\bar{W}(\omega)$ is the Hamming function $\bar{W}_H(\omega)$ of equation (9.26). In this case the stationary-phase point $T_s(\omega)$ of the filter is, from equation (9.33),

$$T_s(\omega) \;=\; \pm \frac{T}{2\pi}\left[\frac{\omega - \omega_c}{B} + \frac{0.46}{0.54}\sin\left(\frac{\omega - \omega_c}{B}\right)\right], \quad \text{for } |\omega - \omega_c| \leqslant \pi B \tag{9.34}$$

and this is approximately equal to the group delay. The constant of integration has been chosen such that $T_s(\omega) = 0$ at the centre frequency ω_c. The function is shown on the right of Figure 9.7, where the lower sign has been chosen in equation (9.34) so that the filter impulse response is a down-chirp. Since $\Omega_i(T_s) = \omega$, the curve can also be regarded as the instantaneous frequency, as a function of time. The input waveform $s(t)$ is the time-reverse of the filter impulse response $h(t)$ and therefore has its instantaneous frequency reversed in time, as shown on the left in Figure 9.7.

For a given weighting function, the output envelope is the same as that obtained for an amplitude-weighted linear chirp, considered above. For example, if the Taylor weighting function $\bar{W}_T(\omega)$ is used, the width of the main peak can be traded-off against the sidelobe rejection, as in the left part of Figure 9.6. However, the mis-match loss is zero, since the filter is matched.

The limitations of the stationary-phase approach, mentioned in the previous section, apply here also. Thus, it is usual to add extensions to $h(t)$ to avoid abrupt truncation, and for small TB-products an adaptation of the reciprocal-ripple design method can be used [295]. With these modifications, the filter is no longer exactly matched to the input waveform, but the resulting mis-match loss is very small. For example [295], a reciprocal-ripple design with $TB = 60$ gave 36 dB sidelobe rejection and a mis-match loss of 0.04 dB, which is typical.

9.3. CHIRP TRANSDUCERS

Having considered the design of chirp waveforms, we now turn to the analysis and design of surface-wave interdigital chirp filters, which are used as matched filters for these waveforms and for passive waveform generation. The present section is concerned with analysis and design of chirp transducers, while the design and performance of devices using these transducers is described in Section 9.4 below. In both of these sections, no restriction is placed on the nature of the chirp waveform involved. Thus the filter impulse response may be any of the types of chirp mentioned in Section 9.2, including weighted and unweighted linear chirps, and non-linear chirps.

In this section we are concerned with chirp transducers such as that shown on the right of the chirp filter of Figure 9.8. The main transducer properties of interest are the frequency response $H_t(\omega)$ and the admittance. The analysis here is based on the quasi-static approach of Chapter 4, assuming that the only acoustic wave present is a piezoelectric Rayleigh wave, and ignoring propagation loss and surface-wave diffraction. Electrode interactions are also ignored, and the reader is reminded that the analysis predicts that the surface-wave velocity within the transducer is the free-surface velocity v_0. In practice the electrodes reduce the velocity a little even when reflections due to interactions are negligible (Section D.2), and this can be allowed for in the analysis below by adjusting the value of the wavenumber k_0. End effects are assumed to be negligible, and this condition can be obtained by adding a few guard electrodes at each end, as explained in Section 4.5.1. However, chirp transducers are usually many wavelengths long, and consequently end effects are usually negligible even if the guards are omitted. For apodised transducers the analysis assumes that dummy electrodes are included. This is usually done in practice, for reasons mentioned in Section 4.7.2. It is assumed that all the electrodes connected to either bus-bar have a uniform potential, and this implies that the metallisation has negligible resistivity, and that the transmission-line effect (Section 9.5.4) is neglected. Second-order effects are considered in Section 9.5. below, which also considers the resulting distortion of the output waveform when the device is used for pulse compression.

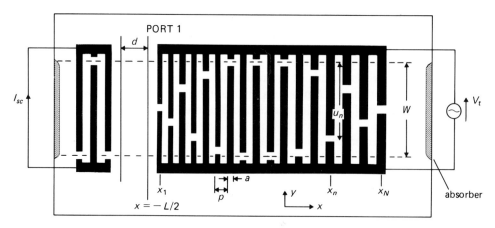

FIGURE 9.8. Single-dispersive interdigital chirp filter. The chirp transducer is the single-electrode ($S_e = 2$) type.

The chirp transducer of Figure 9.8 is typical of designs obtained by synchronous sampling, that is, a chirp waveform with phase $\theta(t)$ is sampled at times corresponding to equal increments of $\theta(t)$. In the case shown there are two samples for period of the waveform, and hence the transducer is a "single-electrode" type, with $S_e = 2$. Multi-electrode designs, with $S_e > 2$, can be obtained by using more than two samples per period, as discussed for constant-pitch transducers in Sections 4.2 and 8.1.3; such designs are often used in order to minimise electrode interaction effects. Synchronous sampling is nearly always used in chirp filters as it generally leads to relatively slow changes of apodisation, reducing second-order effects due to diffraction. In Section 9.3.1 it will be shown that, if synchronous sampling is used, the transducer response can be expressed in terms of an array factor and an element factor, in a manner very similar to the delta-function model of Section 4.1.

9.3.1. Transducer Analysis

We consider first an *unapodised* chirp transducer, such as that shown in Figure 1.5. The electrode pitch p and width a vary along the length. It will be assumed that p and a vary slowly with distance, and that the ratio a/p is constant throughout the transducer. We define $\varrho_e(x)$ as the electrostatic charge density on the electrodes when unit voltage is applied across the transducer, and $\bar{\varrho}_e(\beta)$ as the Fourier transform of $\varrho_e(x)$. The transducer has acoustic ports at $x = \pm\tfrac{1}{2}L$, and when a voltage V_t is applied, at frequency ω, a surface-wave with potential $\phi_{s1}(\omega)$ emerges at port 1, that is, at $x = -\tfrac{1}{2}L$. According to the quasi-static analysis of Chapter 4 this potential is given by equation (4.50), so that

$$\phi_{s1}(\omega) = j\Gamma_s V_t \bar{\varrho}_e(k_0) \exp(-\tfrac{1}{2}jk_0 L), \qquad (9.35)$$

where $k_0 = \omega/v_0$ is the wavenumber for surface waves on a free surface, and Γ_s is a piezoelectric coupling parameter such that $(\varepsilon_0 + \varepsilon_p^T)\Gamma_s = \Delta v/v$, as in equation (4.21).

In Section 4.5 it was shown that the response can be written in a more convenient form if the electrodes are regular, so that p and a are constants. For a chirp transducer, p and a vary along the length. However, a development similar to that of Section 4.5 can be used here. We define $\varrho_{en}(x)$ as the electrostatic charge density on the electrodes when electrode n has unit voltage and all the other electrodes are grounded. Then, according to the charge superposition principle, equation (4.33), the charge density $\varrho_e(x)$ is given by

$$\varrho_e(x) = \sum_{n=1}^{N} \hat{P}_n \varrho_{en}(x), \qquad (9.36)$$

where N is the number of electrodes and $\hat{P}_n$ are the electrode polarities, taking $\hat{P}_n = 1$ if the electrode n is connected to the upper bus and $\hat{P}_n = 0$ if it is connected to the lower bus. In Section 4.5.1 it was shown that, for regular electrodes, $\varrho_{en}(x)$ can be equated with $\varrho_f(x - x_n)$ provided end effects are negligible. Here x_n is the centre location of electrode n, and $\varrho_f(x)$ is the elemental charge density shown in Figure 4.10. In the present case, the electrodes are not regular. However, $\varrho_f(x - x_n)$ gives significant charge densities only in the immediate vicinity of electrode n. Since the pitch p usually varies very little from one electrode to the next, $\varrho_{en}(x)$ will not be substantially affected by the pitch variation, and is therefore given by

$$\varrho_{en}(x) \approx \varrho_f(x - x_n). \tag{9.37}$$

In practice, this equation is usually an excellent approximation. Note that $\varrho_f(x - x_n)$ depends on p, which varies from one electrode to another.

The Fourier transform of $\varrho_f(x)$ is $\bar{\varrho}_f(\beta)$, given by equation (4.88) and shown in Figure 4.11. Equation (4.88) shows that $\bar{\varrho}_f(\beta)$ is a function of the normalised variable βp. Since p varies with position, it is convenient here to change the notation slightly. The pitch in the vicinity of electrode n is denoted by p_n, and the function $\bar{\varrho}_f(\beta)$ for the electrodes in this vicinity is written as $\bar{\varrho}_f(\beta p_n)$. Using equations (9.37) and (9.36), we thus have

$$\bar{\varrho}_e(\beta) = \sum_{n=1}^{N} \hat{P}_n \bar{\varrho}_f(\beta p_n) \exp(-j\beta x_n). \tag{9.38}$$

Evaluating this at $\beta = k_0$ gives $\phi_{s1}(\omega)$, as shown by equation (9.35).

We now consider an *apodised* transducer, such as that shown in Figure 9.8. In this case the surface-wave potential $\phi_{s1}(\omega)$ will vary across the width of the transducer. The general analysis for apodised transducers is given in Section 4.7.2. The frequency response $H_t(\omega)$ of the transducer is defined such that

$$H_t(\omega) = -j\sqrt{\frac{\omega W}{\Gamma_s}} \frac{\langle \phi_{s1}(\omega) \rangle}{V_t}. \tag{9.39}$$

Here it is assumed that a voltage V_t is applied, and $\langle \phi_{s1}(\omega) \rangle$ is the average potential of the surface waves at port 1, taking the average over the aperture W of the transducer. As in Section 4.7.2, the transducer is imagined to be divided into a number of parallel channels, each of which can be regarded as an unapodised transducer. For channel j, the surface-wave potential at port 1 is, from equations (9.35) and (9.38),

$$\phi_{s1}^j(\omega) = j\Gamma_s V_t \exp(-\tfrac{1}{2}jk_0 L) \sum_{n=1}^{N} \hat{P}_n \bar{\varrho}_f(k_0 p_n) \exp(-jk_0 x_n). \tag{9.40}$$

Here the polarity sequence $\hat{P}_n$ will vary from channel to channel. To find the average potential, we integrate with respect to y and divide by W. This is easily done by integrating individual terms in equation (9.40). For electrode n, the break is at a location given by u_n, as shown in Figure 9.8. Thus the nth term in equation (9.40) is uniform over a distance u_n, and zero for the remaining distance $W - u_n$. The average potential $\langle \phi_{s1}(\omega) \rangle$ is therefore obtained by replacing $\hat{P}_n$ by u_n/W in equation (9.40), and substitution into equation (9.39) gives

$$H_t(\omega) = (\omega W \Gamma_s)^{1/2} \exp(-\tfrac{1}{2}jk_0 L) \sum_{n=1}^{N} \frac{u_n}{W} \bar{\varrho}_f(k_0 p_n) \exp(-jk_0 x_n). \tag{9.41}$$

An alternative formulation in terms of gap elements is obtained by the method of Section 4.5.3, and is found to give

$$H_t(\omega) = (\omega W \Gamma_s)^{1/2} e^{-\frac{1}{2}jk_0 L} \sum_{n=1}^{N-1} \frac{u_{n+1} - u_n}{W} \bar{\varrho}_g(k_0 p_n) e^{-jk_0(x_n + p_n/2)} \quad (9.42)$$

where $\bar{\varrho}_g(k_0 p_n)$ is the transform of the elemental charge density for gaps, given by equation (4.96). In both of the above equations, p_n is the pitch in the vicinity of electrode n.

The above results can be simplified if the transducer is designed using synchronous sampling, that is, if it is designed to given an impulse response of the form $a(t) \cos[\theta(t)]$ and the electrode locations correspond to equal increments of $\theta(t)$. This is nearly always the case in practice. Consider a small region of the transducer, containing several periods, and assume that the pitch does not vary much over this region. For synchronous sampling, the electrodes in this region generate surface waves most effectively at a frequency such that the local period $p_n S_e$ equals the wavelength, that is, when $k_0 = 2\pi/(p_n S_e)$. At other frequencies the waves generated by individual periods are not in phase, so the surface-wave amplitude is generally much smaller. Thus, in the function $\bar{\varrho}_f(k_0 p_n)$ we can replace $k_0 p_n$ by $2\pi/S_e$, and this is usually a very good approximation because $\bar{\varrho}_f(k_0 p_n)$ varies slowly with frequency. Furthermore, $\bar{\varrho}_f(2\pi/S_e)$ is a constant, and can therefore be taken outside the summation in equation (9.41). Similarly, $\bar{\varrho}_g(k_0 p_n)$ can be replaced by $\bar{\varrho}_g(2\pi/S_e)$, and thus equation (9.42) becomes

$$H_t(\omega) = (\omega W \Gamma_s)^{1/2} e^{-\frac{1}{2}jk_0 L} \bar{\varrho}_g(2\pi/S_e) \sum_{n=1}^{N-1} \frac{u_{n+1} - u_n}{W} e^{-jk_0(x_n + p_n/2)}. \quad (9.43)$$

This is very similar to the delta-function analysis of Section 4.1, though the element factor, that is, the expression outside the summation, is proportional to $\omega^{1/2}$. The term $(x_n + p_n/2)$ is the location of the nth gap element, midway between the centres of electrode n and electrode $n + 1$.

The above approximation is nearly always adequate for practical purposes. It should be noted that the above refers to the fundamental response. For a harmonic, $k_0 p$ is replaced by $2\pi M/S_e$, where M is the harmonic number.

The transducer response is defined such that the short-circuit response of a two-transducer device is, from equation (4.127),

$$H_{sc}(\omega) \equiv I_{sc}/V_t = H_t^a(\omega) H_t^b(\omega) \exp(-jk_0 d), \quad (9.44)$$

where $H_t^a(\omega)$ and $H_t^b(\omega)$ are the transducer responses, d is the separation between their acoustic ports, and I_{sc} and V_t are shown in Figure 9.8. It is assumed that at least one of the two transducers is unapodised. As in Chapter 4, the transducer admittance can be written $Y_t(\omega) = G_a(\omega) + jB_a(\omega) + j\omega C_t$. For an unapodised transducer the parallel conductance is $G_a(\omega) = |H_t(\omega)|^2$, as shown by equation (4.124), but for an apodised transducer it is necessary to sum the conductances of individual channels to obtain the overall conductance. The capacitance C_t can be obtained by summing the electrostatic charges on all the electrodes connected to one bus. The scattering coefficients for an unapodised transducer are given in Section 4.4.4, and the device response allowing for terminating circuits is given in Section 4.8.

9.3.2. Admittance in the Stationary-phase Approximation

In this section we give an approximate analysis for the parallel conductance $G_a(\omega)$ of a chirp transducer, making use of the stationary-phase approximation introduced in Section 9.2.1. Although the result does not predict the ripples which are given by an accurate analysis, it is very useful in practice and conveniently relates the performance of a chirp transducer to that of a uniform transducer with the same basic electrode structure.

We first consider a uniform transducer with aperture W and with N_{pu} periods. At the fundamental centre frequency ω_c, the parallel conductance $G_{au}(\omega_c)$ can be written

$$G_{au}(\omega_c) = \omega_c N_{pu}^2 W \Gamma_s G_0, \tag{9.45}$$

where G_0 is a constant dependent only on the properties of the substrate and the structure of the electrodes in each period, that is, it depends on S_e and on a/p. This form for $G_{au}(\omega_c)$ is derived in Section 4.6. Comparison with equation (4.118) shows that $G_0 = (\varepsilon_0 + \varepsilon_p^T)^2 \tilde{G}_{a1}$, where $\tilde{G}_{a1}$ is given in Table 4.1. When a voltage V_t is applied to an unapodised transducer, the potential $\phi_s(\omega)$ of the surface wave radiated in either direction is related to the conductance $G_a(\omega)$ by

$$|\phi_s(\omega)|^2 = V_t^2 \Gamma_s G_a(\omega)/(\omega W), \tag{9.46}$$

which follows from power conservation. Thus, in the particular case of a uniform transducer at the centre frequency ω_c we have

$$|\phi_s(\omega_c)| = V_t N_{pu} \Gamma_s G_0^{1/2}. \tag{9.47}$$

We now consider an *unapodised* chirp transducer. For each "period" of the transducer, the surface-wave amplitude generated is obtained by regarding the electrodes as a uniform transducer with one period. At any frequency ω in the band, the surface waves are generated primarily in the region where the periodicity corresponds to the wavelength, so the surface-wave potential can be obtained from equation (9.47). Surface waves generated elsewhere contribute little to the total amplitude, so equation (9.47) may be used for all the periods in the transducer. We define x_m as the location of period m, such that the wave potential ϕ_s due to this period is real if the phase is referred to x_m. The total wave potential generated by the chirp transducer can then be written

$$\phi_{s1}(\omega) \approx V_t \Gamma_s G_0^{1/2} \exp(-\tfrac{1}{2}jk_0 L) \sum_{m=1}^{N_p} \exp(-jk_0 x_m), \tag{9.48}$$

where N_p is the number of periods, so that the number of electrodes is $N = N_p S_e$. The potential is evaluated at port 1, that is, at $x = -\tfrac{1}{2}L$.

To evaluate $\phi_{s1}(\omega)$, it is assumed that the transducer is designed by sampling a non-linear function $\theta(t)$ at points t_m such that $\theta(t_m) = 2m\pi$. The points t_m therefore have unequal spacing. The locations x_m are taken to be $x_m = v_0 t_m$, and hence $k_0 x_m = \omega t_m$. The summation in equation (9.48) can be written

$$S = \sum_{m=1}^{N_p} \exp(-j\omega t_m) = \sum_{m=1}^{N_p} \exp[j\theta(t_m) - j\omega t_m]. \tag{9.49}$$

Here the spacing of the points is given by

$$t_{m+1} - t_m \approx 2\pi/\dot{\theta}(t_m)$$

and the summation can therefore be replaced by the integral

$$S \approx \frac{1}{2\pi} \int_{t_1}^{t_2} \dot{\theta}(t) \exp[j\theta(t) - j\omega t] dt, \qquad (9.50)$$

where the limits are the times corresponding to the two ends of the transducer. Equation (9.50) can be evaluated by the stationary-phase method. Comparing with equations (9.7) and (9.12a), the magnitude of S is

$$|S| \approx \frac{\omega}{2\pi\sqrt{|\mu(T_s)|}}, \qquad (9.51)$$

where $\mu(t)$ is the chirp rate, so that $2\pi\mu(t) = \ddot{\theta}(t)$, and $T_s(\omega)$ is the stationary-phase point, defined as the solution of $\dot{\theta}(T_s) = \omega$. From equation (9.48) the magnitude of the surface-wave potential is $|\phi_{s1}(\omega)| = V_t \Gamma_s |S| G_0^{1/2}$ and, using equation (9.46), the conductance of the unapodised transducer is

$$G_a(\omega) \approx \frac{\omega^3 W G_0 \Gamma_s}{4\pi^2 |\mu(T_s)|}. \qquad (9.52)$$

For an *apodised* chirp transducer, a simple modification can be made. The apodisation is taken to be given by a function $U(t)$, which gives the electrode overlaps at the corresponding position, so that $U(x_n/v_0) = |u_{n+1} - u_n|$. At frequency ω, the surface wave excitation arises primarily from the region corresponding to the stationary-phase point $T_s(\omega)$. Thus, if $U(t)$ varies slowly enough, the conductance is given by equation (9.52) with the aperture taken as $U[T_s(\omega)]$, so that

$$G_a(\omega) \approx \frac{\omega^3 G_0 \Gamma_s U(T_s)}{4\pi^2 |\mu(T_s)|}. \qquad (9.53)$$

Equations (9.52) and (9.53) are very convenient for estimating the conductance, and hence the insertion loss of a device.

The acoustic susceptance $B_a(\omega)$ is the Hilbert transform of $G_a(\omega)$, as shown by equation (4.62). It is usually the case that $G_a(\omega)$ varies slowly with ω and hence $B_a(\omega)$ is small, and can therefore be neglected.

Effective number of periods. It is helpful to compare the conductance $G_a(\omega)$ of a chirp transducer, equation (9.53), with the centre-frequency conductance $G_{au}(\omega_c)$ of a uniform transducer, equation (9.45). To clarify this, we introduce an effective number of periods, $N_{\text{eff}}(\omega)$. This is defined as the number of periods in a uniform transducer, with centre frequency ω, such that its conductance equals that of the chirp transducer at frequency ω. Both transducers are taken to have the same aperture. We thus set equations (9.45) and (9.53) equal, and in equation (9.45) put $\omega_c = \omega$, $N_{pu} = N_{\text{eff}}(\omega)$ and $W = U(T_s)$. This gives

$$N_{\text{eff}}(\omega) = \frac{\omega}{2\pi\sqrt{|\mu(T_s)|}}. \tag{9.54}$$

With this formula, the conductance of a chirp transducer can be found straightforwardly from the centre-frequency conductance of a uniform transducer with the same value of S_e. We can interpret $N_{\text{eff}}(\omega)$ roughly as the number of periods in the chirp transducer that are actively generating surface waves at frequency ω; the remaining periods on either side of the active region behave essentially as a capacitive shunt.

In the case of a linear chirp, $|\mu|$ is a constant equal to B/T, and we find $N_{\text{eff}}(\omega)$ is related to the total number of periods N_p by

$$\frac{N_{\text{eff}}(\omega)}{N_p} = \frac{\omega}{\omega_c\sqrt{TB}}, \tag{9.55}$$

where ω_c is the centre frequency of the chirp. Thus, for $TB = 100$, say, about 10% of the electrodes are active at any particular frequency.

Capacitance and Q-factor. For a uniform transducer, the capacitance C_t is given by equation (4.117), that is, by

$$C_{tu} = WN_{pu}(\varepsilon_0 + \varepsilon_p^T)\tilde{C}_t, \tag{9.56}$$

where N_{pu} is the number of periods and $\tilde{C}_t$ is a constant depending on S_e and a/p, given in Table 4.1 and Figure 4.15. Since this is proportional to N_{pu}, we can regard the contribution of each period as being given by equation (9.56), with $N_{pu} = 1$. The same formula can be applied to each period of a chirp transducer, if the apodisation $U(t)$ is varying slowly. For period m the aperture is $U(t_m)$, where t_m is the solution of $\theta(t_m) = 2m\pi$. The capacitance of the chirp transducer is thus

$$C_t \approx (\varepsilon_0 + \varepsilon_p^T)\tilde{C}_t \sum_{m=1}^{N_p} U(t_m). \tag{9.57}$$

For an unapodised chirp transducer, the capacitance is the same as that of a uniform transducer, equation (9.56), with the same number of periods.

An important result relates the Q-factor of a chirp transducer to that of a uniform transducer with the same acoustic bandwidth. The chirp transducer is taken to be unapodised and to have a linear chirp, so that its centre-frequency conductance $G_a(\omega_c)$ is given by equation (9.52) with $|\mu| = B/T$. The centre-frequency conductance of the uniform transducer is $G_{au}(\omega_c)$, given by equation (9.45). Here N_{pu} is replaced by $\omega_c/(2\pi B)$, so that the acoustic bandwidths of the two transducers are approximately the same. Taking the apertures to be equal, the conductances are found to be in the ratio

$$\frac{G_a(\omega_c)}{G_{au}(\omega_c)} \approx TB. \tag{9.58}$$

The capacitances are given by equations (9.56) and (9.57), and since the chirp transducer has $N_p = T\omega_c/(2\pi)$ periods we have

$$C_t/C_{tu} \approx TB. \tag{9.59}$$

The Q-factor of a transducer is, by definition, $\omega_c C_t/G_a(\omega_c)$. Thus the Q-factors of the chirp and uniform transducers are the same if they have the same acoustic bandwidth. This is of course still true if the apertures are different, since the aperture does not affect the Q-factor. This important result was first noted by Hartmann et al. [296]. It enables the Q-factor of a linear chirp transducer to be estimated directly from the analysis for uniform transducers in Section 4.6. The Q-factor is particularly relevant when the transducer is tuned in order to minimise the conversion loss, as discussed in Section 7.1.1.

9.3.3. Transducer Design

It is assumed here that the transducer is to be designed such that its impulse response is a good approximation to some required response, given by the chirp waveform

$$v(t) = a(t) \cos [\theta(t)], \tag{9.60}$$

where $\theta(t)$ is a non-linear function of time. This waveform will be of finite length, so that $a(t)$ is zero outside the interval occupied by the waveform. If the required response is specified in the frequency domain, $a(t)$ and $\theta(t)$ can be found by transforming the spectrum to the time domain. It may then be necessary to truncate in order to obtain a finite length, and if necessary short extensions can be added to avoid abrupt discontinuities. Owing to the sampled nature of the transducer, the frequency response will include a fundamental and a series of harmonics. It is assumed here that the fundamental component is to correspond with the required response of equation (9.60). In most practical cases the harmonics are irrelevant, or are eliminated by filtering elsewhere.

The design procedure is based on some results for non-uniform sampling, derived in Section A.4 and repeated here for convenience. Suppose that the waveform $v(t)$ is sampled at times t_n, such that

$$\theta(t_n) = 2\pi n/S_e + \theta_0, \tag{9.61}$$

where θ_0 is a constant and n is an integer. Thus, synchronous sampling is assumed, and the integer S_e is the number of samples per period of the waveform. The sampled waveform is

$$v_s(t) = \sum_{n=1}^{N} v(t_n)\delta(t - t_n), \tag{9.62}$$

where N is the number of samples. It is assumed that θ_0 and N are chosen such that $v(t) = 0$ for $t < t_1$ and for $t > t_N$. Section A.4 shows that the *fundamental* componential of $v_s(t)$ is

$$\tilde{v}_s(t) = Ca(t)\dot{\theta}(t)\cos[\theta(t)], \quad (9.63)$$

where C is a constant. This is valid for $S_e > 2$, and for $S_e = 2$ it is valid provided θ_0 is a multiple of π.

Now consider the design of a transversal filter to given an impulse response $v(t)$. The impulse response of the transversal filter has the form

$$h_s(t) = \sum_{n=1}^{N} h_n \delta(t - t_n) \quad (9.64)$$

with t_n given by equation (9.61). The transversal filter has the form shown in Figure 8.3, though here the tap delays t_n are unequally spaced. Comparing with equations (9.62) and (9.63), it can be seen that if the coefficients h_n were equated with $v(t_n)$ the fundamental component of $h_s(t)$ would be proportional to $v(t)$ except for the amplitude distortion due to the term $\dot{\theta}(t)$. To compensate for this distortion, the tap weights must be given by

$$h_n = v(t_n)/\dot{\theta}(t_n) \quad (9.65)$$

so that $h_s(t)$ is a sampled version of a waveform $v(t)/\dot{\theta}(t)$. The fundamental component of $h_s(t)$ will then be proportional to $v(t)$, equation (9.60).

For a *transducer* required to give an impulse response $v(t)$, we consider the design in terms of gap elements, though an alternative method based on electrode elements is very similar. The frequency response is given by equation (9.43), which has the form

$$H_t(\omega) = K\omega^{1/2} \sum_{n=1}^{N} (u_{n+1} - u_n) \exp(-jk_0 x_n'), \quad (9.66)$$

where K is a constant and x_n' is the location of gap n relative to the acoustic output port. Here N is the number of gaps and u_n are the locations of the electrode breaks, as in Figure 9.8. For comparison, the frequency response $H_s(\omega)$ of the transversal filter considered above is the transform of equation (9.64), so that

$$H_s(\omega) = \sum_{n=1}^{N} h_n \exp(-j\omega t_n). \quad (9.67)$$

Comparing with equation (9.66), it is clear that the required gap locations are given by $x_n' = v_0 t_n$, where $v_0 = \omega/k_0$ is the free-surface velocity. The transducer response of equation (9.66) differs from equation (9.67) in that a term $\omega^{1/2}$ is present. In the design, this is easily allowed for by invoking the stationary-phase approximation. The response at frequency ω is mainly determined by the electrodes in the vicinity of gap n, such that $\dot{\theta}(t_n) = \omega$. Thus the $\omega^{1/2}$ term can be compensated by making the gap strength proportional to $h_n/[\dot{\theta}(t_n)]^{1/2}$, so that

$$u_{n+1} - u_n \propto v(t_n)/[\dot{\theta}(t_n)]^{3/2}. \quad (9.68)$$

CHIRP FILTERS AND THEIR APPLICATIONS

A more accurate method of compensation is to transform $v(t)$ to the frequency domain, divide by $\omega^{1/2}$, and then return to the time domain. However, equation (9.68) is usually accurate enough in practice.

To give an example, consider the linear chirp waveform

$$v(t) = a(t) \cos [\theta(t)] = a(t) \cos (\omega_1 t + \pi\mu t^2), \quad \text{for } 0 \leq t \leq T,$$
$$= 0, \quad \text{for other } t. \quad (9.69)$$

This gives instantaneous frequency $\dot{\theta}(t) = \omega_1 + 2\pi\mu t$ and has bandwidth $B = |\mu|T$. From equation (9.61) the sampling times t_n are given by

$$\mu t_n = [f_1^2 + \mu\theta_0/\pi + 2\mu n/S_e]^{1/2} - f_1, \quad (9.70)$$

where $f_1 = \omega_1/(2\pi)$. The gap strengths are given by equation (9.68), so that

$$u_{n+1} - u_n \propto \frac{a(t_n) \cos [\theta(t_n)]}{(\omega_1 + 2\pi\mu t_n)^{3/2}}. \quad (9.71)$$

For $S_e = 2$ the terms $\cos [\theta(t_n)]$ all have the same magnitude, and simply cause an alternation in sign.

It is often necessary to modify the design somewhat to compensate for second-order effects, but this topic is deferred to Section 9.5.

9.4. DESIGN AND PERFORMANCE OF INTERDIGITAL DEVICES

The analysis and design of chirp transducers have been considered in Section 9.3 above. Here we consider the design of a two-transducer chirp filter. This of course pre-supposes that the required response of the device has already been determined. In practice, the starting point is usually the relevant performance requirements of the radar system, for example the output pulse width, sidelobe levels and the receiver mis-match loss. In this case the first task is to design the *waveforms*, that is, the required transmitter waveform and the impulse response of the filter in the receiver. These are designed by the methods of Section 9.2. The devices considered here may have impulse responses that are any of the types of chirp waveform considered in Section 9.2.

Single-dispersive Devices The simplest form of chirp filter is illustrated in Figure 9.8, and comprises one chirp transducer and one uniform transducer. This is called a *single-dispersive* device to distinguish if from devices using two chirp transducers, considered later. Since the single-dispersive device has one transducer unapodised, its short-circuit response is essentially the product of the two transducer responses, as given by equation (9.44). To design the device, the number of electrodes in the uniform transducer is first chosen, such that its response does not vary by more than a few dB's over the required device bandwidth. The chirp transducer is then designed by the method of Section 9.3.3, with a modification to allow for the roll-off given by the uniform transducer. In practice it is often necessary to compensate also for a variety of second-order effects, discussed in Section 9.5. The modifications involved are usually small and can be implemented straightforwardly using the stationary-phase approximation, that is, by modifying the time-domain amplitude $a(t)$ and phase $\theta(t)$. Often, the transducers are tuned in order to reduce the conversion losses, and in this case it is usually necessary to allow for the circuit effect when designing the chirp transducer. However, for chirp transducers it is usually difficult to obtain low conversion losses, because the radiation resistance $R_a(\omega)$ is usually much less than $50\,\Omega$. This is because, at any ω, most of the electrodes act as a capacitive shunt across the active region, as seen in Section 9.3.2. This reduces the radiation resistance, as shown in Section 7.1.2.

The substrate material is usually quartz, chosen because it gives good temperature stability and because electrode interactions are relatively weak. However, for small time-bandwidth products a strongly-piezoelectric material such as lithium niobate is often feasible, and gives lower insertion losses than quartz.

The triple-transit spurious signal can be reduced if necessary by adjusting the terminal impedances "seen" by the transducers. However, it is unusual for the triple-transit signal to cause problems in chirp devices. From the basic nature of the device it is easily seen that, in the *time* domain, the triple-transit response gives a chirp waveform similar to the main response, but stretched in time by a factor of 3. When the device is used as a matched filter the triple-transit response is not matched to the input waveform, and this factor discriminates against the triple-transit response.

Although the device shown in Figure 9.8 is a down-chirp, the device may equally well be an up-chirp, so that the frequency of its impulse response increases with time. This makes no difference to the basic principles. However, for wide-band devices there can be difficulties due to the electrodes scattering surface waves into bulk waves, as explained in Appendix F. For this reason, up-chirp devices are usually constrained to fractional bandwidths less than about 25%, and the bulk wave scattering is then negligible.

In many radar systems a matched pair of chirp devices is required, one to serve as an expander in the transmitter, generating the chirp waveform, and the other to serve as the compressor in the receiver. Usually, one device will be an up-chirp and the other will be a down-chirp. In fact, the chirp polarity can be reversed in the receiver by using a local oscillator with frequency above the band occupied by the R.F. input signal; for this case the two filters will have the same chirp polarity. The expansion process is most efficient if the expander is excited by a short rectangular pulse of carrier, whose length τ is such that its bandwidth is compatible with the device bandwidth. It is usual to design the expander to allow for the roll-off in the spectral amplitude of this input pulse. An important factor is the loss of power due to the expansion process. For a flat-envelope linear chirp it can be shown that, in comparison with C.W. excitation, the expansion process reduces the output power in the ratio of approximately $\tau^2 B/T$, provided $\tau \leqslant 1/B$. Typically, $\tau \approx 1/B$, giving an "expansion loss" of $10\log(TB)\,\text{dB}$. This must be added to the C.W. insertion loss to give the total loss. For large TB the total loss can be prohibitive, and then the transmitter must use active rather than passive generation.

Double-dispersive Devices. For weakly-piezoelectric substrates, such as quartz, the single-dispersive device is generally suitable only if the fractional bandwidth is small. This is due to the difficulty of matching uniform transducers over a wide bandwidth. For example, on *ST*, *X* quartz a matched uniform transducer cannot give more than about 5% bandwidth, as noted in Section 7.1.1; for wider bandwidths the loss is substantial unless a complex matching network is used.

In this situation, a *double-dispersive* design is usually preferrable. As shown in Figure 9.9(a), this design has two chirp transducers. Usually, each transducer contributes about half of the dispersion required of the overall device. One of the transducers is unapodised, so that the device response is given by the product of the transducer responses. Most of the earlier remarks on second-order effects and choice of materials, referring to single-dispersive devices, apply here also.

To design a double-dispersive device, the unapodised transducer is designed first. Typically, the dispersion required for the overall device is halved, and the corresponding phase $\theta_u(t)$ for the unapodised transducer is deduced. The sampling points are then obtained from $\theta_u(t)$ in the usual way. The form of $\theta_u(t)$ is not at all critical, though it is necessary to ensure that the transducer has adequate bandwidth. Once this transducer has been designed, its frequency response can be calculated and divided into the required device response in order to obtain the response required of the apodised transducer. The latter can then be designed by the method of Section 9.3.3.

Figure 9.9(b) shows a *slanted* double-dispersive device [297–299]. Here, waves generated at any point in one transducer travel through relatively few electrodes before emerging on the free surface. Consequently, second-order effects are reduced and, for an up-chirp device, the bulk-wave scattering problem mentioned earlier does not occur. Also, since waves of different frequencies traverse the substrate at different locations, there is the possibility of adding a "phase plate" to correct for phase errors, as done in RAC's (Section 9.6). However, compared with other chirp devices, the

(a) double-dispersive

(b) slanted double-dispersive

FIGURE 9.9. Other types of interdigital chirp filter.

slanted device makes less effective use of the substrate area and is more difficult to fabricate.

Another interesting variant is obtained if one transducer of the double-dispersive device, Figure 9.9(a), is reversed end to end. In this case the device will have one down-chirp transducer and one up-chirp. If the frequency-time curves for the two transducers are the same except for the time-reversal, the overall device will be non-dispersive even though the transducers are dispersive. This configuration can be used effectively to realise a bandpass filter with linear phase, as already mentioned in Section 8.4.5, and for this application the design techniques of Chapter 8 are applicable.

Performance. Interdigital devices typically have bandwidths up to about 500 MHz and dispersion T up to 50 µsec, the latter being limited by the length of substrate required. For large time-bandwidth products the insertion loss can be excessive and errors due to second-order effects can be unacceptable, and consequently the TB-product is generally limited to 1000. Insertion losses (for C.W. signals) are typically 30 to 60 dB, larger values being obtained for larger TB-products and for substrates with weaker piezoelectric coupling.

The accuracy achievable using surface-wave devices is well illustrated by the fact that the time-sidelobes of the compressed output waveform can be as low as 45 dB below the main peak. It will be seen in Section 9.5.1 that this implies a phase accuracy, in the frequency domain, of only a few degrees, while the phase itself varies by typically several thousand degrees over the device bandwidth. The accuracy required becomes more stringent as the required sidelobe level is reduced, but even at low sidelobe levels the performance can be quite close to the theoretical ideal given by the analysis in Section 9.2: sidelobe levels are typically within a few dB's of the ideal, and the output pulse width and the processing gain are also close to ideal. In practice the sidelobe level is not usually minimised because a low sidelobe level implies, even in the ideal case, a relatively wide output pulse and both factors are relevant to the system performance. Typically, the sidelobe suppression is 25 to 40 dB.

A typical example is shown in Figure 9.10, which refers to a matched pair of devices. Here the chirps are linear, with 11 MHz bandwidth and 14 µsec duration, and the compressor has amplitude weighting to suppress the time-sidelobes. The upper photograph shows the two devices, mounted in one package to minimise errors due to temperature differences. Both devices are of the double-dispersive type, with ST, X quartz substrates. The lower photograph on the left shows the expanded pulse, obtained by applying a short rectangular pulse to the expander. The lower right photograph, with the same time scale, is the compressor output when the expanded pulse is applied to it, simulating the output in a radar system for a point target. The width of the output pulse, at the 3 dB points, is 120 nsec, and the largest sidelobe (not visible in the figure) is 35 dB below the peak.

Table 9.1 shows performance data for a variety of chirp devices, and illustrates the wide range of parameters obtainable. For example, bandwidths in the table range from 2.4 to 500 MHz, while the dispersion T ranges from 0.5 to 34 µsec. The first column refers to the devices of Figure 9.10.

Expander

Compressor

Expanded pulse

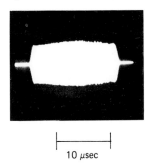

|—————|
10 μsec

Compressed pulse

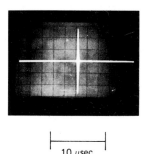

|—————|
10 μsec

FIGURE 9.10. Performance of matched pair of interdigital chirp filters. (Courtesy Plessey Research)

Devices for Frequency-scanning Radar. The last column on Table 9.1 refers to a chirp filter for a frequency-scanning radar, which uses pulse compression in novel way [303–305]. The radar antenna here is a phased array, with the elements driven from a slow-wave structure with the input port at one end. Owing to the slow-wave structure, the relative phases of the signal at the array elements vary quite rapidly with frequency, and this feature is used to form a narrow transmitted beam whose direction depends on frequency. Thus the beam can be scanned rapidly by electronic means. In the particular example here, the elevation scan is achieved by applying a single chirp pulse. For a point target, the returned echo has a centre frequency indicating its elevation. The echo is essentially a short segment of the transmitted waveform, with duration given by the width and sweep rate of the beam at the target elevation. Since this signal is a chirp waveform, processing gain can be obtained by applying it to a matching chirp filter in the receiver. However, the chirp filter must cover the entire bandwidth of the transmitted waveform in order to process echos from targets at all elevations, and in the present case this implies a very large time-bandwidth product of several thousand. The chirp filter also compensates for range errors due to the fact that beams with different elevation are radiated at different times.

TABLE 9.1

Performance examples for interdigital chirp filters. Data refer to a matched pair of devices, unless noted otherwise

Centre frequency (MHz)	70	60	30	60	300	30	300	750	75
Bandwidth B (MHz)*	11	12	5	4	100	2.4	100	500	27
Duration T (μs)	14	8	4	12	5	34	10	0.5	28
TB	153	96	20	49	500	81	1000	250	737
Chirp type[†]	L	L	L	NL	L	L	L[§]	L[§]	NL[§]
Time-sidelobe rejection (dB)	35	40	29	45	32	32	20	13	—
Substrate[‡]	Q	Q	LN	Q	Q	Q	LN	LN	Q
Reference	See text	[300]	[302]	[300]	[293]	[293]	[301, 302]	[338]	See text

* Range of instantaneous frequencies.
[†] L = linear; NL = non-linear.
[‡] Q = quartz; LN = lithium niobate.
[§] Data refer to single device, sidelobe rejection is for ideal input waveform.

CHIRP FILTERS AND THEIR APPLICATIONS

Owing to the large time-bandwidth product required the receiver used seven interdigital chirp filters, multiplexed in frequency. The data in the last column of Table 9.1 refer to the device with the largest time-bandwidth product. The instantaneous frequency for this device was a very non-linear function of time, but the device nevertheless gave a frequency response with phase accurate to within $\pm 10°$ of the response required.

9.5. SECOND-ORDER EFFECTS IN CHIRP FILTERS

A variety of second-order effects are significant in surface-wave chirp filters, and it is important to consider not only the resulting distortion of the device response but also the effect of this distortion on the output waveform of a pulse-compression radar system. The main practical consequences are degradations in the time-sidelobe levels, the pulse width and the signal-to-noise ratio. In many cases, these degradations are primarily associated with slowly-varying phase errors in the frequency response of the compressor or expander, or both. Such errors arise, for example, if there is a temperature change, or if the surface-wave velocity differs from its expected value due to misalignment during fabrication. Similar effects are caused by doppler shifts.

In order to assess these effects, we first consider the effect of a slowly-varying phase error on the output waveform, without considering the physical origin of the error. This is done in Section 9.5.1. The error is expressed as a sum of Legendre polynomials, and it is shown that individual polynomials have quite distinctive effects on the output waveform. It is often the case that errors due to specific second-order effects can be conveniently expressed in terms of these polynomials, and hence the corresponding distortion of the output waveform is readily assessed. This is illustrated in Section 9.5.2, which describes the effects of velocity errors and temperature changes. Doppler shifts are considered in Section 9.5.3, which uses the analysis of Section 9.5.1 for the case of non-linear chirps.

Nearly all of the material in Sections 9.5.1 to 9.5.3 is applicable to reflective array compressors, described later in Section 9.6, as well as to interdigital chirp filters. However, some further second-order effects, specific to interdigital devices, are considered in Section 9.5.4.

9.5.1 Effect of Phase Errors on Compressor Output Waveform

It is assumed here that an ideal chirp waveform is applied to a compressor designed to give an output waveform with well-suppressed time-sidelobes, as described in Sections 9.2.3 and 9.2.4. However, here the compressor is taken to be imperfect, so that its frequency response is $H'(\omega)$ instead of the ideal response $H(\omega)$. If $S(\omega)$ is the spectrum of the input waveform, the ideal output waveform has a spectrum $G(\omega) = S(\omega)H(\omega)$, while the actual output waveform has a spectrum $G'(\omega) = S(\omega)H'(\omega)$. Thus

$$G'(\omega) = G(\omega)H'(\omega)/H(\omega). \qquad (9.72)$$

In practice, phase errors in the compressor response are often much more significant than amplitude errors, and the latter will therefore be ignored here. The phase error will be denoted by $\Delta\phi(\omega)$, so that $H'(\omega)/H(\omega) = \exp[j\Delta\phi(\omega)]$, and the distorted output spectrum becomes

$$G'(\omega) = G(\omega)\exp[j\Delta\phi(\omega)]. \tag{9.73}$$

Here the ideal output spectrum $G(\omega)$ is proportional to $\bar{W}(\omega - \omega_c)$, where $\bar{W}(\omega)$ is the weighting function used to suppress the time-sidelobes, as explained in Sections 9.2.3 and 9.2.4. Note that the input waveform and the ideal filter response need not be considered explicitly. For example, for a given $G(\omega)$ equation (9.73) gives the output distortion irrespective of whether the chirps are linear or non-linear.

Surface-wave devices often exhibit a slowly-varying phase error, considered below, with a rapidly-varying error superimposed. The latter is usually due to either surface wave reflections or inaccuracy in the locations of electrodes or grooves. An indication of the effect of rapid errors is given by "paired-echo" theory [277, 279]. This shows that a small sinusoidal error of the form $\Delta\phi(\omega) = k\sin\omega\tau$, where k and τ are constants, causes spurious time-sidelobes to appear at times $\pm\tau$ relative to the main peak, with relative amplitude $k/2$. However, in practice the rapidly-varying error usually has an irregular form, and hence the paired-echo theory is of limited value. In addition, rapidly-varying errors do not generally cause significant degradations of the output waveform, so they will not be considered further here.

Slowly-varying Phase Errors. Slowly-varying errors can be expressed as simple polynomial functions. In order to distinguish the distortions arising in the compressed output waveform it is important to use orthogonal polynomials, and a suitable choice is the Legendre polynomials $P_n(x)$ defined in Appendix C. These are orthogonal over the interval $-1 \leq x \leq 1$. The phase error is therefore written as

$$\Delta\phi(\omega) = \sum_{n=0}^{\infty} \Phi_n P_n(x), \tag{9.74}$$

where Φ_n are constant coefficients and

$$x = (\omega - \omega_c)/(\pi B). \tag{9.75}$$

This choice for x makes the polynomials orthogonal over the compressor band, that is, over the interval $|\omega - \omega_c| \leq \pi B$. The first four polynomials are $P_0(x) = 1$, $P_1(x) = x$, $P_2(x) = (3x^2 - 1)/2$, and $P_3(x) = (5x^3 - 3x)/2$. The orthogonality relation is

$$\int_{-1}^{1} P_n(x)P_m(x)\,dx = (n+\tfrac{1}{2})^{-1}, \quad \text{for } m = n,$$
$$= 0, \quad \text{for } m \neq n. \tag{9.76}$$

Given some phase error $\Delta\phi(\omega)$, this can be used to evaluate the coefficients Φ_n in equation (9.74); $\Delta\phi(\omega)$ is multiplied by $(n+\tfrac{1}{2})P_n(x)$, and the integral with respect to x from -1 to 1 gives Φ_n. Alternatively, least-squares fitting gives the same results.

Generally, the terms with $n > 3$ in equation (9.74) are found to be of little significance, so here we shall only consider the first four terms. These give rise to distortions of the output waveform in the vicinity of the main peak, and the forms of distortion are quite distinct. The first term, $\Phi_0 P_0(x)$, is a constant, and is rarely of any consequence. The second term, $\Phi_1 P_1(x)$ is proportional to $(\omega - \omega_c)$, and corresponds to a delay $\tau = -\Phi_1/(\pi B)$ of the output waveform. It should be noted that this delay is valid irrespective of the presence of any other terms in the expansion.

To see the effect of a *quadratic* error, suppose that only the $n = 2$ term is present in equation (9.74), so that $\Delta\phi(\omega) = \Phi_2 P_2(x)$. This function is shown on the left in Figure 9.11. The coefficient Φ_2 is the phase error at the band edges. The consequent distortion of the output waveform can be found by using equation (9.73) and transforming to the time domain. The effect of the error depends on the nature of the ideal output waveform, and has been calculated by Klauder et al. [277] for a typical case with an ideal time-sidelobe suppression of 40 dB. The error is found to broaden the main peak and reduce its amplitude, though the levels of the sidelobes are hardly affected. The right part of Figure 9.11 shows the 6 dB width of the peak as a function of Φ_2 and also the reduction of output signal-to-noise ratio; the latter follows directly from the peak amplitude since the output noise level is not affected by the phase error.

For a *cubic* error, suppose that only the $n = 3$ term is present in equation (9.74), so that $\Delta\phi(\omega) = \Phi_3 P_3(x)$. As shown in the upper part of Figure 9.12, this gives a phase error $\pm\Phi_3$ at the band edges. The effect of this on the output waveform is shown in the lower left part of the figure, which shows the envelope of the output for $\Phi_3 = 0$ and for $\Phi_3 = 22.4°$. In this case Hamming weighting has been chosen as an example, so that the ideal output spectrum $G(\omega)$ in equation (9.73) is taken as $\bar{W}_H(\omega - \omega_c)$, equation (9.26). It can be seen that the error causes an unwanted sidelobe to appear before the main peak, and also distorts the shape of the peak. However, the peak broadening is negligible, and there is no significant delay or loss of amplitude. The plot on the lower right in Figure 9.12 shows the spurious sidelobe level as a function of Φ_3.

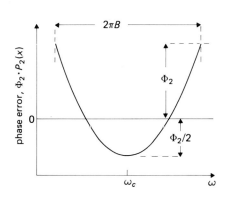 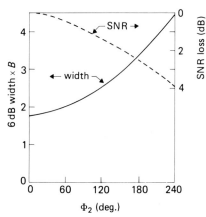

FIGURE 9.11. Effect of quadratic phase error on compressed output waveform. Ideal output has -40 dB sidelobes. From data given by Klauder et al. [277].

250   SURFACE-WAVE DEVICES

Note that quadratic and cubic errors have been considered individually, assuming other terms to be absent. In practice it is often the case that one of these two terms is dominant, and the above conclusions are then valid. If both terms are present with comparable magnitude, the above conclusions will not in general be quantitatively correct.

9.5.2. Velocity Errors and Temperature Effects

These factors cause slowly-varying phase errors in the frequency response of the compressor, and the resulting distortion of the compressed output waveform is readily assessed using the method of Section 9.5.1 above. Velocity errors are caused by, for example, misalignment during fabrication (Section 6.4) or the presence of electrodes (Appendix D). It is assumed here that the velocity error is independent of frequency and is uniform over the surface of the substrate, though these assumptions are not always valid.

It was shown in Section 6.4 that a velocity error or temperature change scales the phase of the device frequency response, so that

$$\phi'(\omega) = \phi[\omega(1 + \varepsilon)], \qquad (9.77)$$

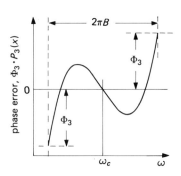

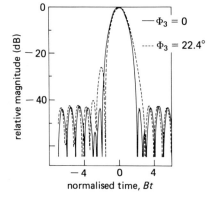
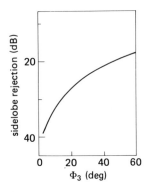

FIGURE 9.12. Effect of cubic phase error on compressed output waveform. Ideal output waveform is Hamming weighted.

where $\phi(\omega)$ and $\phi'(\omega)$ are respectively the ideal and actual phase. There is also a small amplitude change, but this is usually insignificant and is neglected here. The quantity ε is small. For a velocity error, $\varepsilon = -(v' - v)/v$, where v and v' are the ideal and actual velocities. For a temperature change, $\varepsilon = (T' - T)/T$, where T and T' are the delays between two points at the "ideal" temperature and at the actual temperature.

As in Section 9.5.1, the spectrum of the compressed output waveform is $G'(\omega) = G(\omega) \exp[j\Delta\phi(\omega)]$, and here $\Delta\phi(\omega) = \phi'(\omega) - \phi(\omega)$. Since $\varepsilon \ll 1$, we obtain from equation (9.77)

$$\Delta\phi(\omega) \approx \varepsilon\omega \frac{d\phi(\omega)}{d\omega} = -\varepsilon\omega\tau_g(\omega), \tag{9.78}$$

where $\tau_g(\omega)$ is the ideal group delay of the compressor. The phase error therefore depends on the dispersion curve of the device.

(a) Linear Chirps. If the chirp is linear, the ideal group delay of the compressor is approximately

$$\tau_g(\omega) \approx (\omega - \omega_c)/(2\pi\mu) + \tau_c. \tag{9.79}$$

This equation is obtained from the stationary-phase approximation, neglecting the ripples in the response. The parameter τ_c is the delay at the centre frequency ω_c. Clearly, the phase error $\Delta\phi(\omega)$ of equation (9.78) is quadratic; it corresponds to a change of chirp rate from μ to μ', where $\mu' = \mu/(1 + 2\varepsilon)$, and also to a change of the delay pedestal. If the phase error is written as a sum of Legendre polynomials, as in equation (9.74), the coefficients Φ_1 and Φ_2 are found to be

$$\Phi_1 = -\varepsilon B[\pi\tau_c + \tfrac{1}{2}\omega_c/\mu]. \tag{9.80}$$

and

$$\Phi_2 = \mp \tfrac{1}{3}\pi\varepsilon TB, \tag{9.81}$$

where the upper sign applies for $\mu > 0$ and the lower sign for $\mu < 0$. The linear term, with coefficient Φ_1, causes a delay equal to $\varepsilon[\tau_c + \omega_c/(2\pi\mu)]$, while the quadratic term broadens the output pulse and reduces the SNR, as shown for example in Figure 9.11. Usually the quadratic term is more significant, and a typical requirement would be that Φ_2 should not exceed $\pi/2$. This implies that $|\varepsilon| \leq 3/(2TB)$, and hence the tolerances on velocity and temperature errors become more stringent as TB is increased.

Generally, these tolerances are not difficult to meet. For devices on ST, X quartz substrates, the temperature stability is generally sufficient for operation over a range of 100°C or more. For Y, Z lithium niobate the range is more restricted; for example, if $TB = 200$, temperature changes need to be restricted to about ± 20°C. Note that, since velocity changes and temperature changes cause similar effects, a velocity change can be compensated for by changing the device temperature, provided the response is sufficiently sensitive to temperature. This is often feasible for lithium niobate substrates.

For interdigital devices on lithium niobate substrates, difficulties can occur due to the relatively large velocity perturbation, typically 1%, due to electrical loading. A repeatable velocity error can be compensated for by scaling the transducer geometry; however, the velocity depends on the electrode widths, and for large TB-products the required tolerance on the widths can be difficult to achieve. In addition, the variation of electrode pitch along the transducer causes the fractional error in the widths to vary, since narrower electrodes are generally more subject to fabrication errors than wider ones. This causes the device phase error to have a cubic component, and hence the output waveform has a spurious sidelobe which may be unacceptable [302]. This type of error cannot be compensated for by changing the temperature.

(*b*) **Non-linear Chirps.** The methods described above can be applied to non-linear chirps, but for brevity the details will not be given here. For a temperature change or a non-dispersive velocity change, the phase error is given by equation (9.78), and for Hamming weighting the group delay $\tau_g(\omega)$ is given by equation (9.34). The phase error is found to have a cubic component, giving rise to a spurious sidelobe. In consequence, a non-linear chirp is found to be much less tolerant of errors than a linear chirp.

A study of phase errors in non-linear chirps, including surface-wave dispersion due to mechanical loading, is reported by Morgan and Deacon [306].

(*c*) **Matched Pairs of Devices.** In many radar systems a matched pair of chirp filters is used, with one device serving as the expander and the other as the compressor, as explained in Section 9.1. In this case the output waveform is affected by errors in both devices. Usually one device will be an up-chirp and the other a down-chirp, and this has the useful feature that much of the phase error is cancelled out. If the ideal group delays of the expander and compressor are respectively $\tau_e(\omega)$ and $\tau_c(\omega)$, then $\tau_e(\omega) + \tau_c(\omega) = \tau_0$, where τ_0 is a constant equal to the sum of the group delays at the centre frequency. If there is a temperature change or a non-dispersive velocity change, and if the change is the same for both devices, the phase error in either device is given by equation (9.78); thus the total phase error is $\Delta\phi(\omega) = -\varepsilon\omega\tau_0$. This simply gives a delay $\varepsilon\tau_0$, without distorting the output waveform. The velocities or temperatures of the two devices may of course be different, in which case the methods described above give the differences allowable if the distortion of the output waveform is to be kept within prescribed limits. Temperature differences can be minimised by locating the devices close to each other, as illustrated in Figure 9.10, and this usually eliminates the need for temperature regulation, even when lithium niobate substrates are used.

9.5.3. Doppler Shifts

A doppler shift, due to motion of the target relative to the radar system, causes phase errors rather similar to those discussed above and thus generally tends to increase the

CHIRP FILTERS AND THEIR APPLICATIONS

time-sidelobe levels and to reduce the output SNR. These effects are often significant in practical systems, and must therefore be considered when the transmitted chirp waveform is designed. Here we consider the effects for linear and non-linear chirps, showing that the latter are generally more sensitive to doppler shifts.

It is usual to analyse doppler effects by assuming that each frequency component of the receiver input waveform is shifted upwards in frequency by an amount ω_d, the doppler frequency. Thus, if $S(\omega)$ is the spectrum of the input waveform for zero doppler, the spectrum for non-zero doppler is $S'(\omega)$, given by

$$S'(\omega) = S(\omega - \omega_d), \quad \text{for } \omega > 0, \tag{9.82}$$

where the doppler frequency is

$$\omega_d = 2V_r\omega_0/c. \tag{9.83}$$

Here ω_0 is the centre frequency of the transmitted waveform, at R.F., and c is the velocity of light. V_r is the radial velocity of the target relative to the radar, taken to be positive if the range is decreasing. This representation is actually an approximation. Strictly [279, p. 64], the doppler effect contracts or expands the waveform along the time axis, causing the spectrum to expand or contract along the ω-axis. However, the above equations usually give a very good approximation because the waveform bandwidth is much less than the R.F. centre frequency ω_0. Moreover, equation (9.82) is convenient because it is valid at the filter input, in the I.F. section of the receiver, as well as at the receiver input, which is at R.F. As a numerical example, if the R.F. centre frequency is $\omega_0/(2\pi) = 10$ GHz, an aircraft target with a radial velocity of Mach 1, about 330 m/sec, gives a doppler frequency ω_d of about $2\pi \times 22$ kHz.

For zero doppler, the filter output waveform $g(t)$ has a spectrum $G(\omega) = S(\omega)H(\omega)$, where $H(\omega)$ is the filter frequency response. For non-zero doppler, the output spectrum is $G'(\omega) = S'(\omega)H(\omega)$. Thus, using equation (9.82) we have

$$G'(\omega) = G(\omega)\frac{S(\omega - \omega_d)}{S(\omega)}, \quad \text{for } \omega > 0. \tag{9.84}$$

Thus the actual output spectrum is the ideal spectrum multiplied by a complex function $S(\omega - \omega_d)/S(\omega)$. We define $A(\omega)$ and $\phi(\omega)$ as the amplitude and phase of $S(\omega)$, so that $S(\omega) = A(\omega) \exp[j\phi(\omega)]$. Equation (9.84) thus becomes

$$G'(\omega) = G(\omega)\frac{A(\omega - \omega_d)}{A(\omega)} \exp[j\Delta\phi(\omega)], \tag{9.85}$$

where $\Delta\phi(\omega) = \phi(\omega - \omega_d) - \phi(\omega)$. In practice the distortion usually arises primarily from the phase error $\Delta\phi(\omega)$, and since the doppler frequency ω_d is small for practical cases the phase error can be well approximated by

$$\Delta\phi(\omega) = -\omega_d\frac{d\phi(\omega)}{d\omega} \approx \omega_d T_s(\omega), \tag{9.86}$$

where $\phi(\omega)$ is the ideal phase. Here $T_s(\omega)$ is the stationary-phase point of the ideal input waveform, which can be equated with $-d\phi(\omega)/d\omega$ in the stationary-phase approximation, as shown by equation (9.13). If the input waveform is generated by impulsing an expander, the term $-d\phi(\omega)/d\omega$ is equal to the group delay $\tau_g(\omega)$ of the expander.

Doppler Shift for Linear Chirps. For a linear chirp, the spectral phase $\phi(\omega)$ is, in the stationary phase approximation, a quadratic function of ω. The phase $\phi(\omega)$ is given by equation (9.16b) and $T_s(\omega)$ is given by equation (9.15). Using either of these in equation (9.86), equation (9.85) gives

$$G'(\omega) \approx G(\omega) \frac{A(\omega - \omega_d)}{A(\omega)} \exp\left[\frac{j\omega_d(\omega - \omega_c)}{2\pi\mu}\right], \tag{9.87}$$

where μ is the chirp slope of the ideal waveform. For small doppler shifts the amplitude term here is insignificant, while the phase error is equivalent to a delay

$$\tau \approx -\omega_d/(2\pi\mu). \tag{9.88}$$

Thus the doppler shift simply causes a delay of the output waveform, without appreciably distorting it. This is valid irrespective of whether the input waveform or the filter response have amplitude weighting.

If the delay error due to doppler is acceptable it is necessary to consider the amplitude distortion in equation (9.87), and the effect of this depends on the nature of any amplitude weighting in the input waveform or filter response. For example, if the waveform and the filter impulse response both have flat envelopes, it is found [279, p. 134] that the amplitude of the output peak is proportional to $1 - |\omega_d|/(2\pi B)$. This result is intuitively reasonable, since the linear phase error does not affect the peak amplitude, while the bands occupied by the input waveform and the filter response overlap by an amount $2\pi B - |\omega_d|$. If, for example, a 1 dB reduction of output amplitude is acceptable, this corresponds to a doppler frequency of about 11% of the bandwidth. In this case, the output SNR is also reduced by 1 dB, since the noise output power is not affected by the doppler shift.

Doppler Shift for Non-linear Chirps. For a non-linear chirp the phase error $\Delta\phi(\omega)$ of equation (9.86) is more complicated because $T_s(\omega)$ is a non-linear function of ω. To illustrate the effect, we consider the representative case of a Hamming-weighted chirp. For this case the stationary-phase point of the compressor response is given by equation (9.34), and the negative of this is equal to $T_s(\omega)$ in equation (9.86). Equation (9.34) includes a term $\sin(\pi x)$, where $x = (\omega - \omega_c)/(\pi B)$ as before. This is expressed as a sum of Legendre polynomials, so that the phase error of equation (9.86) has the form $\Delta\phi(\omega) = \sum \Phi_n P_n(x)$, as in equation (9.74). Apart from the constant term, the values of the first few coefficients Φ_n are found to be

$$\Phi_1 = \mp 0.6295\omega_d T,$$

$$\Phi_2 = 0,$$

$$\Phi_3 = \pm 0.1570\omega_d T, \tag{9.89}$$

where the upper signs apply when the compressor is an up-chirp, and the lower signs for a down-chirp.

Since $\Phi_2 = 0$, the doppler shift causes no appreciable pulse broadening or loss of SNR. However, the linear term causes a delay $\tau = -\Phi_1/(\pi B)$, which is proportional to ω_d. The cubic term gives a spurious sidelobe, whose level can be obtained from the plot in Figure 9.12. The delay and the sidelobe rejection are shown in Figure 9.13, as functions of ω_d. The delay τ is normalised by expressing it as the product τB. Typically, the sidelobe rejection required would be, say, 30 dB or more, requiring $|\omega_d|T < 1.5$. Thus, if $TB = 100$ for example, the doppler shift must not exceed about $\frac{1}{4}\%$ of the bandwidth B. This is much more stringent than the requirement for linear chirps which, as seen above, can tolerate a doppler shift of typically 10% of the bandwidth. The delay error for a non-linear chirp is usually of no consequence because it is less than the resolution of the system, which is typically $1.5/B$.

Accurate simulations of doppler effects are reported by Newton [295] and by Butler [300]. For these simulations the TB-products were less than 100 and the weighting functions were not the same as the Hamming function, so the analysis give above is, strictly speaking, not applicable. Nevertheless, there is quite reasonable agreement with the results of Figure 9.13.

9.5.4. Other Second-order Effects in Interdigital Devices

As in the case of bandpass filters, chirp filters are usually designed taking account of second-order effects, and it is usual to include a design stage compensating for errors observed experimentally. Several types of error are conveniently compensated for by using the stationary phase approximation, that is, amplitude and phase errors at a particular frequency are compensated by changing the apodisation and position of the electrodes at the corresponding location. This applies to velocity errors, considered above, and to errors due to the circuit effect and surface-wave attenuation.

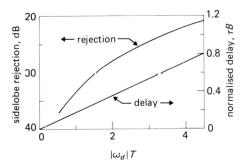

FIGURE 9.13. Effect of doppler shift for Hamming weighted non-linear chirps, in the stationary-phase approximation.

Surface-wave diffraction is often significant, though not if the substrate material is Y, Z lithium niobate, which is a minimal-diffraction orientation. Here again, the stationary-phase approximation gives a useful simplification; for a particular region of the transducer, corrections can be made by considering diffraction only at the frequency equal to the local instantaneous frequency [307]. Diffraction is described in Section 6.2, and design techniques for compensation are discussed in Section 8.4.3.

In long chirp devices, transmission-line effects can be significant [308]. As in a conventional two-wire transmission line, the transducer bus-bars have a distributed inductance L and capacitance C per unit length, and the velocity for electromagnetic waves is $(LC)^{-1/2}$. Typically, L is about $0.6\,\mu H/m$. The capacitance C is primarily associated with the electrodes, and can be large enough to significantly reduce the velocity. The transmission-line effect causes the voltage across the bus-bars to vary along the length of the transducer, and this is sometimes significant even at relatively low frequencies, below 100 MHz. Usually, the transmission-line effect can be adequately suppressed by limiting the transducer aperture, thus minimising the capacitance, though diffraction effects become unacceptable if the aperture is made too small.

9.6. REFLECTIVE ARRAY COMPRESSORS

The Reflective Array Compressor, or RAC, is an alternative type of surface-wave chirp filter, introduced by Williamson and Smith [286, 309–311]. The device relies for its operation on reflection of surface waves by large arrays of grooves, and thus uses physical principles quite different from the interdigital devices described above. In comparison with interdigital devices, RAC's enable much larger time-bandwidth products, up to about 10,000, to be obtained, though the fabrication procedures needed are somewhat more complex and time-consuming.

Section 9.6.1 below discusses the basic nature of the RAC, and makes a more detailed comparison with interdigital devices. Section 9.6.2 considers the analysis and design, including second-order effects, and concludes with a summary of performance. In Section 9.6.3 some modification of the basic form of RAC are discussed, including the use of different types of reflector and different geometries.

9.6.1. Basic Principles

Figure 9.14 shows schematically the commonest type of RAC, using arrays of grooves. Surface waves are generated by a uniform interdigital transducer at one end, and are then reflected by two identical arrays of grooves. Each array reflects the waves through 90°, so that they arrive at a uniform output transducer at the same end as the input. The groove depth is quite small, typically 1% of the wavelength, and consequently the reflection coefficient of any one groove is small. However, at any frequency in the band of the device the overall reflection coefficient is much larger than that of one groove, because reflected waves from many grooves are added coherently. This condition is obtained when the pitch p of the grooves, measured in the propagation direction of the incident waves, is equal or nearly equal to the surface-wave wavelength. In the RAC, the pitch is varied along the length of the

CHIRP FILTERS AND THEIR APPLICATIONS 257

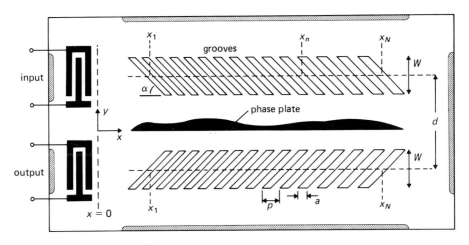

FIGURE 9.14. Reflective Array Compressor (RAC).

device, and consequently waves with different frequencies are reflected at different locations; thus the path length, and hence the delay, varies with frequency. The substrate material is usually Y, Z lithium niobate, where the orientation notation specifies that the input transducer generates waves directed along the Z-axis. This material is favoured because RAC's are particularly suited to large time-bandwidth requirements, where interdigital devices are inapplicable, and for these cases low diffraction and low propagation loss are particularly desirable.

The use of graded-pitch reflective arrays to form dispersive devices was proposed by Sittig and Coquin [312], though not in quite the same configuration as the RAC. A closely related device is the IMCON described by Martin [313], which emerged a few years before the RAC. In this case the wave involved is a non-dispersive bulk shear wave propagating in a parallel-sided steel strip. The two groove arrays are etched on one side of the strip and have the same basic configuration as in the RAC, so that each array reflects the waves through 90°. The waves are launched and received by means of parallel-sided piezoelectric plate transducers bonded on to the end of the strip. The thickness of the strip must be less than about one acoustic wavelength in order to avoid excitation of higher modes, and consequently fabrication considerations limit the IMCON to relatively low frequencies and bandwidths. However, the IMCON can be substantially longer than the RAC, giving larger time dispersion, and large TB-products can be obtained.

As for interdigital devices, amplitude weighting is often necessary in order to suppress the time-sidelobes of the compressed waveform, and to compensate for amplitude variations arising from several causes, for example, the varying groove spacing. In the RAC, amplitude weighting is obtained by varying the depth of the grooves, exploiting the fact that shallow grooves give a reflection coefficient approximately proportional to the depth. This somewhat complicates the fabrication procedure. The grooves are made by exposing the substrate to an ion beam, which slowly etches the surface; to define the geometry, a metal film is deposited first, and the metal is removed at the required groove locations by conventional photolithography. To vary the depths, the ion beam is confined by means of a narrow aperture, and the substrate is drawn past at a varying rate so that different regions are subject to different exposure times.

A useful feature of the RAC is that the phase of the waves propagating between the two groove arrays can be modified by means of a metal film, called the "phase plate", which exploits the velocity reduction due to electrical loading. Since waves of different frequencies traverse the device at different locations, the phase change introduced can be made frequency-dependent by varying the width of the phase plate along its length, as illustrated in Figure 9.14. The phase plate can therefore be used to compensate for phase errors in the frequency response of the device The metallisation is usually of aluminium. Quite commonly, an initial device without a phase plate is fabricated and tested, and a phase plate compensating for the observed phase errors is then designed and fabricated on the device. This enables non-repeatable errors to be compensated, though the fabrication process becomes more complicated.

To ensure that the grooves reflect the waves through 90°, it is necessary to take account of the anisotropy of the substrate. Thus the inclination of the grooves is not generally equal to 45°. From straightforward phase-matching considerations it can be seen that the required inclination, α, is given by

$$\tan \alpha = v_y/v_x. \tag{9.90}$$

Here the x-axis gives the propagation direction of surface waves launched by the input transducer, and α is the groove inclination relative to this axis, as in Figure 9.14; v_x and v_y are the surface-wave phase velocities in the x and y directions. In practical devices the accuracy of the groove inclination is often quite critical. An angular error causes the wavefronts incident on the output transducer to be non-parallel to the transducer electrodes, and hence the output signal level is reduced. Often, an accuracy of 0.1° or better is needed. It is found that velocities calculated from the bulk constants of the material, as in Section 2.3.2, do not give sufficient accuracy, so α-values determined from experimental tests are used for device design. Williamson [310] quotes experimental values for several materials. For example, for Y, Z lithium niobate, $\alpha = 46.82°$ at a temperature of 25°C.

An additional consideration is that the requirement on groove inclination, equation (9.90), must be satisfied with adequate accuracy over a reasonable range of temperatures. Generally, v_x and v_y have different temperature coefficients and, for a particular device, α also varies with temperature because the expansion coefficients in the two directions differ. To satisfy equation (9.90) over a range of temperatures, it is found that the temperature coefficients of delay, $(1/T) \, dT/d\Theta$, should be the same in the x and y directions [310]. These coefficients are generally different, and this applies in particular for ST, X quartz substrates. Thus RAC's on ST, X quartz are quite temperature-sensitive, and so this material is usually avoided. This is in marked contrast to interdigital devices, which give good temperature stability when ST, X quartz is used. For Y, Z lithium niobate the two temperature coefficients of delay are more similar, enabling RAC's to operate successfully over temperature ranges of typically several tens of degrees. The response obtained is of course quite temperature-sensitive because the delays involved vary with temperature; quantitatively, temperature changes cause effects similar to those found in interdigital devices, as discussed in Section 9.5.3. Thus, for individual lithium niobate RAC's temperature regulation is usually mandatory, though for a matched pair comprising an expander and a compressor regulation is often unnecessary.

Comparison with Interdigital Devices. RAC's enable time-bandwidth products up to about 10,000 to be obtained, in contrast to interdigital devices which are limited to *TB*-products below about 1,000. One reason for this is that, for large *TB*-products, the insertion loss of an interdigital device becomes unacceptable because the inactive regions of the transducer act as a capacitive shunt, reducing the impedance. In a RAC, the inactive regions of the groove arrays have little effect and hence the length of the arrays can be increased without appreciably changing the loss. In addition, RAC's are found to give responses less affected by second-order effects, when compared with interdigital devices with similar *TB*-products. This is partly because the RAC structure discriminates against bulk waves generated by the input transducer, since bulk waves reflected by the groove arrays are generally found to have wavefronts not parallel to the electrodes of the output transducer. Further, second-order effects in groove arrays can be made less significant quite simply by reducing the groove depth. The phase plate is another useful feature, enabling phase errors to be reduced as noted above. The RAC also enables a larger time dispersion to be obtained for a given length of substrate, since the surface waves traverse the length of the substrate twice. Thus, a time dispersion up to about 100 μsec can be obtained, while interdigital devices are limited to about 50 μsec. At high frequencies a RAC can be somewhat easier to fabricate than a corresponding interdigital device, because the groove width, approximately $\lambda/(2\sqrt{2})$, is rather larger than the $\lambda/4$ width of the electrodes in a single-electrode transducer; in addition, defects that can cause the electrodes of a transducer to be shorted have little effect on the RAC.

The main disadvantage of the RAC is that a more complex fabrication procedure is needed and, since this is time-consuming, RAC's are relatively costly. Also, since a temperature-stable substrate cannot normally be used, RAC's are generally quite temperature-sensitive.

9.6.2. Analysis and Performance

In analysing RAC's a geometrical complication arises because the wave reflected by a groove in one array illuminates several grooves in the other array, and generally the grooves in the second array are only partially illuminated. In addition multiple reflections can be significant, and there are also some second-order effects associated with the physics of the reflection process. It is convenient to assume initially that all of these effects are negligible, which is the case if the grooves are shallow enough and if the width of the arrays, in the *y*-direction, is only a few surface-wave wavelengths. This gives a formulation convenient for design purposes. The complications mentioned above generally give quite small perturbations and are discussed later.

(a) First-order Analysis. We first consider reflection of surface waves through 90° by a single groove with depth h, assuming that h is much less than the wavelength λ of the incident waves. The amplitude reflection coefficient for this case is known to have the form [311]

$$r_g(\omega) = -2jC\frac{h}{\lambda}\sin(\tfrac{1}{2}ak_x), \tag{9.91}$$

where $k_x = 2\pi/\lambda$ is the wavenumber, a is the groove width measured in the x-direction (the propagation direction of the incident waves), and C is a constant depending only on the substrate material and orientation. This formula has been established experimentally for several substrate materials [311]. For Y, Z lithium niobate, the constant C is found experimentally to be 0.51. The formula is obtained by adding waves reflected at the two edges of the groove, with reflection coefficients $\pm Ch/\lambda$, and the phase of $r_g(\omega)$ is referred to the centre of the groove. The reflection coefficient is maximized when $a = \lambda/2$, which gives $r_g(\omega) = -2jCh/\lambda$. In a RAC, a is usually proportional to the pitch p in Figure 9.14, and since the reflections occur primarily in the region where p equals the wavelength it is a good approximation to take ak_x to be a constant in equation (9.91). Thus $r_g(\omega)$ is proportional to ω.

For convenience, it is assumed here that the two groove arrays in the RAC are identical and have constant width W, as in Figure 9.14. The groove depths h are the same in both arrays, though h will vary along the length of the device. Diffraction and propagation loss are neglected here. The response of the two arrays is described by a function $R(\omega)$, defined as the ratio of the surface-wave amplitude incident on the output transducer to the amplitude of the wave launched by the input transducer. Both amplitudes are measured at the transducer ports, taken to be at $x = 0$. Consider first the wave reflected by groove n of one array, and then by groove n of the other array, where both grooves are centred at $x = x_n$. For this process the contribution to $R(\omega)$ is

$$[r_{gn}(\omega)]^2 \exp(-2jk_x x_n) \exp(-jk_y d),$$

where $r_{gn}(\omega)$ is the reflection coefficient of groove n, $k_y = \omega/v_y$ is the wavenumber in the y-direction and d is the centre-to-centre spacing of the two arrays. If the aperture W were very small, $R(\omega)$ would be given simply by summing the above expression with respect to n. In practice however, the wave reflected by one groove in one array illuminates several grooves in the other array, and the number of grooves illuminated is inversely proportional to the pitch, p. To allow for this we multiply by a factor K/p_n, where K is a constant and p_n is the groove pitch in the vicinity of groove n. Thus

$$R(\omega) \approx \exp(-jk_y d)\sum_{n=1}^{N}(K/p_n)[r_{gn}(\omega)]^2 \exp(-2jk_x x_n), \tag{9.92}$$

where N is the number of grooves in each array. It will be seen later that K can be taken as a constant provided the aperture W is not too large. The overall device has a short-circuit response $H_{sc}(\omega) = H_t^a(\omega) H_t^b(\omega) R(\omega)$, where $H_t^a(\omega)$ and $H_t^b(\omega)$ are the transducer responses.

To simplify further we make use of the stationary-phase approximation. From equation (9.91), the groove reflection coefficient $r_{gn}(\omega)$ is proportional to ωh_n, where h_n is the depth of groove n. At any ω, the reflections occur mainly in the region where the pitch p is approximately equal to the wavelength. Thus, for groove n, ω can be replaced by the instantaneous frequency Ω_n in the region of this groove, where $\Omega_n = 2\pi v_x/p_n$. For convenience we ignore the initial complex exponential in equation (9.92), and also a constant multiplier, giving

$$R(\omega) \propto \sum_{n=1}^{N} \frac{h_n^2 \Omega_n^2}{p_n} e^{-j\omega t_n}, \tag{9.93}$$

where we define times $t_n = 2x_n/v_x$ corresponding to the groove positions. It is assumed that these times are obtained from a time-domain phase function $\theta(t)$, such that

$$\theta(t_n) = 4n\pi. \tag{9.94}$$

This definition is chosen such that the instantaneous frequency $\Omega_i(t) = \dot{\theta}(t)$, evaluated at $t = t_n$, is equal to Ω_n defined above.

Equation (9.93) shows that the response is that of a transversal filter, as in equation (9.67) of Section 9.3.3. Comparing with equations (9.62) to (9.64), it is seen that the fundamental component of the impulse response has the form $a(t)\dot{\theta}(t) \cos[\theta(t)]$, where $a(t)$ is defined such that $a(t_n) = h_n^2 \Omega_n^2/p_n$. It follows that if $\theta(t)$ is the required phase in the time domain, the groove positions are given by $x_n = v_x t_n/2$, where t_n are the solutions of equation (9.94). The frequency response $R(\omega)$ is readily found, approximately, from the stationary-phase approximation. Comparing with equations (9.3) and (9.12a), the magnitude of $R(\omega)$ in the fundamental region is, ignoring a constant multiplier,

$$|R(\omega)| \propto \frac{\omega^4 [h(T_s)]^2}{\sqrt{|\ddot{\theta}(T_s)|}}, \tag{9.95}$$

where $T_s(\omega)$ is the stationary-phase point and the groove depths are expressed by the function $h(t)$, such that $h(t_n) = h_n$. This equation gives the required variation of groove depth. For a linear chirp, $\ddot{\theta}(T_s)$ is a constant. Thus, if $|R(\omega)|$ is required to be flat in the band, as in an expander, h_n should be made proportional to Ω_n^{-2}. In practice it is usually necessary to modify this to allow for the responses of the transducers, and for several second-order effects.

(b) Geometrical Complications.

The above analysis does not properly allow for the fact that waves reflected by a groove in one array illuminate several grooves in the other array. A more accurate analysis can be obtained with the aid of Figure 9.15(a), which illustrates some of the reflections involved. For convenience the

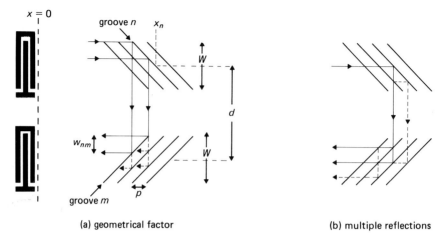

FIGURE 9.15. Factors in RAC analysis.

grooves are here represented as straight lines. It is assumed that the reflection coefficient is small, so that multiple reflections can be neglected and the transmission coefficient of each groove can be taken to be unity. Some surface-wave rays reflected by groove n of the upper array and groove m of the lower array are shown as continuous lines. All such rays have the same phase shift, and are therefore added coherently by the output transducer. Some rays reflected by subsequent grooves in the lower array are indicated by broken lines. Relative to waves reflected by groove m, the reflections from subsequent grooves have phase shifts $k_x p$, $2k_x p$, and so on. Consequently, the lowest frequency at which strong reflections are obtained is the frequency at which the pitch p is approximately equal to the surface-wave wavelength. This condition is generally used in practical devices, though there are also strong reflections at the odd harmonics.

In general, groove n of one array and groove m of the other array will overlap only partially, and in consequence the output beam due to reflection from these grooves has a width less than the width of the groove arrays. This can be seen in Figure 9.15(a), where the two rays shown as continuous lines illustrate the limits imposed by the overlap; thus the output rays, directed toward the output transducer, define the edges of the output beam. Now, the output current produced by a uniform transducer is proportional to the average of the surface-wave amplitude at its input port, as shown in Section 4.7.2, equation (4.126). The output current is therefore proportional to the width w_{nm} of the output beam. To allow for this we define an effective reflection contribution $R_{nm}(\omega)$ proportional to w_{nm}. Taking the two groove arrays to be identical and to have width W, $R_{nm}(\omega)$ is given by

$$R_{nm}(\omega) = \frac{w_{nm}}{W} r_{gn}(\omega) r_{gm}(\omega) \exp[-jk_x(x_n + x_m)] \exp(-jk_y d), \quad (9.96)$$

where $r_{gn}(\omega)$ is the reflection coefficient of groove n in either array. The total response $R(\omega)$ of the groove arrays is obtained by summing $R_{nm}(\omega)$ over both n and m, though the sum over m needs to include only those grooves in the lower array that overlap groove n in the upper array. The output beam width w_{nm} is equal to $W - |x_n - x_m|$ for $|x_n - x_m| < W$, and is zero otherwise; this follows by taking $\alpha = 45°$, a good approximation here.

The above formulation is quite time-consuming to compute, owing to the double summation required. A convenient approximate form for $R(\omega)$ has been derived by Martin [313], making use of the stationary-phase approximation. The result is expressed in terms of a parameter

$$N_{\text{eff}}(\omega) = \frac{\omega}{2\pi\sqrt{2|\ddot{\theta}(T_s)|}}. \tag{9.97}$$

We can regard $N_{\text{eff}}(\omega)$ as the effective number of grooves in either array that are active at frequency ω. Martin's derivation shows that the magnitude of the response is approximately [310, 313]

$$|R(\omega)| \approx |N_{\text{eff}}(\omega) r_g(\omega)|^2 g\left(\frac{W}{\lambda N_{\text{eff}}}\right), \tag{9.98}$$

where $r_g(\omega)$ is the reflection coefficient for a groove in the region active at frequency ω. The function $g(z)$ is a geometrical factor allowing for the fact that reflections from many grooves are involved, and for $z \leq 1$ is well approximated by $g(z) \approx z/\sqrt{2}$. When this is valid, that is, when W is less than the effective length λN_{eff}, the form of equation (9.98) is the same as that of the result derived earlier, equation (9.95). This confirms the validity of the assumption that K can be taken to be a constant in equation (9.92). However, for $W > \lambda N_{\text{eff}}$, that is, for $z > 1$, $g(z)$ increases more slowly with z and the results given earlier are invalid. The slower increase of $g(z)$ with z is due to the fact that a ray propagating in the y-direction encounters grooves whose pitch is varying, so that the phases of the reflected waves are not quite the same. For $z > 1$ the phase differences involved are sufficient to cause a significant reduction in the amplitude of the output beam.

(c) Multiple Reflections. The term "multiple reflections" refers to waves that are successively reflected two or more times in one or both of the groove arrays. This is illustrated in Figure 9.15(b), where rays due to first-order reflections are shown as continuous lines and those due to multiple reflections are shown as broken lines. The multiple reflections, ignored in the analysis described above, are negligible if the groove reflectivity is low enough.

Multiple reflections in one-dimensional structures, such as unapodised transducers or multi-strip couplers, can be analysed by using a transmission matrix for each element and cascading the matrices, as shown in Appendix E. For RAC's, the two-dimensional nature of the problem introduces considerable extra complexity. Otto [314] has developed a two-dimensional analysis for a periodic array of grooves, representing it as a two-dimensional array of cells with each cell small enough for multiple reflections within it to be neglected. A two-dimensional transmission matrix was used for each cell. The application of this method to non-periodic grooves is described by Bloch et al. [315]. Alternatively, Wright and Haus [316] have developed an analysis based on coupled mode theory, and this enables the response to be computed more easily.

In a RAC, it is found that the main effect of multiple reflections is to reduce the amplitude of the frequency response $R(\omega)$ by an amount that varies slowly with frequency [317]. The groove depths can be modified in order to compensate this. However, it is also found [317] that multiple-reflection effects are generally insignificant if the total reflection loss exceeds about 15 dB, that is, if $20 \log |R(\omega)| < -15$. In practical devices this condition is nearly always acceptable, and if so multiple reflections need not be considered.

(d) Other Second-order Effects. The interaction between surface waves and grooves is found to be accompanied by some additional second-order effects, associated with the physical processes involved. The most important effect is termed *stored energy*. It is found that a periodic array of grooves causes a slight reduction of the surface-wave velocity, by an amount dependent on the groove depth h. This phenomenon is attributed to evanescent acoustic fields which store energy in the regions near the groove edges, but do not cause any loss of power. The perturbation can be expressed in terms of a surface-wave phase shift $\Delta\phi = -\hat{B}/2$ at each groove edge. This gives a perturbed wavenumber $k = k_0 + \hat{B}/p$, where k_0 is the free-surface wavenumber and p is the pitch, which in a RAC varies with distance. From experimental measurements [311] the parameter $\hat{B}$ is known to have the form

$$\hat{B} = 2C'(h/\lambda)^2, \tag{9.99}$$

where C' is a material-dependent constant, equal to 4.5 for a RAC on Y, Z lithium niobate. Thus, $\hat{B}$ is proportional to ω^2.

For a RAC, the main consequence of stored energy is that a phase error arises in the device frequency response. This error can be estimated using the stationary-phase approximation [306, 311], though some care is necessary since the perturbed velocity is frequency-dependent and varies with position. For a linear chirp device the error is found to include a cubic term of the form $\Phi_3 P_3(x)$, causing a spurious sidelobe to appear when the device is used for pulse compression, as explained in Section 9.5.1. In large RAC's Φ_3 is often 50° or more. The error can of course of corrected by means of a phase plate, or alternatively the groove positions can be adjusted. Another consequence of stored energy is that a groove array is found to give responses at the even harmonics, not predicted by the first-order analysis.

Propagation loss is often significant in RAC's, and can be allowed for by adjusting the groove depths. The loss for free-surface propagation is described in Section 6.3. It is also found that the grooves cause some additional loss, attributed to excitation of non-coherent surface or bulk waves, called the "non-synchronous scattering loss" [318]. The attenuation coefficient due to this effect is dependent on position. In large RAC's, both types of loss can cause the amplitude of the frequency response to vary by 10 dB or more, so it is important to design the device to compensate for them.

For most RAC's, using Y, Z lithium niobate substrates, diffraction is not

CHIRP FILTERS AND THEIR APPLICATIONS

significant. However, other substrate materials are sometimes used, and it is then usually necessary to design the device compensating for diffraction.

(e) Performance. Figure 9.16 shows the performance of a linear-chirp RAC with 200 MHz centre frequency, using a Y, Z lithium niobate substrate. The bandwidth was 80 MHz and the time dispersion 80 μsec, giving a time-bandwidth product of 6400. The device was designed to have a flat amplitude characteristic in the frequency demain, and to obtain this the groove depths were varied from about 500 to 2000 Å. The overall insertion loss was 38 dB. A phase plate was used to minimise the phase error. Ideally, the phase of the frequency response is a quadratic function of frequency, and the phase error curve in Figure 9.16 shows the result obtained when a quadratic was subtracted from the measured phase. The error is less than $10°$ at most frequencies. It should be remembered that the ideal phase varies very rapidly with frequency; in the present case the variation over the band is some 6×10^5 degrees, so the phase error corresponds to an accuracy of a few parts in 10^5. In laboratory tests this device was used as an expander in conjunction with a second RAC used as a compressor, with the latter having amplitude weighting, and a time-sidelobe suppression of 33 dB was obtained. The compressor had a bandwidth of 40 MHz, half that of the expander, and therefore had a time-dispersion of 40 μsec and a time-bandwidth product of 1600. The difference in the bandwidths is due to the intended application of the devices in a compressive receiver, as explained in Section 9.7 below.

Some examples of RAC and IMCON performance are summarised in Table 9.2, where the fifth column refers to the device described above. For the remaining columns, the time-sidelobe rejection refers to a pair of devices with the same bandwidth. It can be seen that RAC's can give very large time-bandwidth products,

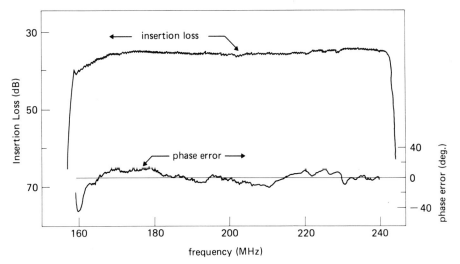

FIGURE 9.16. Frequency response of a RAC with 80 MHz bandwidth and 80 μsec dispersion. (Courtesy Plessey Research)

TABLE 9.2

Performance examples for reflective array compressors. All entries refer to linear-chirp devices

Type[†]	IM	IM	R	R	R	R	R	L	IL	RDA	RDA, IL
Centre frequency (MHz)	20	15	60	1000	200	400	60	1000	150	60	150
Bandwidth, B (MHz)	10	6	6	512	80	180	2.5	500	50	20	50
Duration, T (μsec)	250	100	100	10	80	90	125	1	20	10	20
TB	2500	600	600	5120	6400	16,200	312	500	1000	200	1000
Time-sidelobe rejection, dB	—	48	37	—	33*	—	33	—	30	26	—
C.W. insertion loss (dB)	—	30	34	52	38	56	33	45	47	56	—
Substrate[‡]	Steel	Steel	LN	LN	LN	LN	BGO	LN	LN	LN	LN
Reference	[313]	[319]	[320]	[310]	*	[335]	[321]	[322]	[323]	[324]	[325]

* See text.
[†] R = conventional RAC; IM = IMCON; L = length weighted; IL = in-line RAC; RDA = reflective dot array.
[‡] LN = lithium niobate, BGO = bismuth germanium oxide.

16,200 for one device, and the dispersion T is typically up to 100 μsec. One column refers to devices using bismuth germanium oxide substrates, which give an unusually low surface-wave velocity and thus enable a larger time dispersion to be obtained. The last four columns refer to some novel types of RAC, discussed in the next Section.

9.6.3. Other Types of RAC

While the results discussed above show RAC's to be capable of impressive performance, there are some practical disadvantages. Notably, the time required to etch an array of grooves with varying depth renders the devices relatively expensive, and the devices are quite temperature sensitive. Some modified forms of RAC have been developed to overcome these disadvantages, though with some loss of performance.

The fabrication time required can be reduced substantially if all the grooves have the same depth, since this avoids the need to constrain the ion beam by a narrow aperture. However, this implies that some novel method of weighting is necessary. Figure 9.17 shows two possibilities, for which experimental data are included in Table 9.2. In the length-weighted RAC [322] the groove lengths are varied, and "dummy" grooves at a different angle are included so that the velocity perturbation is uniform. The in-line RAC [323] uses a 3 dB multi-strip coupler to partition the input wave between two arrays of grooves, which reflect it through 180°. Amplitude weighting is obtained by slightly displacing the grooves in one track relative to those in the other track, though length weighting would alternatively be applicable.

Further simplification of the fabrication can be obtained by using metal structures to reflect the waves instead of grooves, so that the etching procedure can be omitted altogether. Metal strips can sometimes be used, but for lithium niobate substrates this is not feasible because the electrical loading causes too strong a perturbation. To overcome this, Solie [324] introduced the "reflective dot array" in which each reflector is divided into a large number of metal segments, known as dots. This also enables weighting to be obtained by varying the density of the dots. Table 9.2 includes some results, and also some data for an in-line version of the dot array device [325].

Continuous metal strips can be used as reflectors on some substrates, as shown by a matched pair of devices on quartz [326] with a TB-product of 500. These devices used 90° reflections and were length-weighted. A time-sidelobe suppression of 28 dB was demonstrated.

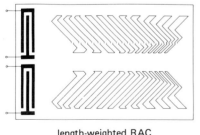
length-weighted RAC

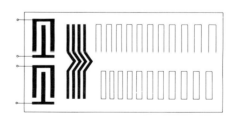
in-line RAC

FIGURE 9.17. RAC's using grooves of uniform depth.

9.7. SPECTRUM ANALYSIS AND OTHER TYPES OF SIGNAL PROCESSING

In Section 9.1 it was seen that chirp filters play a central role in the operation of pulse compression radar systems. In this section we consider a number of other applications, mostly for filters whose impulse responses are linear chirps. The most prominent of these is a system for frequency measurement, known as the *compressive receiver*. It was shown in Section 9.5.3 that, for linear chirps, the main effect of a doppler shift on a pulse-compression system is to delay the output pulse. The compressive receiver uses a similar principle to measure the frequency of a C.W. input signal; the delay of the output pulse is measured in order to determine the input frequency. The delay is a linear function of frequency, and the system is also linear in the sense that a signal comprising several C.W. components can be analysed, giving output pulses corresponding to the individual input components. From this it can be concluded that, for a more general form of input signal, the output waveform must be closely related to the Fourier transform of the input, that is, to its spectrum. In fact, Fourier transformation can be achieved using a system similar to the compressive receiver, though three chirp filters are required rather than two. This enables a complex function to be transformed, presenting the complex output as an amplitude and phase modulated waveform, or as a pair of waveforms. In both the compressive receiver and the Fourier transform system, the time taken to perform the transform is typically a few times $1/\Delta f$, where Δf is the frequency resolution. Thus the transformation is achieved much more rapidly than in a conventional spectrum analyser, which mixes the signal with a slowly-swept local oscillator and applies the output to a narrow-band filter.

The development of such systems has been given considerable impetus by the high performance standards achievable using surface-wave devices, particularly RAC's. However, the principles do not depend on the technology. Thus any of the technologies suitable for chirp filters, mentioned in Section 9.1, would be applicable subject to performance constraints. In fact, the compressive receiver and Fourier transform systems have been studied extensively using digital techniques and using charge-coupled devices, with modifications owing to the fact that sampled data are used.

Section 9.7.1 below gives a descriptive account of the compressive receiver. The basic mathematical relations involved are given in Section 9.7.2, which includes systems for Fourier transformation, and experimental results are discussed in Section 9.7.3. Section 9.7.4 considers a variety of other signal processing functions that can be accomplished using chirp filters, including a variable delay line and a chirp filter system with variable chirp rate.

9.7.1. Compressive Receiver Principles

The principle involved here is closely related to the observation by Klauder *et al.* [277, p. 761] that a pulse compression system can be used to perform Fourier transformation. Subsequently, Darlington [327] showed in detail how the method could be applied to frequency measurement, and Darlington's system is now called

CHIRP FILTERS AND THEIR APPLICATIONS 269

a compressive receiver, or sometimes microscan receiver. Recently, reviews of the topic have been given by Jack et al. [328, 329] and by Roberts et al. [330].

The system is illustrated in Figure 9.18. The input signal, taken to be C.W. with angular frequency Ω, is mixed with a linear-chirp waveform obtained by impulsing an expander, and the resulting output is applied to a compressor with a chirp slope matching that of the expander. For the moment, both chirp filters are taken to have impulse responses with flat envelopes. The lower part of the figure shows the operation of the system in terms of the frequency-time curves of the devices; the curves for the expanded pulse are shown as they appear at the compressor input, and thus shift in frequency according to the value of Ω. It is assumed that the frequency band of the signal at the compressor input overlaps completely the band occupied by the compressor response. To ensure this, the centre frequencies of the two filters must be different and the expander bandwidth must be larger than that of the compressor. In addition, Ω must be constrained to a finite band, equal to the difference between the two device bandwidths. The curves in Figure 9.18 refer to the two extreme values of Ω, which are denoted Ω_1 and Ω_2.

For a particular input frequency, the compressor output waveform has an envelope given approximately by sinc $(\pi B_c t)$, where B_c is the compressor bandwidth. This is similar to the output of an unweighted linear-chirp pulse compression system. However, it can be seen in Figure 9.18 that the expanded pulse sweeps through the band of the compressor at times which vary according to the value of Ω. Thus the time of the output pulse varies with the input frequency, and the relationship is linear because linear chirps are used. The input frequency can therefore be found by detecting the output pulse and measuring its delay relative to the pulse used to excite the expander.

Important parameters are the frequency resolution of the system and the number of resolvable points. If μ is the chirp slope of the expander, the time τ of the output

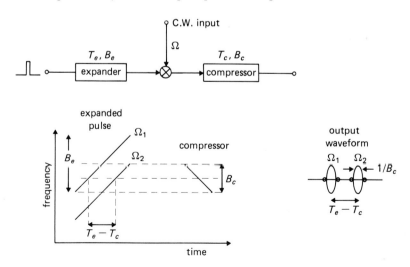

FIGURE 9.18. Compressive receiver.

pulse is given by $\Omega = -2\pi\mu\tau + $ constant. The width of the output pulse is approximately $1/B_c$, and hence the resolution is $\Delta f = |\mu|/B_c = 1/T_c$, where T_c is the length of the compressor impulse response. The range of acceptable input frequencies is $(B_e - B_c)$, where B_e is the expander bandwidth, and hence the number of resolvable points is $(B_e - B_c)T_c$. For technological reasons, it is usually advisable to use the minimum possible value of the expander time-bandwidth product, $T_e B_e$. This value can be deduced by writing $B_c = kB_e$, from which it follows that the number of points is $(k - k^2)T_e B_e$. Differentiating this shows that, for a given number of points, $T_e B_e$ is minimised by taking $k = \frac{1}{2}$, so that the expander bandwidth is twice that of the compressor. The number of points is then $T_e B_e/4$.

Provided the multiplier used is linear with respect to the signal input the system behaves linearly, so that for an input signal comprising several C.W. components the output gives pulses at the corresponding times. However, the output does not in this case give the Fourier transform of the input; it corresponds instead to a function known as the *sliding* Fourier transform, as will be shown in Section 9.7.2 below. In practice, this distinction is not usually very consequential. For the multiplier a balanced mixer is generally used, and for optimum dynamic range one of the inputs, the "local oscillator" input, is driven into saturation. For the compressive receiver the expanded pulse is applied to the local oscillator input, and saturation is acceptable here because this pulse has a flat envelope.

In practical systems it is usual to apply amplitude weighting in the compressor so as to reduce the time-sidelobes of the output pulse, as described for pulse-compression systems in Section 9.2.3. This enables the system to resolve two C.W. components of a signal, with closely similar frequencies but substantially different power levels. In comparison with an unweighted system, this incurs some loss of resolution and a corresponding reduction in the number of points, but the device bandwidths can be increased to compensate for this.

It has been assumed so far that the expander has a bandwidth larger than that of the compressor. An alternative arrangement, with the compressor having the larger bandwidth, operates in a very similar manner. In this case the expander can be weighted to reduce the time-sidelobes. However, this implies that the multiplier used must be linear with respect to both input ports, and this imposes technological constraints which limit the performance obtainable [329]. For this reason, the expander usually has the wider bandwidth in practical compressive receivers.

9.7.2. Analysis of Compressive Receivers and Fourier Transform Systems

Here the compressive receiver is considered in more detail, and in view of the close similarity we also consider systems for Fourier transformation. Both cases are related to a fundamental concept which expresses the Fourier transform in terms of chirp waveforms. Consider a waveform $f(t)$, with Fourier transform $F(\omega)$. Writing τ instead of t, and $2\pi\mu t$ instead of ω, we have

$$F(2\pi\mu t) = \int_{-\infty}^{\infty} f(\tau) \exp(-j2\pi\mu t\tau) \, d\tau \qquad (9.100)$$

Using the identity $2t\tau = t^2 + \tau^2 - (t - \tau)^2$, this becomes [331]

$$F(2\pi\mu t) = \exp(-j\pi\mu t^2) \int_{-\infty}^{\infty} [f(\tau) \exp(-j\pi\mu \tau^2)] \exp[j\pi\mu(t-\tau)^2] d\tau \quad (9.101)$$

This form of the Fourier transform is called *chirp transform*. An equivalent time-discrete form, suitable for digital or charge-coupled devices, is known as the chirp *z* transform [332, 237]. The chirp transform, equation (9.101), can be seen to be equivalent to multiplication by a complex chirp, convolution with another chirp, and finally multiplication by a third chirp. Consequently, this is called the M–C–M scheme. If we define

$$h_e(t) = h_m(t) = \exp(-j\pi\mu t^2),$$
$$h_c(t) = \exp(j\pi\mu t^2), \quad (9.102)$$

then equation (9.101) becomes

$$F(2\pi\mu t) = h_m(t)[\{f(t)h_e(t)\} * h_c(t)]. \quad (9.103)$$

Figure 9.19 shows a representation of this process in terms of chirp filters. The waveforms $h_e(t)$ and $h_m(t)$ are obtained by impulsing chirp filters having $h_e(t)$ and $h_m(t)$ as their impulse responses. The convolution with $h_c(t)$ is obtained by applying the signal to a filter with impulse response $h_c(t)$. The final multiplication by $h_m(t)$ changes the phase of the output waveform, but not its amplitude. Thus, if only the amplitude of the spectrum is required, the final multiplication can be omitted. In this case, the system becomes essentially the same as the compressive receiver of Figure 9.18. The representation in Figure 9.19 is actually a rather naive interpretation of equation (9.103), because in practice the device impulse responses must be real and must have finite lengths. A more detailed analysis allowing for these factors is given below.

A dual form of the system, shown in Figure 9.20, involves convolution followed by multiplication and then another convolution, and is called the C–M–C scheme. The

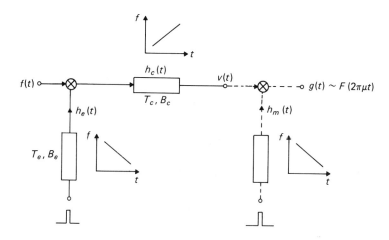

FIGURE 9.19. M–C–M Fourier transform system.

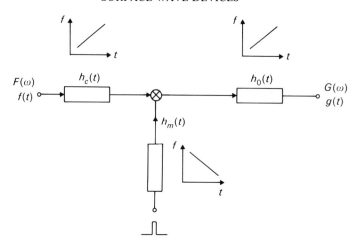

FIGURE 9.20. C–M–C Fourier transform system.

input $f(t)$ is convolved with $h_c(t)$, multiplied by $h_m(t)$, and then convolved with $h_0(t)$. To show that this gives the Fourier transform, we consider the waveforms in the frequency domain. Define $H_c(\omega)$, $H_m(\omega)$ and $H_0(\omega)$ as the filter frequency responses, that is, the transforms of $h_c(t)$, $h_m(t)$ and $h_0(t)$ respectively. The multiplication by $h_m(t)$ becomes, in the frequency domain, a convolution with $H_m(\omega)$, as shown by the convolution theorem. Thus if $F(\omega)$ is the spectrum of the input waveform, the spectrum $G(\omega)$ of the output waveform is

$$G(\omega) = \frac{1}{2\pi} H_0(\omega)[\{F(\omega)H_c(\omega)\} * H_m(\omega)], \qquad (9.104)$$

which is analogous to equation (9.103). To show that this gives the Fourier transform of the input, note that the transform of $\exp(j\pi\mu t^2)$ is proportional to $\exp(-j\alpha\omega^2)$, with $\alpha = 1/(4\pi\mu)$. We can thus take $H_c(\omega) = H_0(\omega) = \exp(-j\alpha\omega^2)$ and $H_m(\omega) = \exp(j\alpha\omega^2)$, ignoring multiplying constants. Substituting these into equation (9.104) gives

$$G(\omega) = f\left(-\frac{\omega}{2\pi\mu}\right), \qquad (9.105)$$

where $f(t)$ is the original input waveform. From the definition of the Fourier integral, it then follows that the output waveform is $g(t) = \mu F(2\pi\mu t)$, which is essentially the same as the output given by the M–C–M scheme.

Analysis for Finite Length Devices (M–C–M scheme). To appreciate the system operation in more detail, we must take account of the fact that the device impulse responses have finite lengths. For brevity, only the M–C–M scheme is considered here. The following account is a summary of that given by Jack et al.

CHIRP FILTERS AND THEIR APPLICATIONS

[328, 329], simplified by taking the filter impulse responses to be centred at $t = 0$. In practice, causality requires each impulse response to be centred at some positive t, so that it is zero for $t < 0$, but this makes no difference to the output waveform except for a constant delay.

In the M–C–M system of Figure 9.19, the compressor output waveform $v(t)$ is

$$v(t) = [f(t)h_e(t)] * h_c(t) \qquad (9.106)$$

and this also gives the output of a compressive receiver. The expander impulse response is taken to have a center frequency ω_e, chirp slope $-\mu$ and duration T_e, so that

$$h_e(t) = \text{rect}(t/T_e) \cos(\omega_e t - \pi\mu t^2). \qquad (9.107)$$

For the compressor, ω_c and T_c are defined similarly, so that

$$h_c(t) = \text{rect}(t/T_c) \cos(\omega_c t + \pi\mu t^2), \qquad (9.108)$$

where the chirp slope is μ. Substituting into equation (9.106) gives

$$v(t) = \int_{-\infty}^{\infty} f(\tau) \text{rect}(\tau/T_e) \cos(\omega_e \tau - \pi\mu\tau^2) \text{rect}\left(\frac{t-\tau}{T_c}\right)$$

$$\times \cos[\omega_c(t-\tau) + \pi\mu(t-\tau)^2] d\tau. \qquad (9.109)$$

Here the product of cosines gives cosines of the sum and difference of the arguments. However, the difference term can be neglected because the waveforms involved can be taken to be band-pass functions; the difference term arises from the negative-frequency part of $h_e(t)$ and the positive-frequency part of $h_c(t)$ and vice versa. Neglecting this term, we have

$$v(t) = \frac{1}{2} \int_{-\infty}^{\infty} f(\tau) \text{rect}(\tau/T_e) \text{rect}\left(\frac{t-\tau}{T_c}\right) \cos(\omega_c t + \pi\mu t^2 - \Omega\tau) d\tau, \qquad (9.110)$$

where Ω is a linear function of t, defined by

$$\Omega(t) = \omega_c - \omega_e + 2\pi\mu t. \qquad (9.111)$$

We now consider separately the case when the expander has the shorter impulse response ($T_e < T_c$), and the case when the expander is longer ($T_e > T_c$). As explained earlier, the latter case is usually chosen for a compressive receiver.

Case 1: Short Expander. It is first necessary to consider the role of the rect functions in equation (9.110). If all the other functions in the integrand are replaced by unity, it is found that, for $T_e < T_c$, the waveform $v(t)$ has a flat region for $|t| < \frac{1}{2}(T_c - T_e)$ and sloping regions of length T_e at each end, so that the total duration of the output waveform is $T_c + T_e$. The flat region corresponds to the time interval when the expanded pulse is entirely "inside" the compressor. Clearly, the system can only be expected to give a Fourier transform for times corresponding to the flat region, so only this region is considered here.

For this region, that is, for $|t| < \frac{1}{2}(T_c - T_e)$, it is found that the presence of the first rect function in equation (9.110) implies that the second rect function can be omitted. We thus find

$$v(t) = \tfrac{1}{4} \exp(j\omega_c t + j\pi\mu t^2) \int_{-\infty}^{\infty} f_1(\tau) \exp(-j\Omega\tau) d\tau + \text{c.c.}, \quad (9.112)$$

where "c.c." indicates complex conjugate and we define

$$f_1(t) = f(t) \operatorname{rect}(t/T_e), \quad (9.113)$$

which is a segment of the input waveform $f(t)$, with length T_e. The integral in equation (9.112) is clearly equal to $F_1(\Omega)$, where $F_1(\omega)$ is the Fourier transform of $f_1(t)$. If we define $\Phi_1(\omega)$ as the phase of $F_1(\omega)$, so that $F_1(\omega) = |F_1(\omega)| \exp[j\Phi_1(\omega)]$, then the compressor output is

$$v(t) = \tfrac{1}{2}|F_1(\Omega)| \cos[\omega_c t + \pi\mu t^2 + \Phi_1(\Omega)], \quad \text{for } |t| < \tfrac{1}{2}(T_c - T_e). \quad (9.114)$$

This shows that the envelope of the output waveform is proportional to the magnitude of the spectrum of $f_1(t)$, given by equation (9.113). The spectral frequency Ω is linearly related to the time-axis of the output waveform, as shown by equation (9.111). Note that, since the system only transforms a segment of input waveform of length T_e, the frequency resolution, in Hz, is approximately $1/T_e$. In addition, the output gives the spectrum only over a finite band of frequencies. This band, of width $\Delta\Omega$, can be deduced from the time interval for which equation (9.114) is valid, substituting into equation (9.111) to give $\Delta\Omega = 2\pi(B_e - B_c)$. The chirp slope μ in the above equations can be positive or negative; in the latter case, the output gives the inverse Fourier transform instead of the forward transform.

Although the input waveform $f(t)$ must be real, the system can in fact transform complex data. This is easily seen by writing $f(t) = f_0(t) \exp(j\omega_0 t) + \text{conjugate}$, where $f_0(t)$ is a complex function and ω_0 is a carrier frequency. If $F_0(\omega)$ is the transform of $f_0(t)$, the transform of $f(t)$ includes a term $F_0(\omega - \omega_0)$. The system can therefore be used to determine $|F_0(\omega - \omega_0)|$, from which $|F_0(\omega)|$ follows.

If the phase of the spectrum is required in addition to its amplitude, an additional multiplication by a chirp waveform $h_m(t)$ is included, as shown in Figure 9.19. This removes the quadratic phase term $\pi\mu t^2$ in the compressor output of equation (9.114). The chirp slope of $h_m(t)$ is $-\mu$, and its duration must be at least $(T_c - T_e)$. If ω_m is the centre frequency of $h_m(t)$, the output waveform has the form

$$g(t) = \tfrac{1}{4}|F_1(\Omega)| \cos[(\omega_c + \omega_m)t + \Phi_1(\Omega)], \quad \text{for } |t| < \tfrac{1}{2}(T_c - T_e),$$

$$(9.115)$$

where it is assumed that a difference-frequency term has been eliminated by appropriate filtering. It is usual to demodulate this waveform using two synchronous detectors in phase quadrature, with reference frequency $\omega_c + \omega_m$. With the phases properly adjusted this gives outputs of the form $|F_1(\Omega)| \cos[\Phi_1(\Omega)]$ and $|F_1(\Omega)| \sin[\Phi_1(\Omega)]$, the real and imaginary parts of $F_1(\Omega)$, as noted by Darlington [327].

CHIRP FILTERS AND THEIR APPLICATIONS   275

Case 2: Long Expander. For $T_e > T_c$, which is usually the case for a compressive receiver, we again deduce the compressor output $v(t)$ from equation (9.110). For this case the rect functions alone give a flat region for $|t| < \frac{1}{2}(T_e - T_c)$. The total length of the output waveform is $T_e + T_c$, as before. For the interval $|t| < \frac{1}{2}(T_e - T_c)$, the first rect function in equation (9.110) can be omitted, giving

$$v(t) = \tfrac{1}{4} \exp(j\omega_c t + j\pi\mu t^2) \int_{-\infty}^{\infty} f_2(\tau, t) \exp(-j\Omega\tau) d\tau + \text{c.c.}, \quad (9.116)$$

where

$$f_2(\tau, t) = f(\tau) \operatorname{rect}\left(\frac{t - \tau}{T_c}\right), \quad (9.117)$$

and $\Omega(t)$ is given by equation (9.111) as before. The integral in equation (9.116) is denoted $F_2(\Omega, t)$, the transform of $f_2(\tau, t)$ from the τ-domain to the Ω-domain. If $\Phi_2(\Omega, t)$ is the phase of $F_2(\Omega, t)$, we have

$$v(t) = \tfrac{1}{2}|F_2(\Omega, t)| \cos[\omega_c t + \pi\mu t^2 + \Phi_2(\Omega, t)], \quad \text{for } |t| < \tfrac{1}{2}(T_e - T_c). \quad (9.118)$$

This is similar to the short-expander case, but here the transform obtained does not quite correspond to the Fourier transform of the input. Transforming equation (9.117) shows that $F_2(\Omega, t)$ can be written

$$F_2(\Omega, t) = \int_{t - T_c/2}^{t + T_c/2} f(\tau) \exp(-j\Omega\tau) d\tau \quad (9.119)$$

This shows that, for any t, the output gives one point in the spectrum of a segment of the input $f(t)$. However, the location of the segment is varying with t, and so equation (9.119) is called the *sliding* Fourier transform. Generally, the Fourier components of an input signal cannot be deduced from the sliding transform. However, if the signal comprises a finite sum of C.W. waveforms, as is often the case for compressive receiver applications, the magnitude of the output given by the sliding transform is the same as that given by the conventional transform; the distinction is then of no significance, provided the spectral phase is not required.

9.7.3. Experimental Compressive Receivers and Fourier Transform Systems

We consider here a few of the many experimental examples described in the literature. RAC's are often used since, for a large number of resolvable points, a large TB-product is needed. With TB-products up to about 10,000, the number of points can be up to about 2500. For such RAC's the time dispersion is typically 20 to 100 μsec, and hence the frequency resolution of the system is typically 20 to 100 kHz.

Figure 9.21 shows the output of a compressive receiver for several C.W. input signals with different frequencies, where the traces have been displaced vertically for clarity. The system used two RAC's, an expander with 80 MHz bandwidth and 34 μsec dispersion, and a compressor with 40 MHz bandwidth and 17 μsec dispersion. The compressor incorporated amplitude weighting, giving a time-sidelobe

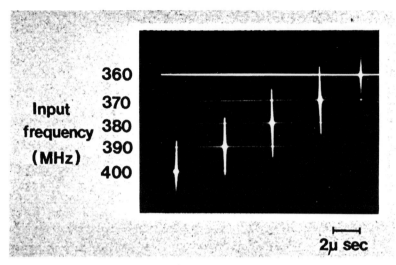

FIGURE 9.21. Performance of a compressive receiver using RAC's (Courtesy, Plessey Research).

suppression of 34 dB. This system enabled signals over a 40 MHz band to be analysed, with a resolution of 75 kHz.

Some other examples are given in References [333–340]. The performance achievable using RAC's is well illustrated by the system of Gerard et al. [335], which transformed signals with up to 60 MHz bandwidth and 60 μsec duration, giving a frequency resolution of 17 kHz. For this system the largest RAC had an impressive time-bandwidth product of 16,200, with other parameters as shown in Table 9.2. A Fourier transform system with digital input and output, incorporating D-to-A and A-to-D convertors, was demonstrated by Dolat et al. [336], and analysed a 10 MHz band with 40 kHz resolution. This was seen as an alternative to an all-digital approach using the Fast Fourier Transform. In comparison, the surface-wave approach offers rapid operation with reduced weight and power consumption, though the accuracy obtainable is not as great. In a later development of this system [337] an automatic temperature compensation technique, using recurrent test signals, eliminated the need for bulky ovens to control the temperature.

Moule et al. [338] developed a wide-band compressive receiver, analysing 250 MHz bandwidth with 4 MHz resolution. This used interdigital devices with 500 MHz bandwidth, shown on Table 9.1. A direct consequence of the wide bandwidth is that the output waveform had a very fine time resolution of 4 nsec. In order to process this, the waveform was detected and entered into a digital shift register clocked at 256 MHz. The data was subsequently read out at a slower rate for further processing.

Williamson et al. [339] describe a satellite-borne processor for demodulation of frequency-shift-keyed communication signals. This system simultaneously handled 100 users, each using a 4-ary code with 4 tones, so that there were 400 tones in all. The tones had a separation of 38 kHz, and were 50 μsec long. In this application the system must process the entire length of each input tone in order to preserve the information content of the signal, so synchronisation is necessary and the system must have a 100% duty factor. A system using four RAC's was used to meet these requirements.

Another application for such systems is in measurement of doppler frequency shifts of radar signals, thus determining target velocities. Generally, the resolution given by surface-wave systems is insufficient for doppler measurements. However for a radar using optical rather than radio-wave propagation the doppler shifts are relatively large, so that surface-wave systems become applicable. For example, a compressive receiver for an infra-red radar [340] had 84 kHz resolution, corresponding to a velocity resolution of 1.6 km/h. Alternatively, for radar systems using radio-wave propagation there is the possibility of contracting the time-scale of the input waveform, thus increasing the doppler shift sufficiently for measurement by a surface-wave system. Roberts *et al.* [341] have demonstrated this by entering the received waveform into a charge-coupled device and then reading it out at a faster rate, suitable for a subsequent surface-wave Fourier transform system. Using a time-compression factor of 1000 in the charge-coupled device, a doppler resolution of 40 Hz was obtained.

Time compression can also be achieved digitally, by entering the waveform into a shift register and then reading out using a faster clock rate. This requires the surface-wave system to have a digital input, as in Dolat's system [336] mentioned above. The additional flexibility introduced by combining surface-wave and digital techniques leads to a variety of novel applications, which have been discussed by Gautier and Tournois [342]. For example, in a sonar system with a phased-array receiver, beam forming can be accomplished by sampling the receiving elements in turn and Fourier transforming the resulting waveform. The rapid processing performed by the surface-wave system enables the beam forming to be done during the time interval corresponding to the range resolution of the sonar system. Several other applications were considered [342], including two-dimensional Fourier transformation.

9.7.4. Other Types of Signal Processing

In addition to Fourier transformation, considered above, surface-wave chirp filters have been used to perform a variety of other signal processing functions. Some of these make use of the Fourier transform system as a building block, giving systems with a higher level of complexity. For example, suppose an input waveform $s(t)$ is Fourier transformed to give $S(\Omega)$, which is multiplied by some waveform $u(t)$ and then inverse transformed to give an output $g(t)$. This configuration, involving two Fourier transform units, can perform a variety of signal processing functions, depending on the waveform $u(t)$. For example, if $u(t)$ is sinusoidal the output is a delayed version of the input, with delay determined by the frequency of $u(t)$, while if $u(t)$ is a linear ramp the output is the time-differential of the input. Atzeni *et al.* [334] demonstrated these functions experimentally. If $u(t)$ is a rectangular pulse with a gating action, so that only a short segment of $S(\Omega)$ is allowed to pass, the system behaves as a variable bandpass filter whose bandwidth and centre frequency are determined by the pulse timing [333, 334, 343]. Alternatively, if the gate is inverted so that a segment of $S(\Omega)$

is suppressed, the system becomes a variable band-stop filter. This principle has been used to demonstrate suppression of narrow-band interference corrupting a wide-band signal [333].

Some signal-processing functions can be accomplished quite simply using the configuration shown in Figure 9.22(a). Here the input signal is passed through two chirp filters in sequence. The first filter has a higher centre frequency than the second, and the signal is up-converted before entering it and down-converted afterwards, using mixers with a C.W. reference at frequency Ω. Using linear chirp devices with matching chirp slopes this system can act as a *variable delay line*, with delay determined by the value of Ω. Figure 9.22(b) illustrates the delays involved for this case. The delay for filter 1, a down-chirp, is shown as it appears between points A and B, before the up-conversion and after the down-conversion. This delay curve therefore moves along the frequency axis as Ω is varied. The bandwidth of filter 1 is made larger than that of filter 2, and Ω is constrained so that the delay curves of the two filters overlap in frequency. It can be seen that the total delay is independent of signal frequency, but its value varies linearly with Ω. A system of this type [344] enabled the delay to be varied by 30 μsec, for signals with bandwidths up to 10 MHz. A similar configuration can be used as a *variable bandpass filter*. In this case the two devices have the same bandwidth, and the bands overlap by an amount dependent on the value of Ω. This principle was demonstrated, in a modified form, by Melngailis et al. [345].

The configuration of Figure 9.22(a) can also be used to give a *variable chirp filter*, that is, a chirp filter whose chirp slope can be varied electronically. In this case the two

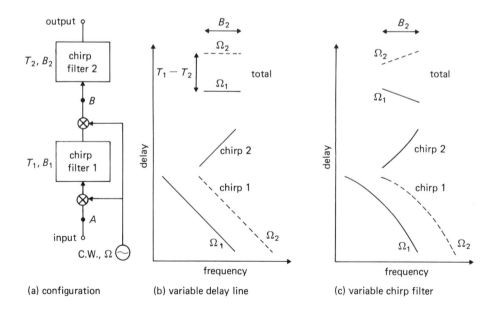

FIGURE 9.22. Signal processing using two chirp filters.

devices have non-linear chirps, with delays varying quadratically with frequency. The delay curves are shown in Figure 9.22(c). The quadratic terms are arranged to cancel so that the total delay is a linear function of frequency. Thus the overall system behaves as a linear chirp filter with bandwidth equal to B_2, the bandwidth of filter 2, and with dispersion T dependent on Ω. The system can even be designed such that the chirp slope can change sign, as shown in the figure. An experimental system [346] had a 100 MHz bandwidth, and enabled the dispersion to be varied from 0 to 1 μsec. It should be noted that systems using the configuration of Figure 9.22 are asynchronous, that is, there is no internal timing so that the signal processing is carried out irrespective of the signal time of arrival.

Chirp filters can also be used to perform *time scaling* [347, 348], using the system of Figure 9.23. Here there are two convolutions with chirp waveforms and two multiplications with chirps. The chirp slopes of the first three chirps are made to satisfy the relation $\alpha + \beta + \gamma = 0$. If $f(t)$ is the input waveform, it can be shown [348, p. 203] that the output $g(t)$ is given by

$$g(t) = f(-\gamma t/\alpha)$$

apart from a constant multiplier. Thus, if α and γ have different signs the output is an expanded or compressed version of the input waveform, with a scaling factor of $|\gamma/\alpha|$. If α and γ have the same sign, the output waveform is also reversed in time. The final multiplication may of course be omitted if the phase of the output waveform is not required.

Many of the above examples have counterparts in optical systems, as discussed by Papoulis [348]. Multiplication by a linear chirp waveform is mathematically equivalent to the action of a lens, while convolution with a chirp is equivalent to optical propagation in the space between two parallel planes. Thus the C–M–C Fourier transform system corresponds to the optical system in which the light distribution arriving at one focal plane of a lens is the Fourier transform of the distribution incident at the other focal plane. The time-scaling system described above corresponds to image formation by a lens; the time-scaling corresponds to the change of image size, related to the distances of the object and image planes from the lens, while time-reversal corresponds to the formation of an inverted image. Papoulis gives several other examples.

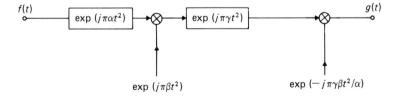

FIGURE 9.23. Time scaling using chirp filters.

Chapter 10

Devices for Spread-Spectrum Communications

In digital communications systems, the term "spread spectrum" refers to a system in which the bandwidth of the transmitted signal is much greater than the bandwidth of the data being transmitted. This feature gives a number of advantages over a conventional system, notably that it provides security and is less sensitive to interference from unwanted signals. Surface-wave devices have been developed for a variety of applications in this area, in particular for correlation of waveforms with large time-bandwidth products, that is, for matched filtering. We have already seen in Chapter 9 that surface-wave chirp filters are used as matched filters in radar systems. The applications in spread-spectrum systems are similar in some respects, but there are some important differences. For example, the waveforms involved are usually phase-shift-keyed (PSK) waveforms rather than chirps, and there is a strong emphasis on programmability, that is, the ability to change the device response electronically.

Another important distinction here lies in the basic physical principles of many of the devices. In previous chapters, nearly all of the devices considered relied on interdigital transducers, multi-strip couplers or reflective groove arrays. In this chapter some new principles are introduced. For example the acoustic convolvers of Section 10.3 exploit the weak acoustic non-linearity of lithium niobate to mix two surface-wave beams, and thus rely on an effect that is avoided in other devices. This gives a very versatile and quite simple method for correlating complex waveforms. Other types of convolver use non-linear interactions in *semiconductors*, using a semiconductor in close proximity to the surface of a piezoelectric and thus exploiting the electric field accompanying the surface wave. Alternatively, the surface wave may propagate on a semiconductor substrate, with a piezoelectric film to generate an electric field. The use of semiconductors in connection with surface waves also introduces a range of other possibilities, including the ability to store an acoustic signal and thus provide an acoustic memory, as described in Section 10.4.2. In fact, semiconductor interactions open up the possibility of an "acoustic chip", in which integrated circuitry and surface-wave devices are combined on the same semiconductor substrate. Some preliminary devices of this type are mentioned in Section 10.2.1. The performance to date has been rather limited owing to

282   SURFACE-WAVE DEVICES

technological difficulties but, in view of the wide range of possibilities such chips would offer, there could well be substantial future developments in this area.

Section 10.1 below gives a brief introduction to the principles of spread-spectrum systems. Section 10.2 is concerned with several linear surface-wave devices, including fixed and programmable matched filters for PSK waveforms. The acoustic convolver is introduced in Section 10.3, which includes devices using surface-wave waveguides in order to improve the efficiency of the non-linear interaction. Section 10.4 is mainly concerned with non-linear semiconductor devices, including convolvers and storage devices. Finally, Section 10.5 gives a brief description of surface-wave oscillators which are included here for convenience, though their applications are not of course limited to spread-spectrum systems.

10.1 PRINCIPLES OF SPREAD-SPECTRUM SYSTEMS

Spread-spectrum communication systems occur in a very wide variety of forms, as described for example in References [349–351]. In this section we review the basic principles briefly. The applications of surface-wave devices in these systems will be described in subsequent sections, and are also discussed in References [352–354].

A common type of spread-spectrum transmitter is illustrated in Figure 10.1. The input data is taken to be a stream of binary digits, each of length T, and these are applied to a balanced modulator. A C.W. waveform is also applied, and the modulator either passes this directly or inverts it, depending on whether the current input digit is a "one" or a "zero". The modulator output is thus a *phase-shift keyed*, or PSK, waveform, with relative phase 0 or 180° corresponding to the data. In a

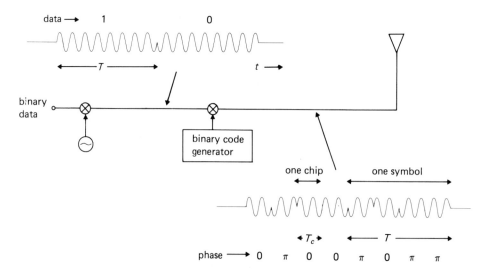

FIGURE 10.1. Spread-spectrum transmitter using PSK waveform.

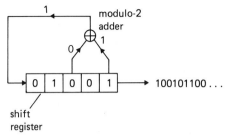

FIGURE 10.2. Pseudo-noise code generator. For the case shown the code repeats after 31 digits.

conventional communication system a waveform of this type would be transmitted, after amplification and up-conversion. In a spread-spectrum system an additional stage of bi-phase modulation is employed, using a binary code generated by a code generator. This second stage introduces phase changes much more frequently than the original data, with the consequence that the signal bandwidth is substantially increased. The waveform corresponding to one data digit, of length T, is called a "symbol" and can be regarded as a sequence of contiguous pulses, each of length T_c and phase 0 or 180°, known as "chips". The spectrum of the output waveform depends in a complex manner on the coding employed; however it is usual to choose the code sequence in a quasi-random manner, and it is then found that the power spectrum of the output has a shape similar to that of an individual chip, and hence its bandwidth is approximately $1/T_c$. In contrast, the waveform prior to code modulation has a bandwidth of approximately $1/T$. Hence the bandwidth has been increased by a factor T/T_c, equal to the number of chips in each symbol. Typically, each symbol will contain 50 to 500 chips. Symbol lengths vary widely according to the application, and are generally in excess of $10\mu\text{sec}$.

The code to be used may be simply stored in a digital memory and read out as required, but a convenient alternative is to use a digital shift register as shown in Figure 10.2. Here two of the register stages are combined by a modulo-2 adder to provide a feedback to the register input; the adder gives a "one" output if its two inputs are different and a "zero" if they are the same. As the register is clocked, a binary code emerges at the right. Provided an appropriate feedback logic is used, a register with n stages can give $2^n - 1$ output digits before repeating, and the code is then called a "pseudo-noise" code. Such a code is found to have quasi-random properties, and is thus effective for spreading the bandwidth of the signal. Furthermore, the number of stages required in the shift register is generally much less than the number of digits in one cycle of the code.

In the receiver the phase changes produced by the code generator must be removed in order to extract the data, and Figure 10.3 shows a receiver using a matched filter for this purpose. The matched filter is the commonest application for surface-wave technology in spread-spectrum systems. Suppose initially that the same code is used for each symbol, as in Figure 10.1, so that the symbols are all the same except that some of them are inverted. The matched filter (Section A.3) has an impulse response corresponding to the time-reverse of one symbol. For each input symbol the filter

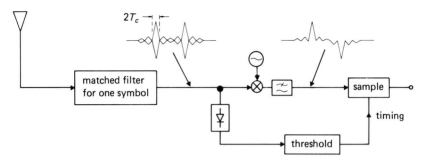

FIGURE 10.3. Spread-spectrum receiver using matched filter.

output typically gives a triangular correlation peak, with basewidth $2T_c$, and a series of relatively small time-sidelobes on either side. The relative phases of the peaks correspond to the data, and this is recovered by demodulating the filter output using a synchronous detector and sampling at the peak times; a comparator is used to determine the polarity of the sampler output, which corresponds to the original data. The required sampling times can be obtained by using an envelope detector followed by a threshold circuit.

Compared with a conventional communication system, a spread-spectrum system offers several advantages. It has increased security, since the signal can only be demodulated if the code is known and its wide bandwidth makes its presence more difficult to detect. The receiver responds preferentially to the waveform that it is matched to, and so is relatively insensitive to narrow-band interference. If the transmitter timing is known, the timing of the narrow correlation peak in the receiver can be used to give a high-resolution measure of the range between the transmitter and receiver. Also, the narrow correlation peak enables the system to reject multi-path signals with different delays, due to reflections from buildings for example. On the other hand, a spread-spectrum system does *not* give an improved signal-to-noise ratio when wide-band noise is present at the receiver input; the output signal-to-noise ratio of the matched filter is $2P^sT/N_i$, as shown by equation (9.1) of Section 9.1, and is therefore independent of the signal bandwidth. All of these features are in strong contrast with the advantages of matched filters in pulse compression radar systems, discussed in Section 9.1.

Before the advent of surface-wave devices, the implementation of matched filters placed heavy demands on existing technology. Consequently another type of receiver, shown in Figure 10.4, has been more commonly used. Here the signal is correlated by an *active correlator*, which multiplies it by a code identical to that used in the transmitter, thus converting each symbol to an unmodulated pulse of length T. The latter is down-converted to baseband and integrated over the symbol duration, and a polarity decision then gives the data. Ideally this receiver gives the same performance as the matched filter receiver described above, and moreover it is easier to implement. However, it has the significant drawback that the code generator must be accurately synchronised with the code of the received signal, typically with an accuracy of $T_c/4$. In general the signal timing will be initially unknown, so the receiver must first perform an acquisition procedure: trial correlations are done with the code timing

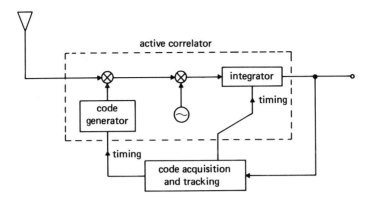

FIGURE 10.4. Spread-spectrum receiver using active correlator.

changed slightly after each symbol length, continuing until a correlation is observed at the output. Typically, the number of trials needed is about twice the number of chips in the symbol, so that the acquisition procedure can be quite lengthy. In contrast, the matched filter receiver does not require this lengthy acquisition procedure, and this is its main advantage.

It was assumed above that the same coding was used for each symbol, but in practice it is quite common to vary the coding in the interest of greater security. In the transmitter, this may be done by using a memory with a library of codes. This does not substantially complicate the operation of an active correlator receiver. For a matched filter receiver a bank of matched filters can be envisaged, but generally the number of codes used would make this approach inconvenient. Thus a *programmable* matched filter, in which the coding can be changed electronically, is preferable. It will be seen later that there are several surface-wave techniques for realising programmable matched filters. An alternative, and often used, method of varying the coding is simply to use a pseudo-noise code generator in the transmitter, with a repetition period much greater than the symbol length.

10.2. LINEAR DEVICES

We have seen above that the the main application for surface-wave devices in spread-spectrum systems lies in matched filtering of PSK waveforms. Section 10.2.1 below describes surface-wave devices for this purpose, including programmable devices. The most important limitations of these devices arise from phase errors due to velocity errors or temperature changes, and these are considered in Section 10.2.2. In addition to PSK waveforms a variety of other waveforms can be considered for spread-spectrum systems, and in Section 10.2.3 we consider surface-wave devices for use with MSK waveforms. Devices for frequency-hopped waveforms are considered in Section 10.2.4.

The devices described here are all linear. In Sections 10.3 and 10.4 below some non-linear devices, particularly convolvers, will be described, and it will be shown that these offer a very different and effective approach for programmable matched filtering.

10.2.1. Matched Filters for PSK Waveforms

We first consider matched filters in which the coding is fixed. It was seen in Section 8.1 that an interdigital transducer can be designed simply by sampling the required impulse response, apart from the minor complication of the presence of the element factor. This principle can be applied to PSK filters, but has the disadvantage that electrode interaction effects are severe if a single-electrode design is used, since the impulse response is usually many wavelengths long [355]; also, this method would require the other transducer to have few electrodes, which could make it difficult to match efficiently if a wide bandwidth were required (Section 7.1.1). For these reasons most devices have used a modified configuration, in which the required impulse response is synthesised by the cooperative action of two transducers. As shown in Figure 10.5, one transducer is uniform with length vT_c, where v is the velocity. The other transducer is essentially an array of short uniform transducers, often called taps, all connected to two bus-bars. The taps have uniform spacing vT_c, and are all identical except that some are inverted in accordance with the required code; this inverts the corresponding contribution to the device output waveform. In response to a short impulse, the uniform transducer at the left generates a rectangular surface-wave pulse which propagates along the device, exciting the taps in sequence, so that a bi-phase PSK waveform is produced as shown in the figure. The principle is very similar to thinning (Section 8.3), but here the distortion due to thinning is compensated by the response of the uniform transducer. The device response is in practice distorted somewhat by the element factor, and by the fact that the taps have finite lengths, but these distortions are usually small enough to be acceptable for spread-spectrum applications.

The performance of these devices is reviewed by Bell *et al.* [356, 357], and further experimental results are given in References [352, 358–364]. The substrate material is usually quartz because good temperature stability is usually needed, as explained below. Most experimental devices had 30 to 150 taps with a chip rate $(1/T_c)$ in the range 5 to 20 MHz, though a chip rate of 200 MHz has been demonstrated [364]. Typically, the devices were coded in accordance with pseudo-noise codes. A common experimental test of the device fidelity is to examine the output waveform when the

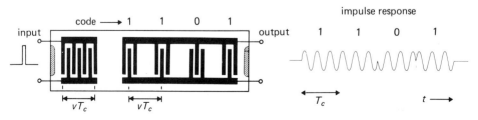

FIGURE 10.5. Fixed-coded matched filter for PSK.

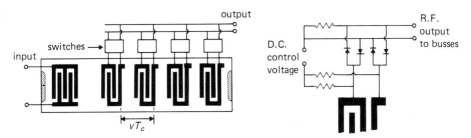

FIGURE 10.6. Programmable matched filter for PSK, with one type of switch circuit shown at right.

device is used to correlate the waveform that it is matched to; the relative levels of the correlation peak and time-sidelobes are sensitive to amplitude and phase errors in individual taps. For typical devices, the peak-to-sidelobe ratios were within 1 or 2 dB of the ideal values.

As already seen in Section 10.1, spread-spectrum systems often change the coding employed, so that a *programmable* matched filter would be required rather than the fixed-coded devices considered above. Programmability can be implemented by connecting the individual taps to a bank of electronic switches, as illustrated in Figure 10.6; each switch has a D.C. control input whose polarity determines the phase (0 or 180°) of the contribution from the corresponding tap. Arrangements of this type have been demonstrated by several authors [365, 366], using either hybrid or integrated circuitry; however, the need to accommodate the physical size of the circuit and the bonding pads for the wire interconnections implies a minimum tap spacing, corresponding to a minimum chip length T_c of typically 100 nsec.

To overcome this limitation, and the inconvenience of a large number of wire bonds, Hickernell et al. [367, 368] have developed a very different approach. Here the surface wave propagates on a *silicon* substrate, and is tapped by an array of field-effect transistors making use of the piezoresistive effect, that is, the change of resistivity accompanying an acoustic strain. The transistor bias can be used to control the amplitude and phase. The wave is launched by an interdigital transducer, with a piezoelectric zinc oxide overlay. This approach enables integrated circuitry to be fabricated in the *same* substrate, eliminating the need for individual wire bonds to each tap; Hickernell's device included a shift register, so that the code required could be read in serially, and a variety of other circuits. The device demonstrated programmable correlation of 31-chip waveforms with 10 MHz chip rate, showing that the approach is practically feasible even though the technology is rather demanding.

This type of technology, in which surface-wave and semiconductor devices are integrated on the same substrate, could well be very significant in the future development of surface-wave devices. It implies that the well known flexibility of integrated circuits can be combined directly, on the same substrate, with almost any type of surface-wave device. However, the technology needed is at present relatively immature, and consequently the results obtained to date are rather limited.

An alternative approach to integration is that of Hagon [369], who demonstrated a programmable PSK filter using a piezoelectric aluminium nitride film on a sapphire substrate. The taps in this case were interdigital transducers, and the switches needed to control the phase were fabricated in a silicon film deposited on the same sapphire substrate. More recently, Grudkowski *et al.* [370] have shown that field-effect transistors on gallium arsenide may be used as programmable surface-wave taps, and clearly these could also be combined directly with integrated circuitry. These two technologies are further possibilities for fully integrated devices, though neither has to date been developed to the extent of the silicon device described above.

10.2.2. Output Waveform and Effect of Phase Errors

It is shown here that, when a PSK waveform is applied to an appropriate surface-wave matched filter, the output waveform is generally quite sensitive to errors in the surface-wave velocity and to temperature changes, and also to doppler shifts. For comparison, we first consider the ideal output waveform.

(a) Ideal Output Waveform. We consider a PSK waveform $s(t)$, with centre frequency ω_0, given by

$$s(t) = \sum_{n=1}^{N} a_n v_c(t - nT_c) \exp(j\omega_0 t) + \text{c.c.}, \quad (10.1)$$

where "c.c." indicates complex conjugate, N is the number of chips and $a_n = \pm 1$ are coefficients corresponding to the coding. The function $v_c(t)$ is the envelope of one chip, given by

$$v_c(t) = \text{rect}(t/T_c), \quad (10.2)$$

which is a rectangular pulse of duration T_c. The impulse response of the filter is taken to be

$$h(t) = \sum_{n=1}^{N} b_n v_c(t - nT_c) \exp(j\omega_0 t) + \text{c.c.}, \quad (10.3)$$

where $b_n = \pm 1$. Apart from a phase constant, which is inconsequential, the filter is matched to the waveform $s(t)$ if the coefficients b_n are given by

$$b_n = a_{N+1-n}. \quad (10.4)$$

We also define $r_c(t)$ as the correlation function of the chip envelope $v_c(t)$, so that

$$r_c(t) = \int_{-\infty}^{\infty} v_c(\tau) v_c(\tau - t) \, d\tau. \quad (10.5)$$

Using equation (10.2), this is found to be a triangular function of height T_c and basewidth $2T_c$.

When the waveform $s(t)$ is applied to the filter, the output waveform $g(t)$ is the convolution of $s(t)$ with $h(t)$, and $g(t)$ is the correlation function of $s(t)$ if the filter is matched. In practice, $s(t)$ and $h(t)$ can be taken to be bandpass functions, and it follows that it is sufficient to convolve the positive-frequency parts of $s(t)$ and $h(t)$ (the terms proportional to $\exp(j\omega_0 t)$), and then add the conjugate. With some re-arrangement this gives

$$g(t) = \sum_{k=2}^{2N} c_k r_c(t - kT_c) \exp(j\omega_0 t) + \text{c.c.} \quad (10.6)$$

where the coefficients c_k are given by

$$c_k = \sum_{n=1}^{k-1} a_{k-n} b_n, \quad \text{for } k \leq N + 1,$$

$$= \sum_{n=k-N}^{N} a_{k-n} b_n, \quad \text{for } k \geq N + 1. \quad (10.7)$$

Thus the output waveform is essentially a sum of delayed versions of the chip correlation function $r_c(t)$, with amplitudes proportional to c_k. Since $r_c(t)$ has duration $2T_c$, the envelope at times $t = kT_c$ is simply $2c_k T_c$. For a matched filter the correlation peak occurs at $k = N + 1$, and using equations (10.4) and (10.7) its amplitude is given by $c_{N+1} = N$. The amplitudes of the time-sidelobes, given by the other c_k, depend on the coding.

(b) Velocity Errors and Temperature Changes.
We now suppose that the response of the surface-wave filter is somewhat different from ideal. The effects of velocity errors and temperature changes were considered in Section 6.4, in terms of a small quantity ε defined as the fractional change of delay, as in equations (6.35) or (6.39). The actual impulse response $h'(t)$ of the device is related to the ideal response $h(t)$ by

$$h'(t) = h\left(\frac{t}{1+\varepsilon}\right) \approx h[t(1-\varepsilon)]. \quad (10.8)$$

This is substituted into equation (10.3) to find $h'(t)$. It can be assumed that ε is too small to have any significant effect on the envelope, and it follows that the only effect of the error is to multiply the waveform by a phase term $\exp(-j\omega_0 \varepsilon t)$. To a good approximation this term can be taken to be constant over the duration of each chip, so that the impulse response becomes

$$h'(t) \approx \sum_{n=1}^{N} b'_n v_c(t - nT_c) \exp(j\omega_0 t) + \text{c.c.}, \quad (10.9)$$

where $b'_n = b_n \exp(-j\omega_0 nT_c \varepsilon)$. This is the same as the ideal response, equation (10.3), except that b_n has been replaced by b'_n. The filter output waveform can therefore be obtained from the earlier analysis for the ideal case, equation (10.6). Thus the amplitudes of the correlation peak and time-sidelobes are given by the c_k of equation (10.7), with b_n replaced by b'_n. For the correlation peak $k = N + 1$ and equation (10.7) is readily summed as a geometric progression, so that the peak amplitude is found to be given by

$$|c_{N+1}| = \left| \frac{\sin(\pi M \varepsilon)}{\sin(\pi M \varepsilon / N)} \right|, \qquad (10.10)$$

where $M = N\omega_0 T_c/(2\pi)$ is the number of cycles in the waveform. The peak amplitude is thus zero for $\varepsilon = \pm 1/M$. A typical requirement is that it should not fall by more than 1 dB, and for $N \gg 1$, which is usually the case, this implies $|\varepsilon| < 0.26/M$. This is generally more stringent than the requirement for linear chirp filters, considered in Section 9.5.2.

Carr et al. [360] give examples for a variety of codes, showing that the error affects the sidelobe amplitudes as well as the correlation peak, though the sidelobe changes are not generally significant if the reduction of the peak is acceptable. The analysis for the temperature sensitivity has been confirmed experimentally for Y, X quartz substrates [360] and for ST, X quartz substrates [357].

(c) Doppler Shifts. The effect of a doppler shift of the input waveform can be found in a similar way. As stated in Section 9.5.3, if $S(\omega)$ is the spectrum of the ideal input waveform, the doppler-shifted waveform can be taken to have a spectrum $S'(\omega) = S(\omega - \omega_d)$, where ω_d is the doppler frequency. Using the shifting theorem, equation (A.11), it is found that the doppler-shifted waveform $s'(t)$ is given by the ideal waveform $s(t)$ of equation (10.1), with ω_0 replaced by $\omega_0 + \omega_d$. There is therefore a phase error $\omega_d t$, and if this is taken to vary little over the duration of one chip we have

$$s'(t) \approx \sum_{n=1}^{N} a'_n v_c(t - nT_c) \exp(j\omega_0 t) + \text{c.c.}, \qquad (10.11)$$

where $a'_n = a_n \exp(j\omega_d n T_c)$. This has the same form as the ideal input waveform, equation (10.1). Thus, when the doppler-shifted waveform is applied to an ideal matched filter, the output peak and sidelobe amplitudes are given by the coefficients c_k of equation (10.7), with a_n replaced by a'_n. The correlation peak amplitude is found to be given by equation (10.10) with ε replaced by ω_d/ω_0, and thus falls to zero for $\omega_d = \pm 2\pi/T$, where $T = NT_c$ is the length of the waveform. A 1 dB reduction of the peak occurs for $\omega_d = \pm 1.6/T$, showing that PSK waveforms are much more doppler sensitive than chirp waveforms (Section 9.5.3). Experimental confirmation is given by Carr et al. [360].

10.2.3. Devices for MSK Waveforms

Although we have assumed so far that the waveform transmitted in a spread-spectrum system is a PSK waveform, there are in fact several other possibilities. Here we consider *minimum-shift keyed* (MSK) waveforms. Compared with PSK, MSK waveforms have spectral sidelobes whose amplitudes fall more rapidly with the distance from the centre frequency, as will be shown below. This feature enables MSK waveforms with different centre frequencies to be spaced more closely in frequency,

An MSK waveform may be written in the form

$$y(t) = \sum_m y_m(t),$$

with

$$y_m(t) = \cos\left(\omega_0 t + d_m \frac{\pi t}{2T_c} + \phi_m\right), \quad \text{for } mT_c \leq t \leq (m+1)T_c \quad (10.12)$$

and $y_m(t) = 0$ for other t. Here $d_m = \pm 1$ are the coding coefficients, T_c is the chip period, ω_0 is the centre frequency and ϕ_m are constants. It is usual to constrain the ϕ_m such that the waveform is continuous for all t, and this requires $\phi_m - \phi_{m-1} = \frac{1}{2}\pi m(d_{m-1} - d_m)$. The MSK waveform is clearly a form of frequency-shift keying, where the two frequencies involved are separated by $\Delta\omega = \pi/T_c$ and the choice of frequency for each chip is determined by the d_m. An example is shown in Figure 10.7.

Some methods for generating an MSK waveform are discussed by Amoroso and Kivett [371] who show, in particular, that it can be generated by applying a *PSK* waveform to a filter with a rectangular impulse response of length T_c. Such a filter is easily realised using surface-wave technology [372, 373]; as shown in Figure 10.7 the filter simply has a uniform transducer with length corresponding to the length of impulse response required, and another short uniform transducer which does not appreciably influence the response. This method enables MSK waveforms to be generated very conveniently.

To show that this method can indeed give an MSK waveform, the input PSK waveform is written as

$$v(t) = \sum_n v_n(t) = \sum_n a_n \, \text{rect} \, (t/T_c - n) \exp (j\omega_1 t) + \text{c.c.}, \quad (10.13)$$

where $v_n(t)$ respresents chip n and $a_n = \pm 1$ gives the PSK coding. The filter impulse response $h(t)$ is a rectangular pulse of length T_c, written as

$$h(t) = \text{rect} \, (t/T_c) \exp (j\omega_2 t) + \text{c.c.} \quad (10.14)$$

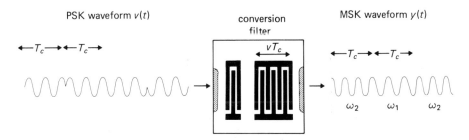

FIGURE 10.7. Conversion of PSK to MSK.

The centre frequencies of these two waveforms are made equal to the two frequencies of the MSK waveform, so that $(\omega_1 + \omega_2)/2 = \omega_0$, the MSK centre frequency, and $\omega_2 - \omega_1 = \pi/T_c$. The filter output waveform $y(t)$ is the convolution of $v(t)$ and $h(t)$, and is written as

$$y(t) = \sum_n g_n(t), \qquad (10.15)$$

where $g_n(t) = v_n(t) * h(t)$ is the filter output due to chip n of the input waveform. Equations (10.13) and (10.14) are used to find $g_n(t)$, and the waveforms are taken to be bandpass functions so that it is sufficient to convolve the positive-frequency terms and then add the conjugate. After some manipulation, this gives

$$g_n(t) = \frac{2a_n T_c}{\pi} (-j)^n \exp(j\omega_0 t) \cos[\tfrac{1}{2}\pi(t/T_c - n)] + \text{c.c.}, \quad \text{for } |t - nT_c| \leq T_c \qquad (10.16)$$

and $g_n(t) = 0$ for other t. Thus, $g_n(t)$ is a waveform of duration $2T_c$, with a cosine-shaped envelope falling to zero at each end. Alternatively, writing the cosine as a sum of exponentials shows that $g_n(t)$ is a sum of two rectangular pulses, with carrier frequencies ω_1 and ω_2. The total output waveform $y(t)$ is the sum of the $g_n(t)$, as in equation (10.15), but it can also be expressed as a sum of the $y_m(t)$ as in equation (10.12). The term $y_m(t)$ is the chip, of length T_c, commencing at $t = mT_c$, and for this time interval only $g_m(t)$ and $g_{m+1}(t)$ are non-zero. We thus find, from equation (10.16),

$$y_m(t) = \frac{T_c}{\pi}[(a_m + a_{m+1})e^{j\omega_1 t} + (-1)^m(a_m - a_{m+1})e^{j\omega_2 t}] + \text{c.c.},$$

$$\text{for } mT_c \leq t \leq (m+1)T_c. \qquad (10.17)$$

Since $a_m = \pm 1$, this expression has the same form as equation (10.12) and is therefore an MSK waveform; the coefficient d_m is 1 when a_m and a_{m+1} are different, and -1 when they are the same. It also follows that $y(t)$ is continuous for all t.

Some further deductions can be made if we assume that the centre frequency is restricted such that $\omega_0 T_c = 2\pi M + \pi/2$, where M is some integer. In this case it follows from equation (10.16) that the waveform can be written

$$y(t) = \frac{2T_c}{\pi} \sum_n a_n g_c(t - nT_c), \qquad (10.18)$$

where

$$g_c(t) = \exp(j\omega_0 t) \cos(\tfrac{1}{2}\pi t/T_c) \text{ rect}(\tfrac{1}{2}t/T_c) + \text{c.c.} \qquad (10.19)$$

It follows that an alternative method of generation is to use a filter whose impulse response has the form of $g_c(t)$ and apply impulses with spacing T_c and with polarities corresponding to the a_n; this method has been demonstrated using a surface-wave

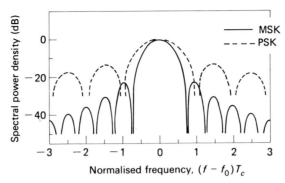

FIGURE 10.8. Spectral power density for one chip of MSK and one chip of PSK.

filter [374]. It is also clear that a matched filter for MSK can be realised by a method similar to the PSK filter of Figure 10.5, except that the uniform input transducer is replaced by an apodised transducer designed to give an impulse response proportional to $g_c(t)$. The output given by this filter in response to an input MSK waveform can be obtained from the analysis of Section 10.2.2 above, replacing $v_c(t)$ by the envelope of $g_c(t)$. It can also be anticipated that, for a waveform containing a large number of chips with coding chosen in a quasi-random manner, the power spectral density will be similar to that of $g_c(t)$. This conclusion is confirmed elsewhere [375]. Figure 10.8 compares the spectrum of $g_c(t)$ with that of a PSK chip, showing that the sidelobes of the MSK waveform fall more rapidly with frequency.

10.2.4. Frequency Hopping

Frequency hopping is a rather different approach to the problems of increasing security and reducing interference effects, and is applicable to both radar and communication systems. In its basic form the transmitter simply makes occasional changes of the waveform centre frequency, according to a pre-arranged pattern known to the receiver. This requires the use of a frequency synthesiser in the transmitter. In the receiver a similar synthesiser can be used, in conjunction with a balanced modulator, to remove the frequency hops.

A simple method of frequency synthesis using surface-wave devices [376, 377] is shown in Figure 10.9. Two chirp filters, with the same chirp slope, are impulsed at slightly different times. The resulting chirp waveforms, which overlap, are mixed together and the sum-frequency term, with constant frequency, is extracted by bandpass filtering. The output frequency is linearly related to the spacing τ of the input pulses, and can therefore be easily varied. By using two such systems and impulsing repetitively, a continuous waveform can be generated. Frequency synthesis can of course be done by more conventional methods, using multipliers and dividers; the main advantage of the surface-wave method is that a very large range of different frequencies can easily be obtained, since chirp filters with large time-bandwidth products are available (Chapter 9). Furthermore, in a communication system frequency hopping is most effective when there are many hops within each data digit, and this requires the frequency changes to be made rapidly and phase coherence to

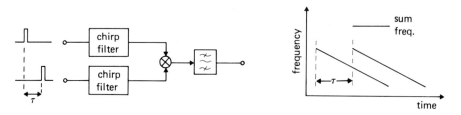

FIGURE 10.9. Frequency synthesis using chirp filters.

be maintained over many hops. These requirements are difficult to meet by conventional methods but can be met by the surface-wave method, which can change the frequency in a few nsec. The phase coherence has been demonstrated by correlating the waveform using a surface-wave convolver [377], described in Section 10.3.

Another surface wave method of frequency synthesis makes use of a filter bank, already described in Section 8.5. A repetitive comb waveform is generated using a stable oscillator, and the filter bank is used to filter the harmonics. An electronic switch is connected to each output of the filter bank so that the required frequency can be selected [378]. This technique gives better spectral purity than the chirp method, but is more limited in the number of frequencies obtainable.

Hunsinger et al. [352, 363] point out that a limited number of frequencies can be obtained simply by impulsing surface-wave devices with different centre frequencies. They used four surface-wave PSK filters, thus generating waveforms employing both PSK modulation and frequency hopping. This method of introducing frequency hopping provides a useful increase of signal bandwidth quite simply.

10.3. ACOUSTIC CONVOLVERS

In Section 10.2 we have considered some devices for correlating PSK and MSK waveforms, including programmable devices since programmability is a common requirement in spread-spectrum systems. The surface-wave convolver [379–382] offers another approach for programmable correlation, with the significant advantage of being much simpler to implement. The convolver differs markedly from all the devices described so far in this book, in that its operation relies on a non-linear effect and thus the basic physical principles are quite different. In this section we consider "acoustic" convolvers, in which the non-linearity involved is a property of the substrate material. In its basic form the acoustic convolver is structurally one of the simplest of surface-wave devices, consisting of two interdigital transducers and a uniformly metallised area between them. Despite this simplicity it performs one of the most sophisticated signal processing operations, correlating complex waveforms with a very high degree of programmability, subject only to constraints on the waveform duration and bandwidth. In addition the convolver is much less sensitive to temperature and velocity changes than most surface-wave devices.

The basic principles common to all convolvers are described in Section 10.3.1 below, and in Section 10.3.2 the performance of the basic acoustic convolver is considered. In recent years many devices have used waveguides to confine the surface-wave energy, thus increasing the energy density and hence the strength of the non-linear effect, and these devices are described in Section 10.3.3. Section 10.3.4 shows how the signal processing operation of a convolver can be analysed using a two-dimensional frequency response.

Some other types of convolver, using non-linear effects in semiconductors, will be described in Section 10.4.

10.3.1. Principles of Non-linear Convolvers

In this section the basic principles of non-linear convolvers are described, illustrating by referring to surface-wave devices in which the non-linearity is a property of the substrate material. While such devices are of most interest here, it should be borne in mind that the principles are in fact quite general. Thus, other sources of non-linearity can be exploited, as will be seen in Section 10.4, and other types of waves, such as bulk acoustic waves, can be used.

We first consider the non-linear interaction of two surface waves in the convolver shown in Figure 10.10(a), which is essentially a simple interdigital delay line. At this stage the two input waveforms $f_1(t)$ and $f_2(t)$ are taken to be C.W. waveforms, with frequencies ω_1 and ω_2 respectively, though more general waveforms will be considered later. For low power levels, such that non-linear effects are insignificant, the corresponding surface-wave amplitudes have the forms

$$u_1(t, x) = A_1 \cos(\omega_1 t - k_1 x)$$

and

$$u_2(t, x) = A_2 \cos(\omega_2 t + k_2 x), \quad (10.20)$$

where A_1 and A_2 are constants, $k_1 = \omega_1/v$ and $k_2 = \omega_2/v$ are the wavenumbers and v is the wave velocity. Attenuation and diffraction are assumed to be negligible. The

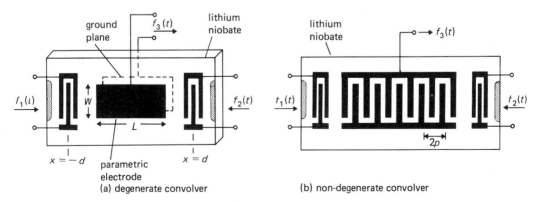

FIGURE 10.10. Surface-wave acoustic convolvers.

convolver makes use of non-linearity to mix the two waves. A variety of non-linear mechanisms can be used, and for the device of Figure 10.10(a) the non-linearity is a property of the substrate material itself. For this reason the device is called an "acoustic" convolver. As already seen in Section 6.3, acoustic non-linearity causes a surface wave to be accompanied by harmonics, and also causes attenuation of the fundamental. The convolver is operated at relatively low power levels, such that the fundamental components of the waves are not significantly affected and are therefore given by equations (10.20). However, the non-linearity will generate terms proportional to the squares $u_1^2(t, x)$ and $u_2^2(t, x)$ of the fundamentals, and a product term proportional to

$$u_1(t, x) u_2(t, x) = \tfrac{1}{2} A_1 A_2 [\cos\{(\omega_1 + \omega_2)t + (k_2 - k_1)x\}$$
$$+ \cos\{(\omega_1 - \omega_2)t - (k_1 + k_2)x\}]. \tag{10.21}$$

For low power levels, any further higher-order terms can be assumed to be negligible.

For a piezoelectric material, each of the various terms present has in general an associated electric field, and the convolver makes use of metal electrodes to selectively sense one of the terms. In the convolver of Figure 10.10(a) a uniform metal film called the "parametric electrode" is used, acting in conjunction with a ground plane; the latter may be provided simply by the metal carrier that the substrate is mounted on. The parametric electrode selectively senses the component of the electric field invariant with x. The sum-frequency component of the product term, equation (10.21), gives such a field when the input frequencies ω_1 and ω_2 are equal. Assuming that ω_1 and ω_2 are non-zero, the only other spatially-invariant fields arise from the difference-frequency components of the squares $u_1^2(t, x)$ and $u_2^2(t, x)$, but these terms have zero frequency and are suppressed because there is no D.C. path through the parametric electrode. Thus, ideally the output voltage arises only from the sum-frequency component of the product term, and exists only when the input frequencies are equal. With $\omega_1 = \omega_2 = \omega$, the output voltage has the form $\tfrac{1}{2} A_1 A_2 \cos(2\omega t)$, and has a frequency equal to twice the input frequency. Note that the output amplitude is proportional to the product of the input amplitudes; a device giving this relation is said to be *bilinear*. This type of surface-wave mixing was first observed by Svaasand [383] using a quartz substrate, though most later devices have used lithium niobate, which gives a stronger interaction.

The device of Figure 10.10(a) is called a *degenerate* convolver because it responds most strongly when the input frequencies are equal. If the frequencies of the C.W. input waveforms are different the sum-frequency component of the product term, equation (10.21), has a spatial periodicity $2\pi/(k_2 - k_1)$. This component can be selectively sensed by introducing this periodicity in the parametric sensor [384]. As shown in Figure 10.10(b) this can take the form of an interdigital array of electrodes with pitch $2p$, giving an output at the sum frequency $(\omega_1 + \omega_2)$ when $\omega_1 - \omega_2 = \pm \pi v/p$. This device is therefore a *non-degenerate* convolver. The parametric sensor is essentially the same as a single-electrode transducer designed for a centre frequency of $|\omega_1 - \omega_2|$, though there is of course no surface wave present at this frequency. Apart from the frequency difference involved the non-degenerate device operates in a manner very similar to the degenerate device. However, it has been found necessary to space the parametric sensor from the surface in order to prevent it causing substantial bulk wave excitation [379], and since this makes the fabrication inconvenient the non-degenerate device has received relatively little attention.

To appreciate the signal processing applications of the convolver it is necessary to consider more general input waveforms. Suppose that input waveforms $f_1(t)$ and $f_2(t)$ are applied to the input transducers of the degenerate convolver of Figure 10.10(a). Assuming that no distortion arises from the transducer responses or from propagation effects, and that the power levels are low enough, the surface-wave amplitudes will be proportional to $f_1(t - x/v)$ and $f_2(t + x/v)$. For simplicity, a delay corresponding to the transducer separation is ignored here. Since the surface-wave amplitudes must be oscillatory it can be concluded that, as in the C.W. case, the output will be mainly due to the product term; this will be justified formally below. The parametric electrode gives an output voltage $f_3(t)$ proportional to the spatial integral of the product, and thus

$$f_3(t) = \int_{-\infty}^{\infty} f_1(t - x/v) f_2(t + x/v) \, dx. \tag{10.22}$$

This assumes that the input waveforms have finite duration so that the product exists only for a finite region of x, and that the parametric electrode extends at least over this region. Writing $\tau = t - x/v$, equation (10.22) becomes

$$f_3(t) = v \int_{-\infty}^{\infty} f_1(\tau) f_2(2t - \tau) \, d\tau. \tag{10.23}$$

This equation is the *convolution* of $f_1(t)$ and $f_2(t)$, apart from the factor of 2 which causes a contraction in the time-scale. Apart from this contraction the convolver output is formally the same as that given by a *linear* filter, even though it relies on a non-linear effect. For a linear filter (Section A.2) the output waveform is given by the convolution of the input waveform with the filter impulse response. In the convolver, one of the input waveforms, called the *reference*, has the role of the "impulse response", and since this can be varied at will the convolver is clearly an exceptionally versatile device. In principle it can behave as any type of linear filter, for example as a bandpass filter or a matched filter for chirp or PSK waveforms, subject only to constraints on the bandwidth and duration of the reference. In practice a convenient method of generating the reference is needed and in consequence the convolver has been used mainly as a matched filter, as discussed in more detail later. The time-contraction given by the convolver arises physically because the two waves have a relative velocity of $2v$, in contrast to the velocity v which applies for a linear surface-wave device; this is also associated with the fact that for C.W. input waveforms the output frequency is twice the input frequency, as seen earlier.

Further insight into the operation of the degenerate convolver be obtained from a frequency-domain analysis. The Fourier transform of the input waveform $f_1(t)$ is denoted $F_1(\omega)$, and since $f_1(t)$ is real we have $F_1(-\omega) = F_1^*(\omega)$. This enables the transform to be written as an integral over positive frequencies. Writing $F_1(\omega) = A_1(\omega) \exp[j\phi_1(\omega)]$, and also substituting $(t - x/v)$ for t, we find

$$f_1(t - x/v) = \frac{1}{\pi} \int_0^\infty A_1(\omega_1) \cos [\phi_1(\omega_1) + \omega_1 t - k_1 x] \, d\omega_1, \quad (10.24)$$

where $k_1 = \omega_1/v$. A similar relation applies for $f_2(t + x/v)$, with $A_1(\omega_1)$ and $\phi_1(\omega_1)$ replaced by $A_2(\omega_2)$ and $\phi_2(\omega_2)$, and with $-k_1$ replaced by $k_2 = \omega_2/v$. The spatial integral required to obtain the output waveform is readily done with the aid of the relation

$$\int_{-\infty}^\infty \cos (a + kx) \, dx = 2\pi\delta(k) \cos a = 2\pi v\delta(\omega) \cos a, \quad (10.25)$$

where a is a constant and $k = \omega/v$. Equation (10.25) follows from the transform of $\cos(\omega_0 t)$, equation (A.27), evaluated to $\omega = 0$. In evaluating the convolver output waveform, it is noted that the surface-wave waveforms must be bandpass waveforms, a restriction imposed by the transducers. Hence $A_1(\omega)$ and $A_2(\omega)$ are zero at $\omega = 0$, and the integrands involved are finite only when k_1 and k_2 are positive and non-zero. It follows from this that the spatial integrals of the linear and square-law terms are identically zero, apart from a D.C. term which is eliminated because there is no D.C. path through the parametric electrode. The output due to the product term, equation (10.22), involves a product of cosines, giving sum- and difference-frequency components. The spatial integral of the latter is found to be identically zero, and the output waveform $f_3(t)$, due only to the sum-frequency term, is found to be

$$f_3(t) = \frac{v}{2\pi} \int_{-\infty}^\infty F_1(\omega) F_2(\omega) \, e^{2j\omega t} \, d\omega. \quad (10.26)$$

From the convolution theorem, equation (A.19), this is seen to be the time-contracted convolution of $f_1(t)$ and $f_2(t)$, agreeing with equation (10.23).

The same method can be used for the case of the non-degenerate convolver of Figure 10.10(b). For illustration, the spatial sensitivity of the parametric sensor can be taken to be sinusoidal, with period $2p$, so that the output due to the product term has the form

$$f_3(t) = \int_{-\infty}^\infty f_1(t - x/v) f_2(t + x/v) \cos (\pi x/p) \, dx.$$

The input waveforms are assumed to be confined to frequency bands which do not overlap, and are such that only the sum-frequency component can have "wavenumber" equal to $2p$. The output waveform is found to be

$$f_3(t) = \frac{v}{4\pi} e^{j\Omega t} \int_0^\infty F_1(\omega) F_2(\omega + \Omega) \, e^{2j\omega t} \, d\omega + \text{conjugate},$$

where $\Omega = \pm \pi v/p$, the sign depending on which of the input waveforms has the higher centre frequency. Thus, $f_3(t)$ is the time-contracted convolution of the input waveforms, but now with a frequency shift of Ω in one of the inputs and in the output.

The Convolver as a Matched Filter. Since the convolver behaves essentially as a linear filter, with the "impulse response" given by the applied reference

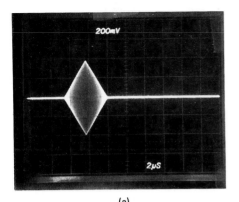 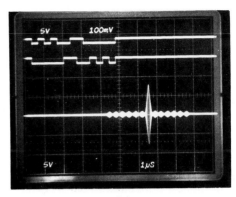

(a) (b)

FIGURE 10.11. Output waveforms produced by a degenerate acoustic convolver similar to that of Figure 10.10(a), for input waveforms that are (a) rectangular pulses; (b) PSK waveforms coded according to the 13-chip Barker code and its time-reverse. In both cases the output waveforms are 4.6 μsec long. (Courtesy J. H. Collins, University of Edinburgh)

waveform, it may in particular be used as a matched filter. Thus the convolver and its reference waveform generator may replace the matched filter in the spread-spectrum receiver illustrated in Figure 10.3. For this application the required reference waveform must correspond to the time-reverse of the ideal input waveform (for one symbol). As we have seen earlier, PSK and MSK waveforms can be generated quite readily, and hence the convolver is particularly suited for matched filtering of these waveforms. Note that the convolver enables the coding to be changed at will, and also the bandwidth and centre frequency if necessary; it is therefore highly programmable.

For illustration, Figure 10.11 shows some output waveforms produced by a degenerate acoustic convolver similar to that of Figure 10.10(a), with an input centre frequency of 110 MHz and a parametric region 4.6 μsec long. In Figure 10.11(a) the inputs are both rectangular pulses 4.6 μsec long, with carrier frequency 110 MHz. The output waveform has a triangular envelope, as given by the convolution of a rectangle with itself, with carrier frequency 220 MHz. The output waveform virtually disappears if either of the two inputs is removed. In Figure 10.11(b) one input is a PSK waveform coded according to the 13-chip Barker code 0101001100000, and the other input is the time-reverse. The binary coding is shown in the upper part of the figure, prior to bi-phase modulation. For this case the ideal output waveform has six sidelobes of equal amplitude on each side of the main peak, and the peak amplitude is 13 times the sidelobe amplitude. It can be seen that the convolver output waveform corresponds very closely to this ideal.

Note that, despite the use of a non-linear mechanism, the convolver processes an input waveform in an essentially linear fashion, as shown by equation (10.23), and this applies even for an input signal accompanied by noise. Ideally, the output signal-to-noise ratio will be the same as that given by a linear matched filter. There is however the limitation that the signal is properly processed only if it arrives at the same, or almost the same, time as the reference is applied, so that the two waveforms overlap only in the parametric region of the device. If the time of arrival of the signal is unknown, an obvious strategy is to apply the reference waveform repetitively. A detailed examination of this process [385] shows that, apart from a time distortion due to the time contraction, an ideal convolver can give *exactly* the same output as a matched filter, irrespective of the signal timing, and this applies even if the signal is accompanied by noise or interference. To obtain this result, the delay along the convolver parametric region must be at least twice the duration of the reference.

Signal Processing Using Idler Wave Generation. An alternative type of signal processing can be obtained by applying input waveforms to one transducer and to the parametric port, instead of to both transducers. With an appropriate choice of frequencies, the non-linearity in this case generates a second surface wave, travelling in the opposite direction to the incident wave. This "idler" wave can be sensed by the input transducer, or by another transducer at the same end. If the input at the parametric port is a short pulse the output waveform is esentially the *time-reverse* of the transducer input waveform. This has been demonstrated using the non-degenerate convolver of Figure 10.10(b) [379, 384], and also using the non-degenerate diode convolver, considered later. For degenerate devices, the use of the idler wave is not usually practicable because of the presence of spurious signals. Time-reversal is of some practical interest since a reference of this form is needed for a convolver using contra-directed input waves. Thus a coded signal can be correlated using a reference of the *same* form if the reference is first time-reversed in another convolver.

10.3.2. Performance of Basic Convolvers

The simplest type of surface-wave convolver is the degenerate acoustic device of Figure 10.10(a). The principles of this device were discussed in the previous section, and here we consider its practical performance. Acoustic convolvers using waveguides to improve the efficiency will be described in Section 10.3.3 below, and convolvers using semiconductors will be described in Section 10.4.

The first convolvers to be demonstrated were similar in principle to the device of Figure 10.10(a), but used bulk acoustic waves rather than surface waves [386]. Convolution using surface waves, with the degenerate and non-degenerate structures of Figure 10.10, was first demonstrated by Luukkala and Kino [384]. In all of these devices the non-linear propagation medium was lithium niobate.

(a) Efficiency of Degenerate Acoustic Devices. As explained in Section 10.3.1, the convolver exploits a non-linear interaction of propagating surface waves to produce an output voltage proportional to the spatial integral of the product. A primary concern is the efficiency of this interaction, and this has been examined theoretically by several authors [387–390]. It is found that, for C.W. input waveforms with the same frequency ω, the open-circuit output voltage at frequency 2ω may be expressed in the form

$$[V_{oc}]_{rms} = \frac{M}{W} \sqrt{P_{s1} P_{s2}}, \qquad (10.27)$$

where P_{s1} and P_{s2} are the surface-wave powers, taken to be small, and W is the width of the parametric electrode, assumed to be equal to the transducer apertures. The constant M depends only on the substrate material and orientation, and is generally defined such that the equation gives the r.m.s. value of the open-circuit voltage. The derivation assumes ideal propagation conditions, so that diffraction, dispersion and attenuation are negligible. Note that the output voltage is independent of the input frequency and of the length of the parametric electrode.

The constant M is related in a complex manner to the linear and non-linear bulk constants of the material [387–390]. In principle this enables M to be calculated for any orientation of interest, but in practice many of the constants required are unknown, or are not sufficiently accurate. It is thus necessary to determine M experimentally, by measuring the output voltages of practical devices, though theoretical predictions for lithium niobate are in fair agreement with experiment [390]. Measurements have been reported for a variety of materials [391], showing that Y, Z lithium niobate gives the relatively large value of $M = 1.2 \times 10^{-4}$ V m/watt. This is the usual choice of substrate material since, in addition to the strong non-linearity, it also gives low attenuation and diffraction spreading. The temperature sensitivity is of little consequence in convolvers, as explained below. The only material known to give a larger M-value is PZT ceramic, but this is generally unacceptable because of the acoustic propagation loss.

The analysis [387, 388] also leads to the conclusion that the convolver output port can be represented by a simple equivalent circuit consisting of a voltage generator V_{oc} in series with a capacitance. The latter is simply the capacitance measured between the parametric electrode and the ground plane. The absence of any resistance here appears to imply that infinite power can be extracted, but in fact the analysis only applies when the interaction is too weak to cause any appreciable reduction of the surface-wave amplitudes, as is usually the case in practice. A small amount of resistance is in fact present because of the resistivity of the parametric electrode, and because the voltage on the electrode causes some excitation of bulk waves.

A more practical measure of the device efficiency is the *bilinearity factor C*, relating the external powers of the input and output waveforms. We define P_1 and P_2 as the available powers of the generators supplying the C.W. input waveforms, and P_0 as the output power delivered to the load. These powers can if appropriate take account of any matching networks, which are usually included in practical devices in order to maximise the efficiency. Since the device is bilinear P_0 is proportional to $P_1 P_2$, and the efficiency can therefore be characterised by defining the bilinearity factor as

$$C = 10 \log_{10}\left(\frac{P_0}{P_1 P_2}\right), \quad (\text{dBm}), \tag{10.28}$$

where P_1, P_2 and P_0 are measured in mW. This definition can be applied to any bilinear device, including several types of convolver, described later, which do not use the acoustic non-linearity; it also applies if the performance is affected by propagation effects such as attenuation or diffraction. Generally, C will depend on the input frequency ω; for an ideal device, C would be independent of ω over the band occupied by the input waveforms to be used. With appropriate matching, the degenerate acoustic convolver of Figure 10.10(a) with an input centre frequency of, say, 100 MHz gives typically $C = -95$ dBm. For practical applications a better efficiency is desirable, though not always essential, and some modified types of convolver giving better efficiencies will be described in Sections 10.3.3 and 10.4.

(b) Second-order Effects. In practice, degenerate convolvers deviate in several ways from the ideal behaviour described above and in Section 10.3.1, though some imperfections can be minimised quite simply [380, 392]. In particular, it is usual to restrict the input frequency band such that it does not overlap the output band, and to use external bandpass filters to reject some unwanted components.

It has been assumed so far that the power levels of the input surface waves are low, so that the non-linearity has a negligible effect on the fundamental waves and causes only square-law and product terms to be generated. At high power levels this will not be true, so that the device will no longer be bilinear; the output amplitude will no longer be proportional to the product of the input amplitudes, and will thus exhibit saturation. In this situation the output voltage is no longer given by equation (10.27), and the bilinearity factor of equation (10.28) is not valid. In practice, saturation is not usually observed at realistic power levels, as might be expected from the low device efficiencies. For example, if C.W. waveforms with $+20$ dBm available power are applied to a device with a typical $C = -95$ dBm, the output power is -55 dBm, which is many orders of magnitude below the input powers.

As in other surface-wave devices, the convolver can be affected by several propagation effects. Consider first the effect of *attenuation*, assuming the wave propagation to be otherwise ideal. We consider the degenerate device of Figure 10.10(a) and allow for the delay due to the transducer separation, which was neglected in the previous section. If waveforms $f_1(t)$ and $f_2(t)$ are applied to the input transducers, and if the power levels are low, the fundamental waves have amplitudes of the form

$$u_1(t, x) \propto f_1(t - x/v - d/v) \exp[-\alpha(d + x)],$$
$$u_2(t, x) \propto f_2(t + x/v - d/v) \exp[-\alpha(d - x)], \quad (10.29)$$

where the transducers are taken to be located at $x = \pm d$ and α is the attenuation coefficient, taken to be independent of frequency. The parametric electrode is taken to be of length L, occupying the region $|x| < \frac{1}{2}L$. The output waveform $f_3(t)$ is found by integrating the product $u_1(t, x)u_2(t, x)$ over this region, giving

$$f_3(t) = \gamma e^{-2\alpha d} \int_{-L/2}^{L/2} f_1(t - x/v - d/v) f_2(t + x/v - d/v) \, dx, \quad (10.30)$$

where γ is a constant. We thus have the important conclusion that the form of the output waveform is not affected by exponential attenuation. This is in marked contrast to a linear surface-wave device, in which attenuation causes distortion of the output waveform. In practice α varies a little with frequency (Section 6.5), causing some distortion, but this is usually very small and has a form such that it can be compensated by external trimming [393].

Equation (10.30), with α constant, can be taken as the form of output waveform given by an *ideal* degenerate convolver. The convolution relation is obtained if the input waveforms have finite durations, less than L/v, and are timed such that the product exists only in the parametric region $|x| < \frac{1}{2}L$. Equation (10.30) is then unaffected if the limits are changed to $\pm \infty$, and with $\tau = t + x/v - d/v$ gives

$$f_3(t) = v\gamma\, e^{-2\alpha d} \int_{-\infty}^{\infty} f_1(2t - 2d/v - \tau)\, f_2(\tau)\, d\tau. \tag{10.31}$$

This is the convolution of $f_1(t)$ and $f_2(t)$, with a time-contraction of 2 and a delay of d/v. Equation (10.31) shows that the output is not distorted by a change of the *velocity* v (due to fabrication errors for example), since this simply changes the delay d/v. In addition *temperature* changes cause no distortion; d and v are both affected, but this simply changes the delay d/v in accordance with the temperature coefficient α_T of Section 6.4. The insensitivity of the convolver to these propagation effects is an important advantage and is particularly relevant when processing PSK waveforms, since PSK filters are very sensitive to velocity and temperature changes (Section 10.2).

Another relevant propagation effect is *dispersion*, which can significantly affect the convolver fidelity. As seen in Section 6.5, dispersion arises from the mass loading due to the parametric electrode, and depends on the electrode thickness. The effect is conveniently expressed in terms of the spectrum of the output waveform. If $V'_0(\omega)$ is the output spectrum, allowing for dispersion, it can be shown that for a degenerate device [393, 394]

$$V'_0(\omega) \approx V_0(\omega)\, \exp\,[-jLk(\omega/2)], \tag{10.32}$$

where $V_0(\omega)$ is the ideal output spectrum, not allowing for a delay $L/(2v)$, and $k(\omega)$ is the surface-wave wavenumber at frequency ω. Equation (10.32) is valid for arbitrary input waveforms. It shows that the phase distortion in the output spectrum, at frequency ω, is the same as the phase distortion for a surface wave at frequency $\omega/2$ travelling the length L of the parametric electrode. Note that, for no distortion, k must be a linear function of ω. This requires the group velocity $v_g = d\omega/dk$ to be independent of ω, though the phase velocity ω/k need not be constant. For PSK waveforms, theoretical simulations [394] lead to the conclusion that the dispersion is generally acceptable if $|dv_g/d\omega|$ is less than $v_g^2 T_c^2/(8L)$. For the basic convolver this requirement is usually satisfied quite easily, though it becomes of some concern for waveguide convolvers, as will be seen in Section 10.3.3.

In many devices the most serious imperfection is *fold-over convolution*, due to the acoustic reflectivity of the transducers. A wave generated at one end of the device is reflected by the transducer at the other end, and the reflected and primary waves are mixed in the parametric region in the usual way. This gives an output signal when only one input waveform is applied, so that the device is not strictly bilinear; thus the bilinearity factor of equation (10.28) is meaningful only if the fold-over convolution is well suppressed. A simple test for this is to measure the output level when C.W. waveforms of the same power are applied to the inputs, and compare with the output level observed when one input is disconnected. Typically, a degenerate device such as that of Figure 10.10(a) gives a fold-over rejection of 25 to 30 dB in this test, and this is not generally acceptable. As in linear devices the transducers could be electrically mis-matched in order to reduce their reflection coefficients (Section 7.1.3), but for convolvers this is not generally attractive as it reduces the efficiency, which is already

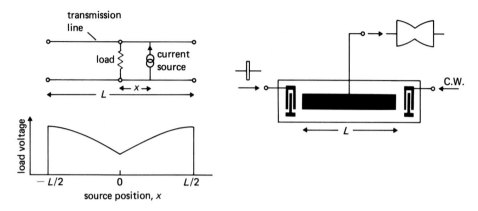

FIGURE 10.12. Transmission-line effect in convolver. Left: analysis. Right: simple experimental test.

rather low. A better method is to adopt a duplicated arrangement, explained later in Section 10.3.3; this typically enables a fold-over rejection of 40 dB to be obtained. The non-degenerate convolver of Figure 10.10(b) gives better fold-over rejection because the product due to the reflected wave and the primary wave does not have a spatial periodicity equal to that of the parametric sensor.

Some devices are significantly affected by electromagnetic wave propagation along the parametric electrode, which is thus found to behave as a *transmission line*. When this is the case, the output voltage due to the product term at location x is found to be dependent on x, so that the device is spatially non-uniform. Thus the ideal form of the output, equation (10.30), is no longer produced even if the product is formed correctly at each x. The effect becomes significant when the length of the parametric electrode exceeds about one quarter of the electromagnetic wavelength, at the output frequency. A simple analysis [395, 396] treats the parametric electrode as a transmission line, with a C.W. current source at the output frequency representing a localised excitation, as in Figure 10.12. The calculated output voltage, as a function of source position, is shown for a transmission line one third of a wavelength long, appropriate for a degenerate device with 260 MHz output frequency and a parametric electrode 25 μsec long [395]. A simple experimental test for this is to apply a C.W. waveform to one input of the convolver, and a short pulse to the other input; as the surface-wave pulse scans along the parametric electrode, the output waveform shows the spatial sensitivity as a function of time. This generally gives good agreement with the analysis, though strictly speaking it is not compatible since the analysis assumes C.W. excitation.

The electrical length of the parametric electrode can be reduced, thus reducing the distortion, by mounting the substrate over a dielectric layer with lower permittivity [395]. Some other methods are mentioned in Section 10.3.3 below, and a more rigorous experimental test is described in Section 10.3.4.

10.3.3. Waveguide Convolvers

In view of the low efficiency of the basic acoustic convolver, several more efficient types of convolver have been developed. In this section we consider a modified type of acoustic convolver, while some other devices are described in section 10.4.

Assuming that the input transducers are well matched and the input power levels are specified, equation (10.27) shows that the only strategy available for increasing the output voltage of a degenerate acoustic convolver is to reduce the width W of the parametric electrode. It is found that a width of only a few wavelengths can be used, and for such a narrow width the electrode behaves as a surface-wave *waveguide*. This is a useful feature since it eliminates the effects of diffraction spreading and beam steering, which would cause severe difficulties if the beam were not guided. On the other hand, the presence of a waveguide introduces some new complications, so we first consider briefly the waveguide behaviour.

A metal strip on a lithium niobate substrate acts as a waveguide because the surface-wave velocity for a uniformly metallised surface is lower than that for a free surface. The velocity reduction is mainly due to electrical loading, and consequently the strip is called a "$\Delta v/v$ waveguide". The behaviour has been analysed by Schmidt and Coldren [397], and experimental results [397, 398] are in good agreement. The waveguide can support a series of propagating modes, each of which is dispersive. The phase and group velocities of the first few modes are shown in Figure 10.13 for a Y, Z lithium niobate substrate. Here the analysis [397] has been used with accurate velocity data obtained from optical probing measurements (Section 6.2.4), since the velocity anisotropy affects the solutions somewhat. The abscissa gives the guide width a divided by λ_0, the wavelength for straight-crested waves propagating in the Z-direction on a free surface. For the present we consider only the continuous curves, referring to a film of zero thickness.

In a convolver it is desirable that only the fundamental mode should be present, since the presence of two or more modes causes distortion. The higher modes can be excluded by operating at frequencies below their cut-off points. However, the excitation of these modes is generally quite weak at higher frequencies, because their transverse amplitude distributions are very different from the distribution of the wave incident at the end of the guide. This is true in particular for the first higher mode, which has an amplitude anti-symmetric about the guide centre line. For this reason, higher-mode excitation is generally small if $a/\lambda_0 < 4$, the cut-off point of the second higher mode. At the same time, since dispersion causes distortion of the convolver response (Section 10.3.2), it is advisable to avoid the region $a/\lambda_0 < 1.5$ where the fundamental is highly dispersive. Thus a suitable range for the guide width is $1.5 < a/\lambda_0 < 4$.

In this region of a/λ_0, the group velocity of the fundamental mode is almost independent of frequency, as shown by the corresponding continuous line in Figure 10.13. This implies that the dispersion causes little distortion of the convolver output waveform, as noted earlier. However, for a finite film thickness the dispersion is increased, as shown by the broken line which refers to a typical experimental value. The added dispersion here is due to mechanical loading, and is acceptable for many applications. In fact, the extra dispersion can be advantageous because it reduces the

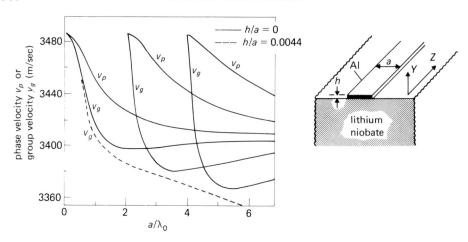

FIGURE 10.13. Phase and group velocities for modes in a $\Delta v/v$ waveguide (Courtesy, Plessey Research).

surface-wave attenuation associated with the non-linearity, as noted in Section 6.3. This is confirmed by the fact that saturation is observed in convolvers using very thin films [399] but not in convolvers using typical film thicknesses, for input power levels of about +20 dBm.

To use a waveguide effectively, some efficient method of launching the narrow surface-wave beams must be found. Narrow-aperture uniform transducers are not generally attractive because their high impedances make them difficult to match efficiently. At high frequencies the effective impedance may be reduced by parasitic capacitance (Section 7.1.2), but this also increases the Q-factor and so limits the bandwidth obtainable. A better approach is to use a wide aperture transducer and compress the beam width using either a multi-strip coupler (Section 5.7) or a waveguide horn [400]. Focussing transducers, in which the electrodes are curved, have also been shown to be effective [401]. Another method is to use a narrow-aperture *chirp* transducer illuminating the end of the waveguide directly, as shown in Figure 10.14. We have already seen, in Section 9.3.2, that the admittance of a chirp transducer is larger than that of a uniform transducer with the same bandwidth and aperture. The use of a chirp thus enables the impedance of a narrow-aperture transducer to be reduced without changing its aperture, as shown experimentally using a time-bandwidth product of about 10 [402]. In a convolver, the dispersion introduced by the transducers is cancelled by arranging one transducer to be an up-chirp and the other a down-chirp, as in Figure 10.14. A further advantage is that the transducer chirps can be made slightly different to compensate for small amounts of dispersion in the waveguide [402, 403].

The first waveguide convolvers, demonstrated by Defranould and Maerfeld [404], used multi-strip beam compressors and gave a bilinearity factor of $C = -71$ dBm, much better than the -95 dBm typical of the basic degenerate convolver. To obtain a suitable capacitance a ground electrode was located on the *upper* surface, adjacent to the waveguide, instead of on the rear surface as in Figure 10.10(a). Here we

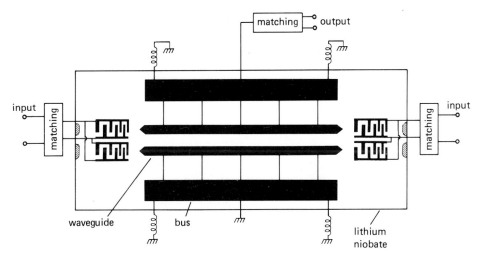

FIGURE 10.14. Waveguide convolver using chirp transducers.

consider as an example the device using chirp transducers [402], shown in Figure 10.14. This device had an input centre frequency and bandwidth of 300 MHz and 120 MHz respectively, and the delay along the parametric region was 16 μsec; thus signals with time-bandwidth products up to 1,920 could be correlated. As shown in the figure, the basic structure is duplicated so that there are two input transducers at each end, and the electrode polarities in one of the four transducers are reversed. This implies that, ideally, the waves launched at one end do not excite the tranducers at the other end; thus the acoustic reflection coefficient is reduced, and the fold-over suppression is improved. The waveguide width was 34 μm, giving $a/\lambda_0 = 3$ at 300 MHz, the input centre frequency. The wide bus-bars, connected at intervals to the waveguides, are included to minimise losses due to the resistance of the narrow waveguides. The small inductors reduce the spatial non-uniformity due to the transmission-line effect [405]; alternatively, it has been shown that the non-uniformity can be reduced by using an external transmission-line network to equalise the phase [406]. A photograph of the device, including simple matching networks, is shown in Figure 10.15.

Figure 10.16 shows the measured output power, for C.W. input waveforms with the same frequency and with 0 dBm power level; this is equal to the bilinearity factor C defined in equation (10.28). At the input centre frequency, 300 MHz, the bilinearity factor was -67 dBm. The graph on the right shows the phase error, that is, the measured phase of the output with a large linear term subtracted. The graph on the left also shows the output powers observed with one of the inputs disconnected; these unwanted components are due to fold-over convolution, and their powers show that the fold-over suppression is typically 38 dB. Similar performance has been demonstrated by devices using horn [406–408] or multistrip [399, 409] beam compressors instead of chirp transducers. The spatial uniformity of the device is considered further in Section 10.3.4 below.

FIGURE 10.15. Photograph of acoustic waveguide convolver, with cover removed. The total length of the package is 12 cm. (Courtesy, Plessey Research).

Required Efficiency and Fold-over Rejection. Assuming typical input power levels of $+20$ dBm the above convolver, with $C = -67$ dBm, gives an output power level of -27 dBm. This figure applies for C.W. input waveforms, and also applies for coded waveforms if the reference corresponds to the time-reverse of the signal and if the waveform durations are equal to the delay along the parametric region. If the output is applied to an amplifier with 240 MHz bandwidth, an output signal-to-noise ratio of about 58 dB is obtained. This implies that the input signal can be reduced by 58 dB before the output disappears below the noise, thus showing a very useful dynamic range. However, a more detailed examination is needed to deduce the efficiency needed for a given application [410]. The main criterion is that, when correlating a signal accompanied by noise, the thermal noise at the convolver output must not appreciably degrade the output signal-to-noise ratio. It is found that the efficiency needed increases with both the bandwidth and the time-bandwidth product

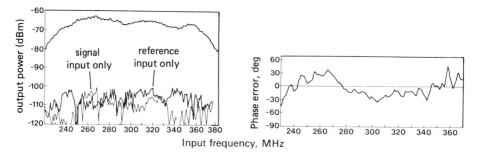

FIGURE 10.16. Performance of the waveguide convolver of Figure 10.15 for C.W. input signals with the same frequency. Left: output powers for 0 dBm input power levels. The upper curve gives the bilinearity factor, C. Right: phase error of output waveform. (Courtesy Plessey Research).

of the signal being correlated; for example, if the signal bandwidth is 120 MHz and the TB-product is 2000, the bilinearity factor needs to exceed about -80 dBm. This requirement is satisfied by the waveguide convolver, but not by the simpler acoustic convolver of Figure 10.10(a).

The level of the fold-over convolution is also an important consideration. A particular consequence of this phenomenon is that a spurious output signal, associated with the reference input, is present irrespective of the level of the signal input. This can obscure the required output signal when the input signal level is low, thus reducing the dynamic range. Analysis of this effect [410] shows that, for a device correlating a coded signal, the requirement on the C.W. fold-over rejection depends on the signal TB-product, and becomes more stringent for *lower* TB products. For the waveguide convolver it is concluded that the C.W. fold-over rejection of 38 dB is adequate for signals with TB-products exceeding 100.

10.3.4. Convolver Fidelity and Frequency Response

We here consider the distortion of the output waveform due to imperfections in the convolver, excluding fold-over convolution. Phenomena such as dispersion, the transmission-line effect and imperfect transducer responses will distort the output waveform, causing for example an increase in the width of the correlation peak and a reduction of signal-to-noise ratio. In addition the peak amplitude will generally depend on the relative timing of the signal and reference inputs, owing to the spatial non-uniformity.

For a *linear* device, any imperfections can be conveniently assessed by considering the frequency response $H(\omega)$, obtained by C.W. measurements. Using $H(\omega)$, the output waveform can be calculated for any input waveform, without explicitly considering the physical processes involved. For a convolver the conventional definition of frequency response is not applicable because the device is not linear. However, it has been shown that a *two-dimensional* frequency response can be used in this case.

To define the convolver frequency response, suppose that C.W. waveforms $\cos(\omega_1 t)$ and $\cos(\omega_2 t)$, with different frequencies, are applied to the two inputs. It is assumed that the device is bilinear, thus neglecting fold-over convolution; this is generally a good approximation for a well-designed device. The output is therefore a sum of two C.W. waveforms at the sum and difference frequencies, and usually the difference-frequency term can be ignored because it is small and is of no interest. The sum-frequency output waveform is written in the form

$$f_3(t) = \tfrac{1}{2} H(\omega_1, \omega_2) \exp[j(\omega_1 + \omega_2)t] + \text{conjugate}, \tag{10.33}$$

where $H(\omega_1, \omega_2)$ is the two-dimensional frequency response, a complex function whose magnitude and phase give the amplitude and phase of the output waveform. It is convenient to define $H(\omega_1, \omega_2)$ for negative frequencies as well as positive, and this is done by defining $H(-\omega_1, -\omega_2) = H^*(\omega_1, \omega_2)$ and $H(\omega_1, \omega_2) = 0$ when ω_1 and ω_2 have different signs. It can be shown [411] that the function $H(\omega_1, \omega_2)$ gives a complete description of the electrical behaviour of any bilinear device, irrespective of the physical processes involved. For arbitrary input waveforms, the output waveform can be deduced from the equation

$$F_3(\omega) = \frac{1}{\pi} \int_{-\infty}^{\infty} F_2(\omega') F_1(\omega - \omega') H(\omega - \omega', \omega') \, d\omega', \qquad (10.34)$$

where $F_1(\omega)$ and $F_2(\omega)$ are the Fourier transforms of the input waveforms, and $F_3(\omega)$ is the transform of the output waveform. It is also possible to define a two-dimensional impulse response as the inverse Fourier transform of $H(\omega_1, \omega_2)$, and the output waveform can then be expressed as a two-dimensional convolution involving the input waveforms [411].

For an ideal degenerate convolver, the frequency response takes the form

$$H(\omega_1, \omega_2) = K \operatorname{sinc} [\tfrac{1}{2}(\omega_1 - \omega_2) T_0] \, e^{-j(\omega_1 + \omega_2) T_d} \qquad (10.35)$$

for positive frequencies, where K is a constant, $T_0 = L/v$ is the delay along the parametric region, and $2T_d = 2d/v$ is the delay corresponding to the separation of the input transducers. Using equation (10.34), it can be shown that this ideal response leads to an integral with the ideal form of equation (10.30), relating the waveforms in the time domain.

Figure 10.17 shows an experimental arrangement for measuring the frequency response, omitting some amplifiers and filters that are necessary in practice. Two signal generators are used to generate C.W. input waveforms at frequencies ω_1, ω_2, and a vector voltmeter measures the amplitude and phase of the convolver output. The phase reference for the voltmeter is obtained by mixing the two input waveforms, giving a waveform at the sum frequency $\omega_1 + \omega_2$. If the input frequencies are the same, so that $\omega_1 = \omega_2 = \omega$, say, the output amplitude is directly related to the bilinearity factor C at frequency ω.

In order to examine the spatial uniformity of the convolver, it is convenient to define another form of the response. This new function, called the *spatial response* $P(\omega, \tau)$, is defined as the inverse Fourier transform of $H(\omega_1, \omega_2)$ with respect to the difference of the two input frequencies, so that [412–414]

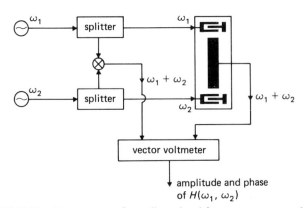

FIGURE 10.17. Measurement of two-dimensional frequency response of convolver.

DEVICES FOR SPREAD-SPECTRUM COMMUNICATIONS 311

$$P(\omega, \tau) = \frac{1}{2\pi} \int_{-\infty}^{\infty} H\left(\frac{\omega + \omega'}{2}, \frac{\omega - \omega'}{2}\right) e^{j\omega'\tau} d\omega'. \qquad (10.36)$$

For an ideal convolver, $P(\omega, \tau)$ is independent of τ for $|\tau| < T_0/2$ and is also independent of ω over the band occupied by the input waveforms. Further analysis [414] shows that an approximate physical interpretation can be obtained by equating τ with x/v; at each x, the product term is in effect applied to a linear filter with frequency response $P(\omega, \tau)$, and the filter outputs, infinite in number, are summed.

Experimentally, the spatial response can be measured using the arrangement of Figure 10.17, varying the input frequencies such that their sum is equal to a constant ω, and transforming the output data with respect to the difference frequency as in equation (10.36). Figure 10.18 shows the amplitude and phase of $P(\omega, \tau)$ for the waveguide convolver described in Section 10.3.3, taking $\omega = 2\pi \times 600\,\text{MHz}$, the output centre frequency. The sharp dips observable in the amplitude are associated with the connections between the waveguides and the bus-bars. This method of measuring spatial uniformity is much superior to the method using a short input pulse, mentioned in Section 10.3.2; the spatial resolution obtainable by the latter method is limited by the fact that short input pulses give low signal-to-noise ratios at the device output. In addition, since short pulses have wide bandwidths the output given by the pulse method is in fact a frequency-domain average of the spatial response, and is therefore less informative.

The observed variation of $P(\omega, \tau)$ with τ implies that, for a device correlating a coded waveform, the amplitude of the output correlation peak will depend on the timing of the input signal. In addition, if the input signal is accompanied by noise, the output noise power will vary with time. However, a detailed study of these effect [412] shows that quite large errors can be tolerated, and it can be concluded that the spatial variations shown in Figure 10.18 would be acceptable for most applications.

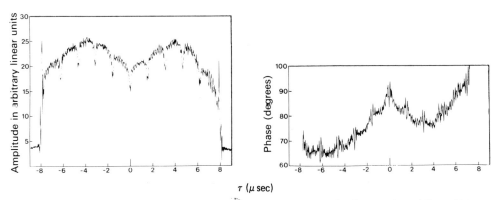

FIGURE 10.18. Amplitude and phase of the spatial response $P(\omega, \tau)$, for the convolver of Figure 10.15. (Courtesy, Plessey Research).

10.4. OTHER NON-LINEAR DEVICES

In addition to the acoustic convolvers described above several other types of convolver have been developed, making use of non-linear effects in semiconductors [381, 382]. Historically, these devices were developed in order to improve on the rather low efficiency of the basic acoustic convolver, and they did indeed give much better efficiencies. However, the later development of the waveguide acoustic convolver has made these novel devices, which are more difficult to fabricate, less attractive. They are nevertheless of some interest, not only because of the performance achieved but also because they have led to the development of several novel signal processing operations. These include the ability to store a reference waveform and to use it to correlate a signal applied later, and also the ability to correlate a very long waveform, with duration much greater than the acoustic delay in the device.

Semiconductor convolvers are described in Section 10.4.1 below, and Section 10.4.2 describes devices for both storage and correlation. Correlation of long waveforms is described in Section 10.4.3.

10.4.1 Semiconductor Convolvers

The surface-wave *diode convolver* [415–417], introduced by Reeder, is shown in Figure 10.19. Here the non-linearity is produced by a set of diodes, each connected to one of a series of uniformly spaced transducers acting as taps. Non-degenerate operation is necessary so that the harmonics of the input waveforms can be rejected by filtering the output. As in the case of the programmable PSK filter, Section 10.2, the need to make many external connections limits the practicable number of taps and their spacing; experimental devices have used up to 150 taps with inter-tap delays down to about 30 nsec, enabling signals with up to 15 MHz bandwidth to be processed. The

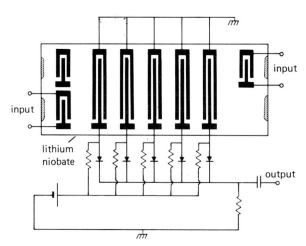

FIGURE 10.19. Diode convolver. The transducer at top left gives the output when idler wave generation is used.

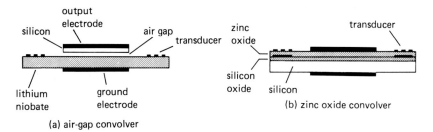

FIGURE 10.20. Semiconductor convolvers.

main advantage is that exceptionally large bilinearity factors, in the region of $-30\,\text{dBm}$, can be obtained. Although saturation is observed at relatively low input power levels, in the region of $0\,\text{dBm}$, the large bilinearity factor leads to a very good dynamic range, up to 80 dB.

The efficiency of the diode convolver is affected by the magnitude of a D.C. bias current applied to each diode, being maximised when each diode is forward biassed such that its low-signal impedance is about equal to the tap impedance. The dependence on bias current can be exploited by using different currents in the individual diodes, and this leads to some additional signal processing operations. For example, a transversal filter with adjustable tap weights can be realised [418], and this can operate as a programmable bandpass filter. Alternatively, using linear chirp waveforms applied to the input transducers, it has been shown [419] that the output can give the Fourier transform of the sequence of bias currents. This contrasts with the chirp filter method of Fourier transformation (Section 9.7) in that the input data are in parallel form rather than serial.

Other devices have introduced a non-linear effect by coupling a surface wave more directly to a semiconductor, and two examples that have been investigated extensively are shown in Figure 10.20. In the *semiconductor air-gap* convolver [381, 420], which is degenerate, a silicon slice is held in close proximity to a lithium niobate substrate on which the contra-directed surface waves propagate. The electric fields associated with the surface waves penetrate into the semiconductor, depleting the surface and giving rise to a potential proportional to the square of the total field. The semiconductor must be held very close to the piezoelectric substrate since, as already seen in Section 3.5, the field decays rapidly above the piezoelectric surface. However, if the semiconductor is too close it is found to cause exponential attenuation of the surface waves, sufficient to reduce the overall efficiency of the convolver. There is therefore an optimum spacing, which in practice is typically $0.5\,\mu\text{m}$. This requirement places considerable demands on the construction of the device, and is its main disadvantage. Usually, the silicon is supported by means of a sparse array of small "posts" on the lithium niobate surface; this does not appreciably perturb the propagating surface waves. The posts are made by ion etching the niobate surface, with the required post areas masked off. Analysis of the device efficiency agrees well with experimental measurements [421, 422]. A device described by Cafarella *et al.* [423] had 100 MHz input bandwidth and $10\,\mu\text{sec}$ interaction length, giving a bilinearity factor of $C = -66\,\text{dBm}$, and several other examples are quoted by Reible [422]. This device can also be used to scan an optical image focussed on to the silicon surface, making the lower ground electrode thin enough to be transparent. Several techniques, reviewed by Kino [381], have been used for this purpose, though they have not been found competitive with other technologies such as charge-coupled devices.

A similar principle is used in the *zinc oxide* convolver [424, 425] of Figure 10.20(b). The piezoelectric zinc oxide film, deposited on a silicon substrate, causes the waves to be accompanied by electric fields which penetrate into the silicon, where the non-linear interaction takes place. The film also enables the surface waves to be generated using conventional interdigital transducers. This technology has the advantage of avoiding the inconvenience of a small air gap, though care is needed in the fabrication of the zinc oxide film [424]. An example is the device of Green and Khuri-Yakub [425], which gave an impressive bilinearity factor of $-44\,\text{dBm}$ and had an input bandwidth of about 5 MHz and a 3.5 μsec interaction length. Several other types of semiconductor convolver are described by Kino [381].

10.4.2. Storage Convolvers

The semiconductor air-gap convolver described above can also be used to store acoustic waveforms by exploiting the phenomenon of charge storage in traps at the semiconductor surface [426, 427]. To obtain reproducible results it has been found advantageous to incorporate diodes in the silicon surface, as in Ingebrigtsen's Schottky-diode device [428] which is illustrated in Figure 10.21(a). Here, a negative voltage applied to the upper electrode causes the diodes to be forward biassed, with a time constant of typically 0.5 nsec. For zero or reversed bias the time constant is

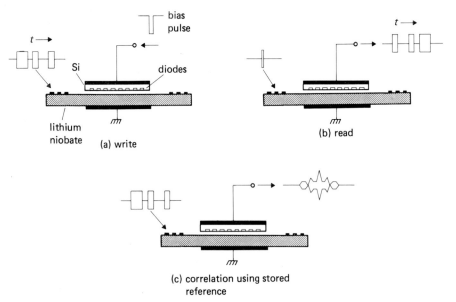

FIGURE 10.21. Air-gap storage convolver.

typically 1 to 100 msec. The waveform $f(t)$ to be stored, which must have a duration not exceeding the acoustic delay along the interaction region, is applied to the transducer at the left. When the surface wave packet is entirely in the interaction region a short bias pulse is applied to the parametric electrode. This forward-biases the diodes so that they rapidly accumulate a charge proportional to the total electric field, which is the sum of the spatially-invariant bias field and the field due to the surface wave. When the bias pulse is turned off the accumulated charges remain, reverse-biasing the diodes so that the decay time for the charges is long. In order that the stored charge pattern should be an accurate representation of the applied waveform the diode spacing must be less than half the surface-wave wavelength (as required by sampling theory), and the length of the bias pulse must be less that one half-period of the surface-wave carrier frequency.

The stored waveform can be read out later by applying a short pulse to the surface-wave transducer, as shown in Figure 10.21(b). The field associated with the travelling surface-wave pulse mixes non-linearly with the field due to the stored charge, and a waveform of the form $f(-t)$, the time-reverse of the stored waveform, emerges at the top electrode. Alternatively a forward output waveform $f(t)$ can be obtained by appling a pulse to the transducer at the right in the figure. The stored signal can be read out very many times without appreciable degradation – over 10^5 times in Ingebrigtsen's case [428].

More generally, if a waveform $s(t)$ is applied to the input transducer at the left, as in Figure 10.21(c), the output waveform $g(t)$ is the cross-correlation of $s(t)$ with the stored waveform $f(t)$, so that

$$g(t) = \int_{-\infty}^{\infty} s(t - \tau)f(-\tau)\,d\tau,$$

where the infinite limits are valid because the duration of $f(t)$ is less than the delay along along the interaction region. The device will therefore correlate a received signal if the stored reference $f(t)$ corresponds to the ideal signal. In contrast with a conventional convolver, the stored reference waveform does not need to be time-reversed and the output waveform is not time-contracted. Moreover, the reference input does not have to be repeated frequently as it does for a conventional convolver (Section 10.3.1). Ingebrigtsen [428] cites a device with 30 MHz input bandwidth and an interaction region 16 μsec long, in which the 3 dB decay time for the stored waveform was about 1 msec. A linear chirp waveform with 30 MHz bandwidth and 1.5 μsec duration was correlated, using a reference stored 1 msec earlier.

Similar devices have use p–n diodes in the silicon in place of Schottky diodes, giving rather longer time constants [429, 430]. Defranould et al. [430] describe a device with a 3 dB storage time exceeding 1 sec, and demonstrate correlation of a chirp waveform with 12 MHz bandwidth and 6 μsec duration using a reference stored 10 msec earlier. Storage has also been obtained using zinc oxide convolvers [431, 432], with a configuration as in Figure 10.20(b) except that Schottky or p–n diodes are added in the silicon surface. For example [432], a device using Schottky diodes with an interaction region about 6 μsec long demonstrated a 3 dB storage time of 7.5 msec.

Several other techniques for storage have been investigated [433]. For example Smythe et al. [434] have developed a silicon air-gap device in which a charge-coupled device is built into the silicon. A complex waveform can be entered into the charge-coupled device, and then serves as a reference to correlate a waveform introduced later in the form of a surface wave. Another method, demonstrated before the advent of the semiconductor devices, uses an electron beam to irradiate the surface of a piezoelectric instantaneously [435]. If a surface wave is present this process gives rise to a static charge distribution on the surface, corresponding to the instantaneous surface-wave amplitude. Application of a second electron-beam pulse at a later time causes generation of a new pair of surface waves, corresponding to the original input waveform and its time-reverse. Storage times of several minutes can be obtained, though the complexity of the electron beam technology makes this method unattractive for practical purposes. It has also been shown that a high-energy *optical* pulse can be used to store a surface-wave waveform on lithium niobate [436]. To read out the information the substrate is illuminated with a low-energy continuous optical beam, and a short surface-wave pulse is applied; the stored waveform then appears as modulation on the output optical beam. This method has been used to store surface-wave signals for remarkably long times, up to several weeks, though the need for short optical pulses with energies in the region of 0.1 J makes it of limited practical interest.

10.4.3. Correlation of Long Waveforms

The devices considered so far all have the limitation that they can only correlate waveforms with duration less than the acoustic delay in the device, and this cannot generally exceed about 50 μsec because of the length of substrate material required. In a spread-spectrum communication system it is necessary to correlate the waveform corresponding to one symbol which, as seen in Section 10.1, represents one binary digit of data, and in practice the symbol length can be much longer than 50 μsec. For this reason some special surface-wave techniques have been considered for correlation of waveforms longer than 50 μsec, and these are the subject of this section.

Most of the techniques considered here differ basically from the devices considered earlier in that the required integration is performed as a function of *time* rather than *space*. We have already seen, in Section 10.1, that the simple active correlator operates in this manner, multiplying the received signal by a corresponding reference waveform and applying the product to an integrator. The main disadvantage of this technique is the need for accurate synchronisation, which leads to long acquisition times. An obvious solution to this problem is to employ a *bank* of active correlators, with the reference timing incremented from one correlator to the next. The multiple delays required for the reference can be provided by a surface-wave tapped delay line, and this is the principle of the "tapped delay line correlator" demonstrated by Darby and Maines [437]. A similar principle was used by Menager and Desormiere [438], using thin strips of n on n^+ silicon as taps. In this case the reference and signal propagate as contra-directed surface waves on the same substrate, and each tap is non-linear and therefore mixes the two waves. Correlation is obtained by integrating each tap output with respect to time. However, both of these devices are limited by the requirement for a large number of external connections, as in the case of the programmable PSK filter (Section 10.2.1).

Another method is provided by the *integrating correlator* [439] in which the three basic functions required – multiple delays, multiplication and time integration – are all incorporated in the same surface-wave device. The structure of the device is essentially the same as that of the air-gap storage convolver, Figure 10.21. In the integrating correlator surface waves are introduced at each end, and each diode mixes the two waves to produce a baseband product which appears as a charge on the diode. The charge accumulates with time, so that when coded input waveforms are used each diode charge gives one point of the required correlation function. The accumulated charge distribution is read out by applying a short surface-wave pulse, using the non-linear interaction of the surface wave with the stored charge. The timing of the output correlation peak corresponds to the timing of the signal applied earlier, relative to the reference waveform.

In practice, the performance of this device has been somewhat restricted by the presence of spurious effects. Nevertheless Reible and Yao [440] have demonstrated correlation of a seqeunce of 40 bursts of signal, each with 10 MHz bandwidth and 6 μsec duration, while Ralston and Stern [441] have correlated a 10 msec waveform with 2 MHz bandwidth using a modified scheme to reduce spurious signals. Smythe and Ralston [442] have developed an integrating correlator using a charge-coupled device in the silicon, and have demonstrated correlation of a 200 μsec waveform with 20 MHz bandwidth. Integrating correlators have also been developed using zinc oxide technology, with a structure as in Figure 10.20(b) except that diodes are incorporated in the silicon; in this way Tuan *et al.* [443] correlated a 1 msec waveform with 5 MHz bandwidth. In all of the above examples, quasi-random PSK waveforms were used.

Some other methods of correlating long waveforms make use of more conventional devices. For example, acoustic convolvers can be cascaded to increase the effective interaction length [444]. A more compact method uses a surface-wave delay line in a recirculation loop, as shown in Figure 10.22. Here the loop input signal is repetitive, with repetition period equal to the delay T_d in the delay line. The loop gain is close to unity, so that successive periods of the input waveform are added coherently. In experimental demonstrations the loop input signal was obtained by correlating segments of PSK waveforms using either a PSK filter [352] or a convolver [445]. In the latter case the convolver input signal need not be repetitive; the reference waveform can be coded such that the output gives a sequence of correlation peaks, all with the

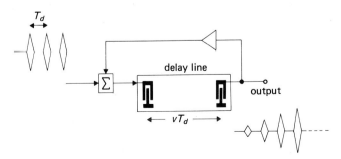

FIGURE 10.22. Recirculation loop for enhancing the SNR of repetitive waveforms.

same phase. The main limitation here lies in the delay line performance; to be effective the delay line frequency response must be very flat in the band of the signal, and spurious responses such as the triple-transit response must be well suppressed. An experimental system [445] with a 35 μsec delay line and a 30 μsec convolver was used to correlate a PSK waveform 2.1 msec long, using 70 circulations in the loop. Using an input waveform with 10 MHz bandwidth and an input signal-to-noise ratio of −41 dB, the output signal-to-noise ratio was +4 dB, within a few dB's of the theoretical ideal.

10.5. OSCILLATORS

In this final section we consider surface-wave oscillators, though since this book is concerned mainly with devices for signal processing this topic is considered only briefly. There are two types of surface-wave oscillator, one using a simple delay line and the other using a resonator. The resonator, which has not been described previously, makes use of reflective arrays and can serve as a narrow-band bandpass filter in addition to its oscillator application. For oscillators the important performance criteria are concerned with the frequency stability, and in this respect surface-wave devices do not perform quite as well as the best bulk crystal oscillators, at the present state of development. However, surface-wave devices can operate directly at frequencies up to 2 GHz. Bulk devices cannot generally operate above about 50 MHz, so that for higher frequencies multiplying circuits are necessary; thus at higher frequencies surface-wave devices can eliminate the need for a multiplier, and this generally leads to more compact devices consuming less power. A comparative review of bulk and surface wave oscillators, including fabrication topics affecting the stability, is given by Lukaszek and Ballato [446].

10.5.1. Delay-line Oscillator

As shown in Figure 10.23, the surface-wave delay line oscillator [447, 448] is essentially a simple interdigital delay line with an external amplifier providing feedback from the output to the input. The amplifier small-signal gain is made to exceed the insertion loss of the delay line, so that the circuit oscillates at a frequency such that the total phase change around the loop is a multiple of 2π. The transducer geometries are both symmetrical, and it follows that the delay line frequency response is non-dispersive, with delay T corresponding to the distance between the transducer centres. Thus the phase change due to the delay line is ωT at frequency ω, and the oscillation frequency must obey the relation

$$\omega T + \phi = 2n\pi, \qquad (10.37)$$

where n is an integer and ϕ is the phase change in the feedback circuit. To ensure that the device can oscillate at only one frequency a narrow-band delay line is used; as shown in Figure 10.23 this is readily obtained by using a long transducer, with length comparable with vT, and it is usually convenient to thin this transducer (Section 8.3). The delay T_1 in the figure is the inter-tap delay multiplied by the number of taps. The

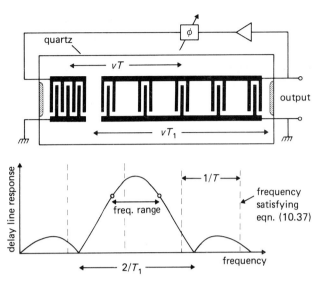

FIGURE 10.23. Delay-line oscillator. Upper: device structure. Lower: frequency selection mechanism. Open circles show typical points for unity loop gain.

amplifier gain is set such that the small-signal loop gain exceeds unity at only one of the frequencies satisfying equation (10.37), as indicated in the lower part of Figure 10.23 where these frequencies are shown by broken lines. It is quite common to include an adjustable phase shifter in the loop to enable the phase ϕ in equation (10.37), and hence the frequency, to be adjusted. This can be used to trim the frequency to compensate for small fabrication errors. Furthermore, if a voltage-controlled phase shifter is used the device becomes a voltage-controlled oscillator, and can generate freqeuncy-modulated waveforms. The substrate material is usually chosen to be ST, X quartz in view of its temperature stability.

10.5.2. Resonators

An alternative type of oscillator uses a surface-wave resonator to stabilise an oscillating circuit, and is thus analogous to the common bulk crystal oscillator mentioned in Chapter 1. A common type of resonator, shown in Figure 10.24(a) consists simply of two parallel reflectors separated by a distance L; this is the basic form for many optical resonators, and for the bulk acoustic resonator. In principle, a surface-wave version of this device could be realised using either tuned interdigital transducers (Section 4.4.4) or multi-strip mirrors (Section 5.5) as the reflectors. However, for these cases the reflection coefficients obtainable are not sufficiently close to unity to give good resonator Q-factors. This is because a small amount of loss arises from the film resistivity, and also from bulk wave excitation associated with the abrupt discontinuity at the edge. A much more effective approach is to use reflective gratings, shown in Figure 10.24(b), as first suggested by Ash [449]. The grating

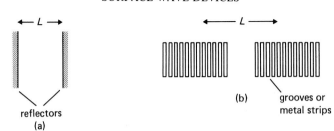

FIGURE 10.24. (a) Basic resonator structure. (b) Surface-wave resonator using reflective gratings.

elements are generally metal strips or grooves, each causing very little perturbation of the wave, and in consequence the loss to bulk waves is very small.

The reflection coefficient of a grating is theoretically given by equation (E.24) of Appendix E, and some examples showing its variation with frequency are given in Figure E.3. The reflection coefficient is close to unity, but only over a narrow frequency band, if $N|r| \gg 1$, where N is the number of elements and r the amplitude reflection coefficient of each element. Typically, $|r|$ will be 0.01 or less, so 100 or more elements are needed. The use of gratings implies that a surface-wave resonator behaves in a rather different manner to the conventional resonator of Figure 10.24(a) since there are now two frequency-selective mechanisms involved – the frequency variation of the grating reflection coefficient, and the cavity resonances. Usually, the device is designed such that all but one of the cavity resonances are suppressed by the frequency variation of the reflection coefficient. Another distinction is that the phase of the grating reflection coefficient varies rapidly with frequency, so that the wave behaves as if it were reflected from a point some distance into the grating; thus the effective cavity length L is considerably larger than the separation of the gratings, as indicated in Figure 10.24(b).

For practical devices, ST, X quartz is usually chosen as the substrate material in view of its temperature stability, and grooves are usually chosen at the reflecting elements since they are found to give better performance than metal strips. The behaviour of grooves reflecting surface waves through 90° has already been considered in Section 9.6.2, in connection with RAC's. For normal incidence on grooves in ST, X quartz the behaviour shows similar features: the groove reflection coefficient is found to be given by equation (9.91) with $C = 0.27$, and stored-energy effects are found to perturb the velocity somewhat in accordance with equation (9.99), with $C' = 17.3$ [450, 451]. It is found that the resonator Q-factor, measured by generating a surface wave outside the resonator, can be remarkably high. For example, at 160 MHz a Q-factor of 27,000 can be obtained [452], while a 1.4 GHz resonator gave a Q-factor of 6000 [453]. These figures are quite typical of well-designed resonators, and are substantially higher than the values obtainable using other technologies suitable for the same frequency range [454]. Comparisons with theoretical predictions show that the Q-factors are quite close to the limit imposed by the acoustic propagation loss which, as seen in Section 6.3, increases with frequency.

DEVICES FOR SPREAD-SPECTRUM COMMUNICATIONS

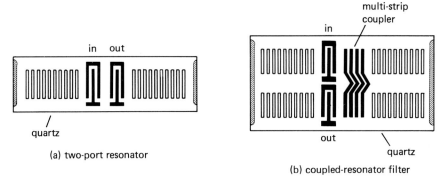

FIGURE 10.25. Resonator configurations.

A practical resonator must of course include one or more transducers to launch and sense the waves. A common arrangement, shown in Figure 10.25(a), uses two transducers between the gratings, forming a two-port resonator. Apart from its use as an oscillator stabilising element, this device can also be regarded as a narrow-band *bandpass filter*, and has been widely investigated in this role [452, 454, 455]. Compared with the interdigital bandpass filter of Chapter 8, the resonator is complementary in that it offers smaller bandwidths, and also has the significant advantage that the insertion loss can be as low as 2 or 3 dB. However, the resonator is somewhat inflexible in that the response is essentially a simple resonance curve, and in particular the stop-band rejection is poor. As in the case of bulk crystal resonators, greater versatility can be obtained by combining several resonators [456], and these may be either separate devices combined in a circuit or several acoustically-coupled resonators on the same substrate. Some experimental examples using the circuit approach [452, 455] show in particular that the pass-band can be flattened and the rejection improved by this method. Acoustic coupling of resonators on the same substrate can be obtained by a variety of methods [454], one of which is illustrated in Figure 10.25(b). Here two resonators on a quartz substrate are coupled by means of a multi-strip coupler; although the coupler cannot efficiently couple two tracks on quartz because of the low piezoelectric coupling (Chapter 5), it can be used effectively in the resonator because only weak coupling is required. A variety of experimental filters are described by Coldren and Rosenberg [454], using several different coupling techniques and up to four coupled resonators.

10.5.3. Oscillator Performance

In both the delay line and the resonator oscillator the short-term and long-term stability are important performance criteria, and any assessment of surface-wave oscillators must take account of their main rival, the well-established bulk crystal oscillator. The stability obtainable using surface-wave techniques is not generally as good as that of the bulk device, but a key advantage is the ability to operate at high frequencies, up to about 2 GHz. Bulk crystals cannot operate in the fundamental mode above about 50 MHz, because they become too fragile. For higher frequencies overtone operation may be used or the output may be applied to a frequency-multiplying circuit, but in both cases the short-term stability is degraded. Thus some care is needed when comparing surface-wave and bulk-wave devices.

The short-term stability is related to noise generated in the circuit, and can be predicted theoretically quite well for both delay line [448, 457] and resonator [457, 458] oscillators. For optimum stability a high Q-factor is needed (or, for a delay line, a long delay), the insertion loss should be low, and the power applied to the device should be relatively high. For resonators it has been found that too high a power level causes migration of the aluminium metallisation and degrades the long term stability [459]; the power level must be restricted to avoid this, though the phenomenon can be mitigated to some extent by adding a small amount of copper. The short-term stability is often expressed in terms of the power spectral density of the oscillator output, relative to the carrier. Results for a 400 MHz delay line device [457] gave a power density of -140 dBc/Hz at 10 kHz from the carrier and a floor of -160 dBc/Hz at 100 kHz and beyond. For frequencies within about 5 kHz of the carrier the stability is affected by "flicker noise", which is dependent on the surface treatment of the crystal and cannot be predicted accurately. Resonators give better stability, as shown by 120 MHz devices giving about -165 dBc/Hz at 10 kHz and -180 dBc/Hz at 100 kHz and beyond; these results are comparable with bulk crystal oscillators [458]. The short-term stability is also affected by vibration [460].

The long-term stability, that is, the ageing, is known to be related to the treatment of the crystal in preparation, mounting and packaging. These factors are not amenable to analysis, but considerable effort has been applied to optimise the performance, as discussed by Parker [461]. With care taken in fabrication, the frequency of a delay line oscillator can be constant within one part per million over one year, and some of Parker's devices are rather better than this. For resonators, ageing rates of about 3 parts per million per year are reported [459].

Trimming of surface-wave oscillators is an important issue because fabrication tolerances do no generally allow the frequency of the initial devices to be more accurate than about 100 parts per million. Here the delay line oscillator has a distinct advantage: the frequency range accessible by external phase shifting is easily increased by increasing the delay-line bandwidth, and hence the design can allow for anticipated tolerances. For resonators the range obtainable by external phase shifting is limited because of the high Q-factors; however it has been found that accurate trimming over quite large ranges can be done by etching the device with an ion beam [461, 462]. Parker points out that some frequency uncertainty is introduced when the device is subsequently encapsulated, but this can be compensated by external phase shifting [461].

The temperature stability of these devices is mainly determined by the substrate. For *ST, X* quartz, the commonest choice, the temperature stability is given in Section 6.4, and is not as good as the stability obtained with bulk crystals. However, other orientations giving better stability have been found [463], including the SST cut (Section 6.5), while considerably better stability can be obtained using surface-skimming bulk waves in quartz, as described in Appendix F. For delay line oscillators there is an interesting alternative technique, in which the stability is improved by using two acoustic tracks at different orientations on the same substrate [464]. The principle has been applied to a digitally-compensated oscillator using an AT-cut quartz substrate [465], demonstrating an impressive stability of 5 parts per million over a temperature range of -13 to $+97°C$.

Appendix A

Fourier Transforms and Linear Filters

The theory of surface-wave devices makes extensive use of Fourier transforms and the theory of linear filters. This appendix summarises the results needed in this context. Many of the results are quoted without proof, and for further details the reader is referred to the many texts available, some of which are listed as References [466–469].

A.1. Fourier Transforms

The Fourier transform of a function $g(t)$ is denoted $G(\omega)$, while the inverse transform of $G(\omega)$ is $g(t)$. The relation is denoted symbolically by

$$G(\omega) \leftrightarrow g(t). \tag{A.1}$$

The transforms are defined by the integrals

$$G(\omega) = \int_{-\infty}^{\infty} g(t)\, e^{-j\omega t}\, dt \quad \text{(forward transform)}, \tag{A.2}$$

$$g(t) = \frac{1}{2\pi} \int_{-\infty}^{\infty} G(\omega)\, e^{j\omega t}\, d\omega \quad \text{(inverse transform)} \tag{A.3}$$

In this appendix, the variables are written as t and ω throughout. The variable t can be taken to refer to time, in which case ω refers to radian frequency. The variables can of course have other meanings; in particular, we sometimes use the spatial coordinate x instead of t and "wavenumber" β instead of ω. In the equations to follow, the use of t and ω specifies whether a forward or inverse transform is involved. As a standard notation, we write time-domain functions using lower-case symbols and the corresponding frequency-domain functions using the corresponding upper-case symbols. Alternatively, an over-bar is used to indicate the forward transform. Thus, the forward transform of $g(t)$ is written $G(\omega)$ or $\bar{g}(\omega)$. In general, $g(t)$ and $G(\omega)$ may both be complex.

A number of useful theorems follow directly from the definitions of equations (A.2) and (A.3). A basic property is that the Fourier transform is linear, so that the transform of the sum of two functions is the sum of their individual transforms. Some useful symmetry theorems are:

(a) If $g(t)$ is even (so that $g(-t) = g(t)$), then $G(\omega)$ is even, and vice versa.
(b) If $g(t)$ is real and even, then $G(\omega)$ is real and even, and vice versa.
(c) If $g(t)$ and $G(\omega)$ are both real, then they are both even functions.

Transforms of conjugates are given by

$$g^*(t) \leftrightarrow G^*(-\omega), \tag{A.4}$$

$$G^*(\omega) \leftrightarrow g^*(-t). \tag{A.5}$$

Thus if $g(t)$ is real we have

$$G(-\omega) = G^*(\omega), \tag{A.6}$$

and if $G(\omega)$ is real we have

$$g(-t) = g^*(t). \tag{A.7}$$

The scaling theorem states that

$$g(t/a) \leftrightarrow |a|G(a\omega), \tag{A.8}$$

where a is a real constant. In particular, with $a = -1$ we have

$$g(-t) \leftrightarrow G(-\omega). \tag{A.9}$$

The shifting theorems are

$$g(t - a) \leftrightarrow e^{-ja\omega}G(\omega), \tag{A.10}$$

$$G(\omega - a) \leftrightarrow e^{jat}g(t), \tag{A.11}$$

and the modulation theorems are

$$g(t)\cos(\omega_0 t) \leftrightarrow \tfrac{1}{2}[G(\omega + \omega_0) + G(\omega - \omega_0)], \tag{A.12}$$

$$g(t)\sin(\omega_0 t) \leftrightarrow \tfrac{1}{2}j[G(\omega + \omega_0) - G(\omega - \omega_0)], \tag{A.13}$$

where ω_0 is a real constant. The differentiation theorems are

$$\frac{d}{dt}g(t) \leftrightarrow j\omega G(\omega), \tag{A.14}$$

$$\frac{d}{d\omega}G(\omega) \leftrightarrow -jtg(t). \tag{A.15}$$

Parseval's theorem states that

$$\int_{-\infty}^{\infty} g_1(t) g_2^*(t) \, dt = \frac{1}{2\pi} \int_{-\infty}^{\infty} G_1(\omega) G_2^*(\omega) \, d\omega, \qquad (A.16)$$

where $G_1(\omega)$ and $G_2(\omega)$ are respectively the transforms of $g_1(t)$ and $g_2(t)$. In particular, if $g_1(t) = g_2(t)$ we have the energy theorem:

$$\int_{-\infty}^{\infty} |g(t)|^2 \, dt = \frac{1}{2\pi} \int_{-\infty}^{\infty} |G(\omega)|^2 \, d\omega. \qquad (A.17)$$

The convolution operation is denoted by an asterisk separating two functions, and is defined by

$$\begin{aligned} g_1(t) * g_2(t) &= \int_{-\infty}^{\infty} g_1(\tau) g_2(t - \tau) \, d\tau \\ &= \int_{-\infty}^{\infty} g_1(t - \tau) g_2(\tau) \, d\tau \\ &= g_2(t) * g_1(t). \end{aligned} \qquad (A.18)$$

Two convolution theorems are

$$g_1(t) * g_2(t) \leftrightarrow G_1(\omega) G_2(\omega), \qquad (A.19)$$

$$G_1(\omega) * G_2(\omega) \leftrightarrow 2\pi g_1(t) g_2(t). \qquad (A.20)$$

Delta Function. The Dirac delta function $\delta(t)$ is defined such that

$$\int_{-\infty}^{\infty} \delta(t) g(t) \, dt = g(0), \qquad (A.21)$$

where the function $g(t)$ is taken to be continuous at $t = 0$. The delta function can be regarded as a function whose value is infinite at $t = 0$ and zero elsewhere. However, it is an example of a "generalised function", and is not a function in the usual sense. Strictly, equation (A.21) has a specialised mathematical interpretation but for most practical purposes it can be regarded as a conventional integral. Some resulting properties are, with a real:

$$\int_{-\infty}^{\infty} \delta(t - a) g(t) \, dt = g(a), \qquad (A.22)$$

$$\delta(t - a) * g(t) = g(t - a), \qquad (A.23)$$

$$\delta(t - a) g(t) = \delta(t - a) g(a), \qquad (A.24)$$

$$\delta(at) = \frac{1}{|a|} \delta(t). \qquad (A.25)$$

If $f(t)$ is a continuous function and is zero at $t = a$, then

$$\delta(t - a) f(t) = 0. \qquad (A.26)$$

Fourier Transforms in the Limit. Strictly speaking, two functions $g(t)$ and $G(\omega)$ can be a Fourier transform pair only if the integrals of equations (A.2) and (A.3) converge. Generally, this requires $g(t)$ to approach zero for $t \to \pm \infty$ and $G(\omega)$ to approach zero for $\omega \to \pm \infty$. However, many functions of practical interest do not satisfy these conditions, an example being the sinusoidal waveform $\cos(\omega_0 t)$. For such cases a special approach can often be used. To illustrate this, consider the output of a practical spectrum analyser when the waveform $\cos(\omega_0 t)$ is applied. This gives two peaks at positions corresponding to the frequencies $\pm \omega_0$. The shapes of the peaks correspond to the response of the spectrum analyser, because the latter has finite resolution while the input waveform comprises components (at $\pm \omega_0$) with vanishing bandwidth. Provided the resolution remains finite (however small), and provided the spectrum analyser can be assumed to be linear with respect to inputs of different frequency, the output is well defined and is easily calculated. Mathematically, it can be shown that this can be expressed in terms of the "Fourier transform in the limit" of the function $\cos(\omega_0 t)$, which is given by

$$\cos(\omega_0 t) \leftrightarrow \pi \delta(\omega + \omega_0) + \pi \delta(\omega - \omega_0). \tag{A.27}$$

The output is then obtained by convolving this transform (with respect to frequency) with a function representing the response of the spectrum analyser for a single-frequency input. Equation (A.27) thus represents a prescription for calculating the output of any linear spectrum analyser with finite resolution, for an input $\cos(\omega_0 t)$; it can also be regarded as the output in the limit when the resolution falls to zero.

Many of the standard Fourier transform formulae, and all those involving the delta function, are valid only in this limiting sense. The formal justification for the method is based on the theory of generalised functions, and is described by Papoulis [466, p. 269] and Bracewell [467, p. 87], for example.

Some Particular Transforms. In the following, a and ω_0 are real constants.

$$\sin(\omega_0 t) \leftrightarrow j\pi[\delta(\omega + \omega_0) - \delta(\omega - \omega_0)], \tag{A.28}$$

$$\exp(-at^2) \leftrightarrow \sqrt{\pi/a} \exp(-\omega^2/4a), \quad \text{for } a > 0, \tag{A.29}$$

$$\exp(\pm jat^2) \leftrightarrow \sqrt{\pi/a} \exp\left[\pm j\left(\frac{\pi}{4} - \frac{\omega^2}{4a}\right)\right], \quad \text{for } a > 0, \tag{A.30}$$

$$\delta(t - a) \leftrightarrow \exp(-ja\omega), \tag{A.31}$$

$$\delta(\omega - \omega_0) \leftrightarrow \frac{1}{2\pi} \exp(j\omega_0 t). \tag{A.32}$$

The function $\text{rect}(x) = 1$ for $|x| < \frac{1}{2}$ and is zero for other x, and gives the transforms

$$\text{rect}(t/a) \leftrightarrow a \, \text{sinc}(\tfrac{1}{2} a\omega), \quad \text{for } a > 0, \tag{A.33}$$

and
$$\text{rect}(\omega/a) \leftrightarrow \frac{a}{2\pi} \text{sinc}(\tfrac{1}{2}at), \quad \text{for } a > 0, \tag{A.34}$$

where $\text{sinc}(x) = (\sin x)/x$.

The function $\text{sgn}(x) = 1$ for $x > 0$ and $\text{sgn}(x) = -1$ for $x < 0$, and gives the transforms

$$1/t \leftrightarrow -j\pi \, \text{sgn}(\omega), \tag{A.35}$$

$$1/\omega \leftrightarrow \tfrac{1}{2}j \, \text{sgn}(t). \tag{A.36}$$

The step function $U(x) = 1$ for $x > 0$ and $U(x) = 0$ for $x < 0$, so that

$$U(x) = \tfrac{1}{2} + \tfrac{1}{2}\text{sgn}(x). \tag{A.37}$$

Using equations (A.32) and (A.36) we have

$$U(t) \leftrightarrow \pi\delta(\omega) + \frac{1}{j\omega}. \tag{A.38}$$

Using the modulation theorems, equations (A.12) and (A.13), we have

$$U(t) \exp(-j\omega_0 t) \leftrightarrow \pi\delta(\omega + \omega_0) - \frac{j}{\omega + \omega_0} \tag{A.39}$$

and using equation (A.9)

$$U(-t) \exp(j\omega_0 t) \leftrightarrow \pi\delta(\omega - \omega_0) + \frac{j}{\omega - \omega_0}. \tag{A.40}$$

Adding the transforms (A.39) and (A.40) gives

$$\exp(-j\omega_0|t|) \leftrightarrow \pi\delta(\omega + \omega_0) + \pi\delta(\omega - \omega_0) + \frac{2j\omega_0}{\omega^2 - \omega_0^2}. \tag{A.41}$$

An infinite train of delta functions transforms into a function of the same form:

$$\sum_{n=-\infty}^{\infty} \delta(t - na) \leftrightarrow \frac{2\pi}{|a|} \sum_{m=-\infty}^{\infty} \delta(\omega - 2\pi m/a). \tag{A.42}$$

From this it follows that

$$\sum_{n=-\infty}^{\infty} e^{jnat} = \frac{2\pi}{|a|} \sum_{m=-\infty}^{\infty} \delta(t - 2\pi m/a), \tag{A.43}$$

which is proved by transforming both sides, making use of equations (A.32) and (A.42).

A.2. Linear Filters

We consider a two-port device with an input waveform $v(t)$ and an output waveform $g(t)$. These waveforms represent physical quantities measured at the two ports, and their precise meaning must be assigned before applying the relations below to a practical device. They usually refer to either voltages or currents. The input $v(t)$ is often defined as the voltage which a specified waveform generator would produce across a matched load.

The term *linear* means that the device obeys superposition with regard to different input waveforms. Suppose that an input waveform $v_1(t)$ gives an output waveform $g_1(t)$, while an input waveform $v_2(t)$ gives an output waveform $g_2(t)$. Then the device is linear if an input waveform $v_1(t) + v_2(t)$ gives an output waveform $g_1(t) + g_2(t)$, irrespective of the forms of $v_1(t)$ and $v_2(t)$. The device is *time-invariant* if an input $v(t - \tau)$ gives an output $g(t - \tau)$ for any input function $v(t)$ and for any value of the delay τ. The term "linear filter" used here refers to a device that is both linear and time-invariant. Most surface-wave devices can be taken to be linear filters provided the power level of the input waveform is not too high; for most practical purposes this is an excellent approximation, though non-linear effects become significant at high power levels (Section 6.3).

It can be assumed here that the input and output waveforms are real. A consequence of linearity and time-invariance is that, if the input $v(t)$ is a real sinusoid, then the output $g(t)$ is also a real sinusoid, with the same frequency but generally with a different amplitude and phase. Thus, if

$$v(t) = \cos \omega_0 t, \quad \text{for } \omega_0 \geqslant 0,$$

then

$$g(t) = A(\omega_0) \cos [\omega_0 t + \phi(\omega_0)], \tag{A.44}$$

where $A(\omega_0)$ and $\phi(\omega_0)$ are functions of frequency but not of time. Since the filter is linear, the input can be written as a sum of two complex exponentials and the output can be regarded as the sum of the responses to these exponentials. We define the *frequency response* $H(\omega)$ such that if

$$v(t) = \exp(j\omega t),$$

then

$$g(t) = H(\omega) \exp(j\omega t). \tag{A.45}$$

This is consistent with equation (A.44) if

$$H(\pm \omega) = A(\omega) \exp [\pm j\phi(\omega)]. \tag{A.46}$$

This defines the frequency response for positive and negative frequencies. Clearly the negative-frequency components are given by

$$H(-\omega) = H^*(\omega). \tag{A.47}$$

Now consider a *general* real input waveform $v(t)$, as in Figure A.1. It is assumed

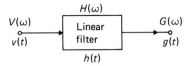

FIGURE A.1. Linear filter, showing notation for input and output waveforms.

that $v(t)$ has a Fourier transform $V(\omega)$ as in equation (A.2), that is, $V(\omega)$ is the *spectrum* of $v(t)$. By equation (A.3), $v(t)$ can be written as an infinite sum of complex exponentials, with coefficients $V(\omega)/2\pi$, and for each component equation (A.45) can be used. The corresponding outputs can be summed to give the total output waveform $g(t)$. This has a spectrum $G(\omega)$ given by equation (A.2), and hence

$$G(\omega) = V(\omega)H(\omega). \tag{A.48}$$

Thus the output spectrum can be obtained for any input waveform once $H(\omega)$ is known. The output waveform $g(t)$ can be obtained by transforming $G(\omega)$. Alternatively, $g(t)$ can be obtained from the *impulse response* $h(t)$, defined as the inverse Fourier transform of the frequency response:

$$h(t) \leftrightarrow H(\omega). \tag{A.49}$$

Applying the convolution theorem of equation (A.19) to equation (A.48) gives

$$\begin{aligned} g(t) &= v(t) * h(t) \\ &= \int_{-\infty}^{\infty} v(\tau) h(t - \tau) \, d\tau, \end{aligned} \tag{A.50}$$

so that the output waveform is the convolution of the input waveform with the device impulse response. In the particular case $v(t) = \delta(t)$, the output waveform is equal to $h(t)$ [by equation (A.23)], and hence $h(t)$ is the device output waveform when a delta function is applied to the input. As expected, $h(t)$ is real; this follows from the symmetry of $H(\omega)$, equation (A.47), and equation (A.6). A practical constraint is that $h(t) = 0$ for $t < 0$, since otherwise the device is not causal.

A particular example is an ideal delay line, which delays the input waveform by an amount τ_0, say, without distortion. In this case $g(t) = v(t - \tau_0)$ and the device responses are

$$h(t) = \delta(t - \tau_0),$$

$$H(\omega) = \exp(-j\omega\tau_0).$$

An important conclusion is that if $H(\omega)$ has a phase term proportional to ω, this causes no distortion of the waveform.

If the input waveform is a random noise waveform, the analysis must be treated statistically [469, p. 312]. We assume here that the input noise is stationary, that is, its statistical properties are independent of time. It follows that the output noise will also be stationary, and a power spectral density can be defined for both the input and output noise waveforms. The power of the input noise waveform is denoted P_i, and is defined by

$$P_i = E\{[v(t)]^2\}, \tag{A.51}$$

where $E\{\ \}$ denotes the statistical expectation value and $v(t)$ is now a statistical ensemble of functions. The expectation value is independent of t because the noise is stationary. A similar definition gives the output noise power P_0, with $v(t)$ replaced by $g(t)$. The input noise has a power spectral density $N_i(\omega)$, and the output noise has a power spectral density $N_0(\omega)$. These spectra are assumed to exist only for positive or zero frequencies. They are related to the noise powers by

$$P_i = \frac{1}{2\pi}\int_0^\infty N_i(\omega)\,d\omega, \tag{A.52}$$

$$P_0 = \frac{1}{2\pi}\int_0^\infty N_0(\omega)\,d\omega, \tag{A.53}$$

and the two spectral densities are related by

$$N_0(\omega) = N_i(\omega)|H(\omega)|^2. \tag{A.54}$$

Physically, $N_0(\omega)$ can be taken as the power per Hz of bandwidth for the output noise spectral components in the immediate vicinity of frequency ω. The same statement can be made about $N_i(\omega)$. This interpretation follows from equations (A.52)–(A.54) by taking the linear filter to be a narrow-band device.

In many cases the input waveform is taken to be *white* noise, which has a spectral density $N_i(\omega)$ independent of frequency. Strictly, this implies that the noise power P_i will be infinite. In all practical cases however, the spectral density will decay at high frequencies, and the power will be finite. White noise can be defined such that its spectral density is constant up to some high frequency, beyond which the filter response $H(\omega)$ can be taken to be negligible. Equations (A.53) and (A.54) can then be used, with $N_i(\omega)$ independent of ω.

A.3. Matched Filtering

We now consider a waveform $s(t)$ of finite length, accompanied by wide-band noise. If this is applied to a linear filter, the output signal-to-noise ratio will depend on the response of the filter. It is shown here that the output signal-to-noise ratio is maximised if the filter is designed such that its impulse response has a specific form, essentially the time-reverse of $s(t)$. Such a filter is called a *matched* filter [277–279, 282]; the terminology refers to the fact that the filter is matched to a specified input waveform $s(t)$, and not to the more familiar meaning referring to the impedances at the terminals. Matched filters are used in pulse-compression radar systems (Section 9.1), where the waveform $s(t)$ represents the received signal reflected from a target, and are also used in spread-spectrum communications (Section 10.1). For present purposes it can be assumed that the noise at the filter input is white, though this is not always the case.

The filter output waveform is a linear sum of a waveform $g(t)$, due to the input waveform $s(t)$, plus random noise due to the noise applied at the input. The power of the output noise is denoted P_0. The output signal power is $[g(t)]^2$, and since this varies with time its maximum value is used when defining the output signal-to-noise power ratio. Denoting the latter by SNR_0, we define

$$\text{SNR}_0 = \frac{[g(t)]^2_{\max}}{P_0}. \tag{A.55}$$

Note that for an oscillatory waveform this refers to the *peak* signal power, not the r.m.s. value; the noise power P_0 is average power. Now, if $H(\omega)$ is the frequency response of the filter and $S(\omega)$ the spectrum of the input waveform $s(t)$, the output waveform $g(t)$ has a spectrum $G(\omega) = S(\omega)H(\omega)$, from equation (A.48). The output signal power can therefore be written

$$[g(t)]^2 = \left| \frac{1}{2\pi} \int_{-\infty}^{\infty} S(\omega)H(\omega)\, e^{j\omega t}\, d\omega \right|^2. \tag{A.56}$$

The input noise is taken to be stationary and white, with spectral power density N_i per Hz of bandwidth, so that N_i is independent of frequency. From equation (A.53), the output noise power is

$$P_0 = \frac{N_i}{4\pi} \int_{-\infty}^{\infty} |H(\omega)|^2\, d\omega. \tag{A.57}$$

We now apply Schwartz's inequality, which states that, if $A(\omega)$ and $B(\omega)$ are complex functions of ω, then

$$\left| \int_a^b A^*(\omega)B(\omega)\, d\omega \right|^2 \leq \int_a^b |A(\omega)|^2\, d\omega \int_a^b |B(\omega)|^2\, d\omega, \tag{A.58}$$

where the equality applies only when $B(\omega) = kA(\omega)$, and k is an arbitrary constant. Assume initially that the signal power $[g(t)]^2$ is maximised at some time t_0, say, so that the output SNR is given by equations (A.55)–(A.57) with $t = t_0$. Using equation (A.58), with $A^*(\omega) = S(\omega)\exp(j\omega t_0)$ and $B(\omega) = H(\omega)$, we find

$$\text{SNR}_0 \leq \frac{1}{\pi N_i} \int_{-\infty}^{\infty} |S(\omega)|^2\, d\omega. \tag{A.59}$$

We define E_s, the energy of the input signal, by

$$E_s = \int_{-\infty}^{\infty} [s(t)]^2\, dt = \frac{1}{2\pi} \int_{-\infty}^{\infty} |S(\omega)|^2\, d\omega, \tag{A.60}$$

where the equality of the two forms is an expression of the energy theorem, given in

Section A.1. We thus have

$$\text{SNR}_0 \leq 2E_s/N_i. \tag{A.61}$$

Thus the maximum value of SNR_0 is simply $2E_s/N_i$. Note that this depends only on the signal energy and the noise spectral density, and not on the form of the input signal. The maximum output SNR is obtained when the equality in equation (A.61) applies, that is, when $B(\omega) = kA(\omega)$, and the filter response $H(\omega)$ then satisfies the condition

$$H(\omega) = kS^*(\omega) \exp(-j\omega t_0). \tag{A.62}$$

A filter satisfying this criterion is a matched filter, matched to the input waveform $s(t)$. The impulse response $h(t)$ of the matched filter is the inverse Fourier transform of $H(\omega)$. Using standard relations from Fourier analysis, this is given by

$$h(t) = ks(t_0 - t). \tag{A.63}$$

This is simply the time-reverse of the signal, delayed by an amount t_0 and multiplied by an arbitrary constant k. Clearly, k must in practice be real.

It was assumed in the above argument that the output power $[g(t)]^2$ was maximised at time $t = t_0$. However, since the maximum value of SNR_0, given by equation (A.61), is independent of t_0, the result is valid irrespective of when the output power is maximised. In practice there is a constraint on t_0 because the filter must be causal, that is, its impulse response $h(t)$ must be zero for $t < 0$. If, for example, the signal $s(t)$ has duration T and commences at $t = 0$, causality requires that $t_0 \geq T$.

As for any linear filter, the output waveform $g(t)$ can be written as the convolution $g(t) = s(t) * h(t)$. Here, this can conveniently be expressed in terms of the *correlation function* of $s(t)$, which is defined as the convolution of $s(t)$ with $s(-t)$. Denoting the correlation function by $c(t)$, we thus have

$$c(t) \equiv s(t) * s(-t) = \int_{-\infty}^{\infty} s(\tau)s(\tau - t)\, d\tau. \tag{A.64}$$

Using equation (A.63) we find

$$g(t) = kc(t - t_0), \tag{A.65}$$

so that $g(t)$ is essentially the delayed correlation function of $s(t)$. By substituting $\tau = \tau' + t$ in equation (A.64) it is readily seen that $c(-t) = c(t)$, so that $c(t)$ symmetric about $t = 0$. Consequently, $g(t)$ is symmetric about the time t_0 where its power is maximised. The spectrum of the output waveform is $G(\omega) = S(\omega)H(\omega)$, and from equation (A.62) we have

$$G(\omega) = k|S(\omega)|^2 e^{-j\omega t_0} = \frac{1}{k}|H(\omega)|^2 e^{-j\omega t_0}. \tag{A.66}$$

A.4. Non-Uniform Sampling

The theory of uniform sampling was given briefly in Section 8.1.2. It was shown that, if a bandpass waveform is sampled using a uniformly-spaced sequence of

FOURIER TRANSFORMS AND LINEAR FILTERS

delta functions, the original waveform can be recovered from the sampled waveform by low-pass filtering, provided the sampling frequency is at least as large as the Nyquist frequency. Here we consider a waveform sampled using delta functions whose spacing varies, and show that a similar result can be obtained. This result is needed for the analysis of chirp filters in Chapter 9. The analysis is based on that given by Tancrell and Holland [79] and others [470, 471].

It is first necessary to establish a relationship concerning delta functions. Consider a monotonic function $u(t)$ which has one zero at $t = t_0$, and whose differential $\dot{u}(t)$ is non-zero at $t = t_0$. We then have [467, p. 95]

$$\delta[u(t)] = \frac{\delta(t - t_0)}{|\dot{u}(t)|}. \tag{A.67}$$

This can be proved by substituting $\delta[u(t)]$ for $\delta(t - a)$ in equation (A.22) and taking u as the independent variable.

Now consider sampling a finite-length continuous waveform $v(t)$. We consider a general case first, and later specialise by taking $v(t)$ to be a chirp waveform. The sampled version of $v(t)$ is $v_s(t)$, a sequence of delta functions at times t_n, given by

$$v_s(t) = v(t) \sum_{n=-\infty}^{\infty} \delta(t - t_n). \tag{A.68}$$

Here the sampling times t_n are to have non-uniform spacing. It is assumed that their values can be obtained from a smooth monotonic non-linear function $\theta(t)$, by solving the equation

$$\theta(t_n) = n\Delta, \quad n = 0, \pm 1, \pm 2, \ldots, \tag{A.69}$$

where Δ is a positive constant. The times t_n thus correspond to uniform increments of $\theta(t)$. With $u(t) = \theta(t) - n\Delta$, we obtain from equation (A.67)

$$\sum_{n=-\infty}^{\infty} \delta(t - t_n) = |\dot{\theta}(t)| \sum_{n=-\infty}^{\infty} \delta[\theta(t) - n\Delta]. \tag{A.70}$$

Here the delta-functions on the right can be expressed as a sum of complex exponentials using equation (A.43), giving

$$v(t) \sum_{n=-\infty}^{\infty} \delta(t - t_n) = v(t) \frac{|\dot{\theta}(t)|}{\Delta} \sum_{m=-\infty}^{\infty} \exp[j2\pi m\theta(t)/\Delta] \tag{A.71}$$

and this is equal to $v_s(t)$, equation (A.68). This shows that the sampled waveform can be expressed as a fundamental, with $m = 0$, plus a series of "harmonics" with other values of m. The fundamental has the same form as the original waveform $v(t)$, except for an amplitude distortion produced by the term $|\dot{\theta}(t)|$. Each "harmonic" is essentially the original waveform multiplied by a chirp waveform. If the original waveform $v(t)$ is band-limited and has finite length, and if the increment Δ is small enough, the frequency band occupied by the fundamental will not overlap the bands occupied by the harmonics. The fundamental component $v(t)|\dot{\theta}(t)|/\Delta$ may then be obtained from the sampled waveform by using a low-pass filter to reject the harmonics.

In surface-wave chirp filters, the waveform $v(t)$ to be sampled represents the required impulse response, and is itself a chirp waveform. Sampling is nearly always done in synchronism with the waveform, that is, at corresponding points in each cycle. This implies that the function $\theta(t)$ which defines the sampling points, equation (A.69), must also be the time-domain phase of the waveform, apart from an additive constant. Thus the original chirp waveform can be written as

$$v(t) = a(t) \cos [\theta(t) + \phi_0], \quad (A.72)$$

where ϕ_0 is an arbitrary constant. The envelope $a(t)$ will have finite length and is taken to be a smooth function, such that $v(t)$ is a band-pass waveform; thus the spectrum of $v(t)$ is non-zero only for $\omega_1 < |\omega| < \omega_2$, where ω_1 and ω_2 are two positive frequencies. We assume here that an integer number of samples is taken for each cycle of the waveform, as is done in practical device design. The number of samples per cycle is denoted by the integer S_e, as in Chapter 9, and in practice $S_e \geqslant 2$. The increment Δ is therefore $2\pi/S_e$, and the sampling times t_n are the solutions of $\theta(t_n) = 2\pi n/S_e$. From equations (A.71) and (A.72) the sampled waveform is

$$v_s(t) = v(t) \sum_{n=-\infty}^{\infty} \delta(t - t_n)$$

$$= \frac{S_e a(t)|\dot{\theta}(t)|}{2\pi} \Bigg[\cos [\theta(t) + \phi_0]$$

$$+ \sum_{m=1}^{\infty} \{\cos [(mS_e + 1)\theta(t) + \phi_0] + \cos [(mS_e - 1)\theta(t) - \phi_0]\} \Bigg].$$

(A.73)

In practice we are mainly interested in the fundamental component of $v_s(t)$, which is denoted $\tilde{v}_s(t)$. For $S_e > 2$ the fundamental is obtained by omitting the terms dependent on m, giving

$$\tilde{v}_s(t) = \frac{1}{2\pi} S_e |\dot{\theta}(t)| a(t) \cos [\theta(t) + \phi_0], \quad \text{for } S_e > 2, \quad (A.74)$$

which is essentially the original waveform, equation (A.72), multiplied by $|\dot{\theta}(t)|$. The harmonic terms which have been omitted have the same form except that the phase $\theta(t)$ in the cosine is replaced by $M\theta(t)$, with the integer $M \geqslant 2$. For $S_e = 2$ the term in equation (A.73) involving $(mS_e - 1)\theta(t)$ contributes to the fundamental when $m = 1$, and we find

$$\tilde{v}_s(t) = \frac{1}{\pi} S_e |\dot{\theta}(t)| a(t) \cos [\theta(t)] \cos \phi_0, \quad \text{for } S_e = 2. \quad (A.75)$$

If ϕ_0 is a multiple of 2π, this is essentially the original waveform multiplied by $|\hat{\theta}(t)|$; in this case the sampling times given by $\theta(t_n) = 2\pi n/S_e$ are at the maxima and minima of the original waveform. For other values of ϕ_0 a constant phase change is introduced, but this is generally inconsequential. However, ϕ_0 must not be an odd multiple of $\pi/2$ because the samples are then at the zeros of the original waveform, and this gives $v_s(t) = 0$.

A.5. Some Properties of Bandpass Waveforms

A waveform $v(t)$ is referred to as a bandpass waveform if its spectrum $V(\omega)$ is finite only in a specified frequency range excluding zero, that is, it is finite only for $\omega_1 < |\omega| < \omega_2$ where ω_1 and ω_2 are two positive frequencies. The impulse response of a linear surface-wave filter is always a bandpass waveform. Here we consider some properties of bandpass waveforms relating to the design of surface-wave transducers, discussed in Section 8.1.3.

It is convenient to represent a bandpass waveform $v(t)$ in terms of its complex envelope $\hat{v}(t)$. To do this, consider the positive-frequency part of the spectrum $V(\omega)$, shifted downward in frequency by an amount ω_r. This base band version of $V(\omega)$ is denoted by $2\hat{V}(\omega)$. The positive-frequency part of $V(\omega)$ is therefore given by

$$V(\omega) = \tfrac{1}{2}\hat{V}(\omega - \omega_r), \quad \text{for } \omega \geq 0. \tag{A.76}$$

Here ω_r is a positive reference frequency whose value is arbitrary except that it is taken to be between ω_1 and ω_2. For an amplitude-modulated waveform, ω_r is usually taken to be the carrier frequency. It is assumed that $v(t)$ is real, which implies that $V(-\omega) = V^*(\omega)$ and hence $V(\omega) = \tfrac{1}{2}\hat{V}^*(-\omega - \omega_r)$ for $\omega \leq 0$. The complex envelope of $v(t)$ is $\hat{v}(t)$, defined as the inverse Fourier transform of $\hat{V}(\omega)$:

$$\hat{v}(t) \leftrightarrow \hat{V}(\omega). \tag{A.77}$$

The waveform $v(t)$ can be related to this by transforming $V(\omega)$, making use of the shifting theorem of equation (A.11), giving

$$v(t) = \tfrac{1}{2}\hat{v}(t)\, e^{j\omega_r t} + \tfrac{1}{2}\hat{v}^*(t)\, e^{-j\omega_r t}, \tag{A.78}$$

where the first term on the right arises from the positive-frequency part of $V(\omega)$, and the second term from the negative-frequency part. We also define $\hat{a}(t)$ and $\hat{\theta}(t)$ as the amplitude and phase of $\hat{v}(t)$, so that

$$\hat{v}(t) = \hat{a}(t) \exp[j\hat{\theta}(t)]. \tag{A.79}$$

The waveform $v(t)$ can then be written as

$$v(t) = \hat{a}(t) \cos[\omega_r t + \hat{\theta}(t)]. \tag{A.80}$$

Thus, $\hat{a}(t)$ is the envelope of the waveform $v(t)$.

Linear Phase in the Frequency Domain. In surface-wave transducer design, where a waveform $v(t)$ is sampled in order to determine the geometry, it is often a requirement that the spectrum $V(\omega)$ should have a phase linear with frequency. It is therefore useful to consider what constraints this imposes on $v(t)$. We define $\phi(\omega)$ as the phase of $V(\omega)$, so that $V(\omega) = |V(\omega)| \exp[j\phi(\omega)]$, and take $\phi(\omega)$ to have the form

$$\phi(\omega) = \theta_c - (\omega - \omega_r)t_0, \quad \text{for } \omega > 0, \tag{A.81}$$

where θ_c and t_0 are arbitrary constants, and ω_r is the reference frequency for the complex envelope. Using equation (A.76), we have in this case

$$\hat{V}(\omega) = |\hat{V}(\omega)| \exp[j(\theta_c - \omega t_0)].$$

The inverse transform of $\hat{V}(\omega)$ is the complex envelope $\hat{v}(t)$. Using the shifting theorem, equation (A.10), we find that the transform of $|\hat{V}(\omega)|$ is given by

$$|\hat{V}(\omega)| \leftrightarrow e^{-j\theta_c} \hat{v}(t + t_0). \tag{A.82}$$

Here the left side is real. Since the transform $g(t)$ of some real function $G(\omega)$ gives $g(-t) = g^*(t)$, it can be concluded that

$$\hat{v}(t_0 - t) = e^{2j\theta_c} \hat{v}^*(t_0 + t).$$

Writing the complex envelope as $\hat{v}(t) = \hat{a}(t) \exp[j\hat{\theta}(t)]$, this gives

$$\hat{a}(t_0 - t)/\hat{a}(t_0 + t) = \exp\{j[2\theta_c - \hat{\theta}(t_0 + t) - \hat{\theta}(t_0 - t)]\}. \tag{A.83}$$

Since $\hat{a}(t)$ is real the right side of this equation must be equal to ± 1, and can be written as $\exp(jn\pi)$. We therefore have

$$\hat{\theta}(t_0 - t) = 2\theta_c - n\pi - \hat{\theta}(t_0 + t) \tag{A.84}$$

and

$$\hat{a}(t_0 - t) = e^{jn\pi} \hat{a}(t_0 + t) = \pm \hat{a}(t_0 + t). \tag{A.85}$$

Thus the envelope $\hat{a}(t)$ is either symmetric or anti-symmetric about $t = t_0$, which is the slope of the phase in the frequency domain, equation (A.81). The phase $\hat{\theta}(t)$ is equal to the constant $\theta_c - n\pi/2$, plus a function anti-symmetric about t_0.

Amplitude Modulation. Another important case for surface-wave transducer design occurs when $v(t)$ is an amplitude-modulated waveform, with no phase modulation. This implies some constraint on the spectrum $V(\omega)$. An amplitude-modulated waveform has the form

$$v(t) = \hat{a}(t) \cos(\omega_r t + \theta_c), \tag{A.86}$$

where θ_c is a constant and the reference frequency ω_r has been chosen to be equal to the carrier frequency. The envelope $\hat{a}(t)$ must of course be real. In this case $\hat{v}(t) = \hat{a}(t) \exp(j\theta_c)$, and hence $\hat{V}(\omega) = \hat{A}(\omega) \exp(j\theta_c)$, where $\hat{A}(\omega)$ is the transform of $\hat{a}(t)$. Since $\hat{a}(t)$ is real we have $\hat{A}(-\omega) = \hat{A}^*(\omega)$, and hence

$$\hat{V}(-\omega) = \hat{V}^*(\omega) \exp(2j\theta_c). \tag{A.87}$$

For positive frequencies, the spectrum $V(\omega)$ of $v(t)$ is given by equation (A.76). Using equation (A.87) we have

$$V(\omega_r - \omega) = V^*(\omega_r + \omega) \exp(2j\theta_c), \quad \text{for } |\omega| < \omega_r. \tag{A.88}$$

If $A(\omega)$ and $\phi(\omega)$ are the spectral amplitude and phase, so that $V(\omega) = A(\omega) \exp[j\phi(\omega)]$, we find, for $|\omega| < \omega_r$,

$$\phi(\omega_r - \omega) = 2\theta_c - n\pi - \phi(\omega_r + \omega) \tag{A.89}$$

and

$$A(\omega_r - \omega) = A(\omega_r + \omega) e^{jn\pi} = \pm A(\omega_r + \omega). \tag{A.90}$$

Thus $A(\omega)$ is either symmetric or anti-symmetric about the carrier frequency ω_r. The phase $\phi(\omega)$ equals a constant $\theta_c - n\pi/2$, plus a function anti-symmetric about ω_r.

Sampling of Amplitude-Modulated Waveforms. It was shown in Section 8.1.2 that a band-limited waveform can be uniformly sampled in a manner such that the original waveform can be recovered by low-pass filtering. In general, a necessary condition for this is that the sampling frequency ω_s must exceed the Nyquist frequency $2\omega_2$, where ω_2 is defined such that the original waveform has negligible spectral energy for $\omega > \omega_2$. Here we show that, for the special case of an amplitude-modulated bandpass waveform, the sampling frequency may be *below* the Nyquist frequency.

Suppose that $v(t)$ is some arbitrary bandpass waveform whose spectrum $v(t)$ is negligible except in the intervals given by $\omega_1 < |\omega| < \omega_2$. Sampling $v(t)$ at times $t = n\tau_s$ gives the sampled version $v_s(t)$, so that

$$v_s(t) = v(t) \sum_{n=-\infty}^{\infty} \delta(t - n\tau_s),$$

as in equation (8.8) of Section 8.1.2. The spectrum of $v_s(t)$ is $V_s(\omega)$, and from equation (8.9) this is given by

$$V_s(\omega) = \frac{\omega_s}{2\pi} \sum_{m=-\infty}^{\infty} V(\omega - m\omega_s), \tag{A.91}$$

where $\omega_s = 2\pi/\tau_s$. If the sampling frequency ω_s is below the Nyquist frequency, there will generally be some overlap of the original spectrum $V(\omega)$ and the image spectra

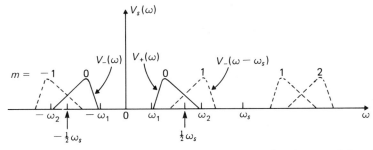

FIGURE A.2. Effect of sampling a bandpass waveform, with sampling frequency below the Nyquist frequency.

due to the sampling, that is, aliasing will occur. This is shown in Figure A.2, where ω_s is assumed to be between $2\omega_1$ and $2\omega_2$ and the terms with $m \neq 0$ are shown by broken lines.

We define $U(\omega)$ as the part of the spectrum $V_s(\omega)$ below frequency ω_s, so that

$$U(\omega) = V_s(\omega), \quad \text{for } |\omega| < \omega_s,$$
$$= 0, \quad \text{for } |\omega| > \omega_s. \tag{A.92}$$

This function can be obtained from $V_s(\omega)$ by low-pass filtering. It is convenient to write

$$V(\omega) = V_+(\omega) + V_-(\omega), \tag{A.93}$$

where $V_+(\omega)$ is the positive-frequency part of $V(\omega)$ and $V_-(\omega)$ is the negative-frequency part. Using equation (A.91), the positive-frequency part of $U(\omega)$ is given by

$$2\pi U(\omega)/\omega_s = V_+(\omega) + V_-(\omega - \omega_s),$$
$$= V_+(\omega) + V_+^*(\omega_s - \omega), \quad \text{for } \omega > 0. \tag{A.94}$$

Here the second form follows because $v(t)$ is real, which implies that $V_-(\omega) = V_+^*(-\omega)$. Now, the original waveform $v(t)$ may be recovered from the sampled waveform if the filtered spectrum $U(\omega)$ is proportional to the original spectrum $V(\omega)$. We therefore consider the case

$$2\pi U(\omega)/\omega_s = KV_+(\omega), \quad \text{for } \omega > 0, \tag{A.95}$$

where K is a constant which may in general be complex. For this to be consistent with equation (A.94) there must be a constraint on $V_+(\omega)$, and therefore on the original waveform $v(t)$. Writing $\omega = \frac{1}{2}\omega_s - \Delta\omega$ we have

$$(K - 1)V_+(\tfrac{1}{2}\omega_s - \Delta\omega) = V_+^*(\tfrac{1}{2}\omega_s + \Delta\omega). \tag{A.96}$$

Considering the case $\Delta\omega = 0$, and assuming that $V_+(\tfrac{1}{2}\omega_s) \neq 0$, this equation shows that $|K - 1| = 1$. We can therefore write $K - 1 = \exp(-2j\alpha_c)$, where the constant α_c is the phase of $V_+(\tfrac{1}{2}\omega_s)$, and equation (A.96) becomes

$$V_+(\tfrac{1}{2}\omega_s - \Delta\omega) = V_+^*(\tfrac{1}{2}\omega_s + \Delta\omega) \exp(2j\alpha_c). \tag{A.97}$$

This equation has the same form as equation (A.88), with ω_r replaced by $\tfrac{1}{2}\omega_s$ and θ_c replaced by α_c. It follows that $v(t)$ must be an amplitude-modulated waveform, with carrier frequency $\tfrac{1}{2}\omega_s$. Comparison with equation (A.86) shows that

$$v(t) = \hat{a}(t) \cos(\tfrac{1}{2}\omega_s t + \alpha_c), \tag{A.98}$$

where the envelope $\hat{a}(t)$ is not determined. We have thus shown that, if an amplitude-modulated waveform is sampled with a sampling frequency ω_s equal to twice the carrier frequency, the sampled waveform has the same spectrum as the original waveform in the region $|\omega| < \omega_s$, except for a complex constant. Equation (A.95) shows that the constant is $K\omega_s/(2\pi)$, and since $K - 1 = \exp(-2j\alpha_c)$ we have

$$K = 2\exp(-j\alpha_c)\cos\alpha_c. \tag{A.99}$$

FOURIER TRANSFORMS AND LINEAR FILTERS

If the sampled waveform is filtered by an ideal low-pass filter, such that all frequency components with $|\omega| > \omega_s$ are rejected, the resulting waveform $u(t)$ is the inverse transform of $U(\omega)$, and this is found to give

$$u(t) = \frac{\omega_s}{\pi} \hat{a}(t) \cos(\tfrac{1}{2}\omega_s t) \cos \alpha_c, \qquad (A.100)$$

which is the same as the original waveform $v(t)$, equation (A.98), except for changes of amplitude and phase.

It should be noted that α_c must not be an odd multiple of $\pi/2$, since this gives $u(t) = 0$. It also gives $K = 0$ and therefore, from equation (A.95), $U(\omega) = 0$. The reason for this is that the sampling points have been taken to be at $t = n\tau_s = 2n\pi/\omega_s$, and for $\alpha_c = \pi/2$ the waveform $v(t)$ is zero at these points. The waveform is usually sampled at the maxima and minima. In this case α_c is a multiple of π and $K = 2$, so that $u(t)$ has the same phase as $v(t)$ and the amplitude of $u(t)$ is maximised.

Appendix B

Reciprocity

The theory of elastic pieozoelectric materials can be used to derive a general reciprocity relation, with a wide range of applications. This appendix discusses the application of the reciprocity relation to transducers consisting of electrodes on the surface of a piezoelectric solid. The main consequence is that the processes of launching and receiving acoustic waves are mathematically related. Thus, if the launching process can be analysed for a particular transducer, reciprocity enables the receiving process to be deduced directly, with little further analysis. We first consider a generalised geometry, and then later discuss surface wave transducers on a half-space.

Throughout this appendix it is assumed that the surface is force free, so that the electrodes cause no mechanical loading. All appropriate quantities are taken to be proportional to exp $(j\omega t)$, with the frequency ω considered constant.

B.1. General Relation for a Mechanically Free Surface

We consider a homogeneous insulating piezoelectric material, whose behaviour is governed by the piezoelectric equations and Newton's laws. Any disturbance in the material gives rise to a displacement $\mathbf{u}$, stress $\mathbf{T}$, potential Φ and electric displacement $\mathbf{D}$, all of which are functions of the coordinates and of time. It is assumed that these functions describe a solution satisfying the constitutive relations of the material, and they will all be assumed to be proportional to exp $(j\omega t)$. A second solution is also assumed, described by the functions $\mathbf{u}'$, $\mathbf{T}'$, Φ' and $\mathbf{D}'$, and these are also proportional to exp $(j\omega t)$. These two solutions are related at each point by an equation called the *real reciprocity relation*, which is written

$$\text{div}\,[\{\mathbf{u}\,.\,\mathbf{T}'\} - \{\mathbf{u}'\,.\,\mathbf{T}\} + \Phi\mathbf{D}' - \Phi'\mathbf{D}] = 0, \qquad (B.1)$$

where the vector $\{\mathbf{u}\,.\,\mathbf{T}'\}$ is defined such that its x_j component is given by

$$\{\mathbf{u}\,.\,\mathbf{T}'\}_j = \sum_{i=1}^{3} u_i T'_{ij}, \quad j = 1, 2, 3.$$

Equation (B.1) is valid provided there are no mechanical or electrical sources within the material, and in particular this means that there must be no free charges. The derivation, given by Auld [32, p. 153], uses the piezoelectric constitutive equations [equations (2.9) and (2.10) of Chapter 2] with Maxwell's equations and Newton's laws. The form given above assumes the electric field to be quasi-static, so that $\mathbf{E} = -\operatorname{grad} \Phi$. The equation is valid even if the material is lossy. Other related reciprocity relations are the complex reciprocity relation given by Auld [32, 472] and the relation give by Lewis [473].

The solid is assumed to be enclosed by a mechanically free surface S, with a vacuum in the space outside this surface. The integral of equation (B.1) over the volume of the solid is related to an integral over the surface S by the divergence theorem

$$\int_V \operatorname{div} \mathbf{A} \, dV = \int_S \mathbf{A} \cdot \mathbf{n} \, dS,$$

where $\mathbf{n}$ is the outward-directed vector normal to the surface, with magnitude unity. For a mechanically free surface, with no forces, it may be shown [32] that $\{\mathbf{u} \cdot \mathbf{T}'\} \cdot \mathbf{n} = \{\mathbf{u}' \cdot \mathbf{T}\} \cdot \mathbf{n} = 0$. Thus, integration of equation (B.1) over the volume of the solid yields

$$\int_S (\Phi \mathbf{D}' - \Phi' \mathbf{D}) \cdot \mathbf{n} \, dS = 0. \tag{B.2}$$

Note that no electrical boundary condition has been applied at this stage.

The normal component of displacement at the surface ($\mathbf{D} \cdot \mathbf{n}$ or $\mathbf{D}' \cdot \mathbf{n}$) can be related to the charge density on a set of electrodes on the surface. The free charges must however be outside the surface S, because the reciprocity relation, equation (B.1), is valid only if there are no sources. To comply with this, the electrodes are assumed to be separated from the surface by an infinitesimal gap. Thus $\mathbf{D} \cdot \mathbf{n}$, the normal component of $\mathbf{D}$, is equal to $-\sigma$, where σ is the charge density on the adjacent electrode. If σ' is the charge density corresponding to the displacement $\mathbf{D}'$, equation (B.2) becomes

$$\int_S (\Phi \sigma' - \Phi' \sigma) \, dS = 0. \tag{B.3}$$

In this equation σ and σ' are the charge densities on the sides of the electrodes adjacent to the piezoelectric. There will also be charges on the vacuum side. The reciprocity argument can be applied to the vacuum as well as the piezoelectric, so that equation (B.3) is valid if σ' and σ are the charge densities on the vacuum side of the electrodes. It follows that the equation is also valid if σ and σ' are the total charge densities, including both the piezoelectric and vacuum sides. In the following description σ and σ' are taken to include the charges on both sides.

B.2. Reciprocity For Two-terminal Transducers

For the type of transducer considered here, each electrode is connected to one of two terminals. Transducers with more than two terminals may be analysed by using the results for two-terminal transducers, in conjunction with the superposition principle.

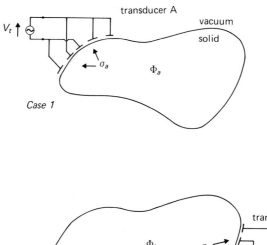

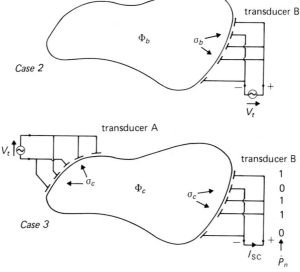

FIGURE B.1. Two-terminal transducers: reciprocity.

It is therefore sufficient to consider two-terminal transducers. The resistivity of the electrodes is assumed to be negligible.

We consider a piezoelectric solid with three different transducer configurations, as shown in Figure B.1. In case 1, transducer A is placed on the surface and a voltage V_t is applied across it. The voltage is proportional to exp ($j\omega t$), with this factor implicit. A charge density σ_a appears on the electrodes, and the potential, specified everywhere, is Φ_a. The transducer may radiate a variety of types of acoustic waves.

In case 2, transducer B is placed on the surface and a voltage V_t is applied, with transducer A absent. A charge density σ_b appears on the electrodes, and the potential is Φ_b. In case 3 both transducers are present. A voltage V_t is applied to transducer A, while transducer B is shorted and produces a current I_{sc}. The potential is Φ_c and the charge density is σ_c, which includes charges on both transducers.

It is assumed that the launching process can be analysed, so that the functions Φ_a, σ_a, Φ_b and σ_b can be found. Reciprocity enables the current produced by the receiving transducer, I_{sc}, to be expressed in terms of these functions. Using reciprocity in the form of equation (B.3) we have

$$\int_S \Phi_c \sigma_b \, dS = \int_S \Phi_b \sigma_c \, dS \tag{B.4}$$

and

$$\int_S \Phi_a \sigma_b \, dS = \int_S \Phi_b \sigma_a \, dS, \tag{B.5}$$

where it is assumed that any fields generated by the leads connecting the electrodes can be ignored. Now, in case 3, Φ_c has the same value at all points on the electrodes of transducer B, because this transducer is shorted and its electrodes are assumed to have zero resistivity. In case 2, σ_b is zero everywhere except on the electrodes of transducer B. We therefore have

$$\int_S \Phi_c \sigma_b \, dS = \text{const.} \int_S \sigma_b \, dS = 0,$$

since the total of all the charges on transducer B must be zero. Equation (B.4) therefore gives

$$\int_S \Phi_b \sigma_c \, dS = 0. \tag{B.6}$$

It is now assumed that, for case 3, the presence of transducer B does not affect the charge density on transducer A so that, on transducer A, $\sigma_c = \sigma_a$. In most practical cases this assumption is amply justified. The coupling between the two transducers can be regarded as of two types, electrostatic and acoustic. Electrostatic coupling causes a "breakthrough" signal to appear at the terminals of the receiving transducer. This signal is usually small by design; thus any perturbation of the charge density on the *launching* transducer due to electrostatic effects associated with the presence of the receiving transducer can be confidently ignored. Acoustic coupling can occur because transducer B can generate acoustic waves when the waves launched by transducer A are incident on it. However, in most of the cases that we are concerned with, this regeneration is small when the receiving transducer is shorted, as it is here. In any case, the effect of the regeneration is usually to produce multiple-transit signals which are easily identified because their delays differ from that of the main signal.

Thus, assuming that the charge density on transducer A is not affected by the presence of transducer B, we have $\sigma_c = \sigma_a$ on transducer A. Define σ_{cb} as the part of σ_c on the electrodes of transducer B, so that $\sigma_{cb} = 0$ except on transducer B. Then $\sigma_c = \sigma_a + \sigma_{cb}$, and equation (B.6) gives

$$\int_S \Phi_b \sigma_{cb} \, dS = -\int_S \Phi_b \sigma_a \, dS. \tag{B.7}$$

To find the current I_{sc} produced by transducer B the electrodes are labelled $n = 1, 2, 3, \ldots$, with areas S_n. The polarity of electrode n is designated $\hat{P}_n$, with a value 1 if it is connected to the positive terminal and 0 if it is connected to the negative terminal, as in Figure B.1. When a voltage V_t is applied to the transducer, as in case 2, the potential of electrode n is $(\hat{P}_n - \frac{1}{2})V_t$. In case 3, the total charge on electrodes connected to the positive terminal is

$$Q_+ = \sum_n \hat{P}_n \int_{S_n} \sigma_{cb} \, dS,$$

and the total charge on electrodes connected to the negative terminal is

$$Q_- = \sum_n (1 - \hat{P}_n) \int_{S_n} \sigma_{cb} \, dS.$$

The current $I_{sc} = j\omega Q_+$, and since the total charge must be zero we have $Q_+ = (Q_+ - Q_-)/2$, so that I_{sc} can be written

$$I_{sc} = j\omega \sum_n \int_{S_n} (\hat{P}_n - \tfrac{1}{2}) \sigma_{cb} \, dS. \tag{B.8}$$

Now, for case 2 we have $\Phi_b = (\hat{P}_n - \tfrac{1}{2}) V_t$ on electrode n of transducer B, since the resistivity is assumed to be zero. Also, σ_{cb} is zero except on these electrodes so, in equation (B.8), $(\hat{P}_n - \tfrac{1}{2})$ may be replaced by Φ_b/V_t. We thus have

$$I_{sc} = \frac{j\omega}{V_t} \int_S \Phi_b \sigma_{cb} \, dS.$$

Finally, using equations (B.7) and (B.5) we have

$$V_t I_{sc} = -j\omega \int_S \Phi_b \sigma_a \, dS = -j\omega \int_S \Phi_a \sigma_b \, dS. \tag{B.9}$$

This equation is the required result, giving the output of the receiving transducer, I_{sc}, in terms of functions obtained by analysis of launching transducers. Analysis of the launching process also gives the impedance of transducer B, so the output voltage and current produced by this transducer can be calculated for any load impedance.

A convenient modification of equation (B.9) is obtained by introducing the functions $\varrho_a = \sigma_a/V_t$ and $\varrho_b = \sigma_b/V_t$. Thus ϱ_a may be defined as the charge density on transducer A when unit voltage is applied, with no other electrodes present on the surface. Equation (B.9) then becomes

$$I_{sc} = -j\omega \int_S \Phi_b \varrho_a \, dS = -j\omega \int_S \Phi_a \varrho_b \, dS. \tag{B.10}$$

B.3. Symmetry of the Green's Function

If the charge density σ is specified at all points on the surface S enclosing the solid, this determines the potential Φ everywhere, apart from a constant term which is ignored here. Since the relationship is linear it may be expressed in terms of a Green's function $G_1(\mathbf{r}; \mathbf{r}')$, defined as the potential at $\mathbf{r}$ due to the charge density at $\mathbf{r}'$, so that

$$\Phi(\mathbf{r}) = \int_S G_1(\mathbf{r}; \mathbf{r}') \sigma(\mathbf{r}') \, dS'. \tag{B.11}$$

The subscript is used to distinguish this from the Green's function $G(x, \omega)$ for a half-space (Chapter 3). Using an argument given by Auld [32, p. 366], we show that the Green's function $G_1(\mathbf{r}; \mathbf{r}')$ is symmetrical. The equations involve values of the potential $\Phi(\mathbf{r})$ only at the surface, so the Green's function is specified with both $\mathbf{r}$ and $\mathbf{r}'$ on the surface.

We consider two solutions, one with potential $\Phi_1(\mathbf{r})$ and charge density $\sigma_1(\mathbf{r})$, and the other with potential $\Phi_2(\mathbf{r})$ and charge density $\sigma_2(\mathbf{r})$. Using reciprocity in the form of equation (B.3) we have

$$\int_S \Phi_1(\mathbf{r})\sigma_2(\mathbf{r})\,\mathrm{d}S = \int_S \Phi_2(\mathbf{r})\sigma_1(\mathbf{r})\,\mathrm{d}S.$$

Using equation (B.11) for the two potentials gives

$$\iint_S \sigma_2(\mathbf{r})\,G_1(\mathbf{r};\,\mathbf{r}')\,\sigma_1(\mathbf{r}')\,\mathrm{d}S'\,\mathrm{d}S = \iint_S \sigma_1(\mathbf{r})\,G_1(\mathbf{r};\,\mathbf{r}')\,\sigma_2(\mathbf{r}')\,\mathrm{d}S'\,\mathrm{d}S.$$

If we interchange $\mathbf{r}$ and $\mathbf{r}'$ in the integral on the right, it becomes the same as the integral on the left except that $G_1(\mathbf{r};\,\mathbf{r}')$ is replaced by $G_1(\mathbf{r}';\,\mathbf{r})$. Since the integrals must be equal for any choice of the charge densities $\sigma_1(\mathbf{r})$ and $\sigma_2(\mathbf{r})$, we conclude that the two forms of the Green's function are equal, that is,

$$G_1(\mathbf{r}';\,\mathbf{r}) = G_1(\mathbf{r};\,\mathbf{r}'). \tag{B.12}$$

The Green's function is therefore symmetrical.

B.4. Reciprocity for Surface Excitation of a Half-Space

We now consider a half-space with its plane force-free surface normal to the z-axis, and assume that there are no variations in the y-direction, so that the potential and charge density are functions of x only. The potential at the surface is denoted $\phi(x)$ and the charge density, which exists only at the surface, is $\sigma(x)$. Equation (B.11) thus becomes

$$\phi(x) = \int_{-\infty}^{\infty} G_1(x;\,x')\sigma(x')\,\mathrm{d}x'. \tag{B.13}$$

For a half-space the relation between $\phi(x)$ and $\sigma(x)$ must be unchanged if the origin for the x-axis is displaced, and hence $G_1(x;\,x')$ depends only on the distance between x and x'. This is expressed by defining a new Green's function $G(x)$, such that

$$G_1(x;\,x') = G(x - x'), \tag{B.14}$$

and hence

$$\phi(x) = \int_{-\infty}^{\infty} G(x - x')\sigma(x')\,\mathrm{d}x' = G(x) * \sigma(x),$$

so that $G(x)$ is the same as the Green's function introduced in Chapter 3, Section 3.4. Comparing equations (B.12) and (B.14) shows that $G(x)$ is symmetrical:

$$G(-x) = G(x). \tag{B.15}$$

It follows that the Fourier transform of $G(x)$, denoted $\bar{G}(\beta)$, is also symmetrical, so that $\bar{G}(-\beta) = \bar{G}(\beta)$. This gives the symmetry of the effective permittivity $\varepsilon_s(\beta)$. The reciprocal of the permittivity is equal to $|\beta|\bar{G}(\beta)$, from equation (3.41), and hence

$$\varepsilon_s(-\beta) = \varepsilon_s(\beta). \tag{B.16}$$

B.5. Reciprocity for Surface Wave Transducers

In general, a transducer on the surface of a half-space may generate surface waves and bulk waves, and will also generate a potential due to electrostatic effects. For such

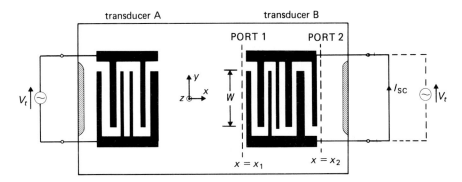

FIGURE B.2. General two-terminal transducers on a plane surface.

cases reciprocity may be used in the form of equation (B.10). However, for devices in which the coupling between transducers is predominantly due to surface waves some useful additional formulae can be derived.

We first consider reception of surface waves by a two-terminal transducer, as shown in Figure B.2. This represents the same situation as case 3 of Figure B.1, but with the transducers on the plane surface of a half-space. The aperture W is assumed to be very large, so that variations in the y-direction can be ignored. From equation (B.10), the short-circuit current produced by transducer B is

$$I_{sc} = -j\omega W \int_{-\infty}^{\infty} \phi_a(x) \varrho_b(x) \, dx. \tag{B.17}$$

Here $\phi_a(x)$ is the potential produced at the surface by transducer A, when a voltage V_t is applied to it, assuming transducer B to be absent. The function $\varrho_b(x)$ is the charge density on transducer B for unit applied voltage, with transducer A absent. Generally, $\phi_a(x)$ and $\varrho_b(x)$ will also be functions of the frequency, ω.

In general, $\phi_a(x)$ includes bulk wave and electrostatic terms in addition to a surface wave term. However, the bulk wave and electrostatic terms decay with distance. We assume that transducer A is generating surface waves, and assume that the transducer separation is large, so that in the region occupied by transducer B the bulk wave and electrostatic terms are negligible in comparison with the surface wave term. Thus, in the region of transducer B, $\phi_a(x)$ has the form

$$\phi_a(x) = \phi_0 \exp(-jk_0 x), \tag{B.18}$$

where k_0 is the free-surface wavenumber for surface waves at frequency ω, and ϕ_0 is a constant. This is substituted into equation (B.17). If $\bar{\varrho}_b(\beta)$ is the Fourier transform of $\varrho_b(x)$, this gives

$$I_{sc} = -j\omega W \phi_0 \bar{\varrho}_b(k_0). \tag{B.19}$$

It is convenient to express this in terms of the surface wave amplitude at the input port, port 1, of the transducer. The input port is defined by a line at $x = x_1$ near the left edge of the transducer, as on Figure B.2. The exact location of this line is immaterial. The incident surface wave potential is given by equation (B.18), and we define ϕ_{i1} as the value of this potential at $x = x_1$, the input port of transducer B. Thus

$$\phi_{i1} = \phi_0 \exp(-jk_0 x_1)$$

and the output current is given by

$$I_{sc} = -j\omega W \phi_{i1} \bar{\varrho}_b(k_0) \exp(jk_0 x_1). \tag{B.20}$$

Note that this is valid even if the transducer couples to bulk waves; it is only necessary to assume that the incident wave has no bulk wave or electrostatic terms.

The same method may be used to deduce the current produced by transducer B when surface waves are incident from the right instead of from the left. In this case equation (B.17) is still valid, but transducer A must be taken to be on the right of transducer B. The input port of transducer B is now port 2, taken to be at $x = x_2$, and the potential of the incident surface wave at this point is denoted ϕ_{i2}. It is found that the current produced by transducer B is

$$I_{sc} = -j\omega W \phi_{i2} \bar{\varrho}_b(-k_0) \exp(-jk_0 x_2). \tag{B.21}$$

Relation Between Launching and Reception. The amplitudes of the surface waves generated by an isolated transducer are derived in section B.6 below, assuming the surface waves to be of the piezoelectric Rayleigh wave type. If $\sigma(x)$ is the charge density on the transducer, with Fourier transform $\bar{\sigma}(\beta)$, the potential of the surface wave radiated in the $-x$ direction is

$$\phi_s(x) = j\Gamma_s \bar{\sigma}(k_0) \exp(jk_0 x), \tag{B.22}$$

where Γ_s is defined in Section B.6. Now suppose that a voltage V_t is applied to the two-terminal transducer B of Figure B.2, with transducer A absent. For unit applied voltage the charge density is $\varrho_b(x)$, with Fourier transform $\bar{\varrho}_b(\beta)$, so that $\bar{\sigma}(k_0) = V_t \bar{\varrho}_b(k_0)$. The transducer generates a surface wave with potential $\phi_s(x)$, and we define ϕ_{s1} as the potential at port 1, so that $\phi_{s1} = \phi_s(x_1)$. We thus have

$$\phi_{s1} = j\Gamma_s V_t \bar{\varrho}_b(k_0) \exp(jk_0 x_1) \tag{B.23}$$

for the wave launched at port 1. For surface waves incident on port 1 the current produced, when the transducer is shorted, is given by equation (B.20). Comparing with equation (B.23) we have

$$\left[\frac{I_{sc}}{\phi_{i1}}\right]_{receive} = -\frac{\omega W}{\Gamma_s} \left[\frac{\phi_{s1}}{V_t}\right]_{launch} \tag{B.24}$$

This equation relates the reception and launching of surface waves at port 1. The reader is reminded that the derivation assumes that there is no mechanical loading, and that the electrodes have zero resistivity. The equation is valid if the transducer couples to bulk waves as well as surface waves; if so, ϕ_{s1} is the surface-wave component of the potential at port 1 for the launching process, while for the receiving process an incident surface wave is assumed, with no bulk wave component.

For port 2 it is readily shown that the same relation, equation (B.24), applies with

ϕ_{i1} and ϕ_{s1} replaced by ϕ_{i2} and ϕ_{s2}, the potentials of incident and launched surface waves at port 2.

B.6. Surface Wave Generation

This section derives the formula for the amplitude of surface waves generated by a transducer, which has already been given in equation (B.22). The formula follows from the nature of the effective permittivity $\varepsilon_s(\beta)$, and therefore assumes that there is no mechanical loading. We also assume that the surface wave is of the piezoelectric Rayleigh wave type, so that $\varepsilon_s(\beta)$ has a zero for $\beta = k_0$, the free-surface wavenumber, and $\varepsilon_s(\beta)$ is real for β close to k_0'.

We consider a set of electrodes on the surface, connected to some electrical network that includes at least one current or voltage source, as in Figure B.3. The electrode edges are parallel to the y-axis, and it is assumed that there are no variations in the y-direction. The surface potential and charge density are $\phi(x)$ and $\sigma(x)$, with Fourier transforms $\bar{\phi}(\beta)$ and $\bar{\sigma}(\beta)$, and by definition these are related by

$$\bar{\phi}(\beta) = \frac{\bar{\sigma}(\beta)}{|\beta|\varepsilon_s(\beta)} \tag{B.25}$$

as in Section 3.2, equation (3.24).

In general the potential $\phi(x)$ includes contributions due to surface waves, bulk waves and electrostatic effects. However, the surface wave contribution may be identified by noting that this is the only term which does not decay in amplitude for large positive or negative values of x, remote from the transducer electrodes. Since the surface is unmetallised outside the transducer, the potential for large positive or negative x must have the form $\exp(-jk_0|x|)$. The Fourier transform of such a function must be infinite at $\beta = \pm k_0$, and hence the right side of equation (B.25) must be infinite at these points. Now, the charge density $\sigma(x)$ is localised in a finite region of x, occupied by the transducer electrodes, and it follows that its

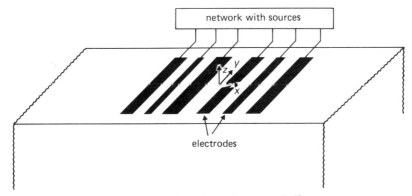

FIGURE B.3. General transducer on a half-space.

transform $\bar{\sigma}(\beta)$ cannot be infinite for any β. The poles of equation (B.25) therefore arise from the zeros of $\varepsilon_s(\beta)$ at $\beta = \pm k_0$.

To evaluate the surface wave potential, it is assumed that the total potential $\phi(x)$ can be written as a sum of two surface wave terms, representing waves of constant amplitude radiated away from the transducer, plus some potential $\phi_1(x)$ which is localised, so that it vanishes for $x \to \pm \infty$. The exact forms assumed for the surface wave terms are not consequential, though they must be defined such that they are harmonic at points remote from the transducer, and they must represent waves propagating away from the transducer. Assuming that the origin of the x-axis is within the transducer, a suitable form for the total potential is

$$\phi(x) = \phi_- U(-x) e^{jk_0 x} + \phi_+ U(x) e^{-jk_0 x} + \phi_1(x), \tag{B.26}$$

where the constants ϕ_- and ϕ_+ are the amplitudes of the two surface wave potentials. $U(x)$ is the step function, equal to unity for $x > 0$ and zero for $x < 0$. From Appendix A, equation (A.40), the transform of the first term is given by

$$U(-x) \exp(jk_0 x) \leftrightarrow \pi\delta(\beta - k_0) + j/(\beta - k_0).$$

This is infinite at $\beta = k_0$. The second term has a similar transform [equation (A.39)], and is infinite at $\beta = -k_0$. Using these and equation (B.25), the charge density has a Fourier transform

$$\begin{aligned}\bar{\sigma}(\beta) &= |\beta|\varepsilon_s(\beta)\bar{\phi}(\beta) \\ &= |\beta|\varepsilon_s(\beta)[\phi_-\pi\delta(\beta - k_0) + j\phi_-/(\beta - k_0) \\ &\quad + \phi_+\pi\delta(\beta + k_0) - j\phi_+/(\beta + k_0) + \bar{\phi}_1(\beta)],\end{aligned}$$

where $\bar{\phi}_1(\beta)$ is the transform of $\phi_1(x)$.

We now evaluate this function at $\beta = k_0$, noting that at this point $\varepsilon_s(\beta) = 0$. In the square bracket, the two ϕ_+ terms and the term $\bar{\phi}_1(\beta)$ can be omitted because at $\beta = k_0$ they are finite or zero. In addition $\varepsilon_s(\beta)\delta(\beta - k_0)$ is zero for all β. Thus for β close to k_0 we only need to consider the remaining term

$$\bar{\sigma}(\beta) = j\phi_-|\beta|\varepsilon_s(\beta)/(\beta - k_0) \tag{B.27}$$

and $\varepsilon_s(\beta)$ can be replaced by the first term of its Taylor expansion:

$$\varepsilon_s(\beta) = -(\beta - k_0)/(k_0 \Gamma_s), \tag{B.28}$$

where the constant Γ_s is defined by

$$\frac{1}{\Gamma_s} = -k_0 \left[\frac{d\varepsilon_s(\beta)}{d\beta}\right]_{k_0},$$

as in Section 3.3. Substituting equation (B.28) into equation (B.27) and evaluating at $\beta = k_0$ gives

$$\phi_- = j\Gamma_s\bar{\sigma}(k_0).$$

Similarly, by evaluating $\bar{\sigma}(\beta)$ at $\beta = -k_0$ we find that $\phi_+ = j\Gamma_s\bar{\sigma}(-k_0)$. Thus, for locations remote from the transducer, such that the localised potential $\phi_1(x)$ in equation (B.26) is negligible, the total potential is given by

$$\begin{aligned}\phi(x) &= j\Gamma_s\bar{\sigma}(k_0)\exp(jk_0x), & \text{for } x \ll 0, \\ &= j\Gamma_s\bar{\sigma}(-k_0)\exp(-jk_0x), & \text{for } x \gg 0,\end{aligned} \quad (B.29)$$

which represents surface waves travelling away from the transducer.

An alternative interpretation is to define a surface wave potential $\phi_s(x)$ for all values of x on the free surface outside the transducer. We can thus write

$$\phi_s(x) = j\Gamma_s\bar{\sigma}(\mp k_0)\exp(\mp jk_0x) \quad (B.30)$$

taking the upper signs for $x > 0$ and the lower signs for $x < 0$. At points remote from the transducer this is equal to the total potential $\phi(x)$. For points close to the transducer the surface wave potential can be taken to be given by equation (B.30), though the total potential $\phi(x)$ also includes the term $\phi_1(x)$ which is due to electrostatic effects and, possibly, bulk wave excitation.

An alternative proof is given by Milsom et al. [93].

Appendix C

Elemental Charge Density for Regular Electrodes

In Chapter 4 it was shown that the properties of transducers using regular electrodes can be conveniently expressed in terms of a function $\bar{\varrho}_f(\beta)$, defined as the Fourier transform of an elemental charge density $\varrho_f(x)$. The main purpose of this appendix is to demonstrate the validity of an analytic expression for $\bar{\varrho}_f(\beta)$. However, before doing this Section C.1 summarises some properties of Legendre functions that are needed in this appendix and elsewhere.

C.1. Some Properties of Legendre Functions

Properties of Legendre functions are given in many texts, in particular by Erdelyi [474], who gives all the properties quoted here.

The Legendre function with variable x and degree v is written as $P_v(x)$. Generally, x and v may be complex, but for analysis of surface-wave devices they are real, and x is in the range $-1 \leqslant x \leqslant 1$. The function may be evaluated using the expansion

$$P_v(x) = \sum_{m=0}^{\infty} a_m, \quad \text{for } |x| \leqslant 1, \tag{C.1}$$

where $a_0 = 1$ and

$$a_m = \frac{(m-1-v)(m+v)(1-x)}{2m^2} a_{m-1}.$$

Some plots of $P_v(x)$, regarded as functions of v, are shown in Figure C.1. The recursion relation is

$$v P_v(x) = (2v-1) x P_{v-1}(x) - (v-1) P_{v-2}(x) \tag{C.2}$$

and the symmetry relation is

$$P_v(x) = P_{-v-1}(x) \tag{C.3}$$

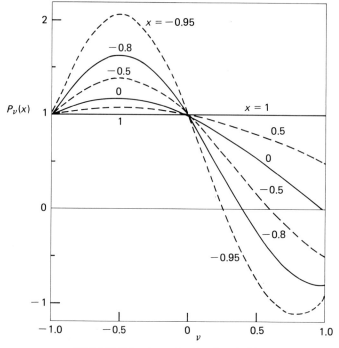

FIGURE C.1. Legendre functions $P_\nu(x)$.

so that, as a function of ν, $P_\nu(x)$ is symmetrical about $\nu = -\tfrac{1}{2}$. From the series of equation (C.1) it can be seen that $P_\nu(-1)$ is infinite if ν is not an integer. For $\nu = -\tfrac{1}{2}$, $P_\nu(x)$ is related to the elliptic integral by

$$P_{-\frac{1}{2}}(x) = 2K(m)/\pi, \qquad m = [(1-x)/2]^{1/2}, \tag{C.4}$$

where $K(m)$ is the complete elliptic integral of the first kind. The Mehler–Dirichlet formula is

$$P_\nu(\cos \Delta) = \frac{1}{\pi\sqrt{2}} \int_{-\Delta}^{\Delta} \frac{\exp[j(\nu + \tfrac{1}{2})\phi]}{\sqrt{\cos\phi - \cos\Delta}} d\phi, \quad \text{for } 0 < \Delta < \pi. \tag{C.5}$$

For positive and negative values of x and ν the Legendre functions are related by

$$P_\nu(x)P_{-\nu}(-x) + P_\nu(-x)P_{-\nu}(x) = \frac{2 \sin \pi\nu}{\pi\nu}, \tag{C.6}$$

as noted by Bløtekjaer et al. [475]. This relation may be proved in two stages. First, using the differentiation formulae [474], it can be shown that the left side of the equation is independent of the variable x. Then the left side is evaluated for $x = 0$ by expressing $P_\nu(0)$ in terms of gamma functions [474] and making use of the properties of gamma functions.

For the particular case when v is an integer, the series of equation (C.1) truncates, so that $P_v(x)$ becomes a polynomial. This is referred to as a Legendre Polynomial, and is often written $P_n(x)$. The relations given above are valid for $P_n(x)$, with v replaced by the integer n. Thus

$$P_0(x) = 1, \tag{C.7}$$

$$P_1(x) = x, \tag{C.8}$$

and, for larger n, $P_n(x)$ may be obtained from the recursion relation, equation (C.2), giving

$$P_2(x) = (3x^2 - 1)/2, \tag{C.9}$$

$$P_3(x) = (5x^3 - 3x)/2, \tag{C.10}$$

and so on. Some Legendre polynomials are shown in Figure C.2. The polynomials are orthogonal over the interval $-1 < x < 1$, and have the symmetry

$$P_n(-x) = (-1)^n P_n(x). \tag{C.11}$$

The Legendre function $P_{-v}(\cos \Delta)$ is expressed as a sum of the polynomials by Dougall's expansion:

$$P_{-v}(\cos \Delta) = \frac{\sin(\pi v)}{\pi} \sum_{n=-\infty}^{\infty} \frac{S_n P_n(-\cos \Delta)}{v + n}, \tag{C.12}$$

where

$$S_n = 1, \text{ for } n \geqslant 0; \quad S_n = -1, \text{ for } n < 0. \tag{C.13}$$

Also, for $0 < \Delta < \pi$,

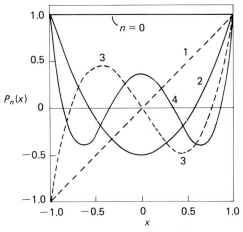

FIGURE C.2. Legendre polynomials $P_n(x)$.

$$\sum_{n=-\infty}^{\infty} P_n(\cos \Delta) \, e^{jn\theta} = \frac{(-1)^m \sqrt{2} \, e^{-j\theta/2}}{\sqrt{\cos \theta - \cos \Delta}}, \quad \text{for } |\theta - 2m\pi| < \Delta,$$

$$= 0, \quad \text{for } \Delta < |\theta - 2m\pi| \leq \pi. \quad \text{(C.14)}$$

This is valid for any value of θ; the integer m is chosen such that $|\theta - 2m\pi| \leq \pi$. Another series valid for $0 < \Delta < \pi$ is

$$\sum_{n=-\infty}^{\infty} S_n P_n(\cos \Delta) \, e^{jn\theta} = \frac{j(-1)^m \sqrt{2} \, \text{sgn}\, (\theta - 2m\pi) \, e^{-j\theta/2}}{\sqrt{\cos \Delta - \cos \theta}},$$

$$\text{for } \Delta < |\theta - 2m\pi| < \pi,$$

$$= 0, \quad \text{for } |\theta - 2m\pi| < \Delta, \quad \text{(C.15)}$$

where S_n is given by equation (C.13) and the function $\text{sgn}\,(x) = 1$ for $x > 0$ and $\text{sgn}\,(x) = -1$ for $x < 0$. Relations similar to equations (C.14) and (C.15) are given by Bløtekjaer et al. [476].

Combining equation (C.14) with the Mehler–Dirichlet formula, equation (C.5), we have, for $0 < \Delta < \pi$,

$$\int_{-\Delta}^{\Delta} \sum_{n=-\infty}^{\infty} P_{n+m}(\cos \Delta) \, e^{-j(n+v)\theta} d\theta = 2\pi P_{m-v}(\cos \Delta). \quad \text{(C.16)}$$

In surface-wave device analysis, this equation is useful when calculating the current flowing into one electrode.

C.2. Elemental Charge Density

The elemental charge density $\varrho_f(x)$ is defined by considering an infinite array of regular electrodes, with width a and pitch p, as in Figure 4.10. The electrodes are deposited on the surface of an anisotropic non-piezoelectric half-space. One of the electrodes is centred at the origin $x = 0$ and has unit voltage applied, while all the other electrodes are grounded. In this situation the charge density on the electrodes is, by definition, the elemental charge density $\varrho_f(x)$. This section shows that the Fourier transform of $\varrho_f(x)$ is given by

$$\bar{\varrho}_f(\beta) = (\varepsilon_0 + \varepsilon_p) \frac{2 \sin(\pi s)}{P_{-s}(-\cos \Delta)} P_n(\cos \Delta), \quad \text{for } n \leq \frac{\beta p}{2\pi} \leq n+1, \quad \text{(C.17)}$$

where $\Delta = \pi a/p$, $s = (\beta p)/(2\pi) - n$, and ε_p is a function of the permittivity components ε_{ij} of the half-space material, given by equation (3.4). The parameter s is in the range $0 \leq s \leq 1$.

To verify equation (C.17) it must be shown that the charge density and the associated potential satisfy Laplace's equation, in a form given by the electrostatic analysis of Section 3.1, and that the boundary conditions are satisfied. Thus the charge density must be zero in the gaps between the electrodes, and the potential on each electrode must be uniform and equal to the applied voltage.

In the x-domain, the charge density $\varrho_f(x)$ is given by the inverse Fourier transform of $\bar{\varrho}_f(\beta)$, so that

$$\varrho_f(x) = \int_{-\infty}^{\infty} \bar{\varrho}_f(\beta) \exp(j\beta x) \, d\beta/(2\pi), \tag{C.18}$$

where $\bar{\varrho}_f(\beta)$ is given by equation (C.17). In view of the form of equation (C.17), this can be integrated from $\beta = 2\pi n/p$ to $\beta = 2\pi(n+1)/p$, and then summed with respect to the integer n. The summation has the form of equation (C.14). We define a normalised x-coordinate by

$$\theta = 2\pi x/p. \tag{C.19}$$

The result is

$$\varrho_f(x) = \frac{\varepsilon_0 + \varepsilon_p}{p} \frac{2\sqrt{2}(-1)^m}{\sqrt{\cos\theta - \cos\Delta}} \Gamma(\theta, \Delta), \quad \text{for } |x - mp| < a/2,$$

$$= 0, \quad \text{for } a/2 < |x - mp| \leqslant p/2, \tag{C.20}$$

where the function $\Gamma(\theta, \Delta)$ is given by

$$\Gamma(\theta, \Delta) = \int_0^1 \frac{\sin(\pi s) \cos[(s - \tfrac{1}{2})\theta]}{P_{-s}(-\cos\Delta)} \, ds. \tag{C.21}$$

Thus $\varrho_f(x)$ satisfies the boundary condition in being zero in the gaps between the electrodes. The form of $\varrho_f(x)$ is shown, for $a/p = \tfrac{1}{2}$, in Figure 4.10.

The associated electric field at the surface has an x-component $E_x(x)$, with Fourier transform $\bar{E}_x(\beta)$. If the potential at the surface is $\phi(x)$, with transform $\bar{\phi}(\beta)$, we have $\bar{E}_x(\beta) = -j\beta\bar{\phi}(\beta)$. The electrostatic analysis in Section 3.1 shows that the charge density and the potential are related, in the β-domain, by equation (3.12), and it follows that

$$\bar{E}_x(\beta) = -j \, \text{sgn}(\beta) \bar{\varrho}_f(\beta)/(\varepsilon_0 + \varepsilon_p). \tag{C.22}$$

Using equation (C.17) for $\bar{\varrho}_f(\beta)$ we have

$$\bar{E}_x(\beta) = -\frac{2j \sin(\pi s)}{P_{-s}(-\cos\Delta)} S_n P_n(\cos\Delta), \quad \text{for } n \leqslant \frac{\beta p}{2\pi} \leqslant n+1, \tag{C.23}$$

where S_n is defined in equation (C.13). Transforming this to the x-domain involves a summation with the form of equation (C.15), and gives the result

$$E_x(x) = \frac{2\sqrt{2}(-1)^m \, \text{sgn}(\theta - 2m\pi)}{p\sqrt{\cos\Delta - \cos\theta}} \Gamma(\theta, \Delta), \quad \text{for } a/2 < |x - mp| \leqslant p/2,$$

$$= 0, \quad \text{for } |x - mp| < a/2, \tag{C.24}$$

where $\Gamma(\theta, \Delta)$ is defined in equation (C.21). Thus $E_x(x)$ is zero on the electrodes, which are therefore equipotentials as required.

To show that the potential $\phi(x)$ is correct at the electrode locations, it is sufficient to evaluate it at the electrode centres $x = mp$. In the β-domain we have $\bar{\phi}(\beta) = j\bar{E}_x(\beta)/\beta$, with $\bar{E}_x(\beta)$ given by equation (C.23). Transforming to the x-domain, the integral involves a summation over n, and for $x = mp$ the summation has the form of Dougall's expansion, equation (C.12). We thus find

$$\phi(mp) = 1, \quad \text{for } m = 0,$$
$$= 0, \quad \text{for } m \neq 0, \qquad (C.25)$$

and hence the electrode potentials are correct. This completes the proof, since all the boundary conditions are satisfied, and Laplace's equation is satisfied in the form given by equation (C.22).

C.3. Net Charges on Electrodes

The total charge per unit length on the electrode centred at $x = mp$ is denoted Q_m, so that

$$Q_m = \int_{mp-a/2}^{mp+a/2} \varrho_f(x)\,dx. \qquad (C.26)$$

These quantities are useful when calculating the capacitance of a transducer. On substituting for $\varrho_f(x)$ from equation (C.20), it is found that the x-integral has the form of the Mehler–Dirichlet formula, equation (C.5), and consequently Q_m can be expressed as

$$Q_m = 2(\varepsilon_0 + \varepsilon_p)\int_0^1 \frac{\sin(\pi s)\cos(2\pi ms)}{P_{-s}(-\cos\Delta)} P_{-s}(\cos\Delta)\,ds. \qquad (C.27)$$

If the metallisation ratio $a/p = \frac{1}{2}$ we have $\cos\Delta = 0$, so that the Legendre functions cancel. The integral is then straightforward, and gives

$$Q_m = \frac{4(\varepsilon_0 + \varepsilon_p)}{\pi(1 - 4m^2)}, \quad \text{for } \cos\Delta = 0. \qquad (C.28)$$

When evaluating the capacitance of uniform transducers, a summation of the Q_m is required. This summation can be obtained from the formula

$$N \sum_{m=-\infty}^{\infty} \cos[2\pi x(mN + n)] = \cos(2\pi nx) \sum_{i=-\infty}^{\infty} \delta(x - i/N), \qquad (C.29)$$

where n and N are integers, and $N \neq 0$. This follows from equation (A.43) of Appendix A, or by Fourier transformation of both sides. Using equation (C.27) for Q_m it is found that, for $N \geq 1$,

$$\sum_{m=-\infty}^{\infty} Q_{mN+n} = \frac{2(\varepsilon_0 + \varepsilon_p)}{N} \sum_{i=0}^{N} \frac{\sin(\pi v)\cos(2\pi nv)}{P_{-v}(-\cos\Delta)} P_{-v}(\cos\Delta) \qquad (C.30)$$

with $v = i/N$.

ELEMENTAL CHARGE DENSITY FOR REGULAR ELECTRODES

Another summation formula is needed for analysis of multi-strip couplers, and is again obtained by using equations (A.43) and (C.27). The formula is

$$\sum_{m=-\infty}^{\infty} Q_m e^{-jkmp} = 2(\varepsilon_0 + \varepsilon_p) \frac{\sin(\pi\mu)}{P_{-\mu}(-\cos\Delta)} P_{-\mu}(\cos\Delta), \qquad (C.31)$$

where μ is defined such that

$$kp = 2\pi(i + \mu)$$

and i is the integer part of $kp/(2\pi)$, so that $0 \leqslant \mu \leqslant 1$.

Appendix D

Floquet Analysis for an Infinite Array of Regular Electrodes

The equations for analysis of multi-strip couplers were derived in Chapter 5 using the Green's function method, making use of the quasi-static approximation. The equations therefore exclude electrode interactions, and in consequence do not predict the presence of any stop bands. In addition, for frequencies outside the stop bands the interactions cause a small change to the surface wave velocity, and this is not correctly predicted by the quasi-static approximation. To rectify these omissions, this appendix gives an account based on the work of Bløtekjaer *et al.* [475, 476], in which the solution is obtained by applying Floquet's theorem and using the effective permittivity discussed in Chapter 3. The permittivity is taken to be adequately represented by Ingebrigtsen's approximation, assuming that the only acoustic wave present is a piezoelectric Rayleigh wave. A similar approach was used by Emtage [477], and an alternative method using perturbation theory [118] gives equivalent results. In the analysis here the surface electric field and charge density are written in terms of space harmonics with coefficients given by Legendre functions, in a manner similar to an earlier analysis of helical waveguides given by Chu [478].

The use of the effective permittivity implies that mass loading due to the electrodes is neglected, and it is also assumed that the electrodes have negligible thickness and resistivity. The electrodes are taken to have a length, in the y-direction, much greater than their width, and the surface waves are assumed to be free of diffraction and to have wavefronts parallel to the y-axis. Thus all quantities, including the surface potential and charge density, can be taken to be invariant with y.

It is a feature of the solution that the equations become very complex when high frequencies are considered. In view of this, the analysis is given here for frequencies $\omega < 2\pi v_0/p$, where v_0 is the free-surface velocity and p the pitch of the electrodes. In most practical cases this range includes the frequencies of interest, and it includes the first stop-band, which occurs for frequencies near $\pi v_0/p$. The analysis for higher frequencies is discussed briefly in Section D.4.

D.1. General Solution for Low Frequencies

For a piezoelectric half-space, the surface potential $\phi(x)$ and charge density $\sigma(x)$ are related by the effective permittivity $\varepsilon_s(\beta)$, which gives the ratio of the Fourier transforms $\bar{\phi}(\beta)$ and $\bar{\sigma}(\beta)$. For the analysis here it is convenient to use the parallel electric field $E_x(x)$ at the surface, rather than the potential $\phi(x)$. Taking $\bar{E}_x(\beta)$ as the transform of $E_x(x)$, we have $\bar{E}_x(\beta) = -j\beta\bar{\phi}(\beta)$. Thus, with $\varepsilon_s(\beta)$ defined by equation (3.24) of Section 3.2, we have

$$\bar{\sigma}(\beta) = j\varepsilon_s(\beta) \text{ sgn }(\beta) \, \bar{E}_x(\beta) \qquad (D.1)$$

with sgn $(\beta) = 1$ for $\beta > 0$ and sgn $(\beta) = -1$ for $\beta < 0$. The solution must satisfy this relation and the boundary conditions that $E_x(x) = 0$ on the electrodes and $\sigma(x) = 0$ in the gaps between the electrodes.

The configuration of the electrodes is shown in Figure 5.1. The electrodes are centred at $x = mp$, with $-\infty \leqslant m \leqslant \infty$, and have width a. For the electrode centred at $x = mp$, the voltage is V_m and the current entering the electrode is I_m. For the analysis here we consider solutions in which these take the forms

$$V_m = V_0 \exp(-j\kappa mp),$$

$$I_m = I_0 \exp(-j\kappa mp), \qquad (D.2)$$

where a term $\exp(j\omega t)$ is implicit. The factor κ has the rôle of a propagation constant. These equations are unaffected if a multiple of $2\pi/p$ is added to κ, and it is convenient to restrict κ to the range

$$0 \leqslant \kappa \leqslant 2\pi/p. \qquad (D.3)$$

This restriction does not affect the generality of the solution.

With the electrode voltages given by equation (D.2), it is reasonable to assume that the electric field has the property $E_x(x + p) = E_x(x) \exp(-j\kappa p)$, and it then follows that $E_x(x) \exp(j\kappa x)$ must be a periodic function, with period p. Thus $E_x(x)$ can be written in the form

$$E_x(x) = \sum_{n=-\infty}^{\infty} E_n \, e^{-j2\pi nx/p} \, e^{-j\kappa x} \qquad (D.4)$$

and similarly the charge density $\sigma(x)$ can be written

$$\sigma(x) = \sum_{n=-\infty}^{\infty} \sigma_n \, e^{-j2\pi nx/p} \, e^{-j\kappa x}, \qquad (D.5)$$

where the coefficients E_n and σ_n are to be determined. These equations have the forms given by Floquet's theorem, which is readily proved for the case of a wave in an unbounded medium with periodic properties [114]. However, here we are concerned with a bounded medium having a periodic boundary condition, and for such cases Floquet's theorem is not always valid [114]. Nevertheless, it will be seen that a solution with the form of equations (D.4) and (D.5) can be obtained for the present case.

In the β-domain, each term in equations (D.4) and (D.5) has the form of a delta function located at $\beta = -(\kappa + 2\pi n/p)$. The ratio of E_n to σ_n is thus given by equation (D.1) for this value of β. In view of equation (D.3), we have sgn $(-\kappa - 2\pi n/p) = -S_n$, with S_n defined by

$$S_n = 1, \text{ for } n \geqslant 0; \quad S_n = -1, \text{ for } n < 0. \qquad (D.6)$$

FLOQUET ANALYSIS FOR AN INFINITE ARRAY OF REGULAR ELECTRODES

Noting also that $\varepsilon_s(-\beta) = \varepsilon_s(\beta)$, we have

$$\frac{\sigma_n}{E_n} = -jS_n\varepsilon_s(\kappa + 2\pi n/p). \tag{D.7}$$

Now, $\varepsilon_s(\beta)$ is taken to be given by Ingebrigtsen's approximation, equation (3.38), which is plotted in Figure 3.2. This function is almost constant except for β close to $\pm k_0$, where $k_0 > 0$ is the wavenumber for surface wave propagation on a free surface at frequency ω. This condition is expressed by writing $\varepsilon_s(\beta) = \varepsilon_s(\infty)$ for all β except for values near $\pm k_0$. We also assume that the frequency is restricted such that

$$0 \leq k_0 < 2\pi/p, \tag{D.8}$$

which implies that $\omega < 2\pi v_0/p$. In view of this restriction, and of equation (D.3), it is found that $\kappa + 2\pi n/p$ can only be close to $\pm k_0$ if $n = 0$ or -1, and so equation (D.7) becomes

$$\frac{\sigma_n}{E_n} = -jS_n\varepsilon_s(\kappa + 2\pi n/p), \quad \text{for } n = 0, -1, \tag{D.9a}$$

$$= -jS_n\varepsilon_s(\infty), \quad \text{for } n \neq 0, -1. \tag{D.9b}$$

If k_0 is close to $2\pi/p$, the additional case $n = 1$ must be included in equation (D.9a); however this is excluded here by assuming that k_0 is at least a few percent below $2\pi/p$.

The solution for σ_n and E_n is found using properties of Legendre functions, given in Appendix C. Consider the functions

$$E'(x) = \sum_{n=-\infty}^{\infty} S_{n-r}P_{n-r}(\cos \Delta)\, e^{-j2\pi nx/p}\, e^{-j\kappa x}$$

$$\sigma'(x) = -j\varepsilon_s(\infty) \sum_{n=-\infty}^{\infty} P_{n-r}(\cos \Delta)\, e^{-j2\pi nx/p}\, e^{-j\kappa x}, \tag{D.10}$$

where $\Delta = \pi a/p$ and a and p are respectively the width and pitch of the electrodes. From equation (C.15) it can be seen that $E'(x)$ is zero at the electrode locations, irrespective of the value of the integer r. From equation (C.14), $\sigma'(x)$ is zero in the gaps between the electrodes, for any r. If $\varepsilon_s(\beta)$ were independent of β, equations (D.9) would be satisfied by equations (D.10), with $r = 0$. However, because $\varepsilon_s(\beta)$ has different values for $n = 0, -1$ it is necessary to add terms with several values of r. It is found that terms with $r = -1, 0, 1$ are sufficient, so that the field and charge density are given by

$$E_x(x) = \sum_{n=-\infty}^{\infty} \sum_{r=-1}^{1} \alpha_r S_{n-r} P_{n-r}(\cos \Delta)\, e^{-j2\pi nx/p}\, e^{-j\kappa x}, \tag{D.11}$$

$$\sigma(x) = -j\varepsilon_s(\infty) \sum_{n=-\infty}^{\infty} \sum_{r=-1}^{1} \alpha_r P_{n-r}(\cos \Delta)\, e^{-j2\pi nx/p}\, e^{-j\kappa x}, \tag{D.12}$$

where the coeffients α_r are to be determined. Since these are linear combinations of equations (D.10), the required boundary conditions are satisfied. We also have, writing $P_n(\cos \Delta)$ as P_n for brevity,

$$\frac{\sigma_n}{E_n} = \frac{-j\varepsilon_s(\infty) [\alpha_{-1}P_{n+1} + \alpha_0 P_n + \alpha_1 P_{n-1}]}{\alpha_{-1}S_{n+1}P_{n+1} + \alpha_0 S_n P_n + \alpha_1 S_{n-1}P_{n-1}} \quad (D.13)$$

and this is required to satisfy equations (D.9). Now, from equation (D.6), it can be seen that $S_{n-1} = S_n = S_{n+1}$ for $n \neq 0$ or -1, and hence equation (D.9b) is already satisfied. The ratios of the coefficients α_r are therefore determined by equation (D.9a). It is convenient to define A_0 and A_{-1} by

$$A_r = \frac{\varepsilon_s(\kappa + 2\pi r/p) + \varepsilon_s(\infty)}{\varepsilon_s(\kappa + 2\pi r/p) - \varepsilon_s(\infty)}, \quad \text{for } r = 0, -1. \quad (D.14)$$

Substituting equation (D.13) into equation (D.9a), it is then found that

$$\frac{\alpha_1}{\alpha_0} = \frac{A_{-1} + \cos \Delta}{A_0 A_{-1} - \cos^2 \Delta},$$

$$\frac{\alpha_{-1}}{\alpha_0} = \frac{A_0 + \cos \Delta}{A_0 A_{-1} - \cos^2 \Delta}, \quad (D.15)$$

where use has been made of the relations $P_{-1} = P_0 = 1$ and $P_{-2} = P_1 = \cos \Delta$.

The effective permittivity $\varepsilon_s(\beta)$ is taken to be given by Ingebrigtsen's approximation, equation (3.38), so that

$$\varepsilon_s(\beta) = (\varepsilon_0 + \varepsilon_p^T) \frac{\beta^2 - k_0^2}{\beta^2 - k_m^2}, \quad (D.16)$$

where k_0 and k_m are respectively the wavenumbers for surface waves on a free surface and on a metallised surface. We thus have $\varepsilon_s(\infty) = \varepsilon_0 + \varepsilon_p^T$ and, using equation (D.14),

$$A_0 = (2\kappa^2 - k_0^2 - k_m^2)/(k_m^2 - k_0^2) \quad (D.17)$$

and

$$A_{-1} = [2(\kappa - 2\pi/p)^2 - k_0^2 - k_m^2]/(k_m^2 - k_0^2). \quad (D.18)$$

Electrode Voltages and Currents. The electrode voltages V_m are found by integrating the field to give the potential, evaluating this at the electrode centres $x = mp$. Using equation (D.4) for the field, the electrode voltages are found to be $V_m = V_0 \exp(-j\kappa mp)$, with V_0 given by

$$V_0 = -\frac{jp}{2\pi} \sum_{n=-\infty}^{\infty} \frac{E_n}{n + s}, \quad (D.19)$$

where we have defined

$$s = \kappa p/(2\pi), \tag{D.20}$$

and, in view of equation (D.3), we have $0 \leqslant s \leqslant 1$. The coefficients E_n in equation (D.19) are identified by comparing equation (D.4) with equation (D.11), and the summation over n can be done by using Dougall's expansion, equation (C.12). We thus have

$$V_0 = -\frac{jp}{2\sin(\pi s)} \sum_{r=-1}^{1} (-1)^r \alpha_r P_{r+s-1}(-\cos \Delta). \tag{D.21}$$

The current I_m entering the electrode centred at $x = mp$ is found by integrating the charge density over the area of the electrode and differentiating with respect to time. The integral over x has limits $x = mp \pm a/2$, and the electrode length, in the y-direction, is W. Using equation (D.5) for the charge density, it is found that $I_m = I_0 \exp(-j\kappa mp)$, with

$$I_0 = \frac{j\omega W p}{2\pi} \int_{-\Delta}^{\Delta} \sum_{n=-\infty}^{\infty} \sigma_n e^{-j(n+s)\theta} \, d\theta, \tag{D.22}$$

where $\theta = 2\pi(x - mp)/p$ and $\Delta = \pi a/p$. The coefficients σ_n can be evaluated by comparing equations (D.5) and (D.12). Equation (D.22) may then be evaluated using equation (C.16) of Appendix C, giving the result

$$I_0 = \omega W p (\varepsilon_0 + \varepsilon_p^T) \sum_{r=-1}^{1} \alpha_r P_{r+s-1}(\cos \Delta), \tag{D.23}$$

where the relation $\varepsilon_s(\infty) = (\varepsilon_0 + \varepsilon_p^T)$ has been used.

Equations (D.21) and (D.23) give the electrode voltages and currents in terms of the coefficients α_r, which are related to the propagation constant κ by equations (D.15), (D.17) and (D.18). If the ratio I_0/V_0 is specified these equations determine the relative values of the α_r and the value of κ, and hence the field and charge density, equations (D.11) and (D.12), are determined. The following sections describe some particular cases.

D.2. Propagation Outside the Stop Band

For propagation on a free surface, the surface-wave wavenumber is $\pm k_0$. Strong coupling to a surface wave is therefore expected only if one or more of the wavenumbers $\kappa \pm 2\pi n/p$ in the Floquet expansion is near to $\pm k_0$. Usually, at most one of the wavenumbers is close to $\pm k_0$. However, in the special case $k_0 \simeq \pi/p$, it is possible to have $\kappa \simeq k_0$ and $\kappa - 2\pi/p \simeq -k_0$, so that coupling occurs for $n = 0$ and for $n = -1$. In this case two surface waves propagating in opposite directions are coupled by the structure, and this leads to the presence of a stop band. This occurs for frequencies close to $\pi v_0/p$. In this section it will be assumed that the frequency is not close to this value. The stop band will be considered in Section D.3.

It can be assumed that, if coupling to surface waves occurs, it occurs for $\kappa \simeq k_0$. The alternative case, when $\kappa - 2\pi/p \simeq -k_0$, is essentially the same solution, and can therefore be neglected. In view of this, we can take $\varepsilon_s(\kappa - 2\pi/p) = \varepsilon_s(\infty)$ and thus from equation (D.14) we have $A_{-1} = \infty$. Equations (D.15) then give $\alpha_{-1} = 0$ and

$$\alpha_0/\alpha_1 = A_0 = (2\kappa^2 - k_0^2 - k_m^2)/(k_m^2 - k_0^2), \tag{D.24}$$

where A_0 has been obtained from equation (D.17). Thus the α_{-1} term in the electric field and charge density, equations (D.11) and (D.12), is not required here. This can also be deduced more directly by noting that equation (D.9b) must be valid for all $n \neq 0$. We now consider separately three cases, in which the electrodes are short-circuited or open-circuited, or have some more general termination.

(a) Shorted Electrodes. If the electrodes are all connected together the voltages V_m must all be zero, and hence $V_0 = 0$. Thus, noting that $\alpha_{-1} = 0$, equation (D.21) gives

$$\alpha_0/\alpha_1 = P_s(-\cos \Delta)/P_{s-1}(-\cos \Delta). \tag{D.25}$$

Using equation (D.24), this gives a relation determining the propagation constant κ, which for this case is denoted k_{sc}. Thus,

$$\kappa^2 = k_{sc}^2 = k_0^2 + \tfrac{1}{2}(k_m^2 - k_0^2)\left[1 + \frac{P_s(-\cos \Delta)}{P_{s-1}(-\cos \Delta)}\right]. \tag{D.26}$$

Since k_m is close to k_0 it can be concluded that k_{sc} is also close to k_0. In the above equation $s = k_{sc}p/(2\pi)$, so that the equation is transcendental. However, since the Legendre functions vary slowly with s, a good approximation is obtained by using $s = k_0 p/(2\pi)$, and the right side is then independent of k_{sc}. The surface-wave velocity in the structure is $v_{sc} = \omega/k_{sc}$. Noting that $k_m \approx k_0$, the velocity is approximately given by

$$v_{sc} \approx v_0 + \tfrac{1}{2}(v_m - v_0)\left[1 + \frac{P_s(-\cos \Delta)}{P_{s-1}(-\cos \Delta)}\right], \tag{D.27}$$

where $v_0 = \omega/k_0$ and $v_m = \omega/k_m$.

(b) Open-Circuit Electrodes. With the electrodes disconnected electrically the currents I_m are zero, and hence $I_0 = 0$. From equation (D.23) we have

$$\alpha_0/\alpha_1 = -P_s(\cos \Delta)/P_{s-1}(\cos \Delta), \tag{D.28}$$

and κ may then be obtained using equation (D.24). For this case κ is denoted by k_{oc}, and we thus have

$$\kappa^2 = k_{oc}^2 = k_0^2 + \tfrac{1}{2}(k_m^2 - k_0^2)\left[1 - \frac{P_s(\cos \Delta)}{P_{s-1}(\cos \Delta)}\right]. \tag{D.29}$$

Here again, k_{oc} must be close to k_0, and a good approximation is obtained by setting $s = k_0 p/(2\pi)$. The surface-wave velocity is $v_{oc} = \omega/k_{oc}$ and is approximately given by

FLOQUET ANALYSIS FOR AN INFINITE ARRAY OF REGULAR ELECTRODES 369

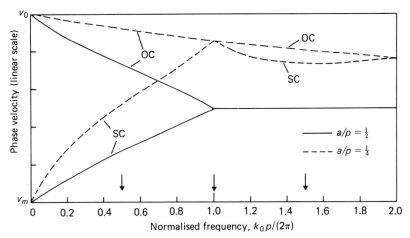

FIGURE D.1. Surface wave velocity in an array of open-circuited (OC) or shorted (SC) electrodes. Arrows indicate locations of stop bands, where the analysis is invalid.

$$v_{oc} \approx v_0 + \tfrac{1}{2}(v_m - v_0)\left[1 - \frac{P_s(\cos \Delta)}{P_{s-1}(\cos \Delta)}\right] \quad (D.30)$$

The velocities v_{sc} and v_{oc} are shown as functions of frequency in Figure D.1, for several metallisation ratios. For $\omega < 2\pi v_0/p$ the velocities are given by equations (D.27) and (D.30), but for higher frequencies the analysis of Section D.4 below must be used. Some experimental results confirming the forms of the curves, for $a/p = \tfrac{1}{2}$ and $0 < \omega < 2\pi v_0/p$, were obtained by Williamson [134] using a Y, Z lithium niobate substrate and an electrostatic probe.

With $s = k_0 p/(2\pi)$, the difference $k_{sc}^2 - k_{oc}^2$ can be obtained by subtracting equation (D.29) from equation (D.26). Using equation (C.6) of Appendix C, this gives

$$k_{sc}^2 - k_{oc}^2 = \frac{k_m^2 - k_0^2}{k_0 p} \frac{2 \sin(\pi s)}{P_{-s}(-\cos \Delta) P_{-s}(\cos \Delta)}. \quad (D.31)$$

Noting that $k_m \approx k_0$, this is in good agreement with the result from the quasi-static analysis, equation (5.23).

(c) General Terminations. The above equations give the propagation constant κ when the ratio I_0/V_0 is either infinite or zero. More generally, I_0/V_0 can take intermediate values. The electrode voltages and currents must of course have the form given by equations (D.2).

The ratio I_0/V_0 is given by equations (D.21) and (D.23), noting that $\alpha_{-1} = 0$ here. The ratio can be expressed as

$$\frac{I_0}{V_0} = 2j\omega W(\varepsilon_0 + \varepsilon_p^T) \sin(\pi s) \frac{P_{s-1}(\cos \Delta)}{P_{s-1}(-\cos \Delta)} \frac{X_+}{X_-}, \tag{D.32}$$

where

$$X_\pm = 1 \pm \frac{\alpha_1}{\alpha_0} \frac{P_s(\pm \cos \Delta)}{P_{s-1}(\pm \cos \Delta)} \tag{D.33}$$

and $s = \kappa p/(2\pi)$. The ratio α_1/α_0 is related to κ by equation (D.24), and hence equation (D.32) gives the ratio I_0/V_0 if κ is specified. Alternatively, the equation gives κ if I_0/V_0 is specified.

The above result may be shown to be in close agreement with the quasi-static analysis. From equation (D.24) it can be seen that α_1/α_0 is small except when κ is close to k_0. Thus, since the Legendre functions vary slowly with s, the functions $X_\pm$ in equation (D.33) can be evaluated approximately by setting $s = k_0 p/(2\pi)$. The ratio of Legendre functions in equation (D.33) is then related to k_{oc} or k_{sc} by equation (D.29) or equation (D.26). This gives

$$X_+ \approx 2(\kappa^2 - k_{oc}^2)/(2\kappa^2 - k_0^2 - k_m^2).$$

For X_- the same result is obtained, with k_{oc} replaced by k_{sc}. Thus I_0/V_0 is given by equation (D.32), with

$$\frac{X_+}{X_-} \approx \frac{\kappa^2 - k_{oc}^2}{\kappa^2 - k_{sc}^2} = 1 - \frac{k_{sc}^2 - k_{oc}^2}{k_{sc}^2 - \kappa^2} \tag{D.34}$$

and with $s = \kappa p/(2\pi)$. Also, $k_{sc}^2 - k_{oc}^2$ is given by equation (D.31). This result agrees well with the quasi-static analysis, equation (5.18), except that the latter has $k_0^2 - \kappa^2$ in the denominator instead of $k_{sc}^2 - \kappa^2$.

D.3. Stop Bands

When k_0 is close to π/p, it is no longer valid to ignore the α_{-1} term in the equations of Section D.1, and it is then found that the propagation constant κ can be complex. In this section we consider the stop band for the two cases when the electrodes are either shorted or open-circuited. It is sufficient to assume that $k_0 \approx \pi/p$, because the solution for other k_0 has already been given in Section D.2.

(a) Shorted Electrodes. For shorted electrodes V_0, given by equation (D.21), is zero. The propagation constant κ must be close to k_0, and since $k_0 \approx \pi/p$ we have $s = \kappa p/(2\pi) \approx \frac{1}{2}$. The Legendre functions vary slowly with s, and so it is a good approximation to set $s = \frac{1}{2}$ in these functions. Equation (D.21) thus gives

$$(\alpha_1 + \alpha_{-1})/\alpha_0 = P_{-1/2}(-\cos \Delta)/P_{1/2}(-\cos \Delta). \tag{D.35}$$

The left side of this equation is related to κ by equations (D.15), (D.17) and (D.18).

To simplify the equations, approximate forms of the functions A_0 and A_{-1} are used. Noting that κ, k_0 and k_m are all close to π/p, equations (D.17) and (D.18) give

$$A_0 \approx (2\kappa - k_0 - k_m)/(k_m - k_0)$$

and

$$A_{-1} \approx (4\pi/p - 2\kappa - k_0 - k_m)/(k_m - k_0). \tag{D.36}$$

The solution for κ is then found by using equations (D.15). It is convenient to define the functions

$$F_\pm(\Delta) = \mp\cos\Delta + P_{1/2}(\pm\cos\Delta)/P_{-1/2}(\pm\cos\Delta) \tag{D.37}$$

and the expression

$$\frac{\Delta v}{v} = (v_0 - v_m)/v_0 \approx p(k_m - k_0)/\pi. \tag{D.38}$$

It is then found that

$$(\kappa - \pi/p)^2 \approx (\omega - \omega_1)(\omega - \omega_{sc})/v_0^2, \tag{D.39}$$

where

$$\omega_1 = \frac{\pi v_0}{p}\left[1 + \frac{1}{2}\frac{\Delta v}{v}(\cos\Delta - 1)\right] \tag{D.40}$$

and ω_{sc} is given by

$$\omega_1 - \omega_{sc} = \frac{\pi v_0}{p}\frac{\Delta v}{v}F_-(\Delta). \tag{D.41}$$

The function $F_-(\Delta)$ is positive, so that $\omega_1 > \omega_{sc}$. Equation (D.39) thus shows that κ is complex for $\omega_{sc} < \omega < \omega_1$, and hence the frequencies ω_{sc} and ω_1 give the two edges of the stop band. These frequencies are shown in Figure D.2, as functions of metallisation ratio. The width of the stop band, $\omega_1 - \omega_{sc}$, is given by equation (D.41) and is shown in Figure D.3.

(b) Open-Circuited Electrodes. For the open-circuit case $I_0 = 0$, and with $s = \frac{1}{2}$ equation (D.23) gives

$$(\alpha_1 + \alpha_{-1})/\alpha_0 = -P_{-1/2}(\cos\Delta)/P_{1/2}(\cos\Delta). \tag{D.42}$$

Using equations (D.36) and (D.15), the solution for κ is

$$(\kappa - \pi/p)^2 \approx (\omega - \omega_1)(\omega - \omega_{oc})/v_0^2, \tag{D.43}$$

where ω_1 is given by equation (D.40) and ω_{oc} is given by

$$\omega_{oc} - \omega_1 = \frac{\pi v_0}{p}\frac{\Delta v}{v}F_+(\Delta). \tag{D.44}$$

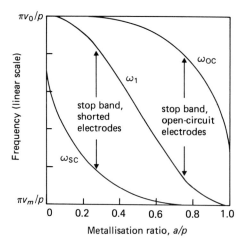

FIGURE D.2. Frequencies of the stop band edges.

with $F_+(\Delta)$ give by equation (D.37). Here $\omega_{oc} > \omega_1$, so κ is complex for frequencies in the range $\omega_1 < \omega < \omega_{oc}$. The frequency ω_{oc} is shown in Figure D.2, and the width of the stop band is shown in Figure D.3.

D.4. Solution at Higher Frequencies

The above sections give the solution for frequencies $\omega < 2\pi v_0/p$, that is, for $0 \leq k_0 < 2\pi/p$. For higher frequencies a similar approach can be used, with the field and charge density given by expressions similar to equations (D.11) and (D.12), but the range of values for r must be increased. Suppose, for example, that $E_x(x)$ and $\sigma(x)$ are given by equations (D.4) and (D.5), but the coefficients E_n and σ_n are now given by

$$E_n = \sum_{r=-R}^{R} \alpha_r S_{n-r} P_{n-r}(\cos \Delta) \tag{D.45}$$

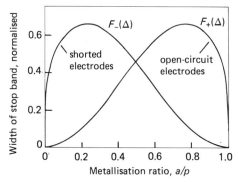

FIGURE D.3. Width of stop band, as a function of metallisation ratio.

and
$$\sigma_n = -j\varepsilon_s(\infty) \sum_{r=-R}^{R} \alpha_r P_{n-r}(\cos \Delta). \qquad (D.46)$$

These equations give
$$\frac{\sigma_n}{E_n} = -jS_n\varepsilon_s(\infty), \quad \text{for } n \geq R,$$
$$\text{or} \quad \text{for } n \leq -R-1.$$

We also have, from equation (D.7),
$$\frac{\sigma_n}{E_n} = -jS_n\varepsilon_s(\kappa + 2\pi n/p), \qquad (D.47)$$

where $0 \leq \kappa \leq 2\pi/p$. Equations (D.45) and (D.46) therefore give a valid solution if $\varepsilon_s(\kappa + 2\pi n/p) = \varepsilon_s(\infty)$ for $n \geq R$ and for $n \leq -R-1$. In the cases of interest, $\varepsilon_s(\beta)$ always approaches a limit for large β, and hence this condition can be satisfied by choosing a large enough value for R. The relative values of the coefficients α_r can then be found by substituting equations (D.45) and (D.46) into equation (D.47). It should be noted that, for $-R-1 < n < R$, the method places no restrictions on the values of $\varepsilon_s(\kappa + 2\pi n/p)$, and is therefore very versatile.

The analysis using this method is given by Bløtekjaer et al. [475, 476]. In view of the complexity of the results, the discussion here is restricted to a consideration of the surface-wave velocities for either shorted or open-circuit electrodes, for frequencies outside the stop bands. For these cases, the analysis can be presented in a somewhat simpler form, similar to that of Datta and Hunsinger [479]. The effective permittivity $\varepsilon_s(\beta)$ is taken to be given by Ingebrigtsen's approximation, and it is a good approximation to assume that $\varepsilon_s(\kappa + 2\pi n/p)$ is equal to $\varepsilon_s(\infty)$ for all except one value of n. This of course excludes the stop bands. It is found that negative values of r are not required, so the coefficients E_n and σ_n can be written as

$$E_n = \sum_{r=0}^{R} \alpha_r S_{n-r} P_{n-r}(\cos \Delta),$$

$$\sigma_n = -j\varepsilon_s(\infty) \sum_{r=0}^{R} \alpha_r P_{n-r}(\cos \Delta). \qquad (D.48)$$

These equations give $\sigma_n/E_n = -jS_n\varepsilon_s(\infty)$ for $n \geq R$ and for $n \leq -1$. It is assumed that $\varepsilon_s(\kappa + 2\pi n/p) = \varepsilon_s(\infty)$ for $n \neq R-1$, which implies that

$$2\pi(R-1)/p \leq k_0 \leq 2\pi R/p, \qquad (D.49)$$

and that, from equation (D.47),
$$\frac{\sigma_n}{E_n} = -jS_n\varepsilon_s(\infty), \quad \text{for } n \neq R-1. \qquad (D.50)$$

Substituting equations (D.48) into equation (D.50), with $n = R-2$, gives $\alpha_{R-1} = -\alpha_R P_{-2}(\cos \Delta)$. Using $n = R-3, R-4, \ldots$ gives the relative values of $\alpha_{R-2}, \alpha_{R-3}, \ldots$. These can be expressed in the form

$$\alpha_{R-m+1} = -\sum_{i=2}^{m} \alpha_{R-m+i} P_{-i}(\cos \Delta), \quad \text{for } 2 \leqslant m \leqslant R. \tag{D.51}$$

This determines the relative values of all the α_r, except for α_0. The additional relation required is obtained by setting $n = R - 1$, using equation (D.47) for σ_n/E_n. This gives

$$\sum_{i=0}^{R-1} \alpha_i P_{R-i-1}(\cos \Delta) = \alpha_R A_{R-1}, \tag{D.52}$$

where A_{R-1} is defined by equation (D.14), taking $r = R - 1$. Using Ingebrigtsen's approximation, equation (D.16), we have

$$A_{R-1} = (2k^2 - k_0^2 - k_m^2)/(k_m^2 - k_0^2), \tag{D.53}$$

where k has been defined by

$$k = \kappa + 2\pi(R - 1)/p. \tag{D.54}$$

When the electrodes are either shorted or open-circuited, it will be found that k is close to k_0, and can therefore be identified as the surface-wave wavenumber.

The electrode voltages and currents are calculated as in Section D.1, giving

$$V_0 = -\frac{jp}{2 \sin (\pi s)} \sum_{r=0}^{R} \alpha_r (-1)^r P_{-r-s}(-\cos \Delta) \tag{D.55}$$

and

$$I_0 = \omega W p(\varepsilon_0 + \varepsilon_p^T) \sum_{r=0}^{R} \alpha_r P_{-r-s}(\cos \Delta), \tag{D.56}$$

where, as before, $s = \kappa p/(2\pi)$ and $0 \leqslant s \leqslant 1$. For shorted electrodes V_0 is set to zero, and with equations (D.51)–(D.54) the wavenumber k is then determined. This wavenumber is denoted k_{sc}. For open-circuited electrodes I_0 is set to zero, and the solution for k is denoted k_{oc}.

As an example, we give the solutions for $R = 2$. Defining

$$B_{\pm}(\Delta) = \cos \Delta + \frac{[P_{s+1}(\pm \cos \Delta) \mp \cos (\Delta) P_s(\pm \cos \Delta)] \cos \Delta}{P_{-s}(\pm \cos \Delta)}$$

the solution for shorted electrodes is

$$k_{sc}^2 = k_0^2 + \tfrac{1}{2}(k_m^2 - k_0^2)[1 - B_-(\Delta)],$$

and for open-circuited electrodes

$$k_{oc}^2 = k_0^2 + \tfrac{1}{2}(k_m^2 - k_0^2)[1 - B_+(\Delta)].$$

The corresponding velocities are shown on Figure D.1.

Appendix E

Electrode Interactions in Transducers

The transducer analysis given in Chapter 4 makes use of the quasi-static approximation, which ignores the perturbation of a propagating surface wave by the electrodes of a shorted transducer. In practice each electrode reflects an incident wave to some extent, and this can be significant if the reflected waves add coherently, that is, if the electrodes are regular and the pitch is close to a multiple of the half-wavelength. The main consequences are that a shorted transducer will reflect surface waves, in contrast with the prediction of the quasi-static analysis, and that its frequency response is distorted. The electrodes also perturb the surface wave velocity, as shown by the analysis in Section D.2. These effects arise from two distinct causes: electrical loading and mechanical loading. For strongly piezoelectric materials, such as lithium niobate, electrical loading can cause severe distortions. However, mechanical loading is not usually very significant in practice, provided the metal film used for the transducer has elastic properties similar to the substrate; this condition is usually met by using an aluminium film.

To analyse electrode interactions in transducers, several types of modified equivalent network models have been used [480, 481]. In particular the crossed-field model, which in its basic form neglects electrode interactions, may be modified by introducing transmission lines to represent surface wave propagation, using different characteristic impedances in the metallised and unmetallised regions [355, 470, 482]. This model does not correctly account for the variation of electrode reflectivity with frequency and with metallisation ratio, though these limitations would be overcome if the transmission line parameters were varied in an appropriate manner. The analysis here is based on a "reflective array model", which assumes that the perturbation due to each electrode can be represented by a transmission matrix related to an effective reflection coefficient, r. The transducer properties can be deduced by cascading the transmission matrices of the electrodes. This approach has the merit of generality, since the properties of the array are related to the reflection coefficient of one electrode without invoking the physical mechanism of the reflection. The method is closely related to the analyses of Skeie [483–486], Emtage [487, 488] and Panasik and Hunsinger [489]. Another method is the generalised Green's function analysis of Milsom et al. [93].

Section E.1 below describes the reflective array model, and Section E.2 uses the Green's function method to deduce the reflection coefficient of one electrode due to electrical loading. The results are brought together in Section E.3, which derives the relevant transducer properties. Mechanical loading is discussed in Section E.4.

Throughout this appendix the electrodes are taken to be regular. End effects, bulk wave excitation and electrode resistivity are ignored.

E.1. Reflective Array Model

We consider an array of identical electrodes with pitch p, as shown in Figure E.1. It is assumed that any electrical connections to the electrodes are all identical, so that the electrodes will all scatter surface waves in the same manner. Here the properties of the array are deduced in terms of the scattering due to individual electrodes. The mechanism of the scattering is not considered at this stage, so the results are valid for both electrical and mechanical loading, and are also applicable to the other types of periodic structure. The analysis for an infinite structure is given in more detail by Collin [490]. An alternative approach, modelling the array as a series of repetitively mismatched transmission lines, was used by Sittig and Coquin [491] to analyse both infinite and finite structures. This gives results equivalent to those presented here.

At some frequency ω, it is assumed that the disturbance in the gaps between the electrodes includes terms of the form $\exp(\pm jk_0 x)$ representing propagating waves, where k_0 is the wavenumber for surface waves on a free surface. Considering one particular electrode, the amplitudes of waves on the left are denoted by c_1 and b_1, and the amplitudes of waves on the right are denoted by c_2 and b_2, as in Figure E.1. These amplitudes are measured at points distant $p/2$ from the centre of the electrode. The terms c_1 and c_2 denote waves propagating to the right, with amplitudes proportional to $\exp(-jk_0 x)$, while b_1 and b_2 denote waves propagating to the left. The amplitudes are linearly related, so that the waves leaving the electrode can be written in terms of incident waves, using a scattering matrix S_{ij}:

$$\begin{bmatrix} b_1 \\ c_2 \end{bmatrix} = \begin{bmatrix} S_{11} & S_{12} \\ S_{12} & S_{11} \end{bmatrix} \begin{bmatrix} c_1 \\ b_2 \end{bmatrix} \tag{E.1}$$

This equation uses the relations $S_{22} = S_{11}$ and $S_{21} = S_{12}$, which follow from the

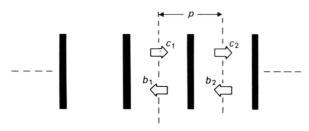

FIGURE E.1. Wave propagation in an infinite array of regular electrodes.

symmetry. If either c_1 or b_2 is zero, S_{11} gives the reflected wave amplitude, while S_{12} gives the transmitted wave amplitude. Defining r as the reflection coefficient of the electrode (referred to its centre) and t as the transmission coefficient, we have

$$S_{11} = r \exp(-jk_0 p) \quad \text{(E.2)}$$

and

$$S_{12} = t \exp(-jk_0 p). \quad \text{(E.3)}$$

If the electrode did not perturb the wave, we would have $r = 0$ and $t = 1$. It should be noted that r and t are defined for an electrode in a periodic array, and their values may be affected by the presence of neighbouring electrodes. This is true in the case of scattering due to electrical loading. Generally, r and t will be frequency dependent.

It is assumed here that there is no loss of power. Setting $b_2 = 0$ and equating the input power to the sum of the output powers, we have

$$|r|^2 + |t|^2 = 1. \quad \text{(E.4)}$$

In addition, if all four of the amplitudes b_1, b_2, c_1 and c_2 are non-zero, power conservation gives

$$rt^* + tr^* = 0. \quad \text{(E.5)}$$

Combining this with equation (E.4) we find

$$r^2 = t^2 - t/t^*, \quad \text{(E.6)}$$

and, using equation (E.4) again, we also have

$$r/t = \pm j|r/t|. \quad \text{(E.7)}$$

Thus the reflection and transmission coefficients are in phase quadrature. Apart from a phase ambiguity of π, the reflection coefficient can be deduced from the transmission coefficient.

For subsequent analysis it is convenient to relate the waves on the right of the electrode to those on the left by means of a transmission matrix T_{ij}, defined such that

$$\begin{bmatrix} c_2 \\ b_2 \end{bmatrix} = \begin{bmatrix} T_{11} & T_{12} \\ T_{21} & T_{22} \end{bmatrix} \begin{bmatrix} c_1 \\ b_1 \end{bmatrix} \quad \text{(E.8)}$$

The coefficients T_{ij} are obtained by re-arranging equation (E.1). Making use of equation (E.6) we find

$$[T] = \begin{bmatrix} 1/\tau^* & r/t \\ -r/t & 1/\tau \end{bmatrix}, \quad \text{(E.9)}$$

where for convenience we define $\tau = S_{12}$, the transmission coefficient for one cell of the periodic structure, so that

$$\tau = t \exp(-jk_0 p). \quad \text{(E.10)}$$

Propagation in an Infinite Array. In a regular array of electrodes the above equations apply to each electrode, and the wave amplitudes c_2 and b_2 on the right of one electrode can be identified with the amplitudes c_1 and b_1 on the left of the subsequent electrode. In an infinite array, the solution corresponds to a propagating wave motion if, for each electrode, c_2 and b_2 are the same as c_1 and b_1 apart from a phase shift. Here the phase shift is taken to be $-\gamma p$, so that γ can be interpreted as the wavenumber. We thus consider solutions in which

$$c_2 = c_1 \exp(-j\gamma p),$$
$$b_2 = b_1 \exp(-j\gamma p), \quad \text{(E.11)}$$

which give a propagating wave if γ is real. Using the transmission matrix, equations (E.8) and (E.9), gives a dispersion relation for γ, and with the aid of equation (E.6) we find

$$2 \cos \gamma p = \frac{1}{\tau} + \frac{1}{\tau^*} \quad \text{(E.12)}$$

This can be re-arranged conveniently if we define θ_t as the phase of the electrode transmission coefficient t, so that

$$t = |t| \exp(j\theta_t), \quad \text{(E.13)}$$

We then have, from equation (E.10), $\tau = |t| \exp[j(\theta_t - k_0 p)]$, and equation (E.12) gives

$$\cos \gamma p = \frac{\cos(k_0 p - \theta_t)}{|t|}. \quad \text{(E.14)}$$

Here k_0, the free-surface wavenumber, is proportional to ω. It is usually the case that θ_t is small and $|t|$ is close to unity, so that for most values of ω there is a real solution for γ, close to $\pm k_0$. However, since $|t| < 1$ the right side of equation (E.14) is greater than 1 or less than -1 when $k_0 p$ is close to a multiple of π, and γ is then complex, giving a stop band. The solution is ambiguous in that equation (E.14) remains valid if a multiple of $2\pi/p$ is added to γ, as expected from the form of equations (E.11). In any interval of width $2\pi/p$ there are two solutions for γ, corresponding to waves propagating in opposite directions. For any particular solution the wave amplitudes c_1 and b_1 are finite, so that the solution involves waves propagating in both directions, even though the overall wave motion is in one direction.

The stop-band edges occur when the right side of equation (E.14) is equal to ± 1. Using Equation (E.4) this condition can be written

$$k_0 p - \theta_t = m\pi \pm \sin^{-1}(|r|), \quad m = 0, \pm 1, \pm 2 \ldots \quad \text{(E.15)}$$

Here $k_0 = \omega/v_0$, where v_0 is the free-surface velocity. Usually $\dot{r}$ and θ_t vary slowly with frequency, and r is small. Thus the width in frequency of each stop band is given by

$$\Delta\omega \approx 2v_0 |r|/p, \quad \text{(E.16)}$$

which is proportional to the electrode reflection coefficient. Within each stop-band, equation (E.14) is satisfied by writing γ as

$$\gamma = M\pi/p + j\alpha, \qquad (E.17)$$

where α is given by

$$\cosh(\alpha p) = (-1)^M \frac{\cos(k_0 p - \theta_t)}{|t|} \qquad (E.18)$$

and the integer M is chosen such that the right side of equation (E.18) is positive.

Propagation in a Finite Array. We now consider an array of N electrodes, as shown in Figure E.2. For analysis of interdigital transducers, given later, it is necessary to evaluate the amplitudes of waves propagating in the structure, and of the reflected wave, when a surface wave is incident at one end. To do this, the transmission matrix of equation (E.9) is used. It is assumed that end effects can be ignored, so that the same transmission matrix can be used for all the electrodes. Thus, for electrode n the wave amplitudes on the two sides are related by

$$\begin{bmatrix} c_n \\ b_n \end{bmatrix} = [T] \begin{bmatrix} c_{n-1} \\ b_{n-1} \end{bmatrix}, \qquad (E.19)$$

where $[T]$ is given by equation (E.9) and $1 \leq n \leq N$. Applying this equation recursively to electrodes 1 to n, we have

$$\begin{bmatrix} c_n \\ b_n \end{bmatrix} = [T]^n \begin{bmatrix} c_0 \\ b_0 \end{bmatrix}, \qquad (E.20)$$

which relates the waves to the right of electrode n to the waves at the left end of the array. Now, for any 2×2 matrix $[T]$, the elements of the matrix $[T]^n$ can be written in terms of the elements of $[T]$, using formulae given in References [491–493]. Denoting the elements of $[T]^n$ by T_{ij}^n, we have in the present case

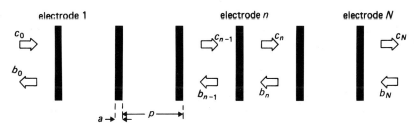

FIGURE E.2. Wave propagation in a finite array of regular electrodes.

$$T_{11}^n = \frac{\sin(n\gamma p) - \tau^* \sin[(n-1)\gamma p]}{\tau^* \sin(\gamma p)}$$

$$T_{12}^n = -T_{21}^n = \frac{r}{t}\frac{\sin(n\gamma p)}{\sin(\gamma p)},$$

$$T_{22}^n = \frac{\sin(n\gamma p) - \tau \sin[(n-1)\gamma p]}{\tau \sin(\gamma p)}, \tag{E.21}$$

where γ is the propagation constant for waves in an infinite array, given by equation (E.14). These equations are readily verified by calculating the product $[T]^n[T]$, which is found to be equal to $[T]^{n+1}$.

If there is only one incident wave, at the left of the structure, we have $b_N = 0$. In this case the ratio b_0/c_0 can be found from equation (E.20), setting $n = N$. Equation (E.20) may then be used to evaluate c_n/c_0 and b_n/c_0, thus giving the wave amplitudes throughout the structure. Thus, for $b_N = 0$ we find

$$\frac{c_n}{c_0} = \frac{\sin(N-n)\gamma p - \tau \sin(N-n-1)\gamma p}{\sin N\gamma p - \tau \sin(N-1)\gamma p} \tag{E.22}$$

and

$$\frac{b_n}{c_0} = \frac{r}{t}\frac{\tau \sin(N-n)\gamma p}{\sin N\gamma p - \tau \sin(N-1)\gamma p}. \tag{E.23}$$

It can be shown from these equations that $|b_0|^2 + |c_N|^2 = |c_0|^2$, so that the power is conserved. If the electrodes were absent we would have $r = 0$ and $t = 1$, and the above equations then give $b_n = 0$ and $c_n = c_0 \exp(-jnk_0 p)$.

For $n = 0$, equation (E.23) gives the ratio b_0/c_0, which is the reflection coefficient of the array. Taking the squared modulus of this gives the power reflection coefficient, which is found to be

$$\left|\frac{b_0}{c_0}\right|^2 = \left[1 + \left|\frac{t}{r}\right|^2 \frac{\sin^2(\gamma p)}{\sin^2(N\gamma p)}\right]^{-1}. \tag{E.24}$$

The amplitude reflection coefficient $|b_0/c_0|$ is plotted as a function of $k_0 = \omega/v_0$ in Figure E.3, taking $|t/r|$ to be independent of frequency, which is usually a good approximation. For $N|r| \ll 1$ the amplitude of the reflected wave is approximately proportional to $\sin(N\gamma p)/\sin(\gamma p)$, which has the form expected when multiple reflections are negligible. For $N|r| \gg 1$ the reflection coefficient is close to unity when k_0 is in the range corresponding to the stop band for an infinite array.

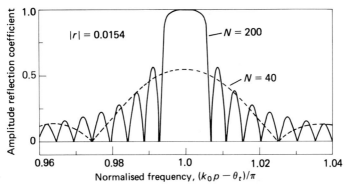

FIGURE E.3. Reflection coefficient of a finite array.

E.2. Electrode Reflection Coefficient Due to Electrical Loading

In this section we evaluate the reflection coefficient r for an electrode in an array of regular electrodes. Mechanical loading is ignored here, so that the reflection coefficient is due only to electrical loading. The analysis involves a particular electrostatic solution, which is considered first for convenience.

(a) Electrostatic Solution for an Incident Surface Wave.

We consider an infinite array of regular electrodes, with pitch p and width a, on the surface of an anisotropic *non-piezoelectric* half-space. It is supposed that, in the absence of the electrode, a surface potential

$$\phi_i(x) = \phi_{i0}\, e^{-j\kappa x} \qquad (E.25)$$

would be present. With the electrodes present, a charge density $\sigma_1(x)$ is generated such that the total potential, which includes the contribution $\phi_i(x)$, is uniform over each electrode. The problem here is to evaluate $\sigma_1(x)$, which will later be seen to give the first-order charge density for a shorted transducer with a surface wave incident on it. The subscript on $\sigma_1(x)$ distinguishes it from other functions required later. The derivation is very similar to the Floquet analysis of Appendix D, and is also described by Skeie [484, 486] and by Datta and Hunsinger [479]. For simplicity, it is assumed here that the constant κ in equation (E.25) is confined to the range

$$0 \leq \kappa \leq 2\pi/p. \qquad (E.26)$$

Taking the electrodes to be centred at $x = np$, the charge density and the x-component of the field at the surface are written as Floquet expansions:

$$\sigma_1(x) = \sum_{m=-\infty}^{\infty} \sigma_m\, e^{-j2\pi m x/p}\, e^{-j\kappa x},$$

$$E_x(x) = \sum_{m=-\infty}^{\infty} E_m\, e^{-j2\pi m x/p}\, e^{-j\kappa x}. \qquad (E.27)$$

Here the total field $E_x(x)$ includes a component $E_i(x)$ corresponding to the potential $\phi_i(x)$ of equation (E.25), so that

$$E_i(x) = jκφ_{i0} \exp(-jκx). \tag{E.28}$$

The remaining part of the field $E_x(x)$ is related to the charge density by the effective permittivity of the material, which in this case is the constant $\varepsilon_0 + \varepsilon_p$, since the material is non-piezoelectric. Thus in this case we have

$$E_m/\sigma_m = jS_m/(\varepsilon_0 + \varepsilon_p), \quad \text{for } m \neq 0 \tag{E.30}$$

and

$$(E_0 - jκφ_{i0})/\sigma_0 = j/(\varepsilon_0 + \varepsilon_p), \tag{E.31}$$

where S_m is defined in equation (D.6). With these equations, the field and charge density may be found using the method given in Appendix D. The coefficients σ_m and E_m are expressed in terms of Legendre polynomials using coefficients α_{-1}, α_0 and α_1, as in equations (D.11)–(D.13). Here, equation (E.30) is satisfied if $\alpha_{-1} = 0$, so only the α_0 and α_1 terms are required. Using equation (E.31) gives α_1, which is found to be

$$\alpha_1 = -jκφ_{i0}/2. \tag{E.32}$$

The value of α_0 depends on the electrical terminations. As in Appendix D, the electrode voltages and currents are $V_n = V_0 \exp(-jκnp)$ and $I_n = I_0 \exp(-jκnp)$, and V_0 and I_0 are given respectively by equations (D.21) and (D.23) with $\alpha_{-1} = 0$. Here we assume that the electrodes are either shorted ($V_0 = 0$) or open-circuited ($I_0 = 0$), and it follows that α_0 is given by

$$\alpha_0 = \pm \tfrac{1}{2} jκφ_{i0} P_s(\pm \cos \Delta)/P_{-s}(\pm \cos \Delta), \tag{E.33}$$

taking the upper signs for open-circuited electrodes and the lower signs for shorted electrodes. Here $s = κp/(2\pi)$ and $\Delta = \pi a/p$.

The charge density $\sigma_1(x)$ is given by equation (D.12), with $\alpha_{-1} = 0$. For subsequent analysis we need the Fourier transform of the charge density on electrode n, evaluated at $\beta = κ$, and this is given by

$$\int_{np-a/2}^{np+a/2} \sigma_1(x) e^{-jκx} dx = -jp(\varepsilon_0 + \varepsilon_p) e^{-2jnκp} [\alpha_0 P_{-2s}(\cos \Delta) + \alpha_1 P_{2s}(\cos \Delta)]. \tag{E.34}$$

This is readily obtained from equation (D.12) with the aid of equation (C.16) of Appendix C.

(b) Second-order Green's Function Analysis. In Chapter 4 the Green's function method was used for transducer analysis, and the quasi-static approximation was used. Here a more accurate analysis is needed. For clarity, the quasi-static equations are summarised first.

For a finite array of electrodes with arbitrary voltages, the total surface potential $\phi(x)$ and the charge density $\sigma(x)$ are related by

$$\phi(x) - \phi_i(x) = [G_e(x) + G_s(x, \omega)] * \sigma(x). \tag{E.35}$$

This is the same as equation (4.18) of Chapter 4, except that a potential $\phi_i(x)$ due to an incident surface wave has been included. $G_e(x)$ and $G_s(x, \omega)$ are respectively the electrostatic and surface-wave Green's functions. For the quasi-static approximation we take $\sigma(x) \approx \sigma_e(x) + \sigma_d(x)$ and neglect a small term $G_s(x, \omega) * \sigma_d(x)$, so that

$$\phi(x) - \phi_i(x) = [G_e(x) + G_s(x, \omega)] * \sigma_e(x) + G_e(x) * \sigma_d(x). \quad (E.36)$$

The electrostatic term $\sigma_e(x)$ is defined such that $G_e(x) * \sigma_e(x) = \phi(x)$ on the electrodes, which implies that $\sigma_e(x)$ vanishes if the electrode voltages are all the same. Defining the acoustic potential

$$\phi_a(x) = G_s(x, \omega) * \sigma_e(x) \quad (E.37)$$

as in Chapter 4, equation (E.36) requires $G_e(x) * \sigma_d(x)$ to be equal to $-\phi_i(x) - \phi_a(x)$ on the electrodes. It follows that the acoustic contribution to the charge density is

$$\sigma_d(x_i) = -\sum_j B_{ij}[\phi_i(x_j) + \phi_a(x_j)], \quad (E.38)$$

where B_{ij} is the electrostatic matrix defined in equation (4.29) and the points x_j exist only on the electrodes, where their spacing is Δx.

The *second-order* solution is obtained by writing

$$\sigma(x) \approx \sigma_e(x) + \sigma_d(x) + \sigma'_d(x), \quad (E.39)$$

where $\sigma_e(x)$ and $\sigma_d(x)$ are the same functions as before. Substituting into equation (E.35), we exclude a small term $G_s(x, \omega) * \sigma'_d(x)$, but include the term $G_s(x, \omega) * \sigma_d(x)$ which was previously omitted. Thus

$$\phi(x) - \phi_i(x) = [G_e(x) + G_s(x, \omega)] * [\sigma_e(x) + \sigma_d(x)] + G_e(x) * \sigma'_d(x). \quad (E.40)$$

We define

$$\phi'_a(x) = G_s(x, \omega) * \sigma_d(x). \quad (E.41)$$

Now, since the potentials on the left of equations (E.36) and (E.40) must be the same on the electrodes, $G_e(x) * \sigma'_d(x)$ must be equal to $-\phi'_a(x)$ on the electrodes, and hence

$$\sigma'_d(x_i) = -\sum_j B_{ij}\phi'_a(x_j). \quad (E.42)$$

This gives the charge density in the second-order approximation. It is convenient to define a potential

$$\phi_s(x) = \phi_i(x) + \phi_a(x) + \phi'_a(x). \quad (E.43)$$

In the unmetallised regions this function includes all the terms proportional to $\exp(\pm jk_0 x)$, and may therefore be interpreted as the surface wave potentials in these regions.

The current entering electrode n is $I_n = I_{en} + I_{an}$, where I_{en} is the electrostatic term due to $\sigma_e(x)$, and can be found by methods given in Chapter 4. The acoustic term I_{an}, due to $\sigma_d(x)$ and $\sigma'_d(x)$, can be written

$$I_{an} = j\omega W \sum_j \hat{p}_n(x_j)[\sigma_d(x_j) + \sigma'_d(x_j)] \Delta x, \quad (E.44)$$

where $\hat{p}_n(x)$ is unity on electrode n and zero elsewhere. Using equations (E.38) and (E.42) and taking the limit $\Delta x \to 0$, this gives

$$I_{an} = -j\omega W \int_{-\infty}^{\infty} \varrho_{en}(x)\phi_s(x)\,dx, \qquad (E.45)$$

where $\phi_s(x)$ is defined by equation (E.43) and $\varrho_{en}(x)$, given by equation (4.32), is the electrostatic charge density on the array when unit voltage is applied to electrode n with the other electrodes grounded. This is very similar to the quasi-static result, equation (4.43).

(c) Reflection Coefficient for Shorted Electrodes.

We now consider the reflection coefficient for one electrode in an array of regular electrodes, with pitch p and width a. From equations (E.43), (E.37) and (E.41), the surface-wave potential $\phi_s(x)$ can be written

$$\phi_s(x) = \phi_i(x) + G_s(x, \omega) * [\sigma_e(x) + \sigma_a(x)]. \qquad (E.46)$$

The function $\phi_i(x)$ represents a surface-wave potential which would exist in the absence of the electrodes, and therefore has the form

$$\phi_i(x) = \phi_{i0} \exp(-jk_0 x), \qquad (E.47)$$

where k_0 is the free-square wavenumber. The surface-wave Green's function $G_s(x, \omega)$ is equal to $j\Gamma_s \exp(-jk_0|x|)$. At any x, the convolution in equation (E.46) gives terms proportional to $\exp(-jk_0 x)$ and $\exp(jk_0 x)$, representing waves propagating in the $+x$ and $-x$ directions respectively. We consider the waves propagating in the $-x$ direction and evaluate these on either side of electrode n. Using the notation of Section E.1, we define b_n as the surface-wave potential at $x = np + p/2$ and b_{n-1} as the potential at $x = np - p/2$. Using equation (E.46), these are related by

$$b_{n-1}\,e^{jk_0 p} - b_n = j\Gamma_s\,e^{jk_0(np+p/2)} \int_n [\sigma_e(x') + \sigma_a(x')]\,e^{-jk_0 x'}\,dx', \qquad (E.48)$$

where the integral is taken over electrode n, so that the limits are $np \pm a/2$. A similar relation can be obtained for waves propagating in the $+x$ direction.

An approximate value for the reflection coefficient is obtained by assuming that, in the vicinity of electrode n, the acoustic potentials $\phi_a(x)$ and $\phi'_a(x)$ in equation (E.43) are much smaller than the incident wave potential $\phi_i(x)$. Thus the wave incident on electrode n from the left, evaluated at $x = np - p/2$, is $\phi_i(np - p/2)$. The electrode reflection coefficient, r, gives the reflected wave amplitude when there is no wave incident from the right, that is, when $b_n = 0$. Referring r to the centre of the electrode, as in Section E.1, we thus have

$$b_{n-1} = r\phi_i(np - p/2)\,e^{-jk_0 p} = r\phi_{i0}\,e^{-jk_0(np+p/2)}, \qquad (E.49)$$

when $b_n = 0$. The reflection coefficient r may therefore be found from the charge density using equation (E.48). In the present case $\sigma_e(x) = 0$ because the electrodes are taken to be shorted. Furthermore, from equation (E.36) the term $\sigma_a(x)$ is the charge density such that the incident potential $\phi_i(x)$ is cancelled at the electrode locations, calculated using the electrostatic Green's function $G_e(x)$. Thus $\sigma_a(x)$ can be identified as the function $\sigma_1(x)$ of Section E.2(a) above, taking $\kappa = k_0$ in order to comply with equation (E.47). It is also necessary to replace ε_p by ε_p^T since the material is now piezoelectric. The integral required in equation (E.48) is given by Equation (E.34), where α_0 and α_1 are respectively given by Equations (E.32) and (E.33); the lower signs are used in equation (E.33) since the electrodes are shorted. The reflection coefficient is thus found to be

$$r \approx -\tfrac{1}{2}j \frac{\Delta v}{v} k_0 p \left[P_{2s}(\cos \Delta) + \frac{P_s(-\cos \Delta)}{P_{-s}(-\cos \Delta)} P_{-2s}(\cos \Delta) \right], \quad (E.50)$$

where the relation $(\varepsilon_0 + \varepsilon_p^T)\Gamma_s = \Delta v/v$ has been used, and $s = k_0 p/(2\pi)$. In view of equation (E.26) this is valid only for $0 \leq k_0 \leq 2\pi/p$, that is, for $\omega \leq 2\pi v_0/p$. The solution for higher frequencies is given by Datta and Hunsinger [479]. Strictly, equation (E.50) applies only if the array is infinite, since this was assumed in the derivation of $\sigma_1(x)$. However, for a finite array the same formula may be used if end effects are neglected.

For an infinite array of electrodes the first stop band occurs for frequencies close to the value where $k_0 p = \pi$, that is, where $s = \tfrac{1}{2}$. At this frequency we find

$$r = -\tfrac{1}{2}j\pi \frac{\Delta v}{v} F_-(\Delta), \quad \text{for } s = \tfrac{1}{2} \quad (E.51)$$

where $F_-(\Delta)$ is defined in Appendix D, equation (D.37). Figure D.3 shows $F_-(\Delta)$ as a function of the metallisation ratio. It is seen that $F_-(\Delta)$, and hence the magnitude of r, has a maximum at $a/p = 0.25$. According to the analysis of Appendix D the width of the stop band is $\omega_1 - \omega_{sc}$, given by equation (D.41), and it follows that $\omega_1 - \omega_{sc} = 2v_0|r|/p$. This agrees with the result obtained from the reflective array model, equation (E.16). The reflection coefficient of equation (E.50) is shown as a function of frequency in Figure E.4, for $a/p = \tfrac{1}{2}$.

(d) Reflection Coefficient for Open-Circuited Electrodes.

This can be deduced in a manner very similar to the above analysis for shorted electrodes. The scattering due to electrode n is described by equation (E.48), but here the electrostatic charge density $\sigma_e(x)$ is non-zero because the electrode voltages are finite. It can be assumed that $\sigma_a(x)$ is negligible in comparison with $\sigma_e(x)$, and that $\sigma_e(x)$ is approximately equal to $\sigma_1(x)$ of Section E.2(a). Equation (E.33) is used for α_0, taking the upper signs. Using the same methods as before, the reflection coefficient is found to be

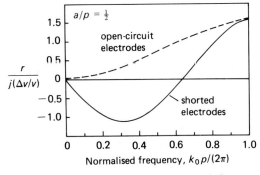

FIGURE E.4. Reflection coefficient for one electrode in a regular array.

$$r \approx -\tfrac{1}{2}j\frac{\Delta v}{v}k_0 p\left[P_{2s}(\cos\Delta) - \frac{P_s(\cos\Delta)}{P_{-s}(\cos\Delta)}P_{-2s}(\cos\Delta)\right]. \quad (E.52)$$

For $s = \tfrac{1}{2}$ we have $r = \tfrac{1}{2}j\pi(\Delta v/v)F_+(\Delta)$, where $F_+(\Delta)$ is defined in equation (D.37) and shown in Figure D.3. The width of the stop band, given by equation (D.44), is again related to the electrode reflection coefficient by equation (E.16). Figure E.4 shows r as a function of frequency.

A finite array of the open-circuited electrodes has the same structure as the basic multi-strip coupler described in Chapter 5, where the reflection of incident waves by the coupler was ignored. However, for a uniform incident wave the reflection coefficient can be obtained from the reflective array model, equation (E.24), using equation (E.52) for r.

E.3. Electrical Loading in Transducers

In this section we consider interactions in an interdigital transducer with regular electrodes. The transducer is taken to be shorted, and to have a surface wave incident on it. The wave amplitudes throughout the transducer can in principle be obtained using the Green's function method of Section E.2. However, owing to the approximations used the resulting expressions do not conserve the power. It is therefore more appropriate to use the reflective array model of Section E.1, taking the electrode reflection coefficient r to be given by the approximate result of equation (E.50). For consistency, the electrode transmission coefficient, t, must be related to r by equations (E.4) and (E.7), where the ambiguity of sign is resolved by noting that t must be close to unity. Since r is imaginary, this implies that t must be real. Consequently, at frequencies where interaction effects are not significant the analysis will predict that the surface-wave velocity within the structure will be the free-surface velocity v_0. This is in fact erroneous, since the velocity will have a value between v_0 and v_m, dependent on the metallisation ratio, as shown in Section D.2. However, the analysis here is readily corrected for the velocity error by adjusting the value of k_0.

(a) Reflection Coefficient. When the transducer is shorted, the polarities of the electrodes have no effect on the reflection coefficient, so the latter is the same as for a simple array of shorted regular electrodes. The reflection coefficient is therefore given by equation (E.24), using equation (E.50) for the electrode reflection coefficient r. Measurements on an array of electrodes can therefore be used to confirm the theoretical expression for r, though care is needed if a large number of electrodes is used since this can give significant propagation losses not allowed for in the reflective array model of Section E.1. For a regular array, significant reflections occur for $s = \tfrac{1}{2}$, that is, when the pitch equals half the wavelength, and for $s = \tfrac{1}{2}$ and $a/p = \tfrac{1}{2}$ equation (E.50) gives $r = -0.718 j \Delta v/v$. For Y, Z lithium niobate, with $\Delta v/v = 2.15\%$, we thus have $|r| = 0.0154$. This is confirmed by measurements on arrays of aluminium electrodes [494, 495]. The variation of r with a/p, shown by equation (E.51) and Figure D.3, is also confirmed experimentally [494]. In addition, it is found that the value of r is independent of the film thickness [495], confirming the assumption that the reflection is predominantly due to electrical, rather than mechanical, loading. The theoretical reflection coefficient for $N = 40$ electrodes with $a/p = \frac{1}{2}$ is shown as a function of frequency in Figure E.3. This is similar to De Vries' measurement [496, p. 276] on a shorted transducer, though the maximum value of 0.55 is somewhat larger than the experimental maximum of 0.44.

(b) Frequency Response. The transducer response is found by a method similar to that described by Skeie and Engan [485]. The analysis here is valid for an apodised transducer, but for convenience we assume initially that the transducer is unapodised. We consider the output current I_{sc} produced when the transducer is shorted and a surface wave is incident. Since the electrodes are shorted the electrostatic charge density $\sigma_e(x)$ is zero, and consequently the current I_n flowing into electrode n consists only of the acoustic term I_{an}, which is given by Equation (E.45). In the present case the electrodes are regular and electrode n is centred at $x = np$, so that $\varrho_{en}(x)$ can be replaced by $\varrho_f(x - np)$, giving

$$I_n = -j\omega W \int_{-\infty}^{\infty} \varrho_f(x - np) \phi_s(x)\,dx, \tag{E.53}$$

where $\phi_s(x)$ is the surface-wave potential, defined by equation (E.43). The elemental charge density $\varrho_f(x - np)$ is small except in the vicinity of electrode n. Now, if the interactions are not too strong the surface-wave amplitude will not vary rapidly with distance, so the waves in the vicinity of electrode n are approximately proportional to $\exp(\pm jk_0 x)$. In the analysis of Section E.1 the waves incident on electrode n are given by c_{n-1}, measured at $x = np - p/2$, and b_n, measured at $x = np + p/2$. Taking these as surface-wave potentials, the total surface-wave potential in the vicinity of electrode n is approximately given by

$$\phi_s(x) \approx c_{n-1} e^{-jk_0(x - np + p/2)} + b_n e^{jk_0(x - np - p/2)}.$$

Substituting into equation (E.53), the integral can be expressed in terms of the Fourier transform of $\varrho_f(x)$, denoted by $\bar{\varrho}_f(\beta)$. Noting that $\bar{\varrho}_f(-\beta) = \bar{\varrho}_f(\beta)$, this gives

$$I_n = -j\omega W \bar{\varrho}_f(k_0)[c_{n-1} + b_n]\exp(-jk_0 p/2), \tag{E.54}$$

The above equation gives the current entering electrode n, for an unapodised shorted transducer of aperture W. We now generalise to an *apodised* transducer, assuming that dummy electrodes are included as in, for example, Figure 8.1. The electrode breaks are taken to be much smaller than W. Since the transducer is shorted, the surface wave amplitudes within it will not be affected by the apodisation, and hence equation (E.54) gives the total current entering the electrodes centred at $x = np$. The current entering from the upper bus-bar, denoted I'_n, is proportional to the length u_n of the electrode connected to this bus-bar, so that

$$I'_n = -j\omega u_n \bar{\varrho}_f(k_0)[c_{n-1} + b_n] \exp(-jk_0 p/2) \qquad (E.55)$$

and the total current flowing in the short-circuit connected to the transducer is

$$I_{sc} = \sum_{n=1}^{N} I'_n, \qquad (E.56)$$

where N is the number of electrodes.

For simplicity it is assumed here that the transducer geometry is symmetrical, and that N is odd. The method given below is readily adapted for the case of an anti-symmetrical geometry, and the two cases can be combined to analyse a more general geometry [485]. The total current is written as

$$I_{sc} = \sum_{n=1}^{(N-1)/2} (I'_n + I'_{N-n+1}),$$

where, for simplicity, the current in the central electrode, with $n = (N+1)/2$, has been omitted. For a symmetrical transducer $u_{N-n+1} = u_n$, and using equation (E.55) we have

$$I_{sc} = -j\omega\bar{\varrho}_f(k_0) e^{-jk_0 p/2} \sum_{n=1}^{(N-1)/2} u_n(c_{n-1} + b_n + c_{N-n} + b_{N-n+1}). \qquad (E.57)$$

Using equation (E.22) for c_n and equation (E.23) for b_n, it is found that this can be written in the form

$$I_{sc}/c_0 = H_0(\gamma) A(\gamma). \qquad (E.58)$$

Here c_0 is the potential of the incident surface wave at $x = p/2$, which can be taken as the location of the acoustic port (the first electrode is centred at $x = p$). The functions on the right are

$$H_0(\gamma) = -2j\omega\bar{\varrho}_f(k_0) e^{-jN\gamma p/2} \sum_{n=1}^{(N-1)/2} u_n \cos(N - 2n + 1)\gamma p/2 \qquad (E.59)$$

and

$$A(\gamma) = \frac{\sin(N+1)\gamma p/2 - \tau(1 - r/t)\sin(N-1)\gamma p/2}{\sin N\gamma p - \tau \sin(N-1)\gamma p} e^{j(N-1)\gamma p/2}. \qquad (E.60)$$

The ratio I_{sc}/c_0 is essentially the transducer frequency response, as can be seen by comparison with equation (4.122). The interaction effects disappear if we put $r = 0$

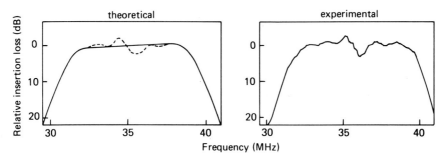

FIGURE E.5. Insertion loss of an interdigital device, showing ripple due to electrode interactions (Courtesy, Plessey Research).

and $\gamma = k_0$. With these values equation (E.60) gives $A(k_0) = 1$, so that $I_{sc}/c_0 = H_0(k_0)$; this is found to be consistent with the quasi-static analysis in Chapter 4, equtions (4.129) and (4.122). When interactions are significant, the function $H_0(\gamma)$ is similar to the first-order response $H_0(k_0)$; the distortion arises mainly from the function $A(\gamma)$. The amplitude distortion is given by $|A(\gamma)|$, and for $\gamma p < 2\pi$ this is close to unity except when γp is close to π.

Figure E.5 shows, on the right, the experimental insertion loss of a device using aluminium electrodes on a Y, Z lithium niobate substrate. The ripple is due to electrode interactions. The device consisted of an apodised single-electrode transducer with $N = 51$ electrodes, and an unapodised double-electrode transducer, the latter giving negligible electrode interactions. The metallisation ratio was 0.5. The continuous theoretical curve shows the response calculated by the quasi-static method of Chapter 4, that is, ignoring electrode interactions. The experimental device was connected directly to the electrical source and load without intervening matching components, and consequently circuit effects are small. The distortion due to interactions can therefore be obtained directly from equation (E.60), using

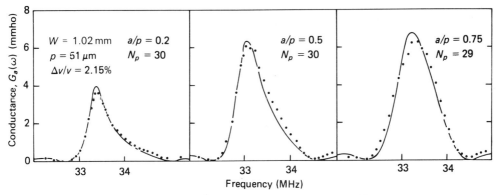

FIGURE E.6. Parallel conductance of single-electrode transducers. Experimental points from Daniel and Emtage [497], with permission.

equation (E.50) for r and taking $\Delta v/v = 2.15\%$. This gives the broken theoretical curve in Figure E.6, agreeing well with the experimental result. Similar experimental and theoretical results were obtained by Skeie and Engan [485].

(c) Conductance of an Unapodised Transducer. Electrode interactions also distort the parallel conductance $G_a(\omega)$ of a transducer. Here we consider only an unapodised transducer, though the method may be extended to the case of an apodised transducer by using parallel channels (Section 4.7.2). For an unapodised transducer, the potential ϕ_{s1} of the wave launched at the acoustic port when a voltage V_t is applied can be obtained by reciprocity, equation (B.24), giving

$$\phi_{s1} = -V_t \frac{\Gamma_s}{\omega W} \frac{I_{sc}}{c_0},$$

where I_{sc}/c_0 is given by equation (E.58). The power carried by this wave is given by equation (3.34) of Chapter 3, and since the transducer is here taken to have a symmetrical geometry an equal power is carried by the wave launched at port 2. The total surface wave power is accounted for by the conductance $G_a(\omega)$, and we thus find

$$G_a(\omega) = \Gamma_s |H_0(\gamma) A(\gamma)|^2/(\omega W). \tag{E.61}$$

This function is shown, for uniform single-electrode transducers, in Figure E.6. The experimental points were obtained by Daniel and Emtage [497], using aluminium transducers on a Y, Z lithium niobate substrate, with metallisation ratios $a/p = 0.2$, 0.5 and 0.75. The number of periods, N_p, was 29 for $a/p = 0.75$ and 30 for the other cases. The aperture W and pitch p were the same for all three transducers. The theoretical curves use equations (E.59) and (E.60) and, because of a small uncertainty in the transducer period, have been displaced along the frequency axis to fit the experimental data. Daniel and Emtage [497] obtained similar theoretical curves. If electrode interactions were negligible the conductance would be an approximately symmetrical function, as shown for example by Figure 4.14. The interactions cause considerable distortion, and the distortion is greatest for $a/p = 0.2$, becoming progressively weaker for $a/p = 0.5$ and 0.75. This is due to the variation of the electrode reflection coefficient r. For $s = \frac{1}{2}$, that is, at the transducer centre frequency, r is proportional to the function $F_-(\Delta)$ shown in equation (E.51) and Figure D.3, and its magnitude is maximised at $a/p = 0.25$. The experimental results therefore confirm the theoretical variation of r with a/p, as well as confirming the analysis for electrode interactions.

E.4. Mechanical Loading

Up to this point we have only considered interactions due to electrical loading. However there are in addition perturbations of a mechanical nature, present even if the substrate is not piezoelectric. There are two types of mechanical perturbation. Firstly, the presence of electrodes on the surface alters the surface geometry, giving a perturbation sometimes called the "topographic effect". This occurs if a set of grooves is etched into the surface. Secondly, an additional perturbation occurs in the case of deposited strips of metal or dielectric material, because the elastic properties differ from those of the substrate; this is often called "mass loading". In practical devices the metal film used for transducers is chosen to have elastic properties similar to the substrate in order to minimise mass loading effects, and usually this implies the choice of an aluminium film. The term "mechanical loading" is used here to include both topographic effects and mass loading.

For metal electrodes on a piezoelectric substrate, both electrical loading and mechanical loading are present. However, for strongly piezoelectric substrates such as lithium niobate, mechanical loading is usually negligible in comparison with electrical loading, and may therefore be ignored as in the analysis above. For weakly piezoelectric substrates, such as quartz, mechanical loading usually dominates. Intermediate cases, where both types of loading are relevant, do not often occur in practice, so here only mechanical loading is considered.

As in the case of electrical loading, the scattering properties of an array of regular electrodes are related to the scattering due to one electrode by the reflective array model of Section E.1, and the relevant transducer properties are given by the analysis in Section E.3. Thus, here it is only necessary to consider the electrode reflection coefficient r and the effective velocity for surface waves in the structure.

The analysis for mechanical loading has been considered by several authors [483, 486, 498–501]. Neglecting piezoelectricity, it is found that to first order the reflection coefficient for one electrode is approximately given by [486, 499, 500]

$$r \approx 2j\alpha_r(h/\lambda) \sin(ak), \tag{E.62}$$

where α_r is a real constant, $k \approx k_0$ is the surface-wave wavenumber, $\lambda = 2\pi/k$ is the wavelength, and a and h are respectively the width and thickness of the electrode. It is assumed that $h \ll \lambda$. The formula applies irrespective of the proximity of any adjacent electrodes. It can be interpreted in terms of a reflection coefficient $\alpha_r h/\lambda$ for the leading edge of the electrode, and an equal and opposite reflection coefficient for the trailing edge. As a function of the width a, the reflection coefficient has a maximum value of $2j\alpha_r h/\lambda$ when $a = \lambda/4$.

For the common experimental case of aluminium electrodes on a ST, X quartz substrate, analysis [486, 500] gives 0.28 or 0.25 for the value of α_r. Experimental measurements [502, 503], using arrays of electrodes with $a \approx \lambda/4$, gave values of 0.25 and 0.33, in reasonable agreement. Further, Marshall [502] showed experimentally that the reflection coefficient is proportional to h, in agreement with equation (E.62), thus confirming that mechanical loading is the main cause of the perturbation. For transducers operating at 100 MHz, say, a typical value of h/λ is 0.01, giving $|r| \approx 0.006$. This is considerably less than the reflection coefficient for aluminium electrodes on lithium niobate, and hence electrode interactions are generally weaker when quartz substrates are used.

Skeie [483] has investigated transducers on Y, Z lithium niobate using gold metallisation. This case was found to give substantial mechanical loading, so that the electrode interactions were considerably stronger than for electrical loading alone.

Thus aluminium is generally preferable to gold.

In addition to reflections, mass loading also causes a phase change of a surface wave transmitted through an electrode. For an array of regular electrodes, and for frequencies outside the stop bands, this is equivalent to a change of surface-wave velocity. According to first-order analysis [501] the velocity varies linearly with the metallisation ratio a/p, and can therefore be calculated straightforwardly from the velocity perturbation due to a continuous film. This gives a velocity perturbation proportional to h/λ. However, Datta and Hunsinger [501] have shown that there is in addition a second-order, or "stored energy", term proportional to $(h/\lambda)^2$. For aluminium electrodes on ST, X quartz this will usually be larger than the first-order term. A 337 MHz double-electrode transducer with 900 Å metallisation gave an experimental velocity perturbation of 0.24%, and the analysis gave good agreement.

Appendix F

Bulk Waves

This appendix gives a brief account of the effects due to unwanted bulk-wave excitation in surface-wave devices, and of some devices that use bulk waves intentionally. It was seen in Chapter 2 that a variety of acoustic modes may propagate in a piezoelectric half-space, including several bulk waves and a variety of surface waves. For interdigital surface-wave devices the substrate material and orientaton are usually chosen such that the only surface wave that can be excited is the piezoelectric Rayleigh wave. However, some excitation of bulk waves is nearly always present. The analysis for the excitation of these waves by interdigital transducers is very complex, but for most practical purposes it is not necessary to consider this in detail; the degradation of device performance is not severe, and generally some simple design features are sufficient to ensure that adequate performance is obtained. Thus a descriptive account illustrating the main practical consequences is sufficient here, and this is given in Sections F.1 and F.2.

Some devices have made use of bulk waves deliberately, and these are described in Section F.3. This includes surface-skimming bulk wave devices, which use interdigital transducers on a substrate chosen such that there is negligible electrical coupling to surface waves. These devices can offer a higher operating frequency, or a better temperature coefficient, than surface-wave devices. We exclude here devices using parallel-plate transducers, such as the bulk wave delay line of Figure 1.1(a).

It is interesting to note that one of the earliest applications for interdigital transducers was in fact concerned with bulk waves rather than surface waves [504]. The device consisted of a block of crystalline quartz with two plane non-parallel surfaces, each bearing an interdigital transducer. The transducers were used to launch and detect bulk waves propagating in the body of the material, and gave a dispersive response owing to the fact that the acoustic path length varied with frequency. The device was proposed as a chirp filter for pulse-compression radar, but has been superseded by the surface-wave interdigital chirp filter (Chapter 9) which is easier to fabricate.

F.1. Bulk Wave Generation by Interdigital Transducers

In a surface-wave device, the general nature of the effects due to bulk-wave excitation can be understood quite readily from a simple phase-matching argument, similar to the delta-function model of Section 4.1 and to the analysis of end-fire antennas. We assume, as is usually the case, that the substrate material and orientation have been chosen such that the only surface-wave mode that can be excited is a piezoelectric Rayleigh wave; this applies, for example, if the substrate is Y, Z lithium niobate or ST, X quartz. Consider a uniform single-electrode launching transducer, illustrated in Figure F.1, and assume initially that the transducer is many periods long. To first order each electrode can be regarded as a source of bulk waves (as well as surface

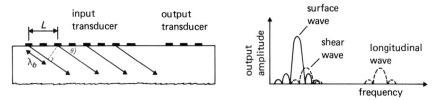

FIGURE F.1. Bulk wave generation by interdigital transducer.

waves), and the total bulk wave power will be relatively large if the waves due to individual electrodes can be added constructively. If L is the transducer period this condition is satisfied by a bulk wave travelling at an angle θ to the surface, such that the bulk-wave wavelength is $\lambda_b = L \cos \theta$, and hence

$$fL = v_b(\theta)/\cos \theta, \tag{F.1}$$

where f is the frequency and $v_b(\theta)$ is the bulk-wave velocity. Generally, $v_b(\theta)$ will vary somewhat with θ because of the substrate anisotropy, but this does not affect the argument. Equation (F.1) refers to the "fundamental" response; there will also be harmonic responses satisfying the relation $M\lambda_b = L \cos \theta$, where M is an odd integer, but for brevity these are ignored here. There will also be bulk waves generated to the left in Figure F.1.

Equation (F.1) shows that coherent bulk wave generation can only occur above a cut-off frequency $f_c = v_b(0)/L$, where $\theta = 0$ and the bulk wave travels parallel to the surface. At lower frequencies the equation has no real solution, but coherent bulk waves can be generated at *all* frequencies above f_c, with the angle θ increasing with frequency and approaching a limit of $\pi/2$ at infinite frequency. This is in contrast with surface-wave excitation, which is significant only over a narrow band if the transducer is many periods long. There will generally be several bulk waves involved, and for each of these the velocity $v_b(0)$ parallel to the surface will be greater than the surface-wave velocity v_s. The cut-off frequencies are therefore all greater than the centre-frequency of the surface-wave response, which is equal to v_s/L.

In practice the above account must be modified a little to allow for the fact that the transducer has finite length. A more detailed examination, treating the electrodes as independent sources, shows that at one frequency the bulk waves are in fact radiated over a range of angles; the polar diagram of the radiated power has a prominent main lobe whose maximum is directed at the angle θ satisfying equation (F.1), and whose width is inversely related to the transducer length. In addition there is no longer a sharp cut-off – for a long finite-length transducer there can in general be some bulk wave excitation below the "cut-off" frequency $f_c = v_b(0)/L$, though the power involved is usually very small.

The radiated bulk waves can be detected experimentally by shining a beam of light through the substrate, parallel to the surface [505–507]. In this way, Schmidt [505] has confirmed experimentally the frequency variation of θ given by equation (F.1), for the several bulk waves generated by a transducer on Y, Z lithium niobate.

In a two-transducer device, the main practical concern is that the bulk waves may cause some unwanted excitation of the output transducer. It is usual practice to roughen the rear surface of the substrate, as indicated in Figure F.1, so that bulk waves incident on it are scattered incoherently and therefore do not significantly excite the output transducer. The unwanted excitation therefore arises primarily from bulk waves travelling almost parallel to the surface, with small values of θ. Assuming that the two transducers have the same period L, it can be expected that the ouput due to a particular bulk wave will be maximised at the frequency for which the period is equal to the wavelength of bulk waves travelling parallel to the surface, that is, at the "cut-off" frequency $f_c = v_b(0)/L$. In fact, we can anticipate a response with a form similar to that of the surface-wave response, but shifted upwards in frequency owing to the higher velocity of the bulk wave. This is indicated on the right in Figure F.1, using broken lines to show responses due to a shear wave and a longitudinal wave, and a continuous line for the surface-wave response. The total response can be taken as the vectorial sum of these three. Note that the shear-wave response is maximised at a frequency quite close to the surface-wave centre frequency because the shear wave velocity is only a little higher than the surface-wave velocity. Thus the shear-wave response generally interferes with the surface-wave response, particularly at frequencies above the centre frequency. On the other hand, the longitudinal-wave velocity is typically twice the surface-wave velocity, giving a response centred at a much higher frequency. This is usually of little consequence because it is easily suppressed by external filtering, though it can interfere with the surface-wave response in very wide band devices.

Practical devices generally show behaviour consistent with the above description, except that the bulk-wave responses are rather distorted in comparison with the surface-wave response. The distortion arises because the bulk waves are not guided by the surface, and because the surface boundary conditions both inside and outside the transducers can considerably perturb the waves; in addition, there are generally two shear-wave components, overlapping in frequency because their velocities are similar. Some experimental results for Y, Z lithium niobate are given by Daniel and Emtage [508] and by Milsom *et al.* [93], showing that the longitudinal-wave response is typically 30 dB below the surface-wave response. The shear wave response, which overlaps the surface-wave response, is difficult to distinguish but appears to be at a similar level. In practice the rejection of the longitudinal-wave response is often rather better than this because the transducers are tuned, thus introducing some external frequency selectivity. The bulk wave contributions can usually be distinguished by applying absorbing material between the transducers, since this attenuates the surface waves much more than the bulk waves.

Another consequence of bulk-wave excitation is a distortion in the parallel conductance $G_a(\omega)$ of a transducer, since $G_a(\omega)$ is proportional to the total acoustic power generated. In addition to the surface-wave contribution, given by the analysis of Chapter 4, the conductance includes bulk-wave contributions extending over a wide frequency band, but mainly confined to frequencies above the surface-wave centre frequency [93, 508]. This affects the device frequency response via the circuit effect, though the consequent distortion is usually small. The presence of bulk-wave radiation in the surface-wave passband also implies an increase in the device insertion loss, but this is not usually significant unless the number of electrodes in the transducer is small. For example, for a transducer with $N_p = 5$ periods on Y, Z lithium niobate, the proportion of the total acoustic power generated in the form of bulk waves is 11%, at the surface-wave centre frequency [93]. For $N_p = 2$, the proportion of bulk wave power is about 25%.

Analysis of Bulk-Wave Effects. While the main features of bulk wave excitation are readily deduced from phase-matching arguments, as discussed above, it is necessary to resort to detailed analysis if quantitative results are required. The transducer analysis given in Chapter 4 neglects bulk waves because it is derived from the approximate Green's function of equation (3.48) in which the bulk-wave contribution $G_b(x_1, \omega)$ was neglected. However, Appendix B gives a reciprocity relation, equation (B.10), which is valid when bulk waves are excited.

Milsom et al. [92, 93] have given a well-developed analysis, based on the Green's function discussed in Section 3.4 with the bulk-wave term included. The Green's function is derived from the effective permittivity $\varepsilon_s(\beta)$ and therefore accounts for all of the acoustic modes that can be excited, including for example Bleustein–Gulyaev waves and leaky surface waves when appropriate. Further discussion is given in Section 3.3. The analysis allows for electrode interactions and for almost arbitrary transducer geometries. Apodisation can be allowed for by channelising the transducer, as in Section 4.7.2. This method is found to give results agreeing well with experiment, though its complexity makes it rather inconvenient to use.

Several other approaches have been developed, making simplifying approximations. In particular, it is usually assumed that electrode interactions are weak so that the waves generated by a launching transducer can be obtained from the electrostatic charge density, as was done for the surface-wave case in the quasi-static analysis of Section 4.3.1. Wagers [509] has considered transducers on a parallel-sided plate, and has obtained impressive agreement between the experimental and theoretical insertion losses of a two-transducer device. However, for most devices the rear surface is roughened and an analysis assuming propagation in a half-space [93, 510–513] is more appropriate. Some analysis for the generated fields of surface and bulk waves, derived from first principles, is given by Yashiro and Goto [510] and Lee [511]. Danicki [512] has introduced a useful approximation for the effective permittivity $\varepsilon_s(\beta)$ to simplify the analysis of bulk wave excitation, in a manner analogous to Ingebrigtsen's approximation for surface-wave excitation (Section 3.3), and this approach was developed further by Josse and Lee [513]. These methods are found to agree very well with experimental measurements of transducer admittances [511–513], even when substantial bulk wave excitation occurs, and they also predict well the measured responses of two-transducer devices [511, 513]. They have been applied in particular to excitation of surface-skimming bulk waves, described in Section F.3 below.

Reduction of Bulk-Wave Effects. Assuming that the rear surface of the substrate is roughened, bulk wave effects are not usually very consequential unless an exacting specification is to be met. The main exception to this occurs for Y, Z lithium niobate substrates, and for this case it is quite common to suppress bulk-wave effects by incorporating a multi-strip coupler, as in Section 5.3. Alternatively, it is sometimes possible to select a material and orientation giving relatively weak bulk-wave coupling, for example the 128°-rotated cut of lithium niobate mentioned in Section 6.5. Some criteria for selecting suitable orientations are discussed by Mitchell and Read [514] and by Milsom *et al.* [515]. It has also been shown that a metal pad on the surface can be beneficial: the pad reduces the surface-wave velocity but generally has little effect on bulk waves, and this feature can be exploited in a two-track device to partially cancel the bulk-wave excitation of the output transducers [516].

F.2. Mode Conversion in Arrays

In Section F.1 we have seen that unwanted bulk waves can be generated by the transducers of a surface-wave device. In some devices, bulk waves can also be generated by an array on the surface converting some of the surface-wave power into the form of bulk waves. This phenomenon can occur for arrays of either metal electrodes or grooves, and can seriously degrade the performance of a surface-wave chirp filter. As before, the main consequences can be appreciated from a simple phase-matching argument.

As shown in Figure F.2(a), we consider a surface wave incident on an array of identical electrodes, with uniform pitch p. The array may represent a transducer or a multi-strip coupler. Assuming that each electrode causes little perturbation of the surface wave, it is readily seen that coherent radiation of a bulk wave at an angle θ to the surface can occur if $fp(v_b(\theta) + v_s \cos \theta) = nv_sv_b(\theta)$, where v_s is the surface-

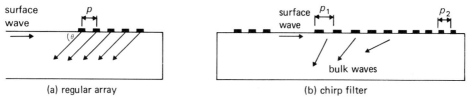

FIGURE F.2. Mode conversion in arrays.

wave velocity and n is an integer. Thus, coherent bulk waves can be produced only at frequencies above a cut-off point f_c for which $\theta = 0$ and $n = 1$, so that

$$pf_c[1 + v_s/v_b(0)] = v_s. \quad (\text{F.2})$$

This equation applies for each of the three bulk waves; to avoid coherent radiation of bulk waves the frequency must be less than the cut-off for slow shear waves, since this is the lowest of the three cut-off frequencies. The same relationships apply if the array consists of grooves rather than electrodes.

Experimental work [517, 518] has shown that quite efficient conversion of surface waves to bulk waves, or vice-versa, can be obtained by using grooves above the cut-off frequency. Conversion coefficients of 1 or 2 dB can be obtained. Some quantitative analysis for the bulk wave excitation is given by Auld [32] and by Haus [519], assuming an isotropic material for simplicity.

In a multi-strip coupler, the first stop-band occurs at the frequency $f = v_s/(2p)$ where p is equal to half the surface-wave wavelength. The cut-off frequency for bulk shear wave generation is a little higher than this, because the shear-wave velocity is a little higher than v_s. Thus, coherent bulk-wave generation cannot occur for a coupler operated below the stop-band frequency, as is usually the case. A periodic array can also be used as a 180° reflector, as in the surface-wave resonator (Section 10.5). In this case the shear-wave cut-off point is a little above the centre frequency $f = v_s/(2p)$, and coherent bulk-wave excitation is of no consequence because the device bandwidth is small. In grooved resonators it has been found that the Q-factor can be reduced by *non-coherent* bulk-wave excitation, but this effect can be reduced substantially by grading the depths of the grooves nearest to the centre of the device [519].

Serious consequences can arise in wide-band chirp filters, as illustrated in Figure F.2(b). Here we consider an up-chirp interdigital device with a short uniform input transducer and a chirp output transducer, taking the latter to be the single-electrode type. If p_1 and p_2 are the electrode pitches at the two ends of the chirp transducer, the surface-wave band extends approximately from $f_1 = v_s/(2p_1)$ to $f_2 = v_s/(2p_2)$. From equation (F.2), coherent bulk-wave excitation can occur in this band if

$$\frac{f_2}{f_1} > \frac{2}{1 + v_s/v_b(0)}. \quad (\text{F.3})$$

The main practical consequence of this is that the surface wave is attenuated. Since the surface wave is detected in a region beyond the region where bulk waves are generated, a corresponding attenuation is seen in the high-frequency part of the device frequency response. To avoid this condition, phase-matching considerations imply that f_2/f_1 should not exceed the right side of equation (F.3), but in practice it is found that significant attenuation occurs only for fractional bandwidths exceeding about 25%. The attenuation can be eliminated by using more than two electrodes per period, or by using the slanted geometry (Section 9.4). The problem does not occur in down-chirp devices, because for this case the bulk wave excitation occurs beyond the point at which the surface waves are detected, so that the attenuation does not significantly affect the device response. The same phenomenon can occur in an up-chirp in-line RAC (Section 9.6.3), which uses arrays of grooves parallel to the wavefronts. This is clearly shown by the frequency response of a device with 33% bandwidth [323].

For a conventional RAC, with angled grooves, the situation is more complicated. The phase-matching conditions for angled reflectors have been studied by Marshall and Paige [520] and by Islam *et al.* [521]. It is found that loss due to bulk wave excitation can be significant in up-chirp grooved RAC's [521]. Measurements using angled metal strips as reflectors on Y, Z lithium niobate showed strong surface-wave attenuation caused by coherent generation of a pseudo-surface wave [520].

F.3. Bulk Wave Devices

In the previous sections we considered the deleterious effects of bulk waves in surface-wave devices. Here we consider devices in which bulk waves are used as the primary acoustic mode, and surface-wave excitation is minimised.

Surface-Skimming Bulk Wave Devices. In the devices considered here, introduced by Lewis [522, 523], interdigital transducers are used to generate bulk waves deliberately. The device structure is essentially the same as that of a surface-wave device, with two interdigital transducers on a plane surface; the bulk wave travels close to the surface between the transducers, and is therefore called a "surface-skimming bulk wave", or SSBW. The configuration is as in Figure F.1 except that the angle θ is small. For practical devices, the substrate material and orientation are chosen such that there is no electrically-coupled surface-wave mode, so that no surface wave is excited. It is also necessary that the bulk wave should not be significantly disturbed by the boundary conditions at the surface. We have seen in Section 2.2.3 that, for an isotropic material, a shear-horizontal bulk wave propagating parallel to the surface is not affected by the surface boundary conditions. A similar solution can sometimes be found for anisotropic piezoelectric materials, and it is found that some orientations give this solution and also give no electrical coupling to surface waves, and are thus suitable for SSBW devices. For example, there are several suitable orientations of quartz [523].

SSBW devices have many features in common with surface-wave devices, including in particular the considerable flexibility obtainable by apodising the transducers. With a suitable choice of substrate the velocity can be 60% higher than that of surface waves, giving the advantage of a higher operating frequency for a given electrode periodicity. Alternatively, for other orientations a very good temperature stability can be obtained, much better than the stability for surface waves on ST, X quartz (Section 6.4). In addition, the SSBW is found to be less affected by surface contamination. Consequently, these waves are particularly suitable for oscillators and for narrow-band bandpass filters. However, the design of these devices is complicated by the fact that the bulk wave is not guided along the surface; thus, the wave diffracts into the bulk of the material, and the disturbance at the surface decays with distance. At large distances a wave generated at $x = 0$ can be expected to have an amplitude proportional to $x^{-1/2}$, as deduced from power conservation [523]. For a device using a short uniform transducer and a long apodised transducer, this attenuation is readily compensated by modifying the apodisation [524]. In practice, the form of the attenuation may be modified by the presence of an additional acoustic mode [513].

Experimental results are given in, for example, references [523–527]. Good temperature stability can be obtained by using $-51°$ rotated Y-cut quartz, with propagation normal to the X-axis; this gives a delay constant within about ± 20 parts per million for a 0 to 100°C temperature range [523, 525]. For high-frequency devices a suitable substrate is $+36°$ rotated Y-cut quartz, again with propagation normal to the X-axis. This gives a velocity of about 5100 m/sec [523], and has been used in delay lines for oscillators with frequencies up to 3.4 GHz [526]. Experimental bandpass filters using several substrate materials and orientations are described by Yen et al. [525].

Bloch et al. [527] have considered SSBW's in rotated Y-cut lithium niobate for application in television bandpass filters, centred at about 40 MHz. It was shown experimentally that the insensitivity to surface contamination implies that the device may be potted in an inert material instead of the usual hermetic encapsulation, thus reducing the cost. The same feature also implies that reflections from the ends of the substrate can be troublesome, since a lossy material on the surface is not an effective absorber. However, it was shown that the effects of reflections could be suppressed by angling the ends of the substrate, giving a device response almost as good as a surface-wave device.

Other Bulk-Wave Devices. Quite recently, Lewis [528] has introduced another interdigital bulk-wave device, shown in Figure F.3. Here the input transducer generates a bulk wave radiating at an angle to the upper surface. The wave is reflected at the lower surface, which must be flat and parallel to the upper surface, and thus reaches the output transducer. There may be several reflections at the lower surface, and Figure F.3 illustrates the case of two reflections. As in SSBW devices it is usual to choose the substrate material such that no surface-wave excitation occurs, and rotated Y-cut orientations of quartz are found to be suitable.

The main attraction of this device is that, for a given transducer periodicity, it enables the operating frequency to be much higher than that of a surface-wave device. This follows directly from equation (F.1), since $\cos \theta$ is less than unity here. Experimental devices [528] have given responses at up to six times the frequencies given by surface-wave devices, showing that for present-day lithography there is the possibility of operating up to about 10 GHz, well beyond the capabilities of surface-wave or SSBW devices. At present the device is insufficiently developed to enable firm conclusions to be drawn on its practicability. However, it has been found

FIGURE F.3. Interdigital bulk-wave filter using reflection at rear surface.

that the frequency response can be predicted quite well [529]. The response is very different from that of a surface-wave device using similar transducers, implying that quite different design techniques are needed. It has also been shown that a temperature stability comparable with the best SSBW devices can be obtained [529].

Another proposed device using bulk waves exploits the coupling of surface waves to bulk waves in a periodic array of grooves [518]. As already noted in Section F.2, this process can be very efficient. Using a parallel-sided plate of lithium niobate, the transmission of surface waves through a groove array on one surface was found to exhibit a series of sharp minima due to coupling to the bulk modes of the plate, for frequencies such that the surface-wave wavelength is approximately equal to the groove pitch (about half the resonance frequency of a surface-wave resonator). It was also shown that the bulk modes generated can be efficiently converted to another surface wave by means of a second groove array, located either on the second surface or on the same surface as the first array. This use of two arrays gives a frequency response with a series of sharp peaks. These methods are therefore potentially applicable to narrow-band bandpass or band-stop filtering.

References

INTRODUCTORY AND GENERAL MATERIAL (Chapter 1)

1. J. H. Eveleth, "A survey of ultrasonic delay lines operating below 100 Mc/s", *Proc. IEEE* **53**, 1406–1428 (1965).
2. G. P. Rodrigue, "Microwave solid-state delay lines", *Proc. IEEE* **53**, 1428–1437 (1965).
3. E. Dieulesaint and D. Royer, *Elastic Waves in Solids*, Masson, 1974 (in French); Wiley, 1980 (in English).
4. W. P. Mason, "Use of piezoelectric crystals and mechanical resonators in filters and oscillators", in W. P. Mason (ed.), *Physical Acoustics*, Vol. 1A, Academic Press, 1964, pp. 335–416.
5. W. J. Spencer, "Monlithic crystal filters", in W. P. Mason and R. N. Thurston (eds.), *Physical Acoustics*, Vol. 9, Academic Press, 1972, pp. 167–220.
6. D. P. Morgan (ed.), *Surface Acoustic Wave Passive Interdigital Devices*, IEE Reprint Series No. 2, Peter Peregrinus, 1976.
7. H. Matthews (ed.), *Surface Wave Filters*, Wiley, 1977.
8. A. A. Oliner (ed.), *Acoustic Surface Waves*, Springer, 1978.
9. R. M. White, "Surface elastic waves", *Proc. IEEE* **58**, 1238–1276 (1970).
10. J. D. Maines and E. G. S. Paige, "Surface acoustic wave components, devices and applications", *Proc. IEE* **120**, 1078–1110 (1973). Also "Surface acoustic wave devices for signal processing applications", *Proc. IEEE* **64**, 639–652 (1976).
11. M. F. Lewis, C. L. West, J. M. Deacon and R. F. Humphryes, "Recent developments in SAW devices", *Proc. IEE* **131A**, 186–215 (1984).
12. J. D. Maines, J. B. G. Roberts and R. C. Williamson (eds.), IEE seminar on "Impact of new technologies in signal processing", IEE Conf. Publ. 144, 1976.
13. J. B. G. Roberts, P. M. Grant and T. W. Bristol (eds.), IEE seminar on "Case studies in advanced signal processing", IEE Conf. Publ. 180, 1979.
14. J. De Klerk and B. R. McAvoy (eds.). Collected papers on "Surface acoustic waves and signal processing", from *IEEE Ultrasonics Symp.*, IEEE, Cat. No. TH0053–9, 1979.
15. T. M. Reeder (ed.), Special issue on "Microwave acoustic signal processing", *IEEE Trans.* **MTT-21**, 161–306 (April, 1973).
16. L. T. Claiborne, G. S. Kino and E. Stern (eds.), Special issue on "Surface acoustic wave devices and applications", *Proc. IEEE* **64**, 579–828 (May, 1976).
17. J. H. Collins and L. Masotti (eds.), Special issue on "Computer-aided design of SAW devices", *Wave Electronics* **2**, 1–304 (July, 1976).
18. R. C. Williamson and T. W. Bristol (eds.), Special issue on "Surface-acoustic-wave device applications", *IEEE Trans.* **SU-28**, 115–228 (May, 1981).
19. R. M. Hays and C. S. Hartmann, "Surface-acoustic-wave devices for communications", *Proc. IEEE* **64**, 652–671 (1976).
20. J. D. Maines, "Surface wave devices for radar equipment", in Matthews [7], pp. 443–476.
21. R. C. Williamson, "Case studies of successful surface-acoustic-wave devices", *IEEE Ultrasonics Symp.*, 1977, pp. 460–468.

22. J. H. Collins and P. M. Grant, "Current and future sub-system modules for intercept receivers", *IEEE Ultrasonics Symp.*, 1980, pp. 204–215.
23. R. M. White and F. W. Voltmer, "Direct piezoelectric coupling to surface elastic waves", *Appl. Phys. Lett.* **7**, 314–316 (1965).
24. W. S. Mortley, British patent 988,102 (May, 1963) and J. H. Rowen, U.S. patent 3,289,114 (Dec., 1963).
25. H. I. Smith, "Fabrication techniques for surface-acoustic-wave and thin-film optical devices", *Proc. IEEE* **62**, 1361–1387 (1974).

ACOUSTIC WAVES (Chapter 2)

26. J. F. Nye, *Physical Properties of Crystals*, Oxford Univ. Press, 1957.
27. W. G. Cady, *Piezoelectricity*, McGraw-Hill, 1946.
28. D. A. Berlincourt, D. R. Curran and H. Jaffe, "Piezoelectric and piezomagnetic materials and their function in transducers", in W. P. Mason (ed.), *Physical Acoustics*, Vol. 1A, Academic Press, 1964, ch. 3.
29. H. F. Tiersten, *Linear Piezoelectric Plate Vibrations*, Plenum Press, 1969.
30. E. Dieulesaint and D. Royer, *Elastic Waves in Solids*, Masson, 1974 (in French); Wiley, 1980 (in English).
31. B. A. Auld, *Acoustic Fields and Waves in Solids*, Vol. I, Wiley, 1973.
32. *Ibid.*, Vol. II.
33. I. A. Viktorov, *Rayleigh and Lamb Waves*, Plenum Press, 1967.
34. T. R. Meeker and A. H. Meitzler, "Guided wave propagation in elongated cylinders and plates", in W. P. Mason (ed.), *Physical Acoustics*, Vol. 1A, Academic Press, 1964, pp. 112–167.
35. A. A. Oliner, "Microwave network methods for guided elastic waves", *IEEE Trans.* **MTT-17**, 812–826 (1969).
36. M. Redwood, *Mechanical Waveguides*, Pergamon, 1960.
37. Lord Rayleigh, "On waves propagated along the plane surface of an elastic solid", *Proc. London Math. Soc.* **17**, 4–11 (1885).
38. W. M. Ewing, W. S. Jardetzky and F. Press, *Elastic Waves in Layered Media*, McGraw-Hill, 1957.
39. L. M. Brekhovskikh, *Waves in Layered Media*, Academic Press, 1960 and 1980.
40. K. Sezawa, "Dispersion of elastic waves propagated on the surface of stratified bodies and on curved surfaces", *Bull. Earthquake Res. Inst. Tokyo* **3**, 1–18 (1927).
41. K. Kanai, "On the M2 waves (Sezawa waves) — parts 1 and 2", *Bull. Earthquake Res. Inst. Tokyo* **29**, 39–48 (1951) and **33**, 275–281 (1955).
42. H. M. Mooney and B. A. Bolt, "Dispersion characteristics of the first three Rayleigh modes for a single surface layer", *Bull. Seism. Soc. Am.* **56**, 43–67 (1966).
43. A. E. H. Love, *Some Problems of Geodynamics*, Cambridge Univ. Press, 1911; Dover, 1967.
44. G. W. Farnell and E. L. Adler, "Elastic wave propagation in thin layers", in W. P. Mason and R. N. Thurston (eds.), *Physical Acoustics*, Vol. 9, Academic Press, 1972, ch. 2.
45. J. J. Campbell and W. R. Jones, "A method for estimating optimal crystal cuts and propagation directions for excitation of piezoelectric surface waves", *IEEE Trans.* **SU-15**, 209–217 (1968).
46. T. C. Lim and G. W. Farnell, "Search for forbidden directions of elastic surface wave propagation in anisotropic crystals", *J. Appl. Phys.* **39**, 4319–4325 (1968).
47. J. J. Campbell and W. R. Jones, "Propagation of piezoelectric surface waves on cubic and hexagonal crystals", *J. Appl. Phys.* **41**, 2796–2801 (1970).
48. G. W. Farnell, "Properties of elastic surface waves", in W. P. Mason and R. N. Thurston (eds.), *Physical Acoustics*, Vol. 6, Academic Press, 1970, pp. 109–166.
49. G. W. Farnell, "Elastic surface waves", in Matthews [7], pp. 1–53.
50. G. W. Farnell, "Types and properties of surface waves", in Oliner [8], pp. 13–60.
51. A. J. Slobodnik, E. D. Conway and R. T. Delmonico, "Microwave acoustics handbook, Vol. 1A, Surface wave velocities", Air Force Cambridge Res. Labs., AFCRL-TR-73-0597, Oct. 1973.

REFERENCES

52. IEEE Standard 176 on Piezoelectricity, *IEEE Trans.* **SU-31**, March 1984, part II. Also, T. R. Meitzler, *IEEE Trans.* **SU-31**, 135–136 (1984).
53. J. L. Bleustein, "A new surface wave in piezoelectric crystals", *Appl. Phys. Lett.* **13**, 412–413 (1968).
54. Y. V. Gulyaev, "Electroacoustic surface waves in solids", *Soviet Phys. JETP Lett.* **9**, 37–38 (1969).
55. C. C. Tseng, "Piezoelectric surface waves in cubic crystals", *J. Appl. Phys.* **41**, 2270–2276 (1970).
56. G. Koerber and R. F. Vogel, "SH-mode piezoelectric surface waves on rotated cuts", *IEEE Trans.* **SU-20**, 9–12 (1973).
57. H. Engan, K. A. Ingebrigtsen and A. Tonning, "Elastic surface waves in alpha quartz: observation of leaky surface waves", *Appl. Phys. Lett.* **10**, 311–313 (1967).
58. T. C. Lim and G. W. Farnell, "Character of pseudo-surface waves on anisotropic crystals", *J. Acoust. Soc. Am.* **45**, 845–851 (1969).
59. K. Yamanouchi and K. Shibayama, "Propagation and amplification of Rayleigh waves and piezoelectric leaky waves in lithium niobate", *J. Appl. Phys.* **43**, 856–862 (1972).
60. L. P. Solie, "Piezoelectric waves on layered substrates", *J. Appl. Phys.* **44**, 619–627 (1973).
61. K. M. Lakin, "Perturbation theory for electromagnetic coupling to elastic surface waves on piezoelectric substrates", *J. Appl. Phys.* **42**, 899–906 (1971).

ELECTRICAL EXCITATION (Chapter 3)

62. K. A. Ingebrigtsen, "Surface waves in piezoelectrics", *J. Appl. Phys.* **40**, 2681–2686 (1969).
63. C. S. Hartmann and B. G. Secrest, "End effects in interdigital surface wave transducers", *IEEE Ultrasonics Symp.* 1972, pp. 413–416.
64. C. A. A. J. Greebe, P. A. van Dalen, T. J. N. Swanenburg and J. Wolter, "Electric coupling properties of acoustic and electric surface waves", *Phys. Rep., Phys. Lett. C* **1C**, 235–268 (1971).
65. R. F. Milsom, N. H. C. Reilly and M. Redwood, "Analysis of generation and detection of surface and bulk acoustic waves by interdigital transducers", *IEEE Trans.* **SU-24**, 147–166 (1977).
66. R. F. Milsom, M. Redwood and N. H. C. Reilly, "The interdigital transducer", in Matthews [7], pp. 55–108.
67. K. M. Lakin, "Perturbation theory for electromagnetic coupling to elastic surface waves on piezoelectric substrates", *J. Appl. Phys.* **42**, 899–906 (1971).
68. B. A. Auld, *Acoustic Fields and Waves in Solids*, Vol. II, Wiley, 1973.
69. B. A. Auld and G. S. Kino, "Normal mode theory for acoustic waves and its application to the interdigital transducer", *IEEE Trans.* **ED-18**, 898–908 (1971).
70. G. S. Kino and T. M. Reeder, "A normal mode theory for the Rayleigh wave amplifier", *IEEE Trans.* **ED-18**, 909–920 (1971).
71. J. Henaff and M. Feldmann, "Monolithic acoustoelectric amplifier using pseudo-surface waves", *Appl. Phys. Lett.* **24**, 447–449 (1974).
72. K. Bløtekjaer, K. A. Ingebrigtsen and H. Skeie, "A method for analysing waves in structures consisting of metal strips on dispersive media", *IEEE Trans.* **ED-20**, 1133–1138 (1973).
73. M. R. Daniel and J. De Klerk, "An apparatus for measuring the velocity and energy flow direction of ultrasonic surface waves", *Rev. Sci. Instr.* **41**, 1463–1465 (1970).
74. K. Yamanouchi and K. Shibayama, "Propagation and amplification of Rayleigh waves and piezoelectric leaky surface waves in $LiNbO_3$", *J. Appl. Phys.* **43**, 856–862 (1972).
75. G. S. Kino, "Acoustoelectric interactions in acoustic-surface-wave devices", *Proc. IEEE* **64**, 724–748 (1976).
76. O. Yamazaki, T. Mitsuyu and K. Wasa, "ZnO thin-film SAW devices", *IEEE Trans.* **SU-27**, 369-379 (1980).
77. F. S. Hickernell, M. D. Adamo, R. V. De Long, J. G. Hinsdale and H. J. Bush, "SAW programmable matched filter signal processor", *IEEE Ultrasonics Symp.*, 1980, pp. 104–108.
78. G. S. Kino and R. S. Wagers, "Theory of interdigital couplers on non-piezoelectric substrates", *J. Appl. Phys.* **44**, 1480–1488 (1973).

INTERDIGITAL TRANSDUCER ANALYSIS (Chapter 4)

79. R. H. Tancrell and M. G. Holland, "Acoustic surface wave filters", *Proc. IEEE* **59**, 393–409 (1971).
80. R. H. Tancrell, "Principles of surface wave filter design", in Matthews [7], pp. 109–164.
81. C. S. Hartmann, D. T. Bell and R. C. Rosenfeld, "Impulse model design of acoustic surface wave filters", *IEEE Trans.* **MTT-21**, 162–175 (1973).
82. T. W. Bristol, W. R. Jones, P. B. Snow and W. R. Smith, "Applications of double electrodes in acoustic surface wave device design", *IEEE Ultrasonics Symp.*, 1972, pp. 343–345.
83. H. Engan, "Surface acoustic wave multi-electrode transducers", *IEEE Trans.* **SU-22**, 395–401 (1975).
84. W. R. Smith, H. M. Gerard, J. H. Collins, T. M. Reeder and H. J. Shaw. "Analysis of interdigital surface wave transducers by use of an equivalent circuit model", *IEEE Trans.* **MTT-17**, 856–864 (1969).
85. W. R. Smith, "Experimental distinction between crossed-field and in-line three-port circuit models for interdigital transducers", *IEEE Trans.* **MTT-22**, 960–964 (1974).
86. W. R. Smith, "Basics of the SAW interdigital transducer", *Wave Electronics* **2**, 25–63 (1976).
87. H. M. Gerard, "Principles of surface wave filter design", in Oliner [8], pp. 60–96.
88. W. R. Smith, H. M. Gerard and W. R. Jones, "Analysis and design of dispersive interdigital surface-wave transducers", *IEEE Trans.* **MTT-20**, 458–471 (1972).
89. W. S. Jones, C. S. Hartmann and T. D. Sturdivant, "Second order effects in surface wave devices", *IEEE Trans.* **SU-19**, 368–377 (1972).
90. W. R. Smith and W. F. Pedler, "Fundamental and harmonic frequency circuit-model analysis of interdigital transducers with abritrary metallisation ratios and polarity sequences", *IEEE Trans.* **MTT-23**, 853–864 (1975).
91. B. A. Auld and G. S. Kino, "Normal mode theory for acoustic waves and its application to the interdigital transducer", *IEEE Trans.* **ED-18**, 898–908 (1971).
92. R. F. Milsom, "Bulk wave generation by the IDT", *Wave Electronics* **2**, 64–81 (1976).
93. R. F. Milsom, N. H. C. Reilly and M. Redwood, "Analysis of generation and detection of surface and bulk acoustic waves by interdigital transducers", *IEEE Trans.* **SU-24**, 147–166 (1977).
94. R. F. Milsom, M. Redwood and N. H. C. Reilly, "The interdigital transducer", in Matthews [7], pp. 55–108.
95. D. P. Morgan, "Quasi-static analysis of generalised SAW transducers using the Green's function method", *IEEE Trans.* **SU-27**, 111–123 (1980).
96. S. Ramo, J. R. Whinnery and T. van Duzer, *Fields and Waves in Communication Electronics*, Wiley, 1965, pp. 154, 314.
97. R. C. Peach, "A general approach to the electrostatic problem of the SAW interdigital transducer", *IEEE Trans.* **SU-28**, 96–105 (1981).
98. C. S. Hartmann and B. G. Secrest, "End effects in interdigital surface wave transducers", *IEEE Ultrasonics Symp.*, 1972, pp. 413–416.
99. H. Skeie and A. Rønnekliev, "Electrostatic neighbour and end effects in weighted surface wave transducers", *IEEE Ultrasonics Symp.*, 1976, pp. 540–542.
100. K. R. Laker, E. Cohen and A. J. Slobodnik, "Electric field interactions within finite arrays and design of withdrawal weighted SAW filters at fundamental and higher harmonics", *IEEE Ultrasonics Symp.*, 1976, pp. 317–321.
101. A. L. Lentine, S. Datta and B. J. Hunsinger, "Analysis of nonperiodic SAW transducers using a circuit model", *IEEE Trans.* **SU-27**, 328–333 (1980).
102. A. Papoulis, *The Fourier Integral and its Applications*, McGraw-Hill, 1962.
103. A. L. Nalamwar and M. Epstein, "Immittance characterisation of acoustic surface-wave transducers", *Proc. IEEE* **60**, 336–337 (1972).
104. B. Lewis, P. M. Jordan, R. F. Milsom and D. P. Morgan, "Charge and field superposition methods for analysis of generalised SAW interdigital transducers", *IEEE Ultrasonics Symp.*, 1978, pp. 709–714.
105. S. Datta, B. J. Hunsinger and D. C. Malocha, "A generalised model for periodic transducers with arbitrary voltages", *IEEE Trans.* **SU-26**, 235–242 (1979).
106. S. Datta and B. J. Hunsinger, "Element factor for periodic transducers", *IEEE Trans.* **SU-27**, 42–44 (1980).

107. H. Engan, "Excitation of elastic surface waves by spatial harmonics of interdigital transducers", *IEEE Trans.* **ED-16**, 1014–1017 (1969).
108. H. Skeie, "Analysis and synthesis of surface acoustic wave filters", ELAB report STF 44A79106, University of Trondheim, 1979.
109. D. P. Morgan, B. Lewis and J. G. Metcalfe, "Fundamental charge distributions for surface-wave interdigital transducer analysis", *Electronics Lett.* **15**, 583–585 (1979).
110. R. H. Tancrell and R. C. Williamson, "Wavefront distortion of acoustic surface waves from apodised interdigital transducers", *Appl. Phys. Lett.* **19**, 456–459 (1971).
111. R. S. Wagers, "Analysis of finite-width interdigital transducer excitation profiles", *IEEE Trans.* **SU-26**, 105–111 (1979).
112. G. L. Matthaei, D. Y. Wong and B. P. O'Shaughnessy. "Simplifications for the analysis of interdigital surface-wave devices", *IEEE Trans.* **SU-22**, 105–114 (1975).

MULTI-STRIP COUPLERS (Chapter 5)

113. F. G. Marshall and E. G. S. Paige, "Novel acoustic-surface-wave directional coupler with diverse applications", *Electronics Lett.* **7**, 460–464 (1971).
114. C. Elachi, "Waves in active and passive periodic structures: a review", *Proc. IEEE* **64**, 1666–1698 (1976).
115. C. Maerfeld and P. Tournois, "An acoustoelectric multistrip amplifier", *IEEE Ultrasonics Symp.*, 1972, pp. 171–176.
116. C. Maerfeld and P. Tournois, "Perturbation theory for the surface-wave multistrip coupler", *Electronics Lett.* **9**, 115–116 (1973).
117. C. Maerfeld, K. Gordon and G. W. Farnell, "Resistive losses in acoustic surface wave multistrip couplers", *IEEE Trans.* **SU-22**, 358–368 (1975).
118. C. Maerfeld, "Multistrip couplers", *Wave Electronics* **2**, 82–110 (1976).
119. K. Bløtekjaer, K. A. Ingebrigtsen and H. Skeie, "A method for analysing waves in structures consisting of metal strips on dispersive media", *IEEE Trans.* **ED-20**, 1133–1138 (1973).
120. K. Bløtekjaer, K. A. Ingebrigtsen and H. Skeie, "Acoustic surface waves in piezoelectric materials with periodic metal strips on the surface", *IEEE Trans.* **ED-20**, 1139–1146 (1973).
121. F. G. Marshall, C. O. Newton and E. G. S. Paige, "Theory and design of the surface acoustic wave multistrip coupler", *IEEE Trans.* **MTT-21**, 206–215 (1973).
122. I. M. Mason, J. Chambers and P. E. Lagasse, "Spatial harmonic analysis of the multistrip coupler", *IEEE Ultrasonics Symp.*, 1973, pp. 159–162.
123. G. W. Farnell and C. Maerfeld, "Modes in multistrip couplers", *IEEE Ultrasonics Symp.*, 1976, pp. 480–485.
124. J. M. Deacon, J. H. Heighway and J. A. Jenkins, "Multistrip coupler in acoustic surface wave filters", *Electronics Lett.* **9**, 235–236 (1973).
125. R. H. Tancrell, "Improvement of an acoustic-surface-wave filter with a multistrip coupler", *Electronics Lett.* **9**, 316–317 (1973).
126. F. G. Marshall, C. O. Newton and E. G. S. Paige, "Surface acoustic wave multistrip components and their applications", *IEEE Trans.* **MTT-21**, 216–225 (1973).
127. T. I. Browning and F. G. Marshall, "Compact 130 μs surface acoustic wave delay line using improved MSC reflecting trackchangers", *IEEE Ultrasonics Symp.*, 1974, pp. 189–192.
128. C. W. Chapman and T. W. Bristol, "The analysis and design of broad-band fractional efficiency multistrip couplers, *IEEE Trans.* **SU-23**, 174–184 (1976).
129. L. P. Solie, "A SAW bandpass filter technique using a fanned multistrip coupler", *Appl. Phys. Lett.* **30**, 374–376 (1977).
130. M. Feldmann and J. Henaff, "Design of multistrip arrays", *IEEE Ultrasonic Symp.*, 1977, pp. 686–690.
131. C. Maerfeld and G. W. Farnell, "Nonsymmetrical multistrip coupler as a surface-wave beam compressor of large bandwidth", *Electronics Lett.* **9**, 432–434 (1973).

PROBING AND DIFFRACTION (Sections 6.1, 6.2)

132. A. J. Slobodnik, P. H. Carr, and A. J. Budreau, "Microwave frequency acoustic surface wave loss mechanisms on lithium niobate", *J. Appl. Phys.* **41**, 4380–4387 (1970).
133. B. A. Richardson and G. S. Kino, "Probing of elastic surface waves in piezoelectric media", *Appl. Phys. Lett.* **16**, 82–85 (1970).
134. R. C. Williamson, "Improved electrostatic probe for measurement of elastic surface waves", *IEEE Trans.* **SU-19**, 436–441 (1972). Also *IEEE Ultrasonics Symp.*, 1972, pp. 323–327.
135. G. I. Stegeman, "Optical probing of surface waves and surface wave devices", *IEEE Trans.* **SU-23**, 33–63 (1976).
136. E. G. H. Lean and C. G. Powell, "Optical probing of surface acoustic waves", *Proc. IEEE* **58**, 1939–1947 (1970).
137. A. J. Slobodnik, "Materials and their influence on performance", in Oliner [8], pp. 225–303.
138. W. S. Goruk, P. J. Vella and G. I. Stegeman, "Optical probing measurements of surface wave generation and reflection in interdigital transducers on lithium niobate", *IEEE Trans.* **SU-27**, 341–354 (1980).
139. R. Adler, A. Korpel and P. Desmares, "An instrument for making surface waves visible", *IEEE Trans.* **SU-15**, 157–161 (1968).
140. H. Engan, "Phase sensitive laser probe for high-frequency surface acoustic wave measurements", *IEEE Trans.* **SU-25**, 372–377 (1978). Also *ibid.*, **SU-29**, 281–283 (1982).
141. R. L. Whitman, L. J. Laub and W. J. Bates, "Acoustic surface displacement measurements on a wedge-shaped transducer using an optical probe technique", *IEEE Trans.* **SU-15**, 186–189 (1968).
142. R. M. De La Rue, R. F. Humphryes, I. M. Mason and E. A. Ash, "Acoustic-surface-wave amplitude and phase measurements using laser probes", *Proc. IEE* **119**, 117–126 (1972).
143. L. Bergstein and T. Zachos, "A Huygen's principle for uniaxially anisotropic media", *J. Opt. Soc. Am.* **56**, 931–937 (1966).
144. M. S. Kharusi and G. W. Farnell, "Diffraction and beam steering for surface-wave comb structures on anisotropic substrates", *IEEE Trans.* **SU-18**, 34–42 (1971).
145. I. M. Mason and E. A. Ash, "Acoustic-surface-wave beam diffraction on anisotropic substrates", *J. Appl. Phys.* **42**, 5343–5351 (1971).
146. G. W. Farnell, "Types and properties of surface waves", in Oliner [8], pp. 13–60.
147. M. G. Cohen, "Optical study of ultrasonic diffraction and focusing in anisotropic media", *J. Appl. Phys.* **38**, 3821–3828 (1967).
148. R. D. Weglein, M. E. Pedinoff and H. Winston, "Diffraction spreading of surface waves on $LiNbO_3$", *Electronics Lett.* **6**, 654–656 (1970).
149. N. R. Ogg, "A Huygen's principle for anisotropic media", *J. Phys. A* **4**, 382–388 (1971).
150. J. C. Crabb, J. D. Maines and N. R. Ogg, "Surface-wave diffraction on $LiNbO_3$", *Electronics Lett.* **7**, 253–255 (1971).
151. J. D. Maines, G. L. Moule and N. R. Ogg, "Correction of diffraction errors in acoustic-surface-wave pulse compression filters", *Electronics Lett.* **8**, 431–433 (1972).
152. R. F. Milsom, "A diffraction theory for SAW filters on non-parabolic high-coupling orientations", *IEEE Ultrasonics Symp.*, 1977, pp. 827–833.
153. R. A. Waldron, "Power transfer factors for nonuniformly irradiated interdigital piezoelectric transducers", *IEEE Trans.* **SU-19**, 448–453 (1972).
154. B. A. Auld, *Acoustic Fields and Waves in Solids*, Wiley, 1973, Vol. I.
155. M. J. Lighthill, "Studies on magneto-hydrodynamic waves and other anisotropic wave motions", *Phil. Trans.* **252A**, 397–430 (1960).
156. G. W. Farnell, "Properties of elastic surface waves", in W. P. Mason and R. W. Thurston (eds.), *Physical Acoustics*, Vol. 6, Academic Press, 1970, pp. 109–166.
157. M. Abramowitz and I. A. Stegun, *Handbook of Mathematical Functions*, Dover, 1968.
158. T. L. Szabo and A. J. Slobodnik, "The effect of diffraction on the design of acoustic surface wave devices", *IEEE Trans.* **SU-20**, 240–251 (1973).
159. D. Murray and E. A. Ash, "Precision measurement of SAW velocities", *IEEE Ultrasonics Symp.*, 1977, pp. 823–826.

160. R. C. Peach, "General diffraction analysis for surface acoustic wave filters", *IEEE Ultrasonics Symp.*, 1983.
161. T. L. Szabo and A. J. Slobodnik, "Diffraction compensation in periodic apodised acoustic surface wave filters", *IEEE Trans.* **SU-21**, 114–119 (1974).
162. D. Penunuri, "A numerical technique for SAW diffraction simulation", *IEEE Trans.* **MTT-26**, 288–294 (1978).
163. E. B. Savage, "Fast computation of SAW filter responses including diffraction", *Electronics Lett.* **15**, 538–539 (1979).
164. R. C. Peach, N. H. Doggett, F. S. McClemont, A. Katsellis and A. J. Dyer, "Diffraction analysis and correction of narrow band SAW transversal filters", *IEEE Ultrasonics Symp*, 1981, pp. 58–62.
165. W. A. Radasky and G. L. Matthaei, "Fast computation of diffraction in general anisotropic media by use of the geometrical theory of diffraction", *IEEE Trans.* **SU-30**, 78–84 (1983).

OTHER PROPAGATION EFFECTS, AND MATERIALS (Sections 6.3–6.5)

166. H. J. Maris, "Attenuation of ultrasonic surface waves by viscosity and heat conduction", *Phys. Rev.* **188**, 1308–1311 (1969).
167. K. Dransfeld and E. Salzmann, "Excitation, detection and attenuation of high-frequency elastic surface waves", in W. P. Mason and R. N. Thurston (eds.), *Physical Acoustics*, Vol. VII, Academic Press, 1970, pp. 219–272.
168. J. J. Campbell and W. R. Jones, "Propagation of surface waves at the boundary between a piezoelectric crystal and a fluid medium", *IEEE Trans.* **SU-17**, 71–76 (1970).
169. J. W. Gibson and P. H. E. Meijer, "Nonlinearly generated harmonics and attenuation of Rayleigh waves on crystalline quartz", *J. Appl. Phys.* **45**, 3288–3295 (1974).
170. B. A. Auld, *Acoustic Fields and Waves in Solids*, Wiley, 1973, Vol. II.
171. Y. Nakagawa, K. Yamanouchi and K. Shibayama, "Control of nonlinear effects in acoustic surface waves", *J. Appl. Phys.* **45**, 2817–2822 (1974).
172. E. G. Lean and C. G. Powell, "Non-destructive testing of thin films by harmonic generation of dispersive Rayleigh waves", *Appl. Phys. Lett.* **19**, 356–359 (1971).
173. R. C. Williamson, "Problems encountered in high-frequency surface-wave devices", *IEEE Ultrasonics Symp*, 1974, pp. 321–328.
174. K. L. Davis and V. L. Newhouse, "Mixing between noncollinear surface elastic waves", *IEEE Trans.* **SU-22**, 33–38 (1975).
175. J. D. Maines, E. G. S. Paige, A. F. Saunders and A. S. Young, "Simple technique for the accurate determination of delay-time variations in acoustic surface wave structures", *Electronics Lett.* **5**, 678–680 (1969).
176. J. F. Dias, H. E. Karrer, J. A. Kusters, J. H. Matsinger and M. B. Schulz, "The temperature coefficient of delay-time for X-propagating acoustic surface waves on rotated Y-cuts of alpha quartz", *IEEE Trans.* **SU-22**, 46–50 (1975).
177. M. B. Schulz, B. J. Matsinger and M. G. Holland, "Temperature dependence of surface acoustic wave velocity on alpha-quartz", *J. Appl. Phys.* **41**, 2755–2765 (1970).
178. S. J. Kerbel, "Design of harmonic surface acoustic wave oscillators without external filtering and new data on the temperature coefficient of quartz", *IEEE Ultrasonics Symp.*, 1974, pp. 276–281.
179. D. T. Bell and L. T. Claiborne, "Phase code generators and correlators", in Matthews [7], pp. 307–346.
180. J. Temmyo, I. Kotaka, T. Inamura and S. Yoshikawa, "Precise measurement of SAW propagation velocity on lithium niobate", *IEEE Trans.* **SU-27**, 218–219 (1980).
181. A. J. Slobodnik, E. D. Conway and R. T. Delmonico, "Microwave acoustics handbook, Vol. 1A, Surface wave velocities", Air Force Cambridge Res. Labs., AFCRL-TR-73-0597, Oct. 1973.
182. M. B. Schulz and J. H. Matsinger, "Rayleigh wave electromechanical coupling constants", *Appl. Phys. Lett.* **20**, 367–369 (1972).
183. K. Shibayama, K. Yamanouchi, H. Sato and T. Meguro, "Optimum cut for rotated Y-cut $LiNbO_3$ crystal used as the substrate of acoustic-surface-wave filters", *Proc. IEEE* **64**, 595–597 (1976).

184. B. K. Sinha and H. F. Tiersten, *Appl. Phys. Lett.* **34**, 817–819 (1979). Also A. Ballato, T. Lukaszek, D. F. Williams and F. Y. Cho, *IEEE Ultrasonics Symp.*, 1981, pp. 346–349.
185. A. J. Slobodnik, J. H. Silva, W. J. Kearns and T. L. Szabo, *IEEE Trans.* **SU-25**, 92–97 (1978). Also P. H. Carr, T. E. Fenstermacher, J. H. Silva, W. J. Kearns and M. R. Stiglitz, *IEEE Ultrasonics Symp.*, 1976, pp. 459–461.
186. R. G. Pratt, G. Simpson and W. A. Crossley, *Electronics Lett.* **8**, 127–128 (1972).
187. A. J. Slobodnik and T. L. Szabo, *J. Appl. Phys.* **44**, 2937–2941 (1973).
188. M. F. Lewis, "Temperature compensation techniques for SAW devices", *IEEE Ultrasonics Symp.*, 1979, pp. 612–622.
189. B. Lewis, N. M. Shorrocks and R. W. Whatmore, *IEEE Ultrasonics Symp.*, 1982, pp. 389–393.
190. R. M. O'Connell and P. H. Carr, *IEEE Trans.* **SU-24**, 376–384 (1977). Also P. H. Carr, J. H. Silva´ and B. H. T. Chai, *IEEE Ultrasonics Symp.*, 1981, pp. 328–331.
191. T. W. Grudkowski, G. K. Montress, M. Gilden and J. F. Black, *IEEE Trans.* **MTT-29**, 1348–1355 (1981).
192. H. Takeuchi, Y. Ito, S. Jyomura, K. Nagatsuma and S. Ashida, *IEEE Ultrasonics Symp.*, 1980, pp. 400–409.
193. D. Penunuri and K. M. Lakin, "Surface acoustic wave velocities for isotropic metal films on selected cuts of $Bi_{12}GeO_{20}$, quartz, Al_2O_3 and $LiNbO_3$", *IEEE Trans.* **SU-21**, 293–295 (1974).
194. K. L. Davis and J. F. Weller, "SAW attenuation in metal film coated delay lines", *IEEE Ultrasonics Symp.*, 1979, pp. 659–662.
195. K. Tsubouchi, K. Sugai and N. Mikoshiba, *IEEE Ultrasonics Symp.*, 1981, pp. 375–380.
196. T. E. Parker and H. Wichansky, "Temperature-compensated surface-acoustic-wave devices with SiO_2 film overlays", *J. Appl. Phys.* **50**, 1360–1369 (1979).

DELAY LINES AND MULTI-PHASE TRANSDUCERS (Chapter 7)

197. W. R. Smith, H. M. Gerard, J. H. Collins, T. M. Reeder and H. J. Shaw, "Design of surface wave delay lines with interdigital transducers", *IEEE Trans.* **MTT-17**, 865–873 (1969).
198. C. S. Hartmann, D. T. Bell and R. C. Rosenfeld, "Impulse model design of acoustic surface wave filters", *IEEE Trans.* **MTT-21**, 162–175 (1973).
199. D. P. Morgan, J. M. Hannah and J. H. Nash, "Broadband surface-acoustic-wave delay lines using weakly piezoelectric substrates", *Wave Electronics* **3**, 97–105 (1977).
200. T. M. Reeder, W. R. Shreve and P. L. Adams, "A new broadband coupling network for interdigital surface wave transducers", *IEEE Trans.* **SU-19**, 466–470 (1972).
201. T. M. Reeder and W. R. Sperry, "Broad-band coupling to high-Q resonant loads", *IEEE Trans.* **MTT-20**, 453–458 (1972).
202. W. R. Smith, "Key tradeoffs in SAW transducer design and component specification", *IEEE Ultrasonics Symp.*, 1976, pp. 547–552.
203. J. Temmyo and S. Yoshikawa, "Aluminium evaporation conditions for SAW interdigital transducers", *IEEE Trans.* **SU-27**, 219–221 (1980).
204. K. M. Lakin, "Electrode resistance effects in interdigital transducers", *IEEE Trans.* **MTT-22**, 418–424 (1974).
205. A. J. Slobodnik, "Materials and their influence on performance", in Oliner [8], pp. 225–303.
206. M. F. Lewis, "Triple-transit suppression in surface-acoustic-wave devices", *Electronics Lett.* **8**, 553–554 (1972).
207. W. J. Tanski and H. Van De Vaart, "Technique for triple-transit suppression in surface-acoustic-wave delay lines", *Electronics Lett.* **15**, 312–313 (1979).
208. D. J. Gunton, "Combined triple-transit and bulk wave suppression in S.A.W. delay lines", *Electronics Lett.* **16**, 149–150 (1980).
209. L. A. Coldren and H. J. Shaw, "Surface-wave long delay lines", *Proc. IEEE* **64**, 598–609 (1976).
210. I. M. Mason, E. Papadofrangakis and J. Chambers, "Acoustic-surface-wave disc delay lines", *Proc. IEEE* **64**, 610–612 (1976).

211. D. H. Warne, J. J. Purcell and D. P. Morgan, "Cascadable tapped delay line module", *IEEE Ultrasonics Symp.*, 1981, pp. 53–57.
212. A. J. Slobodnik and J. H. Silva, "Ultraflat frequency response delay modules at 800 MHz", *IEEE Trans.* **SU-27**, 379–383 (1980).
213. J. H. Silva, A. J. Slobodnik and W. J. Kearns, "A 2.2 GHz SAW delay line fabricated by direct optical projection", *IEEE Trans.* **SU-26**, 312–313 (1979).
214. J. Temmyo, Y. Sakakibara, K. Komatsu, M. Oda and S. Yoshikawa, "Performance of 0.4 μm linewidth SAW interdigital transducers fabricated by scanning electron beam lithography", *IEEE Trans.* **SU-27**, 383–385 (1980).
215. H. Engan, "High frequency operation of surface-acoustic-wave multielectrode transducers", *Electronics Lett.* **10**, 395–396 (1974).
216. R. Almar, B. Lewis and E. G. S. Paige, "SAW interdigital transducers with harmonic responses insensitive to variations in finger widths", *IEEE Trans.* **SU-30**, 51–54 (1983).
217. C. S. Hartmann, W. S. Jones and H. Vollers, "Wide band unidirectional surface wave transducers", *IEEE Trans.* **SU-19**, 378–381 (1972).
218. R. C. Rosenfeld, C. S. Hartmann and R. B. Brown, "Low loss unidirectional acoustic surface wave filters", *Ann. Symp. Frequency Control*, 1974, pp. 299–303.
219. R. C. Peach and C. Dix, "A low loss medium bandwidth filter on lithium niobate", *IEEE Ultrasonics Symp.*, 1978, pp. 509–512.
220. D. C. Malocha, "Quadrature 3-phase unidirectional transducer", *IEEE Trans.* **SU-26**, 313–315 (1979).
221. R. S. Krimholtz, G. L. Matthaei and B. E. Hoiem, "Acoustic-surface-wave interdigital hybrid-junction transducers", *IEEE Trans.* **SU-21**, 23–32 (1974).
222. K. Yamanouchi, F. M. Nyffeler and K. Shibayama, "Low insertion loss acoustic surface wave filter using group-type unidirectional interdigital transducer", *IEEE Ultrasonics Symp.*, 1975, pp. 317–321.
223. D. C. Malocha and B. J. Hunsinger, "Tuning of group type unidirectional transducers", *IEEE Trans.* **SU-26**, 243–245 (1979).
224. R. B. Brown, "Electrical matching of unidirectional surface wave devices", *IEEE MTTS Intl. Microwave Symp.*, 1975, pp. 359–361.
225. K. Yamanouchi, T. Meguro and J. K. Gautam, "Low-loss GHz range SAW filter using group-type unidirectional transducer – new GUDT and new phase shifter", *IEEE Ultrasonics Symp.*, 1982, pp. 212–217.
226. D. C. Malocha, *IEEE Ultrasonics Symp.*, 1981, pp. 83–88.
227. W. S. Ishak and W. R. Shreve, *ibid.*, pp. 7–12.
228. B. R. Potter, *ibid.*, pp. 13–16.
229. D. P. Morgan, B. Lewis and R. G. Arnold, "Charge superposition analysis of multiphase SAW transducers", *IEEE Ultrasonics Symp.*, 1980, pp. 317–321.
230. G. W. Farnell, C. K. Jen and E. L. Cambiaggio, "Multiphase interdigital transducers and the crossed-field model", *IEEE Trans.* **SU-27**, 184–195 (1980).

BANDPASS FILTERS (Chapter 8)

231. P. Hartemann and E. Dieulesaint, "Acoustic surface wave filters", *Electronics Lett.* **5**, 657–658 (1969).
232. R. H. Tancrell and M. G. Holland, "Acoustic surface wave filters", *Proc. IEEE* **59**, 393–409 (1971).
233. H. E. Kallmann, "Transversal filters", *Proc. IRE* **28**, 302–310 (1940).
234. C. R. Hewes, R. W. Brodersen and D. D. Buss, "Applications of CCD and switched capacitor filter technology", *Proc. IEEE* **67**, 1403–1415 (1979).
235. F. F. Kuo and J. F. Kaiser, *System Analysis by Digital Computer*, Wiley, 1966.
236. L. R. Rabiner, J. H. McClellan and T. W. Parks, "FIR digital filter design techniques using weighted Chebyshev approximation", *Proc. IEEE* **63**, 595–610 (1975).
237. L. R. Rabiner and B. Gold, *Theory and Application of Digital Signal Processing*, Prentice-Hall, 1975.
238. R. F. Mitchell, "Basics of SAW filter design: a review", *Wave Electronics* **2**, 111–132 (1976).
239. R. H. Tancrell, "Analytic design of surface wave bandpass filters", *IEEE Trans.* **SU-21**, 12–22 (1974).

240. R. H. Tancrell, "Principles of surface wave filter design", in Matthews [7], pp. 109–164.
241. A. J. Slobodnik, T. L. Szabo and K. R. Laker, "Miniature surface-acoustic-wave filters", *Proc. IEEE* **67**, 129–146 (1979).
242. F. J. Harris, "On the use of windows for harmonic analysis with the Discrete Fourier Transform", *Proc. IEEE* **66**, 51–83 (1978).
243. H. D. Helms, "Nonrecursive digital filters: design methods for achieving specifications on frequency response", *IEEE Trans.* **AU-16**, 336–342 (1968).
244. K. W. Moulding and D. W. Parker, "A technique for SAW bandpass filter design", *IEEE Ultrasonics Symp.*, 1974, pp. 168–171.
245. A. J. De Vries, "Surface wave bandpass filters", in Matthews [7], pp. 263–305.
246. J. H. McClellan and T. W. Parks, "A uniform approach to the design of optimum FIR linear-phase digital filters", *IEEE Trans.* **CT-20**, 697–701 (1973).
247. J. H. McClellan, T. W. Parks and L. R. Rabiner, "A computer program for designing optimum FIR linear phase digital filters", *IEEE Trans* **AU-21**, 506–526 (1973).
248. H. Skeie, "Analysis and synthesis of surface acoustic wave filters", University of Trondheim, ELAB Report STF44 A79106, 1979.
249. P. M. Jordan and B. Lewis, "A tolerance-related optimised synthesis scheme for the design of SAW bandpass filters with arbitrary amplitude and phase characteristics", *IEEE Ultrasonics Symp.*, 1978, pp. 715–719.
250. T. Kodama, K. Sato and Y. Uemura, "SAW vestigal sideband filter for TV broadcasting transmitter", *IEEE Trans.* **SU-28**, 151–155 (1981).
251. H. W. Bode, *Network Analysis and Feedback Amplifier Design*, Van Nostrand, 1945. Also, E. A. Guillemin, *Synthesis of Passive Networks*, Wiley, 1957, p. 256.
252. A. Papoulis, *The Fourier Integral and its Applications*, McGraw-Hill, 1962, p. 206.
253. O. Herrmann and H. W. Schuessler, "Design of nonrecursive digital filters with minimum phase", *Electronics Lett.* **6**, 329–330 (1970).
254. M. Feldmann and J. Henaff, "Design of SAW filter with minimum phase response", *IEEE Ultrasonics Symp.*, 1978, pp. 720–723.
255. C. S. Hartmann, "Weighting interdigital surface wave transducers by selective withdrawal of electrodes", *IEEE Ultrasonics Symp.*, 1973, pp. 423–426.
256. B. Lewis, P. M. Jordan, R. F. Milsom and D. P. Morgan, "Charge and field superposition methods for analysis of generalised SAW interdigital transducers", *IEEE Ultrasonics Symp.*, 1978, pp. 709–714.
257. R. S. Wagers, "Phase error compensation in finger withdrawal transducers", *IEEE Ultrasonics Symp.*, 1974, pp. 418–421.
258. K. R. Laker, E. Cohen, T. L. Szabo and J. A. Pustaver, "Computer-aided design of withdrawal-weighted SAW bandpass filters", *IEEE Trans.* **CAS-25**, 241–251 (1978).
259. T. L. Szabo and A. J. Slobodnik, "Diffraction compensation in periodic apodised acoustic surface wave filters", *IEEE Trans.* **SU-21**, 114–119 (1974).
260. E. B. Savage and G. L. Matthaei, "A study of some methods for compensation for diffraction in SAW IDT filters", *IEEE Trans.* **SU-28**, 439–448 (1981).
261. T. Kodama, "Broad-band compensation for diffraction in surface acoustic wave filters", *IEEE Trans.* **SU-30**, 127–136 (1983).
262. R. M. Hays and C. S. Hartmann, "Surface-acoustic-wave devices for communications", *Proc. IEEE* **64**, 652–671 (1976).
263. O. Yamazaki, T. Mitsuyu and K. Wasa, "ZnO thin-film SAW devices", *IEEE Trans.* **SU-27**, 369–379 (1980).
264. D. L. Ash, "An improved high-performance TV receiver", *Proc. IEEE* **70**, 1345–1357 (1982).
265. R. J. Murray and S. Neylon, "The design, fabrication and performance limitations of narrow-band fast cut-off surface wave filters", *IEEE Ultrasonics Symp.*, 1978, pp. 482–485.
266. W. S. Drummond and S. A. Roth, "Application of high performance SAW transversal filters in a precision measurement instrument", *IEEE Ultrasonics Symp.*, 1982, pp. 494–499.
267. A. J. Slobodnik, T. E. Fenstermacher, W. J. Kearns, G. A. Roberts, J. H. Silva and J. P. Noonan, "SAW Butterworth contiguous filters at UHF", *IEEE Trans.* **SU-26**, 246–253 (1979).

268. J. M. Deacon and J. Heighway, "SAW filters: some case histories", *Proc. IEE* **127F**, 107–117 (1980).
269. D. C. Malocha and B. J. Hunsinger, "Capacitive tap weighted SAW transducers", *IEEE Trans.* **SU-24**, 293–301 (1977).
270. H. Engan, "Interdigital transducer techniques for specialised frequency filters", *Wave Electronics* **2**, 133–154 (1976).
271. R. C. Williamson, "Properties and applications of reflective-array devices", *Proc. IEEE* **64**, 702–710 (1976).
272. F. Sandy and T. E. Parker, "Surface acoustic wave ring filter", *IEEE Ultrasonics Symp.*, 1976, pp. 391–396.
273. L. P. Solie, "The development of high performance RDA devices", *IEEE Ultrasonics Symp.*, 1979, pp. 682–686.
274. L. A. Coldren and R. L. Rosenberg, "Surface-acoustic-wave resonator filters", *Proc. IEEE* **67**, 147–158 (1979).
275. A. J. Slobodnik, G. A. Roberts, J. H. Silva, W. J. Kearns, J. C. Sethares and T. L. Szabo, "Switchable SAW filter banks at UHF", *IEEE Trans.* **SU-26**, 120–126 (1979).
276. L. P. Solie and M. D. Wohlers, "Use of a SAW multiplexer in FMCW radar system", *IEEE Trans.* **SU-28**, 141–145 (1981).

PULSE COMPRESSION AND INTERDIGITAL CHIRP FILTERS (Sections 9.1–9.5)

277. J. R. Klauder, A. C. Price, S. Darlington and W. J. Albersheim, "The theory and design of chirp radars", *Bell Syst. Tech. J.* **39**, 745–808 (1960).
278. M. J. Skolnik, *Introduction to Radar Systems*, McGraw-Hill, 1962.
279. C. E. Cook and M. Bernfeld, *Radar Signals*, Academic Press, 1967.
280. E. C. Farnett, T. B. Howard and G. H. Stevens, "Pulse compression radar", in M. I. Skolnik (ed.), *Radar Handbook*, McGraw-Hill, 1970, pp. 20.1–20.37.
281. J. D. Maines, "Surface wave devices for radar equipment", in Matthews [7], pp. 443–476.
282. G. L. Turin, "An introduction to matched filters", *IRE Trans.* **IT-6**, 311–329 (1960).
283. G. L. Turin, "An introduction to digital matched filters", *Proc. IEEE* **64**, 1092–1112 (1976).
284. P. Hartemann and E. Dieulesaint, "Intrinsic compensation of sidelobes in a dispersive acoustic delay line", *Electronics Lett.* **5**, 219–220 (1969).
285. R. H. Tancrell, M. B. Schulz, H. H. Barrett, L. Davies and M. G. Holland, "Dispersive delay lines using ultrasonic surface waves", *Proc. IEEE* **57**, 1211–1213 (1969).
286. R. C. Williamson and H. I. Smith, "Large time-bandwidth product surface-wave pulse compressor employing reflective gratings", *Electronics Lett.* **8**, 401–402 (1972).
287. I. N. Court, "Microwave acoustic devices for pulse compression filters", *IEEE Trans.* **MTT-17**, 968–986 (1969).
288. C. Lardat, C. Maerfeld and P. Tournois, "Theory and performance of acoustical dispersive surface wave delay lines", *Proc. IEEE* **59**, 355–368 (1971).
289. M. Abramowitz and I. A. Stegun, *Handbook of Mathematical Functions*, Dover, 1968, p. 301.
290. T. T. Taylor, "Design of line-source antennas for narrow beamwidths and low sidelobes", *IRE Trans.* **AP-3**, 16–28 (1955).
291. G. W. Judd, "Technique for realising low time sidelobe levels in small compression ratio chirp waveforms", *IEEE Ultrasonics Symp.*, 1973, pp. 478–481.
292. G. A. Armstrong and M. B. Butler, "Engineering design and evaluation of SAW pulse compression filters with low time sidelobes", *Radio and Electron. Eng.* **46**, 221–228 (1976).
293. G. A. Armstrong, "The design of SAW dispersive filters using interdigital transducers", *Wave Electronics* **2**, 155–176 (1976).
294. C. E. Cook and J. Paolillo, "A pulse compression pre-distortion function for efficient sidelobe reduction in a high power radar", *Proc. IEEE* **52**, 377–389 (1964).
295. C. O. Newton, "Non-linear chirp radar signal waveforms for surface acoustic wave pulse compression filters", *Wave Electronics* **1**, 387–401 (1976).

296. C. S. Hartmann, D. T. Bell and R. C. Rosenfeld, "Impulse model design of acoustic surface-wave filters", *IEEE Trans.* **MTT-21,** 162–175 (1973).
297. R. H. Tancrell and P. C. Meyer, "Operation of long surface wave interdigital transducers", *IEEE Trans.* **SU-19,** p. 405 (1972).
298. J. D. Maines, G. L. Moule, N. R. Ogg and G. Rich, "Improvements in surface-acoustic-wave pulse compression filters", *Electronics Lett.* **9,** 156–157 (1973).
299. B. R. Potter and C. S. Hartmann, "Surface acoustic wave slanted device technology", *IEEE Trans.* **SU-26,** 411–418 (1979).
300. M. B. N. Butler, "Radar applications of SAW dispersive filters", *Proc. IEE* **127F,** 118–124 (1980).
301. H. M. Gerard, W. R. Smith, W. R. Jones and J. B. Harrington, "Design and applications of highly dispersive acoustic surface wave filters", *IEEE Trans.* **MTT-21,** 176–186 (1973).
302. H. M. Gerard, "Surface wave interdigital electrode chirp filters", in Matthews [7], pp. 347–380.
303. K. Milne, "The combination of pulse compression with frequency scanning for three-dimensional radars", *Radio and Electronic Eng.* **28,** 89–106 (1964).
304. P. Bradsell, "The Plessey AR3D frequency scanning three-dimensional radar", IEE Conf. Publ. 180, "Case studies in advanced signal processing", 1979, pp. 151–154.
305. I. W. Hammer, "Frequency-scanned arrays", in M. I. Skolnik (ed.), *Radar Handbook*, McGraw-Hill, 1970, pp. 13.1–13.27.
306. D. P. Morgan and J. M. Deacon, "Dispersive propagation in surface-acoustic-wave pulse compression filters", *Electronics Lett.* **13,** 741–742 (1977).
307. J. D. Maines, G. L. Moule and N. R. Ogg, "Correction of diffraction errors in acoustic-surface-wave pulse compression filters", *Electronics Lett.* **8,** 431–433 (1972).
308. J. Heighway, D. W. Tarrant and C. H. Oxley, "Simple approach to the design of interface networks for acoustic-surface-wave filters", *Electronics Lett.* **8,** 642–643 (1972).

REFLECTIVE ARRAY COMPRESSORS AND CHIRP SIGNAL PROCESSING (Sections 9.6, 9.7)

309. R. C. Williamson and H. I. Smith, "The use of surface elastic wave reflection gratings in large time-bandwidth pulse compression filters", *IEEE Trans.* **MTT-21,** 195–205 (1973).
310. R. C. Williamson, "Reflection grating filters", in Matthews [7], pp. 381–442.
311. J. Melngailis, R. C. Williamson, J. Holtham and R. C. M. Li, "Design of reflective-array surface wave devices", *Wave Electronics* **2,** 177–198 (1976).
312. E. K. Sittig and G. A. Coquin, "Filters and dispersive delay lines using repetitively mismatched ultrasonic transmission lines", *IEEE Trans.* **SU-15,** 111–119 (1968).
313. T. A. Martin, "The IMCON pulse compression filter and its applications", *IEEE Trans.* **MTT-21,** 186–194 (1973).
314. O. W. Otto, "Multiple reflections in acoustic surface wave reflective arrays", *IEEE Trans.* **SU-22,** 251–257 (1975).
315. P. D. Bloch, E. G. S. Paige and L. Solymar, "Reflection of surface acoustic waves from arrays of strips", *IEEE Ultrasonics Symp.*, 1978, pp. 639–642.
316. P. V. Wright and H. A. Haus, "A closed-form analysis of reflective-array gratings", *IEEE Ultrasonics Symp.*, 1980, pp. 282–287.
317. H. M. Gerard, O. W. Otto and R. D. Weglein, "Development of a broadband reflective array 10,000:1 pulse compression filter", *IEEE Ultrasonics Symp.*, 1974, pp. 197–201.
318. O. W. Otto and H. M. Gerard, "Non-synchronous scattering loss in surface-acoustic-wave reflective array compression filters", *J. Appl. Phys.* **49,** 3337–3340 (1978).
319. T. A. Martin, "Low-sidelobe IMCON pulse compression", *IEEE Ultrasonics Symp.*, 1976, pp. 411–414.
320. P. C. Meyer, R. H. Tancrell and J. H. Matsinger, "Long time delay reflective surface wave dispersive delay line", *IEEE Ultrasonics Symp.*, 1973, pp. 498–499.
321. V. S. Dolat and R. C. Williamson, "BGO reflective array compressor (RAC) with 125 μsec dispersion", *IEEE Ultrasonics Symp.*, 1975, pp. 390–394.

REFERENCES

322. H. M. Gerard and G. W. Judd, "500 MHz bandwidth RAC filter with constant groove depth", *IEEE Ultrasonics Symp.*, 1978, pp. 734–737.
323. R. E. Chapman, R. K. Chapman, D. P. Morgan and E. G. S. Paige, "Weighted in-line reflective array devices", *IEEE Ultrasonics Symp.*, 1979, pp. 696–700.
324. L. P. Solie, "Reflective dot array devices", *IEEE Ultrasonics Symp.*, 1977, pp. 579–584.
325. R. C. Woods, "Dispersive delay lines using 180° reflecting metal dot arrays", *IEEE Ultrasonics Symp.*, 1982, pp. 88–91.
326. G. W. Judd and J. L. Thoss, "Use of apodised metal gratings in fabricating low cost quartz RAC filters", *IEEE Ultrasonics Symp.*, 1980, pp. 343–347.
327. S. Darlington, "Demodulation of wide band low-power FM signals", *Bell Syst. Tech. J.* **43**, 339–374 (1964).
328. M. A. Jack and E. G. S. Paige, "Fourier transform processors based on surface acoustic wave chirp filters", *Wave Electronics* **3**, 229–247 (1978).
329. M. A. Jack, P. M. Grant and J. H. Collins, "The theory, design and applications of surface acoustic wave Fourier-transform processors", *Proc. IEEE* **68**, 450–468 (1980).
330. J. B. G. Roberts, G. L. Moule and G. Parry, "Design and application of real time spectrum analyser systems", *Proc. IEE* **127F**, 76–91 (1980).
331. L. Mertz, *Transformations in Optics*, Wiley, 1965, p. 94.
332. L. R. Rabiner, R. W. Schafer and C. M. Rader, "The chirp z transform algorithm and its application", *Bell. Syst. Tech. J.* **48**, 1249–1292 (1969).
333. R. M. Hays, W. R. Shreve, D. T. Bell, L. T. Claiborne and C. S. Hartmann, "Surface wave transform adaptable processor system", *IEEE Ultrasonics Symp.*, 1975, pp. 363–370.
334. C. Atzeni, G. Manes and L. Masotti, "Programmable signal processing by analogue chirp-transformation using SAW devices", *IEEE Ultrasonics Symp.*, 1975, pp. 371–376.
335. H. M. Gerard, P. S. Yao and O. W. Otto, "Performance of a programmable radar pulse compression filter based on a chirp transformation with RAC filters", *IEEE Ultrasonics Symp.*, 1977, pp. 947–951.
336. V. S. Dolat, M. B. Schulz, L. A. Veilleux, G. R. McCully and R. C. Williamson, "High-performance hybrid SAW chirp Fourier transform system", *IEEE Ultrasonics Symp.*, 1978, pp. 527–532.
337. D. R. Arsenault and V. S. Dolat, "Self-aligning bilateral chirp-transform system", *IEEE Ultrasonics Symp.*, 1980, pp. 220–225.
338. G. L. Moule, R. A. Bale and T. I. Browning, "A 1 GHz bandwidth SAW compressive receiver", *IEEE Ultrasonics Symp.*, 1980, pp. 216–219.
339. R. C. Williamson, V. S. Dolat, R. R. Rhodes and D. M. Boroson, "A satellite-borne SAW chirp-transform system for up link demodulation of FSK communication signals", *IEEE Ultrasonics Symp.*, 1979, pp. 741–747.
340. D. R. Arsenault and V. S. Dolat, "Compact multiple-channel SAW sliding-window spectrum analyser", *IEEE Ultrasonics Symp.*, 1981, pp. 220–225.
341. J. B. G. Roberts, R. Eames, D. V. McCaughan and R. F. Simons, "A processor for pulse-doppler radar", *IEEE J. Solid State Circuits* **SC-11**, 100–104 (1976).
342. H. Gautier and P. Tournois, "Very fast signal processors as a result of the coupling of surface acoustic wave and digital technologies", *IEEE Trans.* **SU-28**, 126–131 (1981).
343. J. D. Maines, G. L. Moule, C. O. Newton and E. G. S. Paige, "A novel SAW variable-frequency filter", *IEEE Ultrasonics Symp.*, 1975, pp. 355–358.
344. V. S. Dolat and R. C. Williamson, "A continually variable delay line system", *IEEE Ultrasonics Symp.*, 1976, pp. 419–423.
345. J. Melngailis and R. C. Williamson, "Filter with bandwidth continuously variable from 5 to 100 MHz", *IEEE Ultrasonics Symp.*, 1977, pp. 965–968.
346. G. L. Moule, C. O. Newton and E. G. S. Paige, "Performance of a surface acoustic wave variable slope chirp filter", *IEEE Ultrasonics Symp.*, 1977, pp. 611–614.
347. W. J. Caputi, "Stretch–a time-transformation technique", *IEEE Trans.* **AES-7**, 269–278 (1971).
348. A. Papoulis, *Systems and Transforms with Applications in Optics*, McGraw-Hill, 1968.

SPREAD SPECTRUM AND LINEAR SPREAD-SPECTRUM DEVICES
(Sections 10.1, 10.2)

349. R. C. Dixon, *Spread Spectrum Systems*, Wiley, 1976.
350. J. K. Holmes, *Coherent Spread Spectrum Systems*, Wiley, 1982.
351. C. E. Cook, F. W. Ellersick, L. B. Milstein and D. L. Schilling (eds.), Special issue on "Spread-spectrum communications", *IEEE Trans.* **COM-30,** 817–1069 (May, 1982).
352. B. J. Hunsinger, "Spread spectrum processors", *Ultrasonics* **11,** 254–262 (1973).
353. M. G. Unkauf, "Surface wave devices in spread spectrum systems", in Matthews [7], pp. 477–509.
354. R. D. Haggarty, E. A. Palo and W. M. Bridge, "SAW processors versus competing technologies for spread spectrum data link applications", *IEEE Ultrasonics Symp.*, 1978, pp. 567–572.
355. W. S. Jones, C. S. Hartmann and T. D. Sturdivant, "Second order effects in surface wave devices", *IEEE Trans.* **SU-19,** 368–377 (1972).
356. D. T. Bell, J. D. Holmes and R. V. Ridings, "Application of acoustic surface wave technology to spread spectrum communications", *IEEE Trans.* **MTT-21,** 263–271 (1973).
357. D. T. Bell and L. T. Claiborne, "Phase code generators and correlators", in Matthews [7], pp. 307–346.
358. S. T. Costanza, P. J. Hagon and L. A. MacNevin, "Analog matched filter using tapped acoustic surface wave delay line", *IEEE Trans.* **MTT-17,** 1042–1043 (1969).
359. P. A. DeVito, P. H. Carr, W. J. Kearns and J. H. Silva, "Encoding and decoding with elastic surface waves at 10 megabits per second", *Proc. IEEE* **59,** 1523–1525 (1971).
360. P. H. Carr, P. A. DeVito and T. L. Szabo, "The effect of temperature and doppler shift on the performance of elastic surface wave encoders and decoders", *IEEE Trans.* **SU-19,** 357–367 (1972).
361. G. W. Judd, W. R. Jones and T. W. Bristol, "Improved tapping transducer geometry for surface-wave phase coded delay lines", *IEEE Ultrasonics Symp.*, 1972, pp. 373–376.
362. C. F. Vasile and R. LaRosa, "1000-bit surface-wave matched filter", *Electronics Lett.* **8,** 479–480 (1972) and **9,** 495–496 (1973).
363. B. J. Hunsinger, A. G. Burke, H. J. Bush, W. L. Richards and J. N. Entzminger, "Coherent FH/PN synthesiser using surface wave techniques", *IEEE Trans.* **SU-21,** 289–293 (1974).
364. P. H. Carr, R. D. Colvin and J. H. Silva, "Encoding and decoding at 1 GHz with SAW tapped delay lines", *IEEE Ultrasonics Symp.*, 1979, pp. 757–760.
365. E. J. Staples and L. T. Claiborne, "A review of device technology for programmable surface-wave filters", *IEEE Trans.* **MTT-21,** 279–287 (1973).
366. R. D. Lambert, P. M. Grant, D. P. Morgan and J. H. Collins, "Programmable surface acoustic wave devices using hybrid microelectronic components", *Radio and Electronic Eng.* **44,** 343–351 (1974).
367. F. S. Hickernell, D. E. Olson and M. D. Adamo, "Monolithic surface wave transversal filter", *IEEE Ultrasonics Symp.*, 1977, pp. 615–618.
368. F. S. Hickernell, M. D. Adamo, R. V. DeLong, J. G. Hinsdale and H. J. Bush, "SAW programmable matched filter signal processor", *IEEE Ultrasonics Symp.*, 1980, pp. 104–108.
369. P. J. Hagon, F. B. Micheletti and R. N. Seymour, "Integrated programmable analog matched filters for spread spectrum applications", *IEEE Ultrasonics Symp.*, 1973, pp. 333–335.
370. T. W. Grudkowski, G. K. Montress, M. Gilden and J. F. Black, "Integrated circuit compatible surface acoustic wave devices on gallium arsenide", *IEEE Trans.* **MTT-30,** 1348–1355 (1981).
371. F. Amoroso and J. A. Kivett, "Simplified MSK signaling technique", *IEEE Trans.* **COM-25,** 433–441 (1977).
372. W. R. Smith, "SAW filters for CPSM spread spectrum communication", *IEEE Ultrasonics Symp.*, 1977, pp. 524–528.
373. M. F. Lewis, P. J. Lowe and W. G. Picken, "MSK SAW filter to complement to-day's SAW convolver", *IEEE Ultrasonics Symp.*, 1982, pp. 256–261.
374. D. C. Malocha, J. H. Goll and M. A. Heard, "Design of a compensated SAW filter used in a wide spread MSK waveform generator", *IEEE Ultrasonics Symp.*, 1979, pp. 518–521.
375. S. A. Gronemeyer and A. L. McBride, "MSK and offset QPSK modulation", *IEEE Trans.* **COM-24,** 809–819 (1976).

376. B. J. Darby and J. M. Hannah, "Programmable frequency-hop synthesisers based on chirp mixing", *IEEE Trans.* **SU-28**, 178–185 (1981).
377. P. M. Grant, D. P. Morgan and J. H. Collins, "Generation and correlation of digitally controlled coherent frequency-hopped waveforms using surface acoustic wave devices", *Proc. IEEE* **64**, 826–828 (1976).
378. A. J. Budreau, A. J. Slobodnik and P. H. Carr, "A review of SAW-based direct frequency synthesisers", *IEEE Trans.* **MTT-30**, 686–693 (1982).

ACOUSTIC CONVOLVERS (Section 10.3)

379. G. S. Kino, S. Ludvik, H. J. Shaw, W. R. Shreve, J. M. White and D. K. Winslow, "Signal processing by parametric interactions in delay-line devices", *IEEE Trans.* **MTT-21**, 244–263 (1973).
380. D. P. Morgan, "Signal processing using non-linear convolvers", *Ultrasonics* **12**, 74–83 (1974).
381. G. S. Kino, "Acoustoelectric interactions in acoustic-surface-wave devices", *Proc. IEEE* **64**, 724–748 (1976).
382. E. A. Ash, "Fundamentals of signal processing devices", in Oliner [8], pp. 97–185.
383. L. O. Svaasand, "Interaction between elastic surface waves in piezoelectric materials", *Appl. Phys. Lett.* **15**, 300–302 (1969).
384. M. Luukkala and G. S. Kino, "Convolution and time inversion using parametric interactions of acoustic surface waves", *Appl. Phys. Lett.* **18**, 393–394 (1971).
385. D. P. Morgan, J. H. Collins and J. G. Sutherland, "Asynchronous operation of an analogue convolver", *Proc. IEEE* **60**, 1556–1557 (1972). Also *IEEE Ultrasonics Symp.*, 1972, pp. 296–299.
386. C. F. Quate and R. B. Thompson, "Convolution and correlation in real time with non-linear acoustics", *Appl. Phys. Lett.* **16**, 494–496 (1970).
387. M. Luukkala and J. Surakka, "Acoustic convolution and correlation and the associated nonlinearity parameters in lithium niobate", *J. Appl. Phys.* **43**, 2510–2518 (1972).
388. E. A. Kraut, T. C. Lim and B. R. Tittmann, "Application of nonlinear interactions in ferroelectric ceramics to microwave signal processing", *IEEE Trans.* **SU-19**, 247–255 (1972).
389. V. V. Shtykov, I. M. Mason and M. D. Motz, "On the anisotropy of guided acoustic surface wave degenerate convolution", *IEEE Trans.* **SU-22**, 131–136 (1975).
390. A. K. Ganguly and K. L. Davis, "Theory of degenerate SAW elastic convolver", *IEEE Ultrasonics Symp.*, 1979, pp. 715–720. Also, *J. Appl. Phys.* **51**, 920–926 (1980).
391. T. C. Lim, E. A. Kraut and R. B. Thompson, *Appl. Phys. Lett.* **20**, 127–129 (1972). Also, J. Chambers, I. M. Mason and C. W. Turner, *Electronics Lett.* **8**, 314–316 (1972). Also, M. D. Motz and I. M. Mason, *Electronics Lett.* **10**, 83–84 (1974).
392. M. D. Motz, J. Chambers and I. M. Mason, "Suppression of spurious signals in a degenerate SAW convolver", *IEEE Ultrasonics Symp.*, 1973, pp. 152–154.
393. E. L. Adler and J. H. Cafarella, "The effect of acoustic dispersion and attenuation on SAW convolver performance", *IEEE Ultrasonics Symp.*, 1980, pp. 1–4.
394. D. P. Morgan, "Effect of dispersion in surface-acoustic-wave convolvers", *IEEE Trans.* **SU-22**, 274–277 (1975).
395. D. P. Morgan and J. M. Hannah, "Electromagnetic effects in surface-acoustic-wave convolvers", *Wave Electronics* **1**, 129–138 (1975). Also, *IEEE Ultrasonics Symp.*, 1974, pp. 333–336.
396. E. L. Adler, "Electromagnetic long-line effects in surface wave convolvers", *IEEE Ultrasonics Symp.*, 1980, pp. 82–87.
397. R. V. Schmidt and L. A. Coldren, "Thin film acoustic surface waveguides on anisotropic media", *IEEE Trans.* **SU-22**, 115–122 (1975).
398. H. Engan, "A phase sensitive laser probe for pulsed SAW measurements", *IEEE Trans.* **SU-29**, 281–283 (1982).
399. H. Engan, K. A. Ingebrigtsen and A. Rønnekleiv, "Design of SAW convolver for processing MSK modulated waveforms", *Electronics Lett.* **16**, 908–909 (1980).
400. N. K. Batani and E. L. Adler, *IEEE Ultrasonics Symp.*, 1974, pp. 114–116. Also, R. A. Becker and D. H. Hurlburt, "Wide band lithium niobate elastic convolver with parabolic horns", *IEEE Ultrasonics Symp.*, 1979, pp. 729–731.

401. J. B. Green and G. S. Kino, "SAW convolvers using focused interdigital transducers", *IEEE Trans.* **SU-30**, 43–50 (1983).
402. D. P. Morgan, D. H. Warne, P. N. Naish and D. R. Selviah, "Monolithic SAW convolvers using chirp transducers", *IEEE Ultrasonics Symp.*, 1981, pp. 186–191. Also D. P. Morgan, D. H. Warne and D. R. Selviah, *Electronics Lett.* **18**, 80–81 (1982).
403. W. R. Smith, *IEEE Ultrasonics Symp.*, 1980, pp. 59–64.
404. P. Defranould and C. Maerfeld, *Electronics Lett.* **10**, 209–210 (1974). Also P. Defranould and C. Maerfeld, "A SAW planar piezoelectric convolver", *Proc. IEEE* **64**, 748–751 (1976).
405. J. H. Goll and R. C. Bennett, "Reactive output tuning of high BT product SAW convolvers", *IEEE Ultrasonics Symp.*, 1978, pp. 44–47.
406. I. Yao, "High performance elastic convolver with parabolic horns", *IEEE Ultrasonics Symp.*, 1980, pp. 37–42. Also, *IEEE Ultrasonics Symp.*, 1981, pp. 181–185.
407. B. J. Darby, D. J. Gunton, M. F. Lewis and C. O. Newton, "Efficient miniature SAW convolver", *Electronics Lett.* **16**, 726–728 (1980).
408. B. J. Darby, D. Gunton and M. F. Lewis, *IEEE Ultrasonics Symp.*, 1980, pp. 53–58. Also A. M. Hodge and M. F. Lewis, *IEEE Ultrasonics Symp.*, 1982, pp. 113–118.
409. H. Gautier and C. Maerfeld, "Wide band elastic convolvers", *IEEE Ultrasonics Symp.*, 1980, pp. 30–36.
410. J. H. Cafarella, *Proc. SPIE* **209** (Optical Signal Processing for C3I, 1979), pp. 53–56. Also, D. P. Morgan, D. R. Selviah, D. H. Warne and J. J. Purcell, *"Radar 82" conf.*, IEE Conf. Publ. No. 216, pp. 321–325 (1982).
411. D. P. Morgan, "General analysis of bilinear SAW convolvers", *Electronics Lett.* **17**, 265–267 (1981).
412. D. P. Morgan, "Analysis of amplitude and phase errors in bilinear convolvers", *IEEE Ultrasonics Symp.*, 1981, pp. 196–201.
413. D. R. Selviah, D. H. Warne and D. P. Morgan, "Spatial uniformity measurement of SAW convolvers", *Electronics Lett.* **18**, 837–839 (1982).
414. D. P. Morgan, D. R. Selviah and D. H. Warne, "Spatial uniformity of SAW convolvers", *IEEE Ultrasonics Symp.*, 1982, pp. 143–148.

OTHER NON-LINEAR DEVICES (Section 10.4)

415. T. M. Reeder and M. Gilden, *Appl. Phys. Lett.* **22**, 8–10 (1973). Also T. M. Reeder, *Electronics Lett.* **9**, 254–256 (1973).
416. T. M. Reeder and T. W. Grudkowski, "Large bandwidth diode convolver using exceptionally uniform silicon-on-sapphire diode array", *Electronics Lett.* **11**, 530–532 (1975).
417. D. P. Morgan, "Signal processing using the SAW diode convolver", *IEEE Ultrasonics Symp.*, 1973, pp. 138–141.
418. T. W. Grudkowski and T. M. Reeder, *IEEE Ultrasonics Symp.*, 1977, pp. 710–714.
419. G. K. Montress and T. M. Reeder, "A high performance SAW/hybrid component Fourier transform convolver", *IEEE Ultrasonics Symp.*, 1978, pp. 538–542.
420. C. W. Lee and R. L. Gunshor, *Appl. Phys. Lett.* **20**, 288–290 (1972).
421. H. Gautier and G. S. Kino, "A detailed theory of the acoustic wave semiconductor convolver", *IEEE Trans.* **SU-24**, 23–33 (1977).
422. S. A. Reible, "Acoustoelectric convolver technology for spread-spectrum communications", *IEEE Trans.* **SU-28**, 185–195 (1981).
423. J. H. Cafarella, W. M. Brown, E. Stern and J. A. Alusow, "Acoustoelectric convolvers for programmable matched filtering in spread-spectrum systems", *Proc. IEEE* **64**, 756–759 (1976).
424. G. S. Kino, "Zinc oxide on silicon acoustoelectric devices", *IEEE Ultrasonics Symp.*, 1979, pp. 900–910.
425. J. B. Green and B. T. Khuri-Yakub, "A 100 μm beam width ZnO on Si convolver", *IEEE Ultrasonics Symp.*, 1979, pp. 911–914.
426. E. Stern and R. C. Williamson, "New adaptive signal-processing concept", *Electronics Lett.* **10**, 58–59 (1974).

427. A. Bers and J. H. Cafarella, "Surface state memory in surface acoustoelectric correlator", *Appl. Phys. Lett.* **25**, 133–135 (1974).
428. K. A. Ingebrigtsen, "The Schottky diode acoustoelectric memory and correlator — a novel programmable signal processor", *Proc. IEEE* **64**, 764–769 (1976).
429. C. Maerfeld, P. Defranould and P. Tournois, "Acoustic storage and processing devices using p–n diodes", *Appl. Phys. Lett.* **27**, 577–578 (1975).
430. P. Defranould, H. Gautier, C. Maerfeld and P. Tournois, "P–N diode memory correlator", *IEEE Ultrasonics Symp.*, 1976, pp. 336–347.
431. H. C. Tuan, J. E. Bowers and G. S. Kino, "Theoretical and experimental results for monolithic SAW memory correlators", *IEEE Trans.* **SU-27**, 360–369 (1980).
432. R. L. Thornton and G. S. Kino, "Monolithic ZnO on Si Schottky diode storage correlator", *IEEE Ultrasonics Symp.*, 1980, pp. 124–128.
433. R. A. Becker, R. W. Ralston and P. V. Wright, "Wide band monolithic acoustoelectric memory correlators", *IEEE Trans.* **SU-29**, 289–298 (1982).
434. D. L. Smythe, R. W. Ralston and E. Stern, "A CCD-programmable SAW matched filter", *IEEE Ultrasonics Symp.*, 1979, pp. 767–770.
435. A. G. Bert, B. Epsztein and G. Kantorowicz, "Signal processing by electron-beam interaction with piezoelectric surface waves", *IEEE Trans.* **MTT-21**, 255–263 (1973).
436. N. J. Berg, B. J. Udelson and J. N. Lee, "Long duration acousto-optic memory correlator", *IEEE Ultrasonics Symp.*, 1977, pp. 500–504.
437. B. J. Darby and J. D. Maines, "The tapped delay line active correlator: a neglected SAW device", *IEEE Ultrasonics Symp.*, 1975, pp. 193–196.
438. O. Menager and B. Desormiere, "Surface acoustic wave tapped correlator using time integration", *Appl. Phys. Lett.* **27**, 1–2 (1975).
439. K. A. Ingebrigtsen and E. Stern, "Coherent integration and correlation in a modified acoustoelectric memory correlator", *Appl. Phys. Lett.* **27**, 170–172 (1975).
440. S. A. Reible and I. Yao, "An acoustoelectric burst-waveform processor", *IEEE Ultrasonics Symp.*, 1980, pp. 133–138.
441. R. W. Ralston and E. Stern, "Four-wave interactions in acoustoelectric integrating correlators", *IEEE Ultrasonics Symp.*, 1979, pp. 761–766.
442. D. L. Smythe and R. W. Ralston, "An improved SAW time-integrating correlator with CCD readout", *IEEE Ultrasonics Symp.*, 1980, pp. 14–17.
443. H. C. Tuan, P. M. Grant and G. S. Kino, "Theory and application of zinc-oxide-on-silicon monolithic storage correlators", *IEEE Ultrasonics Symp.*, 1978, pp. 38–43.
444. J. M. Hannah, D. P. Morgan, J. H. Nash and W. M. Morris, "Code synchronisation system using cascaded surface acoustic wave convolvers", *IEEE Ultrasonics Symp.*, 1978, pp. 557–560.
445. D. P. Morgan and J. M. Hannah, "Surface wave recirculation loops for signal processing", *IEEE Trans.* **SU-25**, 30–42 (1978).

OSCILLATORS (Section 10.5)

446. T. Lukaszek and A. Ballato, "What SAW can learn from BAW: implications for future frequency control, selection and signal processing", *IEEE Ultrasonics Symp.*, 1980, pp. 173–183.
447. J. D. Maines, E. G. S. Paige, A. F. Saunders and A. S. Young, "Simple technique for the accurate determination of delay-time variations in acoustic surface wave structures", *Electronics Lett.* **5**, 678–680 (1969).
448. M. F. Lewis. "The surface acoustic wave oscillator – a natural and timely development of the quartz crystal oscillator", *28th Annual Frequency Control Symp.*, 1974, pp. 304–314.
449. E. A. Ash, "Surface wave grating reflectors and resonators", *IEEE Microwave Theory and Techniques Symp.*, 1970.
450. R. C. M. Li and J. Melngailis, "The influence of stored energy at step discontinuities on the behaviour of surface-wave gatings", *IEEE Trans.* **SU-22**, 189–198 (1975).
451. J. Melngailis, R. C. Williamson, J. Holtham and R. C. M. Li, "Design of reflective-array surface wave devices", *Wave Electronics* **2**, 177–198 (1976).

452. D. T. Bell and R. C. M. Li, "Surface-acoustic-wave resonators", *Proc. IEEE* **64**, 711–721 (1976).
453. W. J. Tanski, "GHz SAW resonators", *IEEE Ultrasonics Symp.*, 1979, pp. 815–823.
454. L. A. Coldren and R. L. Rosenberg, "Surface-acoustic-wave resonator filters", *Proc. IEEE* **67**, 147–158 (1979).
455. W. J. Tanski, "Multipole SAW resonator filters: elements of design and fabrication", *IEEE Ultrasonics Symp.*, 1981, pp. 100–104.
456. G. L. Matthaei, E. B. Savage and F. Barman, "Synthesis of acoustic-surface-wave resonator filters using any of various coupling mechanisms", *IEEE Trans.* **SU-25**, 72–84 (1978).
457. T. E. Parker, "$1/f$ phase noise in quartz delay lines and resonators", *IEEE Ultrasonics Symp.*, 1979, pp. 878–881.
458. L. Penavaire, D. Seguines, C. Lardat, J. J. Bonnier, J. Y. Chevalier and Y. Besson, "A 120 MHz resonator stabilised oscillator with high spectral purity", *IEEE Ultrasonics Symp.*, 1980, pp. 256–259.
459. W. R. Shreve, R. C. Bray, S. Elliot and Y. C. Chu, "Power dependence of aging in SAW resonators", *IEEE Ultrasonics Symp.*, 1981, pp. 94–99.
460. T. E. Parker and J. Callerame, "Sensitivity of SAW delay lines and resonators to vibration", *IEEE Ultrasonics Symp.*, 1981, pp. 129–134.
461. T. E. Parker, "Precision surface acoustic wave oscillators", *IEEE Ultrasonics Symp.*, 1982, pp. 268–274.
462. P. S. Cross and W. R. Shreve, "Frequency trimming of SAW resonators", *IEEE Trans.* **SU-29**, 231–234 (1982).
463. M. F. Lewis, "Temperature compensation techniques for SAW devices", *IEEE Ultrasonics Symp.*, 1979, pp. 612–622.
464. T. I. Browning and M. F. Lewis, "A novel technique for improving the temperature stability of SAW/SSBW devices", *IEEE Ultrasonics Symp.*, 1978, pp. 474–477.
465. A. J. Slobodnik, R. D. Colvin, G. A. Roberts and J. H. Silva, "A digitally compensated SAW oscillator", *IEEE Ultrasonics Symp.*, 1981, pp. 135–138.

REFERENCES FOR APPENDICES

466. A. Papoulis, *The Fourier Integral and its Applications*. McGraw-Hill, 1962.
467. R. M. Bracewell, *The Fourier Transform and its Applications*. McGraw-Hill, 1965.
468. A. Papoulis, *Systems and Transforms with Applications in Optics*. McGraw-Hill, 1968.
469. A. Papoulis, *Signal Analysis*, McGraw-Hill, 1977.
470. W. R. Smith, H. M. Gerard and W. R. Jones, "Analysis and design of dispersive interdigital surface-wave transducers", *IEEE Trans.* **MTT-20**, 458–471 (1972).
471. C. Atzeni and L. Masotti, "Linear signal processing by acoustic surface-wave transversal filters", *IEEE Trans.* **MTT-21**, 505–519 (1973).
472. B. A. Auld, "Application of microwave concepts to the theory of acoustic fields and waves in solids", *IEEE Trans.* **MTT-17**, 810–811 (1969).
473. J. A. Lewis, "The effect of driving electrode shape on the electrical properties of piezoelectric crystals", *Bell Syst. Tech. J.* **40**, 1259–1280 (1961).
474. A. Erdelyi, *Higher Transcendental Functions*, McGraw-Hill, 1953.
475. K. Bløtekjaer, K. A. Ingebrigtsen and H. Skeie, "Acoustic surface waves in piezoelectric materials with periodic metal strips on the surface", *IEEE Trans.* **ED-20**, 1139–1146 (1973).
476. K. Bløtekjaer, K. A. Ingebrigtsen and H. Skeie, "A method for analysing waves in structures consisting of metal strips on dispersive media", *IEEE Trans.* **ED-20**, 1133–1138 (1973).
477. P. R. Emtage, "Self-consistent theory of interdigital transducers", *J. Acoust. Soc. Am.* **51**, 1142–1155 (1971).
478. C. M. Chu, "Propagation of waves in helical waveguides", *J. Appl. Phys.* **29**, 88–99 (1958).
479. S. Datta and B. J. Hunsinger, "An analytical theory for the scattering of surface acoustic waves by a single electrode in a periodic array on a piezoelectric substrate", *J. Appl. Phys.* **51**, 4817–4823 (1980).
480. R. F. Milsom, M. Redwood and N. H. C. Reilly, "The interdigital transducer", in Matthews [7], pp. 55–108.

REFERENCES

481. T. Aoki and K. A. Ingebrigtsen, "Equivalent circuit parameters of interdigital transducers derived from dispersion relations for surface acoustic waves in periodic metal gratings", *IEEE Trans.* **SU-24**, 167–178 (1977).
482. W. R. Smith, "Experimental distinction between crossed-field and in-line three port circuit models for interdigital transducers", *IEEE Trans.* **MTT-22**, 960–964 (1974).
483. H. Skeie, "Electrical and mechanical loading of a piezoelectric surface supporting surface waves", *J. Acoust. Soc. Am.* **48**, 1098–1109 (1970).
484. H. Skeie, "Mechanical and electrical reflections in interdigital transducers", *IEEE Ultrasonics Symp.*, 1972, pp. 408–412.
485. H. Skeie and H. Engan, "Second-order effects in acoustic surface wave filters: design methods", *Radio and Electron. Eng.* **45**, 207–220 (1975).
486. H. Skeie, "Analysis and synthesis of surface acoustic wave filters", ELAB report STF44 A79106, University of Trondheim, 1979.
487. P. R. Emtage, "Self-consistent theory of interdigital transducers", *J. Acoust. Soc. Am.* **51**, 1142–1155 (1972).
488. P. R. Emtage, "Description of interdigital transducers", *J. Appl. Phys.* **43**, 4486–4489 (1972).
489. C. M. Panasik and B. J. Hunsinger, "Scattering matrix analysis of surface acoustic wave reflectors and transducers", *IEEE Trans.* **SU-28**, 79–91 (1981).
490. R. E. Collin, *Field Theory of Guided Waves*, McGraw-Hill, 1960.
491. E. K. Sittig and G. A. Coquin, "Filters and dispersive delay lines using repetitively mismatched ultrasonic transmission lines", *IEEE Trans.* **SU-15**, 111–119 (1968).
492. S. Ramo, J. R. Whinnery and T. Van Duzer, *Fields and Waves in Communication Electronics*, Wiley, 1965, p. 611.
493. L. Storch, "The transmission matrix of N alike cascaded networks", *AIEE Trans. Communications and Electronics* **73**, 616–618 (1955).
494. G. L. Matthaei, F. Barman and E. B. Savage, "SAW reflecting arrays", *Electronics Lett.* **12**, 556–557 (1976). Also F. Barman and G. L. Matthaei, *Wave Electronics* **4**, 251–272 (1983).
495. P. S. Cross, "Properties of reflective arrays for surface acoustic resonators", *IEEE Trans.* **SU-23**, 255–262 (1976).
496. A. J. DeVries, "Surface wave bandpass filters", in Matthews [7], pp. 263–305.
497. M. R. Daniel and P. R. Emtage, "Distortion of the central resonance in long interdigital transducers", *Appl. Phys. Lett.* **20**, 320–322 (1972).
498. B. K. Sinha and H. F. Tiersten, "Variational analysis of the reflection of surface waves by arrays of reflecting strips", *J. Appl. Phys.* **47**, 2824–2832 (1976).
499. D. A. Simons, "Reflection of Rayleigh waves by strips, grooves and periodic arrays of strips or grooves", *J. Acoust. Soc. Am.* **63**, 1292–1301 (1978).
500. S. Datta and B. J. Hunsinger, "First-order reflection coefficient of surface acoustic waves from thin-strip overlays", *J. Appl. Phys.* **50**, 5661–5665 (1979).
501. S. Datta and B. J. Hunsinger, "An analysis of energy storage effects on SAW propagation in periodic arrays", *IEEE Trans.* **SU-27**, 333–341 (1980).
502. F. G. Marshall, "Surface acoustic wave resonators constructed of aluminium on ST quartz for use in high stability feedback oscillators", *IEEE Ultrasonics Symp.*, 1975, pp. 290–292.
503. C. Dunnrowicz, F. Sandy and T. Parker, "Reflection of surface waves from periodic discontinuities", *IEEE Ultrasonics Symp.*, 1976, pp. 386–390.
504. W. S. Mortley, "Pulse compression by dispersive gratings on crystal quartz", *Marconi Rev.*, No. 159, 273–290 (1965).
505. R. V. Schmidt, *Appl. Phys. Lett.* **17**, 369–371 (1970).
506. R. V. Schmidt, "Excitation of shear elastic waves by an interdigital transducer operated at its surface wave centre frequency", *J. Appl. Phys.* **43**, 2498–2501 (1972).
507. P. D. Bloch and N. G. Doe, "An automated optical probe for the rapid visualisation of acoustic fields inside microwave acoustic devices", *J. Phys. E* **15**, 307–309 (1982).
508. M. R. Daniel and P. R. Emtage, "Acoustic radiation by interdigitated grids on lithium niobate", *J. Appl. Phys.* **43**, 4872–4875 (1972).
509. R. S. Wagers, "Plate mode coupling in acoustic surface wave devices", *IEEE Trans.* **SU-23**, 113–127 (1976).

510. K. Yashiro and N. Goto, *IEEE Trans.* **SU-25**, 146–153 (1978).
511. D. L. Lee, "Excitation and detection of surface skimming bulk waves on rotated *Y*-cut quartz", *IEEE Trans.* **SU-27**, 22–30 (1980).
512. E. Danicki, "New theory of SSBW devices", *IEEE Ultrasonics Symp.*, 1980, pp. 235–239.
513. F. Josse and D. L. Lee, "Analysis of the excitation, interaction and detection of bulk and surface acoustic waves on piezoelectric substrates", *IEEE Trans.* **SU-29**, 261–273 (1982).
514. R. F. Mitchell and E. Read, *IEEE Trans.* **SU-22**, 264–270 (1975).
515. R. F. Milsom, R. J. Murray, I. Flinn and M. Redwood, "Ultra-low bulk orientations of lithium niobate for SAW TV filters", *Electronics Lett.* **17**, 89–91 (1981).
516. R. La Rosa and C. F. Vasile, *Electronics Lett.* **8**, 478–479 (1972).
517. R. F. Humphryes and E. A. Ash, "Acoustic bulk-surface-wave transducer", *Electronics Lett.* **5**, 175–176 (1969).
518. D. -P. Chen, J. Melngailis and H. A. Haus, "Filters based on conversion of surface acoustic waves to bulk plate modes in gratings", *IEEE Ultrasonics Symp.*, 1982, pp. 67–71.
519. H. A. Haus, "Bulk scattering loss of SAW grating cascades", *IEEE Trans.* **SU-24**, 259–267 (1977).
520. F. G. Marshall and E. G. S. Paige, "Mode conversion in surface-acoustic-wave reflective arrays", *Electronics Lett.* **10**, 137–138 (1974).
521. M. N. Islam, H. A. Haus and J. Melngailis, "Bulk radiation by surface acoustic waves propagating under a grating", *IEEE Trans.* **SU-31**, 123–134 (1984).
522. T. I. Browning and M. F. Lewis, *Electronics Lett.* **13**, 128–130 (1977).
523. M. F. Lewis, "Surface skimming bulk waves, SSBW", *IEEE Ultrasonics Symp.*, 1977, pp. 744–752.
524. T. I. Browning, D. J. Gunton, M. F. Lewis and C. O. Newton, "Bandpass filters employing surface skimming bulk waves", *IEEE Ultrasonics Symp.*, 1977, pp. 752–756.
525. K. H. Yen, K. F. Lau and R. S. Kagiwada, "Recent advances in bulk acoustic wave devices", *IEEE Ultrasonics Symp.*, 1979, pp. 776–785.
526. K. F. Lau, K. H. Yen, R. S. Kagiwada and A. M. Kong, "High frequency temperature stable SBAW oscillators", *IEEE Ultrasonics Symp.*, 1980, pp. 240–244.
527. P. D. Bloch, E. G. S. Paige and M. W. Wagg, "Design considerations for surface skimming bulk wave devices: reflection, suppression and encapsulation", *IEEE Trans.* **SU-31**, 77–83 (1984).
528. M. F. Lewis, "High frequency acoustic plate mode device employing interdigital transducers", *Electronics Lett.* **17**, 819–821 (1982).
529. M. A. Goodberlet and D. L. Lee, "The excitation and detection of surface-generated bulk waves", *IEEE Trans.* **SU-31**, 67–76 (1984).

INDEX

Absorber, 7
Acoustic bandwidth, 159, 162
Acoustic convolver, 294–311
 using waveguide, 305–311
Acoustic potential, 73, 74
Acoustic waves, 1, 19–37
 in anisotropic materials, 29–37
 in isotropic materials, 19–29
 see also bulk waves, longitudinal waves, shear waves, surface waves
Active correlator, 284, 316
Active electrodes, 85
Active generation, 218, 242
Admittance:
 of transducer, 77, 89–95, 98
 of chirp transducer, 235, 236–239
Admittance matrix of device, 104
Ageing of oscillators, 322
Air loading, 144
Aliasing, 189, 340
Aluminium layer:
 effect on surface waves, 26, 147, 150, 154
 resistivity of, 167
Aluminium nitride, 155, 287
Aluminium phosphate (berlinite), 154
Amplification, 54
Amplitude-modulated waveform, 190, 191, 192, 338
 sampling of, 191, 339–341
Angular spectrum of plane waves, 132–134, 140
Anisotropic material:
 elasticity in, 15–19
 layered, 36–37
 orientation of, 30
 waves in, 29–37
 see also materials
Apodisation loss, 103

Apodised transducer, 8, 61–64
 admittance of, 98
 delta-function analysis, 61–64
 design of, 63, 192–200
 frequency response, 99–100, 387–390
 impulse response, 9, 63
 as transversal filter, 185–187
 see also interdigital transducer
Applications, 13
 for chirp filters, 9, 213, 216, 218, 268, 277–279
 in pulse-compression radar, 9, 214–219
 in spread-spectrum communications, 283, 291, 293, 294, 299
 in television, 14, 207, 208
 see also channelised receiver, compressive receiver, Fourier transform systems, frequency hopping, frequency measurement, frequency-shift keying, frequency synthesis, minimum-shift-keyed waveforms, oscillator
Array factor, 59, 85, 99
 for gap elements, 89, 100
Attenuation, in convolvers, 302
 see also propagation loss
Auld's reciprocity relation, 344

Bandpass filter, 9, 183–211
 bank of , 126, 209–211, 294
 using bulk waves, 4, 399–401
 using chirp transducers, 209, 244
 circuit effect in, 204–205
 using crystal resonator, 4
 design of, 192–200, 205–206
 equi-ripple, 198
 L–C type, 183, 207, 208
 minimum phase, 199–200
 monolithic crystal filter, 4
 using multi-phase transducers, 176

with multi-strip coupler, 124–126, 203
performance of, 206–208
using reflective array, 209
using resonator, 4, 209, 321
second-order effects in, 205–206
using surface-skimming bulk waves, 399–400
types of, 203, 208–209
variable, 277, 278, 313
using zinc oxide, 55, 208
Bandpass waveform, 188, 189, 190, 221, 337–341
Bandstop filter, 278, 401
Barker code, 299
Beam compression, 126–127, 306
Beam forming, 277
Beam steering, 43, 134–136, 139, 151
Berlinite, 154
Bilinear device, 296, 309
Bilinearity factor, 301, 302, 303
Bismuth germanium oxide, 36, 154, 267
Bleustein-Gulyaev wave, 34, 36, 46
Bragg cell:
 using bulk waves, 131
 using surface waves, 6
Broadcasting, see television
Bulk waves, 2
 in chirp filters, 242, 398–399
 devices using, 2–4, 219, 394, 399–401
 and effective permittivity, 46, 396–397
 and Green's function, 51
 and interdigital transducers, 66, 80, 153, 393–401
 mode conversion of, 397–399, 401
 and multi-strip coupler, 118, 119, 398
 in reflective array compressor, 264, 399
 surface-skimming, 399–400
 in surface-wave devices, 393–399
 see also longitudinal wave, plane wave, shear wave
Bus-bar, 57

Cadmium sulphide, 34
Capacitance of transducer, 77, 87, 93–95, 238
Cascading, 172, 317
Causality, 78, 199, 334
Ceramics, 154, 301
Channels, 97
Channelised receiver, 210
Charge-coupled device, 187, 268, 315–316, 317
Charge density, 40, 41, 44, 50
 acoustic, 69, 73
 electrostatic, 40, 41, 69–72, 76
  evaluation of, 72

elemental (for electrodes), 84–87, 358–360
elemental (for gaps), 89, 185
in quasi-static approximation, 69
Charge superposition, 71, 88
Chips, 283
Chip rate, 286
Chirp filter, 9–11, 213–279
 bulk waves in, 242, 398–399
 using bulk waves, 219, 394
 design of, 241–244
 diffraction in, 144, 256
 double-dispersive, 243–244
 for frequency synthesis, 293
 group delay of, 222, 231
 matched pair of, 219, 242, 244, 252
 optical counterpart, 279
 performance of, 244
 second-order effects in, 247–252, 255–256
 short-circuit response, 235
 for signal processing, 268–279
 single-dispersive, 241–242
 technologies for, 219, 394
 temperature effects in, 250–252
 transmission-line effects in, 256
 variable, 278
 weighting of, 224–231
 see also chirp transducer, chirp waveform, compressive receiver, compressor, expander, pulse-compression radar, reflective array compressor
Chirp rate, 220
Chirp transducer, 231–241
 admittance of, 236–239
 analysis of, 233–239
 in bandpass filter, 209, 244
 capacitance of, 235, 238
 conductance of, 235, 236–239
 in convolver, 306
 delta-function analysis, 235
 design, 239–241
 effective number of periods, 237–238
 frequency response, 234–235
 Q-factor, 238–239
 and sampling, 233, 235, 335
 scattering coefficients, 235
 second-order effects in, 247–252, 255–256
 slanted, 243, 398
 transmission-line effect in, 256
 see also chirp filter, chirp waveform
Chirp transform, 271
Chirp waveform, 215, 217–231
 active generation of, 218
 bandwidth of, 223
 doppler effect, 218, 252–255

INDEX

down-chirp, 220
linear, 218, 220, 222–229, 254
non-linear, 218, 229–231, 254
passive generation of, 218, 242
phase errors in, 247–255
reciprocal-ripple design, 228, 231
sampling of, 191, 233, 235, 239, 335–337
spectrum of, 221–223
stationary-phase analysis, 221–224
  limitations of, 223, 228, 231
up-chirp, 220
weighting of, 224–227, 229–231
weighting functions for, 226–227
see also chirp filter, instantaneous frequency, pulse-compression radar, time-sidelobes
Circuit effect, 65, 100, 162
  in bandpass filter, 204–205
Circuit factor, 102, 104
Code, **see** Barker code, pseudo-noise code
Communications, 14
  spread-spectrum, 11, 14, 281, 282–285
Complex envelope, 337
Compression ratio, 10, 216
Compressive receiver, 14, 268–277
Compressor, 216
  see also chirp filter
Conductance:
  of transducer, 77, 89–93
  of chirp transducer, 235, 236–239
Conversion coefficient, 81, 169
Conversion loss, 164, 169
  and triple transit, 169
  see also insertion loss
Convolution, 297, 298, 327, 331
  see also convolver
Convolution theorems, 327
Convolver, 281, 294–316
  acoustic, 294–311
    second-order effects in, 302–304
    using waveguide, 305–308
  with air gap, 313–315
  attenuation in, 302
  with beam compression, 306, 307
  bilinearity factor of, 301, 302, 303
  cascaded, 317
  with chirp transducer, 306
  degenerate, 296, 297
  diode, 300, 312–313
  dispersion in, 303, 305, 306
  dynamic range, 308, 309
  efficiency, 300–301, 305, 308
  fidelity, 309–311
  fold-over convolution in, 303–304, 307, 309
  frequency response, 309–310
  ideal, 302, 310, 311
  as matched filter, 298–300
  non-degenerate, 296, 298, 304
  propagation effects in, 302–303
  with repetitive reference, 300
  saturation in, 302, 306, 313
  second-order effects in, 302–304
  using semiconductor, 54, 312–316
  for signal processing, 294, 297, 299, 312, 313
  spatial response, 310–311
  spatial uniformity, 304, 307, 310–311
  with storage, 314–316
  temperature effects in, 303
  and time-reversal, 300
  transmission-line effect in, 304, 307
  using waveguide, 305–311
  using zinc oxide, 314, 315
Correlation function, 288, 334
  see also matched filter
Correlation of long waveforms, 316–318
Correlation peak, 216, 284, 289, 290
Correlator, integrating, 317
Coupled-wave analysis, 108, 146, 264
Coupler, **see** multi-strip coupler
Crossed-field model, 66, 375
Cross-overs, 173, 175
Crystalline material, **see** anisotropic material, materials
Crystal oscillator, 3, 318, 321
  see also oscillator
Crystal resonator, 3, 29
Cut, 31

Delay line, 7, 123, 124, 157–173
  using bulk waves, 2–3
  cascading, 172
  disk, 172
  dispersive, 9, 29
    see also chirp filter
  ideal, 331
  for long delay, 124, 172
  in oscillator, 318
  in recirculation loop, 317–318
  strip, 219
  tapped, 11, 172
    see also PSK filter
  variable, 277, 278
  wrap-around, 172
Delta function, 327, 335
Delta-function model, 57–64, 89, 100
  for chirp transducer, 235
Differentiation theorems, 326
Diffraction of light, 6, 130, 131
Diffraction of surface waves, 132–144

anglar-spectrum-of-plane-waves analysis, 132–134
  compensation for, 206, 256
  and design, 191, 200, 209
  in far-field region, 139
  Green's function analysis, 132, 139
  minimal-diffraction orientations, 136–137, 153, 154
  in multi-strip coupler, 118
  in parabolic approximation, 137–140, 151
  scaling, 137, 139, 151
  for two transducers, 140–144
Digital filter, 187, 193, 197
Digital processing, 219, 268, 276, 277
Diode convolver, 300, 312–313
Disc delay line, 172
Dispersion of surface waves:
  in chirp filters, 252, 264
  in convolver, 303, 305, 306
  due to film (layer), 26, 27, 36, 154, 155
  on "free" surface, 146
  in regular electrodes, 368–369
  see also stored energy
Dispersive delay line, 9, 29
  see also chirp filter
Displacement, acoustic, 15
  for Rayleigh wave, 23, 32, 33
Dolph-Chebyshev window, 196, 227
Doppler effect, 252–254
  for chirp waveforms, 254–255
  for PSK waveform, 290
Doppler shift, measurement of, 277
Dots, 209, 267
Double-electrode transducer, 65, 89
Dougall's expansion, 357
Down-chirp waveform, 220
Driving networks, 174, 179–180
  see also matching
Dummy electrodes, 97, 232
Dynamic range, of convolver, 308–309

Earthquakes, 2
Echo trap, 124
Effective permittivity, 41, 44–49
  for Bleustein-Gulyaev waves, 46
  and bulk waves, 46, 396–397
  Ingebrigtsen's approximation, 48, 52
  and launching transducer, 351–353
  for lithium niobate, 46
  for piezoelectric layer, 55
  for plane above surface, 52–53
  symmetry of, 348
Elastic waves, **see** acoustic waves
Elasticity, 15–19

Electrical loading, 36, 64, 305, 381, 386
Electrode interactions, 64–65, 66, 67, 201
  analysis of 375–392
Electrode polarity, 71, 76
Electrode resistivity, 117, 166–167
Electrostatic charge density, 40, 41, 69–72, 76
Electrostatic probe, 129, 137, 369
Elemental charge density, 84, 85, 89, 185, 358–360
Element factor, 85
  for gaps, 59, 88, 100, 185
End effects, 66, 85, 88, 232
  transverse, 100
Energy theorem, 327
Energy velocity, 135
Equation of motion, 17, 18
  for isotropic material, 19
Equi-ripple response, 198
Equivalent circuit of transducer, 77, 158–159, 160
Expander, 218, 242, 252
Expansion of material, 148
Expansion loss, for chirp, 242
Extensions, of chirp, 228, 229, 231

Fabrication, 12–13
  of RAC, 257–258, 259
Far-field region, 132, 139
Film, **see** aluminium layer, gold layer, layer, piezoelectric film, zinc oxide
Filter, **see** bandpass filter, bandstop filter, chirp filter, digital filter, linear filter, matched filter, minimum-shift-keyed waveforms, PSK filter, transversal filter
Filter bank, 126, 209–211, 294
Finite-impulse-response filter, 187, 197
Flicker noise, 322
Floquet's theorem, 108, 110, 363, 364
Focussing transducer, 306
Fold-over convolution, 303–304, 307, 309
Fourier transforms, 325–329
  sliding, 270, 275
Fourier transform systems, 268, 270–278
  applications of, 276–277
  using diode convolver, 313
Fraunhofer region, 132
Frequency, **see** instantaneous frequency
Frequency hopping, 293–294
Frequency measurement, 14, 210, 268
Frequency response:
  of convolver, 309–310
  of interdigital device, 97, 99, 101–105, 120
  of interdigital transducer, **see** interdigital transducer
  of linear filter, 330

INDEX

Frequency-scanning radar, 245, 247
Frequency-shift keying, 276
  see also minimum-shift-keyed waveforms
Frequency synthesis, 210, 293–294
Fresnel distance, 139
Fresnel integrals, 139, 141
Fresnel region, 132
Fused quartz, see quartz, fused

Gallium arsenide, 154, 288
Gating sidelobes, 228
Generalised function, 327, 328
Gold layer, 26, 27, 391
Green's function, 50–52
  for bulk waves, 51
  in diffraction analysis, 132
  electrostatic, 51
  symmetry of, 347–348
  for surface waves, 50
Grooves, 10, 260–264, 267, 320
  fabrication of, 257–258
  mode conversion by, 397–399, 401
Group delay of chirp filter, 222, 231
Group-type transducer, 175–181
Guard electrodes, 66, 85

Hamming weighting, 226, 227
High-frequency techniques, 155, 172–173, 399, 400
Hilbert transform, 78, 92, 199
Hooke's law, 16
Horn, 306, 307
Huygen's principle, 132
Hybrid junction transducer, 174

Idler wave, 300, 312
Image responses, 189, 200
IMCON, 29, 219, 257, 265
Impulse model, 57
Impulse response, 331
  of transducer, 9, 63, 189
Ingebrigtsen's approximation, 48–49
In-line RAC, 267, 399
Insertion loss, 8, 14, 102
  and triple-transit, 8, 168–170
Instantaneous frequency, 220, 222
  or linear chirp, 222
Integrating correlator, 317
Integration with semiconductors, 281, 287
Interactions, see electrode interactions
Interdigital transducer, 5, 6, 57–67, 75–105
  admittance of, 77, 89–95, 98
  analysis methods, 66–67
  and bulk waves, 66, 93, 393–401

  capacitance of, 77, 87, 93
  channels in, 97
  conductance of, 77, 91–93, 97, 390, 396
  conversion coefficient of, 81, 164, 169
  crossed-field model, 66, 375
  delta-function model, 57–64, 85, 89, 100
  and diffraction, 137, 138, 140–144, 206
  double-eletrode, 65, 89
  dummy electrodes in, 97
  electrode interactions in, 64, 67, 375–392
  electrode resistivity in, 166–167
  equivalent circuit, 77, 158–159, 160
  focussing, 306
  frequency response, 59, 96
    for apodised transducer, 9, 62, 99, 100, 387–390
  Green's function analysis, 67
  guard electrodes in, 66, 85
  harmonics, 59, 60, 91, 172–173
    strengths of, 87, 93, 95
  hybrid junction, 174
  impulse model, 57
  impulse response, 9, 63, 189
  launching surface waves, 76, 81, 96, 99 351–353
    with regular electrodes, 57–60, 62, 85
  matching, 160–162
    and loss, 162–165
  multi-electrode, 65, 233
    see also uniform transducer
  neighbour effect in, 66
  with parasitics, 165–168
  with piezoelectric layer, 54–55, 287, 314
  Q-factor, 94, 159, 162, 164
  quasi-static analysis, 67, 75–82
  radiation resistance of, 160–161, 168
  receiving surface waves, 60, 78–82, 96, 98, 99
  reciprocity for, 79, 348–351
  reflection coefficient, 64–65, 80–82, 101, 386–387
    with regular electrodes, 83–95, 99
  and sampling, 187–191
  scattering coefficients, 80–82
  second-order effects in, 64–67
  single-electrode, 65, 89
  standing wave ratio, 165
  susceptance (acoustic) 77, 78, 92
  thinned, 200–201
  transmission coefficient, 81
  as transversal filter, 185–187
  weighting of, 201–202, 209
    see also apodised transducer
  withdrawal-weighted, 66, 201–202
  voltage standing wave ratio, 165

see also apodised transducer, array factor, chirp transducer, element factor, multi-phase transducer, transducer, triple-transit signal, unidirectional transducer, uniform transducer
Isotropic material, 19
  waves in, 19–29

Kaiser window, 195–196

L–C filter, 183, 207, 208
Lamb waves, 28
Lamé constants, 19
Layer, effect on surface waves, 26, 36, 154–155
  see also aluminium layer, gold layer, piezoelectric film, zinc oxide
Layered Rayleigh wave, 26, 219
Leaky surface wave, 36
Legendre functions, 355–358
Legendre polynomials, 248, 357–358
Linear chirp, 218, 220, 222–229
  bandwidth of, 223
  doppler effect, 254
  spectrum, 223
  weighting of, 224–229
  see also chirp waveform
Linear filter, 6, 330–332
Lithium niobate, 3, 5, 152, 153
  with aluminium layer, 26, 147, 154
  beam steering in, 137
  bulk waves in, 46, 153, 395–396, 397, 401
  diffraction in, 137, 139, 140
  effective permittivity, 46
  non-linear effects in, 146–147, 296, 300, 301, 316
  propagation loss, 144
  pseudo-surface wave in, 36, 399
  in reflective array compressor, 257, 258, 260, 264, 265
  rotated orientation (128°), 153, 397
  surface waves in, 33, 34, 46, 153
  temperature coefficient, 148, 153
  velocity accuracy, 140, 150–151
  Y, Z orientation, 33, 46, 140, 152, 153
    propagation effects in, 144, 146, 148, 150
Lithium tantalate, 153–154, 155
Lithium tetraborate, 154
Lithography, 12–13
Loading, see electrical loading, electrode interactions, mass loading, mechanical loading, stored energy, topographic effect
Longitudinal waves, 1, 20, 30
  in surface-wave devices, 395
Love waves, 27, 219

Mass loading, 36, 391

Matched filter, 10, 332–334
  for chirp waveform, 216, 219, 220
    see also chirp filter
  in communication systems, 283–285
  convolver as, 298–300
  for minimum-shift-keyed waveform, 293
  for phase-shift-keyed waveform, 286–290
  in radar systems, 215–219
    see also correlation of long waveforms
Matched pair, 219, 242, 244, 252
Matching of transducers, 100, 160–162
Materials, 151–155
  see also aluminium nitride, anisotropic material, berlinite, bismuth germanium oxide, ceramics, gallium arsenide, isotropic material, lithium niobate, lithium tantalate, lithium tetraborate, quartz, zinc oxide
Matrix notation for tensors, 18
Mechanical loading, 45, 375, 390–392
Mehler-Dirichlet formula, 356
Memory, see storage
Metallisation ratio, 83, 95
Metallised surface, 31, 32, 46
Microscan receiver, see compressive receiver
Minimal-diffraction orientations, 136–137, 153, 154
Minimum-phase filter, 199–200
Minimum-shift-keyed (MSK) waveforms, 290–293, 299
Mirror, 123–124
Misalignment, 135, 150, 151
Mis-match loss, 227, 231
Mixing of surface waves, 6, 147, 296
  see also acoustic convolver
Modes, of multi-strip coupler, 107, 113, 114, 117, 126
Mode conversion, 397–399, 401
Modulation theorems, 326
Monolithic crystal filter, 4
Multi-electrode transducer, 65
  see also uniform transducer
Multi-phase transducer, 173–181
  analysis of, 176–181
  performance of, 176
Multi-strip coupler, 5, 107–127,
  beam compression, 126–127
  bulk waves in, 118, 119, 398
  diffraction in, 118
  echo trap, 124
  electrode resistance in, 117, 123, 124, 127
  in filter bank, 126, 210–211
  in interdigital device, 118–120, 203
  mirror, 123–124
  modes in, 107, 113, 114, 117, 126

narrow-band, 124–126
reflection coefficient, 118, 386
in reflective array compressor, 267
in resonator, 321
scattering matrix, 121
second-order effects in, 117–118
stop bands in, 108, 114, 118, 363, 370–372
track-changer, 124
in unidirectional transducer, 120–123

Near-field region, 132, 134, 139
Neighbour effect, 66
Noise, 331–334
Non-destructive testing, 2
Non-linear chirp, 218, 229–231
doppler effect, 254–255
see also chirp waveform
Non-linear effects, 145–147
in convolvers, 6, 281, 295–296, 301, 312, 313
Normal mode theory, 49, 55
Nyquist frequency, 188, 339

Optical image scanning, 314
Optical probes, 130–132, 140, 145
Orientation of material, 30, 36
see also misalignment
Oscillator:
using crystal resonator, 3, 318, 321, 322
using surface-skimming bulk waves, 399, 400
using surface waves, 11, 318–323

Packaging, 13
Paired echos, 248
Parallel-plate transducer, 3
Parallel plate, waves in, 28, 29, 400–401
Parasitic (stray) components, 100, 165–168
Parseval's theorem, 326
Partial waves, 22, 26, 31, 42
Passive generation, 218, 219, 242
see also expander, matched pair
Performance of surface-wave devices, 13–14
Periodic structure, 108, 363–374, 376–380
reflection coefficient, 380, 386–387
see also multi-strip coupler, reflective array model
Permittivity tensor, 17, 18, 31
see also effective permittivity
Perpendicular diffraction delay line, 219
Perturbation theory, 37, 66, 108
Phase plate, 243, 258, 259, 265
Phase shift keyed (PSK) waveform, 218, 282, 286–290
doppler effect, 290
matched filter for, 11, 14, 283, 286–290, 299

see also correlation of long waveforms
Photolithography, 12–13
Piezoelectric ceramics, 154
Piezoelectric coupling, 31, 33
across gap, 52
constant (k^2), 153
Piezoelectric film, 154–155
see also zinc oxide
Piezoelectric Rayleigh wave, 33, 46
power of, 47–48
Piezoelectric tensor, 18, 31
Piezoelectricity, 3, 5, 17
Piezoresistive effect, 287
Plane waves:
in anisotropic material, 30
in isotropic material, 1, 19–21, 25
Plate, see parallel plate
Port, of transducer, 76
Posts, 313
Power of surface wave, 47–48
Poynting vector, 135–136
Probing, 6, 129–132
Processing gain, 217
Programmable devices, see convolver, variable devices
Programmable PSK filter, 287–288
Propagation effects, 129–151
see also beam steering, diffraction, dispersion, non-linear effects, propagation loss, temperature effects, velocity accuracy
Propagation loss, 129, 144–145, 151
in metal film, 154
Pseudo-noise code, 283
Pseudo-surface wave, 35, 399
PSK filter, 11, 14, 283, 286–290
programmable, 287–288
Pulse-compression radar, 10, 214–219
see also chirp filter, chirp waveform
Pure mode direction, 33, 135
Pure waves, 30, 33
PZT, 154, 301

Q-factor:
of transducer, 94, 159, 239
of resonator, 3, 320, 322
Quadrature three-phase transducer, 174
Quartz, crystalline, 153
with aluminium layer, 26, 150, 154
bulk waves in, 393, 399, 400
leaky surface waves in, 36
non-linear effects in, 147, 296
propagation loss in, 144–145
pseudo-surface waves in, 35
Rayleigh waves in, 35

in reflective array compressor, 258, 267
SST orientation, 153
ST cut, 144, 148–149, 150, 153
temperature coefficient, 148–149, 150, 153, 258
Quartz, fused, 19, 24
with gold layer, 26, 27
plane waves in, 21
Rayleigh waves in, 24
see also silicon oxide
Quasi-longitudinal wave, 30
Quasi-shear wave, 30
Quasi-static approximation, 67–70, 75, 108, 109

Radar, 14, 211, 277
frequency-scanning, 245, 247
pulse-compression, 9, 10, 214–219
see also chirp filter, chirp waveform
radiation resistance, 160–161
Rayleigh wave:
in isotropic material, 21–24
and Lamb waves, 28
layered, 26, 36
piezoelectric, 33, 35, 36, 46
pure, 33
non-piezoelectric, 32
see also leaky surface wave, lithium niobate, pseudo-surface wave, quartz
Reciprocity, 343–353
Auld's relation, 343–344
for surface-wave transducers, 79, 348–351
Reciprocal ripple design, 228–229, 231
Recirculation loop, 317–318
Reflective array, 209
see also reflective array compressor, reflective array model, resonator
Reflective array compressor (RAC), 10, 13, 14, 256–267
analysis, 259–264
bulk waves in, 264, 399
in compressive receiver, 275, 276, 277
using dot array, 267
fabrication, 257–258, 259
in-line, 267, 399
length weighted, 267
with metal reflectors, 267
non-synchronous scattering loss, 264
performance, 265–267
phase plate in, 258
second-order effects in, 261–265
short-circuit response, 261
temperature effects in, 258–259, 267
Reflective array model, 376–380
Reflective dot array, 209, 267
Regular electrodes, 83, 84, 87, 89, 109, 375

waves in, 363–374, 375–386
Remez exchange algorithm, 197
Resistivity:
of aluminium film, 167
of electrodes, 117, 166–167
Resonator:
crystal, 3, 4, 29
surface-wave, 11, 209, 319–321, 398
Rigidity, 19

Sagittal plane, 21
Sampling, 187–189
of amplitude-modulated waveform, 191, 339–341
of chirp waveform, 191, 239–240, 336–337
non-uniform, 192, 334–337
synchronous, 233, 235, 336
Sapphire, 3, 155
Scaling theorem, 326
Scattering coefficients:
of electrode, 376–377, 385, 386, 391
of multi-strip coupler, 115, 117, 121
of transducer, 80–82
Schwartz's inequality, 333
Seismology, 2, 25
Semiconductor convolver, 312–316
Semiconductor substrate, 55
see also gallium arsenide, silicon
Sezawa modes, 27
Shape factor, 207
Shear-horizontal (SH) waves, 25, 27, 29, 34, 399
Shear waves, 2, 20, 21, 25, 33
in anisotropic materials, 30
in quartz, 35
in surface-wave devices, 395
Shifting theorems, 326
Short-circuit response:
of interdigital device, 97, 99, 102, 104, 105, 120
of reflective array compressor, 261
Sidelobes, **see** time-sidelobes
Signal processing, 6, 8
see also chirp filter, compressive receiver, convolver, diode convolver, Fourier transform systems, matched filtering, pulse-compression radar, spread-spectrum communications
Signal-to-noise ratio, 215, 217, 227, 284, 308, 318, 332–334
Silicon, 37, 287, 313–317
Silicon oxide, 155, 173, 313
see also quartz
Single-electrode transducer, 65, 89, 91, 95, 163, 164
and sampling, 191, 192

electrode interactions in, 64–65, 389–390
Slanted chirp transducer, 243, 398
Sliding Fourier transform, 275
Slowness, 133
Slowness curve, 133
Sonar, 277
Space harmonics, 108, 363
Spectrum analysis, 268–277
   see also filter bank
Split-electrode transducer, see double-electrode transducer
Spread-spectrum communications, 11, 14, 282–285
ST-cut quartz, 144, 148–149, 150, 153
Standing-wave ratio, 165
Stationary-phase approximation, 221–222
   limitations of, 223, 228–229, 231
Stationary-phase point, 221
Step function, 329
Stiffness tensor, 16, 17, 18, 31
   for isotropic material, 19
   in matrix notation, 18
Stop bands, 108, 378
   in multi-strip coupler, 108, 114, 118
   in regular electrodes, 370–372, 385, 386
Storage, 314–316
Storage convolver, 314–316
Stored energy, 264, 320
Straight-crested wave, 132
Strain, 15
Stray components, see parasitic components
Stress, 16, 17
Strip delay line, 219
Substrate, 7
   see also materials
Superposition, see charge superposition, linear filter
Surface-skimming bulk waves (SSBW) 399-400
Surface waves, 2, 21
   in anisotropic materials, 32–35
   in isotropic materials, 21–24, 25–27
   technology for, 4–6
   see also Bleustein Gulyaev waves, leaky surface waves, Love waves, pseudo-surface waves, Rayleigh waves, Sezawa modes
Susceptance of transducer (acoustic), 77, 78, 92
Symbol, in communications, 283
Synchronous sampling, 233, 235, 239, 336

Tapped delay line, 11, 172
   see also PSK filter
Tapped delay line correlator, 316
Taylor weighting, 227
Television, 3, 14, 207, 208, 400

Temperature effects, 129, 147–150, 153–154
   in bandpass filter, 206
   in chirp filter, 250–252
   in convolver, 303
   in oscillators, 322–323, 399, 400, 401
   in PSK filters, 289–290
   reduction of, 13, 155, 218–219, 322–323, 399, 401
   in reflective array compressor, 247, 258–259, 267
   in two-transducer devices, 149
Thinning, 200–201, 286, 318
Three-phase transducer, 173–174, 176, 179
Three-transducer scheme, 170–171
Time-invariant device, 330
Time scaling, 279
Time-sidelobes, 216, 247, 248, 249
   reduction of, 218, 224–231
Time reversal, 279, 300
Topographic effect, 391
Track-changer, 124
Transducer:
   for bulkwaves, 3, 399–400
   parallel-plate, 3
   reciprocity for, 344–347, 348–351
   for surface waves, 6
   see also interdigital transducer
Transform, see Fourier transform, Hilbert transform
Transmission-line effects, 256, 304, 307
Transmission matrix:
   of electrode, 377
   of periodic structure, 379–380
Transversal filter, 125, 185–187, 240
Transverse end effect, 100
Transverse waves, see shear waves
Traps, 203
Triple-transit signal, 8, 64, 100
   analysis of, 102, 104, 105, 120, 168–170
   in bandpass filter, 176, 205, 207, 208
   in chirp filter, 242
   and insertion loss, 168–170
   reduction of, 170–171
Tuning, see matching
Turn-over temperature, 149, 150

Unidirectional transducer, 170
   multi-phase type, 173–181
   using multi-strip coupler, 120–123
Uniform transducer, 8, 57, 59, 89–95, 158–171
   bandwidth of, 158–165
   with parasitic components, 165–168
Up-chirp waveform, 220

Variable devices, **see** bandpass filter, chirp filter, convolver, delay line, programmable PSK filter
Velocity, energy, 135
Velocity of surface waves, 23, 26, 33, 35
   accuracy of, 150–151
   in lithium niobate, 33, 150–151
   measurement of, 140, 151–152, 369
   in regular electrodes, 369
   temperature coefficient of, 148
Voltage standing-wave ratio (VSWR), 165
Vibration of oscillators, 322

Waveguide, 5, 305–306
Waveguide convolver, 305–311
Waves, **see** acoustic waves, surface waves
Wave vector, 20, 133

Wedge delay lines, 219
Weighting, of interdigitial transducer, 201–202, 209
   see also apodised transducer, linear chirp, non-linear chirp, withdrawal weighting
Weighting functions for chirp waveforms, 226–227
White noise, 332
Window functions, 193–196
Withdrawal weighting, 66, 176, 201–202, 203
Wrap-around delay line, 172

Zeros, in filter responses, 183, 199
Zinc oxide, 3, 6, 37, 55, 154
   in bandpass filter, 208
   in convolvers, 314, 315
   in PSK filters, 287